ME: Workshop Conference on Integration of Endocrine and Non Endocrine Mechanisms in the Hypothalamus (1969 : Stresa, Italy)

THE HYPOTHALAMUS

Edited by

L. MARTINI, M. MOTTA, AND F. FRASCHINI

PROCEEDINGS OF THE WORKSHOP CONFERENCE ON
INTEGRATION OF ENDOCRINE AND NON ENDOCRINE
MECHANISMS IN THE HYPOTHALAMUS
HELD IN STRESA, ITALY, MAY 1969

SPONSORED BY
NATO SCIENTIFIC AFFAIRS DIVISION, BRUSSELS

1970

ACADEMIC PRESS, NEW YORK AND LONDON
A Subsidiary of Harcourt Brace Jovanovich, Publishers

ACADEMIC PRESS, INC.
111 Fifth Avenue, New York, New York 10003

United Kingdom Edition published by
ACADEMIC PRESS, INC. (LONDON) LTD.
24/28 Oval Road, London NW1

LIBRARY OF CONGRESS CATALOG CARD NUMBER: 70-117100

Second Printing, 1974

PRINTED IN THE UNITED STATES OF AMERICA

List of Contributors

P. R. ADIGA, McGill University Clinic, Royal Victoria Hospital, Montreal, Canada

R. M. BALA, Division of Endocrinology and Metabolism, McGill University Clinic, Royal Victoria Hospital, Montreal, Canada

S. BALAGURA, Department of Psychology, The University of Chicago, Chicago, Illinois

C. BARKER JØRGENSEN, Laboratory of Zoophysiology A, University of Copenhagen, Copenhagen, Denmark

F. A. BEACH, Department of Psychology, University of California, Berkeley, California

J. C. BECK, Division of Endocrinology and Metabolism, McGill University Clinic, Royal Victoria Hospital, Montreal, Canada

A. BÉRAULT, Department of Sub-Cellular Physiology, Collège de France, Paris, France

R. BURGUS, Department of Physiology, Baylor University College of Medicine, Houston, Texas

B. CECCARELLI, Department of Pharmacology, University of Milan, Milan, Italy

F. CLEMENTI, Department of Pharmacology, University of Milan, Milan, Italy

B. A. CROSS, Department of Anatomy, Medical School, Bristol, England

N. DAFNY, Laboratory of Neurophysiology, Department of Neurology, Hadassah University Hospital, Jerusalem, Israel

M. P. DE LA LLOSA, Department of Sub-Cellular Physiology, Collège de France, Paris, France

L. DESCLIN[1], Department of Pathology, Free University of Brussels, Brussels, Belgium

D. DE WIED, Rudolf Magnus Institute for Pharmacology, Medical School, University of Utrecht, Utrecht, The Netherlands

[1] Deceased.

J. P. DUPOUY, Department of Comparative Physiology, Faculty of Sciences, Paris, France

R. G. DYER, Department of Anatomy, Medical School, Bristol, England

W. FELDBERG, The National Institute for Medical Research, Mill Hill, London, England

S. FELDMAN, Laboratory of Neurophysiology, Department of Neurology, Hadassah University Hospital, Jerusalem, Israel

K. A. FERGUSON, Ian Clunies Ross Animal Research Laboratory, Division of Animal Physiology, Parramatta, Australia

J. FLAMENT-DURAND, Department of Pathology, Free University of Brussels, Brussels, Belgium

B. FLERKÓ, Department of Anatomy, University Medical School, Pécs, Hungary

J. T. FITZSIMONS, The Physiological Laboratory, University of Cambridge, Cambridge, England

P. FRANCHIMONT, Department of Clinical Medicine and Medical Pathology, University of Liège, Liège, Belgium

F. FRASCHINI, Department of Pharmacology, University of Milan, Milan, Italy

K. FUXE, Department of Histology, Karolinska Institute, Stockholm, Sweden

W. F. GANONG, Department of Physiology, School of Medicine, University of California, San Francisco Medical Center, San Francisco, California

A. GELOSO-MEYER, Department of Comparative Physiology, Faculty of Sciences, Paris, France

J. GLOWINSKI, Department of Neurophysiology, Collège de France, Paris, France

R. GUILLEMIN, Department of Physiology, Baylor University College of Medicine, Houston, Texas

J. D. HAHN, Main Laboratory of Schering AG, Berlin, West Germany

G. W. HARRIS, Department of Anatomy, University of Oxford, Oxford, England

T. HÖKFELT, Department of Histology, Karolinska Institute, Stockholm, Sweden

A. JOST, Department of Comparative Physiology, Faculty of Sciences, Paris, France

M. JUTISZ, Department of Sub-Cellular Physiology, Collège de France, Paris, France

B. KERDELHUÉ, Department of Sub-Cellular Physiology, Collège de France, Paris, France

C. KUDO, Division of Endocrinology and Metabolism, McGill University Clinic, Royal Victoria Hospital, Montreal, Canada

J. J. LEGROS, Department of Clinical Medicine and Medical Pathology, University of Liège, Liège, Belgium

L. MARTINI, Department of Pharmacology, Universities of Milan and Perugia Milan and Perugia, Italy

V. MAZZI, Institute of Comparative Anatomy, University of Turin, Turin, Italy

S. M. MCCANN, Department of Physiology, University of Texas Southwestern Medical School, Dallas, Texas

J. M. MCKENZIE, McGill University Clinic, Royal Victoria Hospital, Montreal, Canada

B. MESS, Department of Anatomy, University Medical School, Pécs, Hungary

M. MOTTA, Department of Pharmacology, University of Milan, Milan, Italy

F. NEUMANN, Main Laboratory of Schering AG, Berlin, West Germany

C. S. NICOLL, Department of Physiology-Anatomy, University of California, Berkeley, California

G. C. OLIVIER, Division of Endocrinology and Metabolism, McGill University Clinic, Royal Victoria Hospital, Montreal, Canada

J. L. PASTEELS, Department of Histology, School of Medicine, University of Brussels, Brussels, Belgium

M. PICKFORD, Department of Physiology, Edinburgh University, Medical School, Edinburgh, Scotland

F. PIVA, Department of Pharmacology, University of Milan, Milan, Italy

G. RAISMAN, Department of Human Anatomy, University of Oxford, Oxford, England

K. RETIENE, Department of Endocrinology, Center of Internal Medicine, Johann Wolfgang Goethe University, Frankfurt/Main, West Germany

N. W. RODGER, Department of Medicine Endocrinology and Metabolism, St. Joseph's Hospital, London, Canada

C. H. SAWYER, Department of Anatomy, University of California, Los Angeles, California

J. P. SCHADÉ, Netherlands Central Institute for Brain Research, Amsterdam, The Netherlands

N. B. SCHWARTZ, Department of Physiology, University of Illinois College of Medicine, Chicago, Illinois

C. C. D. SHUTE, The Physiological Laboratory, University of Cambridge, Cambridge, England

P. G. SMELIK, Department of Pharmacology, Medical Faculty, Free University of Amsterdam, Amsterdam, The Netherlands

S. H. SOLOMON, McGill University Clinic, Royal Victoria Hospital, Montreal, Canada

H. STEINBECK, Main Laboratory of Schering AG, Berlin, West Germany

F. S. STUTINSKY, Department of General Physiology, University of Strasbourg, Strasbourg, France

K. C. SWEARINGEN, Department of Physiology-Anatomy, University of California, Berkeley, California

L. TIMA[1], Department of Pharmacology, University of Milan, Milan, Italy

R. J. WURTMAN, Department of Nutrition and Food Science, Massachusetts Institute of Technology, Cambridge, Massachusetts

A. ZANCHETTI, Institute of Cardiovascular Research, University of Milan, Milan, Italy

M. ZANISI, Department of Pharmacology, University of Milan, Milan, Italy

[1] Permanent address: Department of Anatomy, University Medical School, Pécs, Hungary.

Preface

It is now almost thirty years since the first well-known symposium (The Hypothalamus and Central Levels of Autonomic Function. Res. Publ. Ass. nerv. ment. Dis., 1940, XX. Williams & Wilkins Company) on the general functions of the hypothalamus was held in New York in December 1939. Since that time particular aspects of hypothalamic function have been discussed at numerous meetings, but very few attempts have been made to piece together, into a single and total picture, the various roles played by this part of the brain. The present symposium which will discuss " The Integration of Endocrine and Non Endocrine Mechanisms in the Hypothalamus " owes much to its organizers for their ideas, initiative and energy in formulating a meeting on this subject.

It is of considerable interest to compare the programs of the meetings held thirty years apart in 1939 and 1969. In 1939 much basic groundwork on the anatomy of the primate hypothalamus, on the comparative anatomy, embryology, fiber connections, and angioarchitecture of this part of the brain were discussed. Since that time relatively little information has been added in these fields although the relationship of the limbic system and the reticular activating system to the hypothalamus have been found to be exciting areas of work. However detailed knowledge concerning the fiber connections of the different hypothalamic nuclei and neuron pools still remains an extensive area of ignorance. Much may be hoped for from the recent application of electron-microscopy and histochemistry to this problem. The present findings utilizing these techniques will probably receive much discussion at the present meeting, as will topics in the closely related fields of neurosecretion, monoaminergic systems, catecholamine metabolism, brain and pineal indoles and cholinergic pathways in respect to the hypothalamus.

In general terms, the functions of the basal diencephalon may be given as those of a mechanism which integrates and coordinates endocrine activity, autonomic nervous activity and behavioral responses. In the 1939 meeting less than half a dozen papers out of thirty-five dealt with the endocrine and behavioral functions of the hypothalamus although, in retrospect, some of these now seem to be " classics. " In this 1969 meeting two-thirds of the thirty-eight papers are on these topics. The mechanisms controlling the secretion of both posterior and anterior pituitary hormones, the releasing factors and inhibiting factors, the possible role of long and short-loop feedback mechanisms, and the control of rhythmic phenomena and sexual behavior, all form exciting topics for discussion. Much interest also centers on the inductive role of hormones during fetal life, whereby the development of the brain (probably hypothalamic-preoptic region) is affected so that the appropriate regulation is eventually exerted over sexual rhythms and behavior patterns.

Seminars, working parties, association meetings, society meetings, symposia, conferences, and congresses are of frequent occurrence these days. Many have been held in the last few years on neuroendocrine relations. This present meeting in Stresa will certainly deal with recent work in this field, but will also contribute great service in putting these findings into a general physiological context.

Special thanks are extended to Dr. F. Naftolin for his supervision of the papers contributed by scientists whose native tongue is not English and to Dr. F. Piva for his careful checking of manuscript references.

G. W. Harris, C.B.E., F.R.S.

Department of Human Anatomy
University of Oxford
Oxford, U. K.

January, 1970

Contents

Some Aspects of the Neural Connections of the Hypothalamus

G. Raisman

I. Introduction

It is a general idea of the neuroanatomist that the various cytoarchitectonic regions of the central nervous system have characteristic afferent and efferent fiber connections and distinctive functional properties. Although it seems likely that this sort of generalization can be applied to the hypothalamus, the information at present available affords only an incomplete correlation of cell grouping, fiber connections, and functional localization (Harris, 1960). The well-defined nuclei of the hypothalamus lie in the medial region, while, with certain important exceptions, most of the fiber connections have been traced only as far as the lateral hypothalamus, a region in which cell bodies are scattered diffusely among the fascicles of ascending and descending fibers of various lengths, the whole system being referred to as the medial forebrain bundle. Within such a diffusely arranged area, conventional anatomical techniques are at a disadvantage, and it is still largely a matter of speculation as to what extent the lateral hypothalamic area may contain discrete functional and anatomical systems and what may be the degree of interaction between such systems.

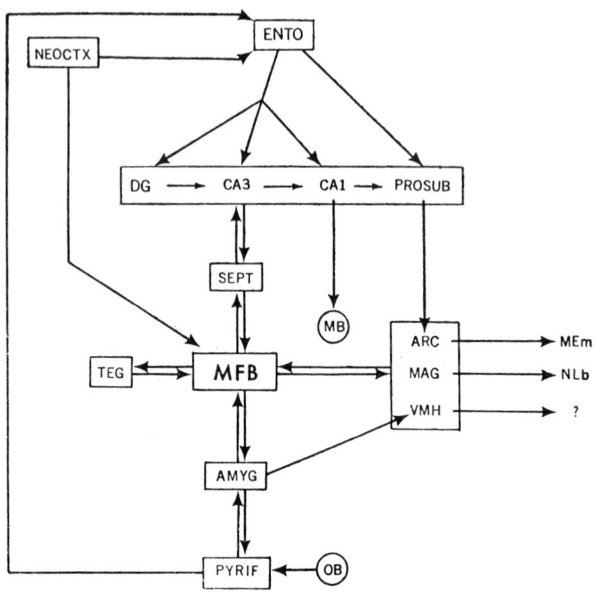

FIG. 1. A schematic diagram of the main hypothalamic fiber connections. AMYG, amygdala; ARC, arcuate nucleus; CA 1, CA 3, fields of hippocampus; DG, dentate gyrus; ENTO, entorhinal area; MAG, magnocellular hypothalamic nuclei; MB, mammillary body; MEm, median eminence; MFB, medial forebrain bundle; NEOCTX, neocortex; NLb, neural lobe; OB, olfactory bulb; PROSUB, prosubiculum; PYRIF, pyriform cortex; SEPT, septum; TEG, midbrain tegmentum; VMH, ventromedial hypothalamic nucleus.

Little is known of the detailed interrelationships of the medial and lateral hypothalamus. Guillery (1957) has demonstrated degeneration of fibers in the medial hypothalamus after lateral hypothalamic lesions, suggesting a general direction of relay from lateral to medial. Furthermore, the dendrites of medial hypothalamic nuclei, such as the ventromedial nucleus, extend for considerable distances into the lateral hypothalamic area and are thus in a position to sample activity in that region (Millhouse, 1969). The present brief account of hypothalamic connections will be limited to some systems (Fig. 1) which have been studied recently with degeneration techniques, and brief reference will be made to results obtained by the fluorescent technique for biogenic amines. Interconnections between the hypothalamus and the thalamus, subthalamus, or striatum, which were described in normal material but have not been examined with recent critical degeneration techniques, will not be further discussed.

II. Afferent Connections

A. Source

Massive and functionally important fiber tracts ascend to the hypothalamus from the brain stem. These connections are of diverse origin, and are distributed principally to the lateral hypothalamus, where analysis of their mode of termination is difficult. The formaldehyde-induced fluorescence method shows that this is an important route by which catecholamine- and serotonin-containing fibers are distributed to the forebrain (Andén *et al.*, 1966). Of the fiber pathways from the forebrain to the hypothalamus, the best understood are those arising in limbic and associated areas (the hippocampus, amygdala, and pyriform cortex). It is clear that fibers from neocortical areas (and in particular the frontal cortex) reach the region of the medial forebrain bundle, but there is no convincing demonstration of a projection from a neocortical area into the medial hypothalamus in subprimate species (Lundberg, 1960; Raisman, 1966a). In view of the importance of a projection from the neocortex to the hypothalamus as a possible route by which affective states could elicit unconscious endocrine and autonomic responses, this is a system which merits further investigation anatomically.

It is convenient to consider the afferents from the pyriform cortex, the olfactory tubercle, and the amygdala as a group (Fig. 1). A massive fiber system arises from all levels of the pyriform cortex and sweeps forwards and medially through the amygdala (from which it recruits further fibers). Emerging from the amygdala in a medial direction, as the so-called ventral amygdalo-fugal pathway, the fibers enter the medial forebrain bundle in a broad band, principally at the preoptic or anterior hypothalamic level. After lesions of this system, degeneration may be traced back through the medial forebrain bundle for some considerable distance and at premammillary levels, specific sites of termination occur in two cell masses called the *nuclei gemini* (Lundberg, 1962). This system is supplemented by another projection pathway, the stria terminalis, which arises specifically in the amygdaloid complex. The strial fibers emerge from the dorsal aspect of the amygdala and turn forward in the floor of the lateral ventricle to the level of the septum where (after distributing fibers to the bed nucleus of the stria, the anterior commissure, and other regions) a great many fibers turn caudally and enter the hypothalamus; these fibers are of special interest in that they terminate in the medial hypothalamic region (see below).

The hippocampal complex, and its dependency the septal nuclear group, are an important source of hypothalamic afferents (Fig. 1). The details of the projection pathways of the individual hippocampal fields are quite complex (Raisman *et al.*, 1966), and for the present purposes it is convenient to make some generalizations which do, however, ignore some of the finer details. The regio superior of the hippocampus (field CA 1) sends fibers through the dorsal fornix (and, from its caudal part, through the fimbria); these fibers are largely directed into the postcommissural fornix which passes through the hypothalamus as a relatively compact bundle and terminates in the mammillary nuclei (as well as sending fibers directly to the anterior thalamic nuclei). It is difficult to decide to what extent the system postcommissural fornix-mammillary body is linked to the rest of the the hypothalamus functionally. Although it is impossible to exclude that the fornix column may not distribute fibers to the adjacent hypothalamus (the so-called perifornical nucleus; see Nauta, 1956), such a demonstration is not as yet convincing (see Guillery, 1956). Moreover, the cells of the mammillary nuclei display a dendritic pattern of the " closed " nuclear type (Millhouse, 1969), indicating a minimum of direct interaction between the mammillary complex and the adjacent posterior hypothalamus (see, however, Guillery, 1957).

On the other hand, the regio inferior of the hippocampus (principally field CA 3) has an intimate relationship with the hypothalamus. The fibers from the regio inferior are distributed through the fimbria and run predominantly in the precommissural fornix, a massive group of fibers which account for about half of the total hippocampal outflow and which radiate out through the septofimbrial and medial and lateral septal nuclei, giving rise to many terminals. In a degeneration study at the electron microscope level, it was found that a lesion of the fimbria gave rise to degeneration of 35% of the total number of axodendritic terminals in the medial septal nucleus, and 56% in the lateral septal nucleus (Raisman, 1969a). The more ventral members of this fiber system extend through the medial septal nucleus into the diagonal band nucleus, and further caudally into the medial forebrain bundle where degeneration may be traced back through the lateral preoptic area and as far as anterior hypothalamic levels.

While this projection from the hippocampus directly into the medial forebrain bundle is not large, lesions of the septum cause (in addition to degeneration in the precommissural fornix, which is damaged *en passant*) considerably more hypothalamic degeneration which is not only denser but can be traced back through the medial forebrain bundle for almost the full rostrocaudal extent of the hypothalamus (for a review see Raisman, 1966b). It seems reasonable to regard this system as a septal relay reinforcing the hippocampal influence over the medial forebrain bundle.

As far as can be concluded at present, neither the fields of the regio superior nor of the regio inferior of the hippocampus send fibers into the medial hypothalamus. However, a detailed analysis of the differential projection of the individual hippocampal fields revealed that a special area, which lies at the junction of the posterior part of field CA 1 and the subiculum (the prosubiculum of Lorente de Nó's classification), gives origin specifically to the fibers of the medial cortico-hypothalamic tract. The general importance of the prosubiculum in the internal economy of the hippocampal complex is suggested by the direction of conduction in the intrinsic fiber connections (Fig. 1), viz: dentate gyrus → regio inferior → regio superior → prosubiculum (for references see Raisman *et al.*, 1965). In the first part of their course the fibers of the medial cortico-hypothalamic tract travel in the ventral tip of the fimbria, but when they reach the level of the anterior commissure, they detach themselves from the medial aspect of the fornix column and run vertically down in close apposition to the wall of the third ventricle to reach the dorsal aspect of the suprachiasmatic nucleus. From this point they run caudally along the base of the brain at the ventral angle of the third ventricle as far as the rostral pole of the arcuate nucleus, at which level it becomes difficult to trace them further with the light microscope (Nauta, 1956; Raisman *et al.*, 1966), although electron microscope studies (see below) reveal a dense termination in the arcuate nucleus.

B. Mode of Termination

Only two of the major afferent pathways are known to terminate in the medial hypothalamus: the stria terminalis in the ventromedial nucleus and the medial cortico-hypothalamic tract in the arcuate nucleus. Most of the afferent fiber systems discussed have been traced only as far as the lateral hypothalamus. Within this diffuse region, degeneration studies do not distinguish readily between terminals as opposed to fibers of passage, and our understanding of the intrinsic organization of this region as well as its relationship with the medial hypothalamus is incomplete. For this reason, most interest will be directed towards the mode of termination of fibers in the medial hypothalamus while at the same time acknowledging that our present inability to unravel the intricacies of the medial forebrain bundle by no means detracts from its functional or anatomical importance.

The question of afferents to the medial hypothalamus has been a subject of much controversy, and it has been amply demonstrated how an artefactitious granularity of silver staining in this region can mimic degeneration (Cowan and Powell, 1956). Until recently, studies employing various modifications of the Nauta method have on the whole shown disappointingly little degeneration in the medial hypothalamus. For example, in an investigation of the projection of the pyriform cortex and amygdala, degenerating fragments were traced back to the whole extent of the medial and lateral preoptic areas, although caudal to the anterior hypothalamic area, the degeneration was restricted to the lateral hypothalamus, reaching only as far medially as the lateral border of the ventromedial nucleus (Cowan *et al.*, 1965). As terminal degeneration was well impregnated in other sites (such as the mediodorsal thalamic nuclei), it seemed safe to accept as genuine the absence of staining of degeneration in the medial hypothalamus. However, the amygdaloid projections have been recently reexamined by Heimer and Nauta (1969), using a modified silver stain which permits the impregnation of degenerating terminals (Heimer and Peters, 1968), and in their material it was clear that the stria terminalis fibers do extend back in the medial hypothalamus as far as the ventral premammillary nucleus. The ventromedial nucleus is surrounded by a dense ring of degenerating terminals whose existence has been confirmed by the electron microscope.

With this in mind the projection of the medial cortico-hypothalamic tract has been reexamined at the electron microscope level after lesions of the fimbria. Whereas in the previous light microscope material it was only possible to trace degenerating fibers with certainty as far as the rostral pole of the arcuate nucleus, electron microscopy of the tract up to this point revealed degenerating myelinated fibers but no appreciable numbers of degenerating terminals. However, electron micrographs of the rostral part of the arcuate nucleus itself show a density of terminal degeneration which is quite remarkable. This degeneration has been assessed quantitatively in relation to the normal population of terminals in the nucleus (Field, personal communication).

C. Sensory Pathways

It is well established that the secretory functions of the pituitary are sensitive to factors in the external environment, and this is strikingly demonstrated by the influence of exteroceptive stimuli on reproductive behavior in many species. It is therefore of some interest to decide to what extent the various afferent pathways to the hypothalamus may represent routes over which specific sensory information could gain access to the neural mechanisms controlling pituitary secretion. Of all the sensory systems the hypothalamic access routes are best understood in the case of olfaction. Odor plays a vital role in the reproductive function in many mammalian species (Parkes and Bruce, 1961) and it is clear that the olfactory structures of the brain are privileged in having a massive route of access to the medial and lateral hypothalamus (for references see Raisman, 1966a). The olfactory bulb receives the fibers of the olfactory nerves and projects to the olfactory tubercle, the pyriform cortex, and also to the corticomedial group of amygdaloid nuclei, from which areas arise two of the most extensive projections to the hypothalamus, the ventral amygdalo-fugal pathway and the stria terminalis.

In the case of visual stimuli, the situation is much less clear, and what follows must be regarded as a speculative rather than factual account. Despite reports from time to time of a direct retinal projection to the hypothalamus, recent critical studies using various degeneration techniques have consistently failed to reveal a direct visual projection to the hypothalamus in mammals (for references see Kiernan, 1967). In considering the effects of light upon mammalian reproductive physiology it may be necessary to recognize two different types of effects. In the first instance, there is the well-known effect of the general intensity of ambient light and the duration and timing of the lightdark periods on such parameters as gonadal size in hamsters. In such reactions it is now known that the pineal gland (with its innervation through the superior cervical ganglion) is vitally important, and it is possible that the pineal may exert its effects upon pituitary function by means of a humoral link rather than by hypothalamic afferent fiber pathways (Wurtman *et al.*, 1968). On the other hand, where the visual input is more complex (involving the recognition of a suitable mate or of the appropriate behavior pattern) it seems likely that such information, in mammals, may reach the hypothalamus only after circuitous routes, probably involving the neocortex. While the status of direct neocortico-hypothalamic projections is uncertain, it is clear that the hippocampal projections to the hypothalamus could provide an indirect route for neocortical influence upon the hypothalamus. Through the intrinsic relays within the neocortex, information from the primary sensory areas gains access via association areas to the entorhinal area (Cragg, 1965), from which arise the major afferents to the hippocampus. It seems reasonable to suggest that a primary function of the hippocampal formation could be to channel information from widely spaced neocortical areas down into the hypothalamus. In this context it is interesting that the olfactory system, despite its own rather direct hypothalamic connections, also has access to this route by means of a massive fiber projection from the pyriform cortex to the entorhinal area.

This common use of the route through the entorhinal area, hippocampus, and septum, with its complex and highly organized hypothalamic projections, suggests that this system may offer some generalized form of neuronal processing by which information from widely different areas is converted into the form necessary for hypothalamic control. Sensory stimuli are registered in the central nervous system by rapid unitary electrical events in neuronal circuits where highly complex information can be expressed in terms of the activity of very small numbers of single units. On the other hand, the hypothalamus exerts its endocrine effects by altering the level of secretion of posterior pituitary hormones or of releasing factors and these events would occur over a much more extended time course, involving minutes if not hours. It is tempting to speculate that the hippocampus may be instrumental in converting momentary events in a few units into a form required for effective control of tonic hypothalamic functions, either by prolonging the time course or by amplifying the effects of single units as a massive discharge phenomenon.

III. Efferent Connections

One of the characteristic features of hypothalamic organization is the system of neurons which either secrete or control the secretion of hypophyseal hormones. These neurons are the last stage at which all nervous information has to be integrated prior to the ultimate endocrine response, and therefore they represent the final common pathway of the endocrine system in a manner comparable to the spinal and bulbar motoneurons in the somatic efferent system. The hypothalamus possesses, in addition, efferent fibers which are not directly concerned with the hypophysis. These fibers run chiefly in the medial forebrain bundle and their origins are less well-defined than the endocrine effector groups. It is convenient to divide them into ascending and descending fibers.

The descending group runs caudally into the midbrain tegmentum, reaching for varying distances into the brain stem. These fibers are important as potential routes over which the hypothalamus can exert its effects upon the autonomic nervous system, although direct monosynaptic connections to regions containing either sympathetic or parasympathetic motoneurons have not been demonstrated. The functions of the ascending hypothalamic efferents in the medial forebrain bundle are less understood, although the hypothalamo-septal relay probably acts as a feedback upon the hypothalamic input from the hippocampus and septum. Some selected features of the efferent connections of the hypothalamus will now be considered.

A. Posterior Pituitary

The system of fibers running from the so-called magnocellular hypothalamic nuclei (the supraoptic and paraventricular nuclei of mammals) to the posterior pituitary is the best understood of the hypothalamic efferent systems (for a review see Sloper, 1966). The cells of the supraoptic and paraventricular nuclei send their axons through the internal layer of the median eminence and the infundibular stalk to terminate in a palisade around the capillaries of the pars nervosa. The hormones of the posterior pituitary are synthesized in the cells of the magnocellular nuclei and are transported down the axons to be released into the blood stream in the pars nervosa. Special features which have contributed to the successful elucidation of this system are the ready accessibility of the pituitary stalk for operative interference, and the peculiar chemical properties of the neurosecretory material or its carrier protein, which permit the use of selective staining; electron microscopy has revealed that the cell bodies, axons, and terminals of the neurosecretory neurons have characteristic large dense core vesicles (of 2000 to 3000 Å diameter) which distinguish them from all other known systems in this region.

Morphologically, the neurosecretory cells are neurons, although highly specialized, and electron micrographs show abundant afferent synapses upon the cell bodies and also upon the proximal parts of the axons; the source of these afferent fibers is unknown, although a decrease in the formaldehyde-induced fluorescence of the supraoptic neuropil occurs after lesions in the region of the ventral tegmental area of the midbrain, suggesting that the fibers may ascend along a route through the medial forebrain bundle (Fuxe and Hökfelt, 1967).

B. Median Eminence

By analogy with the magnocellular nuclei and the pars nervosa it has been suggested that hypothalamic neurons of the so-called parvicellular system (Szentágothai, 1964) send their axons down into the median eminence where they form pericapillary palisades around the primary portal capillaries in the external layer; releasing factors synthesized in the cell bodies would be carried down these axons to be released into the portal hypophyseal circulation by means of which they are brought to act upon the anterior pituitary. There is evidence from Golgi material (see Szentágothai, 1964) that axons directed into the external zone of the median eminence arise in the arcuate nucleus, and Fuxe and Hökfelt (1967) have shown a retrograde increase in formaldehyde-induced fluorescence in this nucleus after lesions of the median eminence. As the arcuate nucleus is, from many standpoints (see Flament-Durand, 1965) the area most likely to be concerned in elaborating gonadotropin releasing factors, the evidence for a parvicellular neurosecretory system is therefore strongest for gonadotropin control, and the massive projection of the prosubiculum through the medial cortico-hypothalamic tract to the arcuate nucleus (see above) may be expected to play some key role in reproductive behavior.

Experimental investigation of the parvicellular neurosecretory system suffers from the practical disadvantages that well-defined lesions are difficult to place and that there is no specific staining procedure for any of the releasing factors. It is therefore not clear whether hypothalamic areas other than the arcuate nucleus contribute fibers to the median eminence, nor how heterogeneous is the population of cells in the arcuate nucleus itself. Furthermore, although detailed quantitative electron microscopy of the median eminence (see Rodríguez, 1969, in the toad) has suggested a number of different classes of nerve endings, the nerve terminals in general exhibit a continuous spectrum of sizes of vesicular inclusions. This makes it difficult to investigate the idea that the different releasing factors are represented by morphologically distinct categories of nerve endings. At the same time the disposition of the tanycyte ependymal cell processes and their characteristic endings upon the capillaries of the primary portal plexus suggests that the ependyma has something more than a passive role in the control of the pars distalis (see Leveque *et al.*, 1966).

C. Ascending Hypothalamic Efferents

Among the hypothalamic efferent systems running in the medial forebrain bundle, the ascending fibers to the septum will be selected as an example, since this system has been studied recently at a quantitative level with the electron microscope (Raisman, 1969a). Of the large and probably heterogeneous group of fibers ascending through the medial forebrain bundle, some undoubtedly arise at midbrain levels, although lesions at successively more rostral levels along the course of the hypothalamus cause progressively heavier degeneration, indicating that the system recruits fibers during its passage through the hypothalamus (Guillery, 1957). An important contingent of fibers leave the medial forebrain bundle at the preoptic level and run through the diagonal band into the septum, where they terminate principally in the medial septal nucleus.

In view of the well-known dense fiber projection from the hippocampus to the septum, it was decided to make a quantitative electron microscope comparison of the terminal degeneration in the medial septal nucleus after lesions of these two fiber systems, in order to discover by what means the two projections might differ in their mode of termination. It was found that the hippocampal fibers to the septum terminate exclusively in axodendritic synapses, accounting for some 35% of the total population of axodendritic synapses in the medial septal neuropil. The hypothalamic projection fibers account for another 20% of the axodendritic synapses, but are distinguished from the hippocampal projection fibers by giving rise to about 25% of the axosomatic synapses of the region. Moreover, the terminals of the two systems also differ in their content of dense-core vesicles (of 1000 Å diameter): the majority (around two-thirds) of the hypothalamic terminals contain two or more dense-core vesicles, whereas over 90% of the hippocampal terminals contain either one or no dense-core vesicles. These results illustrate how the use of the electron microscope permits the analysis of fiber systems whose degeneration patterns can be indistinguishable by conventional light microscope degeneration studies.

D. Plasticity

Two of the hypothalamic efferent systems demonstrate considerable powers of reconstruction after damage. It has long been known that if the terminals of the supraoptico- and paraventriculo-hypophyseal tracts are removed by hypophysectomy, the tracts reconstruct themselves in such a way as to form a miniature posterior pituitary in the region of the median eminence. For one to two weeks after hypophysectomy, an accumulation of neurosecretory material distends the axons in the internal zone of the median eminence. Subsequently, neurosecretory fibers grow out from the internal zone of the median eminence and invade the external zone where they form characteristic pericapillary plexuses (Stutinsky, 1957). This rearrangement of the previously sectioned axons brings them into contact with the circulation, and experiments with osmotic stress show that this system is capable of regulating the release of vasopressin in a manner analogous to the normal intact posterior pituitary (Moll and De Wied, 1962).

In the septal neuropil a different type of plasticity has been investigated by taking advantage of the fact that two fiber systems which converge upon the medial septal nucleus can be destroyed selectively by lesions, in the medial forebrain bundle for the hypothalamic fibers, or in the fimbria for the hippocampal fibers (Raisman, 1969b).

In animals with chronic lesions of the fimbria the septal neuropil shows a permanent change, consisting of an overall increase in the number of axon terminals which (in a single plane of section) contact more than one postsynaptic profile. It seems possible that this phenomenon represents the reoccupation of deafferented postsynaptic sites by the formation of supernumerary contacts by the surviving axon terminals within the region.

Short-term lesions of the medial forebrain bundle in these animals with chronic fimbrial section show that a proportion of the terminals thus affected are derived from hypothalamic efferent fibers, implying that under these circumstances the hypothalamic projection fibers are capable of reinnervating deafferented hippocampal terminal sites. Conversely, in animals with chronic lesions of the medial forebrain bundle (which result in loss of some 25% of the axosomatic terminals in the medial septal nucleus) it can be shown that the fimbrial fibers, which normally do not have axosomatic terminals, now give rise to a consistent proportion of axosomatic terminals. This implies that the axosomatic terminal sites, once deprived of their hypothalamic input, are capable of inducing the formation of new contacts from a fiber system (such as the fimbria) which does not normally innervate them. Such observations as these suggest that the neuropil of the central nervous system may be capable of considerable plastic remodelling following injury.

IV. Conclusions

Some of the main hypothalamic connections are summarized in Fig. 1. The afferent connections of the hypothalamus comprise ascending fibers from the brain stem and descending fibers from the forebrain. The forebrain areas projecting to the hypothalamus form two basic groups: (1) the hippocampal complex (and septum) probably act as a common channel for sensory and neocortical input to the hypothalamus, and (2) the pyriform cortex, olfactory tubercle, and amygdala provide a privileged route of access of olfactory information to the hypothalamus. Both systems have massive terminations in the medial hypothalamus, the hippocampal complex in the arcuate nucleus and the amygdala in the ventromedial nucleus.

The two main categories of efferent hypothalamic connections are: (1) the endocrine effector fibers and (2) projections through the medial forebrain bundle. The endocrine effector fibers include the axons of the magnocellular neurosecretory nuclei which convey the posthypophyseal hormones to the pars nervosa and the axons of the parvicellular nuclei (such as the arcuate nucleus) which are assumed to carry releasing factors to the neurohemal area of the external zone of the median eminence. Hypothalamic projection fibers in the medial forebrain bundle comprise two groups: a descending fiber group, which is probably the first link in a multisynaptic route for hypothalamic influence upon autonomic motoneurons, and an ascending fiber group to the septum, olfactory tubercle, and other basal forebrain areas.

Acknowledgment

This study was supported in part by U.S. Public Health Service Grant No. T1 NBO5591 to Dr. S. L. Palay at Harvard Medical School, awarded by the National Institute of Neurological Diseases and Stroke.

References

Andén, N. E., Dahlström, A., Fuxe, K., Larsson, K., Olson, L., and Ungerstedt, U. (1966). *Acta Physiol. Scand.* **67**, 313.

Cowan, W. M., and Powell, T. P. S. (1956). *J. Anat.* **90**, 188.

Cowan, W. M., Raisman, G., and Powell, T. P. S. (1965). *J. Neurol. Neurosurg. Psychiat.* **28**, 137.

Cragg, B. G., (1965). *J. Anat.* **99**, 339.

Flament-Durand, J. (1965). *Endocrinology* **77**, 446.

Fuxe, K., and Hökfelt, T. (1967). *In* " Neurosecretion " (F. Stutinsky, ed.). Springer-Verlag, Berlin, pp. 165–177.

Guillery, R. W. (1956). *J. Anat.* **90**, 350.

Guillery, R. W. (1957). *J. Anat.* **91**, 91.

Harris, G. W. (1960). *In* " Handbook of Physiology " (J. Field, H. W. Magoun and V. E. Hall, eds.). Section 1, Vol. II. American Physiological Society, Washington, pp. 1007–1038.

Heimer, L., and Nauta, W. J. H. (1969). *Brain Res.* **13**, 284.

Heimer, L., and Peters, A. (1968). *Brain Res.* **8**, 337.

Kiernan, J. A. (1967). *J. Comp. Neurol.* **131**, 405.

Leveque, T. F., Stutinsky, F., Porte, A., and Stoeckel, M. E. (1966). *Z. Zellforsch.* **69**, 381.

Lundberg, P. O. (1960). *Acta Physiol. Scand.* **49**, *Suppl.* 171, 1.

Lundberg, P. O. (1962). *J. Comp. Neurol.* **119**, 311.

Millhouse, O. E. (1969). *Brain Res.* **15**, 341.
Moll, J., and De Wied, D. (1962). *Gen. Comp. Endocrinol.* **2**, 215.
Nauta, W. J. H. (1956). *J. Comp. Neurol.* **104**, 247.
Parkes, A. S., and Bruce, H. M. (1961). *Science* **134**, 1049.
Raisman, G. (1966a). *Brit. Med. Bull.* **22**, 197.
Raisman, G. (1966b). *Brain* **89**, 317.
Raisman, G. (1969a). *Exp. Brain Res.* **7**, 317.
Raisman, G. (1969b). *Brain Res.* **14**, 25.
Raisman, G., Cowan, W. M., and Powell, T. P. S. (1965). *Brain* **88**, 963.
Raisman, G., Cowan, W. M., and Powell, T. P. S. (1966). *Brain* **89**, 83.
Rodríguez, E. M. (1969). *Z. Zellforsch.* **93**, 182.
Sloper, J. C. (1966). *In* "The Pituitary Gland" (G. W. Harris and B. T. Donovan, eds.). Vol. III. Butterworths, London, pp. 131–239.
Stutinsky, F. S. (1957). *Arch. Anat. Microscop. Morphol. Exp.* **46**, 94.
Szentágothai, J. (1964). *Progr. Brain Res.* **5**, 135.
Wurtman, R. J., Axelrod, J., and Kelly, D.E. (1968). "The Pineal." Academic Press, New York.

Fine Structure of Rat Hypothalamic Nuclei

F. Clementi and B. Ceccarelli

I. Introduction

The control of both posterior and anterior lobes of the pituitary gland by hypothalamic structures is well established and is supported by numerous endocrinological, histochemical, histological, and ultrastructural data. It is known that the neurosecretory granules found in the neurohypophysis are synthesized by the neurons of the paraventricular and supraoptic nuclei, and that the endings of these neurons constitute the functional part of the neurohypophysis (Palay, 1960; Nemetschek-Gansler, 1965; Bern and Knowles, 1966; Zambrano and De Robertis, 1966a,b, 1967a; Herlant, 1967; Monroe, 1967; Bargmann, 1968; Ginsburg, 1968). It is also well established that the anterior pituitary gland is controlled by hypothalamic structures; furthermore, specific hypophysiotropic substances (the releasing and inhibiting factors) are synthesized and released by the hypothalamic neurons (Everett, 1964; McCann and Dhariwal, 1966;

Davidson, 1966; Pecile and Müller, 1966; Mess and Martini, 1968; Szentágothai *et al.*, 1968).

Since data concerning the fine morphology of mammalian hypothalamic nuclei are still very limited, additional knowledge of their ultrastructure could be of great importance for clarifying their functions.

The two types of hypothalamic nuclei, the " magnocellular " and the " parvicellular, " will be considered here; in particular the ultrastructure of supraoptic and paraventricular nuclei (magnocellular nuclei) and of anterior, ventromedial and arcuate nuclei (parvicellular nuclei) will be dealt with. The morphological aspect of the median eminence and of the pars tuberalis will also be described; their significance in relation to the present knowledge in the field of endrocrinology will be discussed. Lastly, the fine structure of blood capillaries in the hypothalamus and the median eminence and their relationship with the surrounding nervous tissue will be considered.

II. Technical Considerations

The ultrastructural study of the hypothalamus poses two main problems: fixation and localization of nuclei.

We have found that immersion fixation in any fixative always gives poor results. Therefore, perfusion appears to be the most suitable procedure (Ceccarelli and Pensa, 1968). Mixed aldehyde fixatives are preferable; the best results were obtained using a mixture of 1% formaldehyde and 2.5% glutaraldehyde in 0.12 *M* phosphate buffer with pH 7.3–7.6 (Karnovsky, 1965). Postfixation in osmium tetroxide and the embedding were performed as usual without any particular changes in routine. Fixation of the animals is always performed at the same time of the day to avoid structural modifications due to the cyclic functions of many hypothalamic nuclei (Fraschini and Motta, 1967).

Localization of hypothalamic nuclei in specimens embedded for electron microscopy is an equally important problem. A useful procedure is to embed halves of the hypothalamus and then to cut serial thick sections and examine them under the optic microscope. The localization of the nuclei is facilitated by using the stereotaxic atlas of the rat brain as a guide (König and Klippel, 1963).

III. Paraventricular and Supraoptic Nuclei

The paraventricular and supraoptic nuclei of the mammalian hypothalamus are involved in the synthesis, transport, and release of oxytocin and vasopressin. This was first hypothesized by Bargmann and Sharrer (1951) on the basis of their histological studies, and has been confirmed by more recent studies (Ginsburg, 1968). It seems probable that the two hormones, vasopressin and oxytocin, are synthesized in different neurons. According to Olivecrona (1957), Brooks *et al.* (1966), Nibelink (1961), and Lederis (1961), oxytocin is synthesized primarily by the paraventricular nuclei, while vasopressin is synthesized primarily by the supraoptic nuclei (Sokol and Valtin, 1967). Furthermore, separation of the vasopressin-containing granules from those containing oxytocin has been obtained by means of density gradient centrifugation (Bindler *et al.*, 1967). Although these differences in the localization of oxytocin and vasopressin are established, there are no particular differences between the anatomical structures of the neurosecretory cells of the supraoptic and paraventricular nuclei. The magnocellular neurons of these two nuclei are large cells containing a central, round nucleus and a prominent nucleolus (Fig. 1). Very often the cells of the supraoptic nucleus are in close contact with blood capillaries; the walls of these capillaries appear to follow the periphery of the neurons. In the cytoplasm there is always a very large and elaborate Golgi apparatus which occupies a large area around the nucleus (Figs. 1–3). This apparatus is formed by elongated cisternae and numerous vesicles, some of which contain a dense substance (Fig. 2). When the size of the latter increases, they detach from the Golgi and are then called secretory, or elementary, granules. These are enclosed by a three-layered membrane of a diameter ranging from 900–1900 Å and are scattered randomly in the cytoplasm, mainly in the Golgi zone (Figs. 2, 3); they are very similar to those found in the axons of the hypothalamo-neurohypophyseal tract and in the neural lobe of the pituitary gland.

In the Golgi area there are also numerous dense bodies that have a polymorphic appearance and are limited by a unit membrane. These bodies contain a granular dense substance as well as some electron-lucent droplets that are probably lipid in nature. These bodies also contain acid phosphatase and can be considered to be

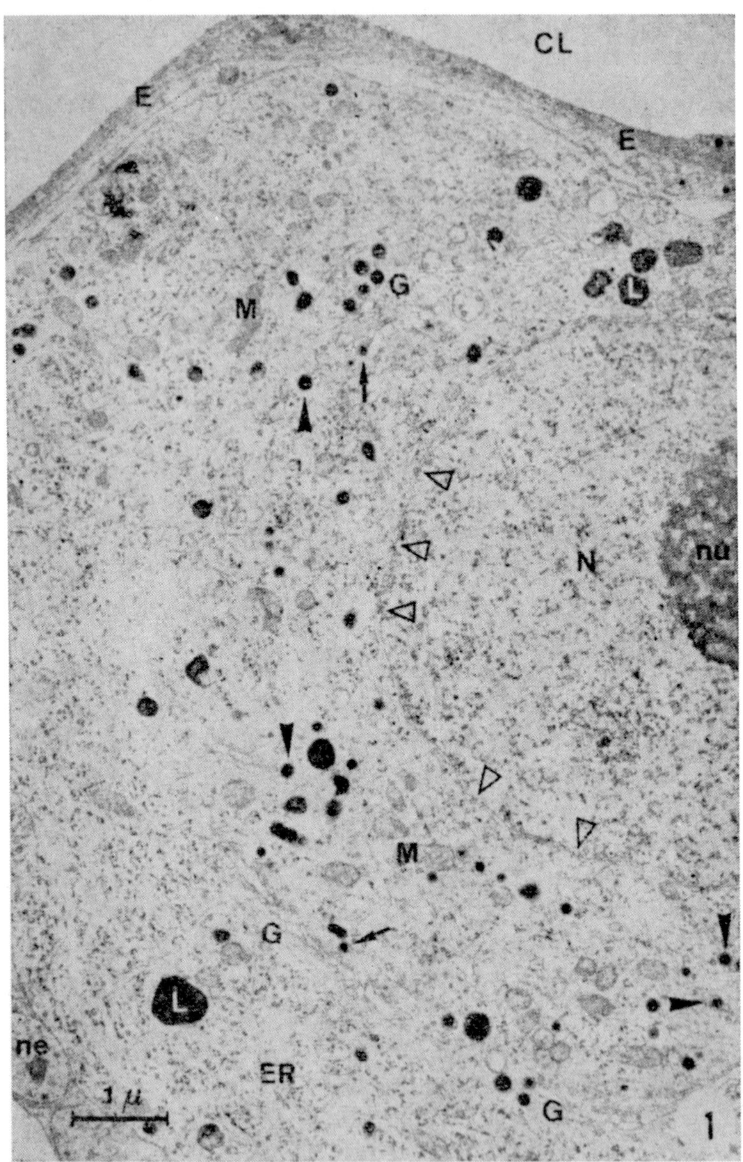

FIG. 1. Supraoptic nucleus. These cells have a large, central nucleus (N) with a well developed nucleolus (nu) and numerous nuclear pores on the nuclear envelope (▷). In the cytoplasm the Golgi apparatus (G) is very well developed and constitutes nearly a continuous belt around the nucleus. In the Golgi area numerous elementary granules (▶) and forming granules (—▶) are present. In the periphery of the cells the rough endoplasmic reticulum (ER) is abundant and contains numerous polysomes. Lysosomes (L) are also numerous. At the periphery of the cells, nerve endings (ne) are present. This photograph also shows the particular relationship between cells and capillaries. The capillary endothelium (E) is very near to the neurosecretory cell and follows its borders; CL, capillary lumen; M, mithocondria.

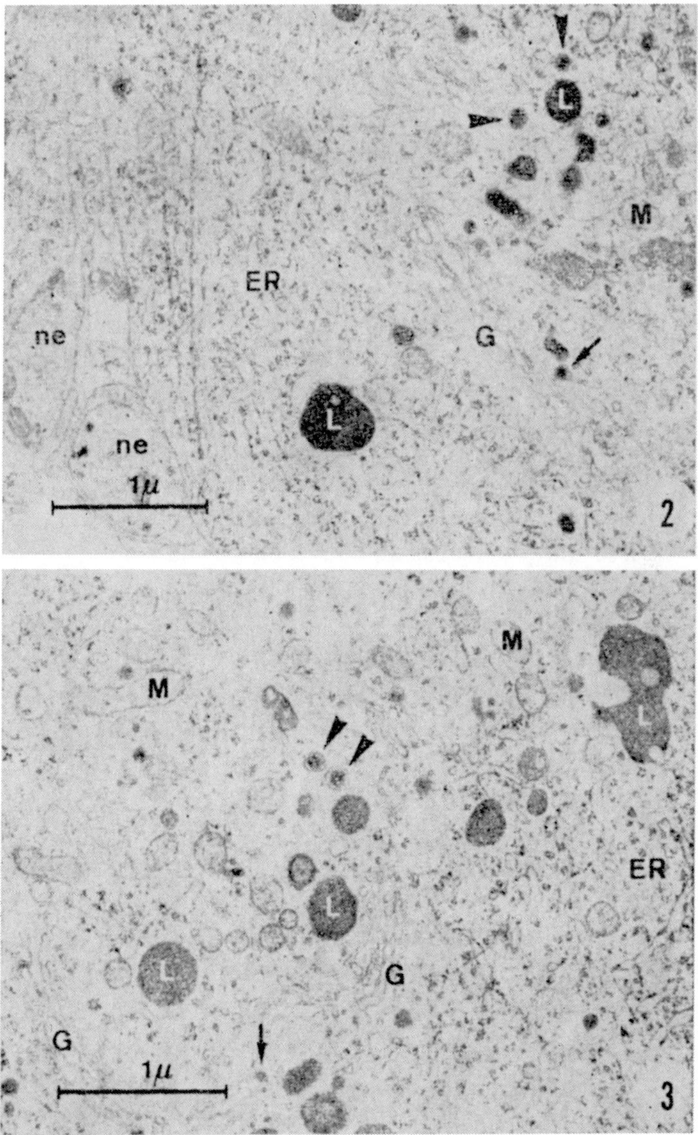

FIG. 2. Supraoptic nucleus. Higher magnification of the neurosecretory cell shown in Fig. 1. A large Golgi apparatus (G) is present with a secretory granule being formed in one cisterna (—▶). A well-developed endoplasmic reticulum (ER) is grouped between the Golgi zone and the cell membrane. On the left corner two nerve ending profiles (ne) are present, containing the usual population of synaptic vesicles and dense-core granules.

FIG. 3. Paraventricular nucleus. The cytoplasm of this neurosecretory cell is particularly similar to that of the cell shown in Figs. 1 and 2. A large and elaborate Golgi apparatus containing several secretory granules (▶), an abundant endoplasmic reticulum, and numerous lysosomes (L) are present.

lysosomes (Osinchak, 1964). Between the Golgi area and the cell membrane there is an abundant endoplasmic reticulum formed by flattened sacs and covered by numerous polysomes (Figs. 1–3). Only in a few neurons is the rough endoplasmic reticulum formed by very dilated cisternae containing a filamentous material. Zambrano and De Robertis (1966b) have shown that the nature of this filamentous substance is proteinaceous and that puromycin inhibits its formation (Zambrano and De Robertis, 1967b). They considered this material to be a secretory protein in its dilute state, ready to be packaged by the Golgi apparatus. In agreement with these data, biochemical studies of incubated hypothalamic and median eminence tissue *in vitro* have shown that an inactive precursor of vasopressin is synthesized on ribosomal RNA; the synthesis of this precursor is inhibited by puromycin (Sachs and Takabatake, 1964).

From these morphological and biochemical data it is possible to postulate that the inactive precursor of vasopressin, a protein, is synthesized on the polysomes of the endoplasmic reticulum, transported to the Golgi apparatus, and there incorporated into the secretory granules. Vasopressin is probably released inside the granules by the precursor during the transport of the secretory granules from the body of the cell to the periphery (Sachs *et al.*, 1967).

The hormonal polypeptides present in the secretory granules are bound to a protein called neurophysin. This protein was first isolated from the neurohypophysis by Van Dyke *et al.*, (1941), subsequently purified by Acher *et al.* (1956), and shown to be very specific, since it binds only oxytocin and vasopressin (Ginsburg *et al.*, 1966; Ginsburg, 1968). The data available indicate that this carrier protein is different from the precursor protein of vasopressin. Therefore it is interesting to note that in these secretory granules there are active hormones, two types of proteins, and possibly other inactive peptides (Ramchandran and Winnick, 1957). It is not known which of these substances is responsible for the dense appearance typical of the granules in the electron microscope; this explains the difficulty of deriving conclusions about the activity of neurosecretory cells solely on the basis of their morphological appearance (Lederis, 1969).

The input neuron pathways of the neurosecretory nuclei are rather complicated. The afferent fibers make synapses mainly with the soma of the cells, but axodendritic or axoaxonic synapses are also found in this area. These nerve endings are similar to those present in other regions of the hypothalamus and contain mainly clear synaptic vesicles of 500 Å diameter and a few dense granules (Fig. 2). Some of the nerve endings, however, contain only clear synaptic vesicles (Fig. 2). Acetylcholine has been clearly established as one of the mediators in the supraoptic and paraventricular nuclei. It seems probable that the cholinergic receptors are located on the neurosecretory cells themselves. These conclusions are based primarily on the presence of cholinesterase in the supraoptic and paraventricular nuclei (Pepler and Pearse, 1957; Abrahams *et al.*, 1957) and on the effects of acetylcholine and acetylcholinesterase inhibitors on the neurosecretory cells (Ginsburg, 1968). In view of what is known about other parts of the nervous system, it might be postulated that the nerve endings containing only clear synaptic vesicles are cholinergic in nature. However, noradrenaline is also present in very high quantities in the nerve fibers penetrating into these nuclei, while the cells of supraoptic and paraventricular nuclei apparently do not contain amines (Konstantinova, 1967; Odake, 1967); there is evidence indicating that the nerve endings containing dense-core granules may be adrenergic terminals (Hökfelt, 1968).

Our observations confirm that two types of neurons are present in the paraventricular nucleus: the neurosecretory cells and smaller cells similar to those present in the anterior and suprachiasmatic nuclei. This has also been suggested by histological procedures. The existence of the small size cells might represent a morphological basis for the finding of Mess and his associates (1967), who have suggested that the paraventricular region might be responsible for the synthesis of Follicle-Stimulating Hormone Releasing Factor (FSHRF). It might be difficult to accept that the neurosecretory cells of the paraventricular nucleus are involved in the synthesis of releasing factors.

IV. Anterior Hypothalamus

In the anterior part of the hypothalamus there are primarily two nuclei: the anterior nucleus and the suprachiasmatic nucleus. There are essentially no differences between the neurons present in these two nuclei. The cells are smaller than those of the neurosecretory neurons, they are round and scattered within the nuclei, and they

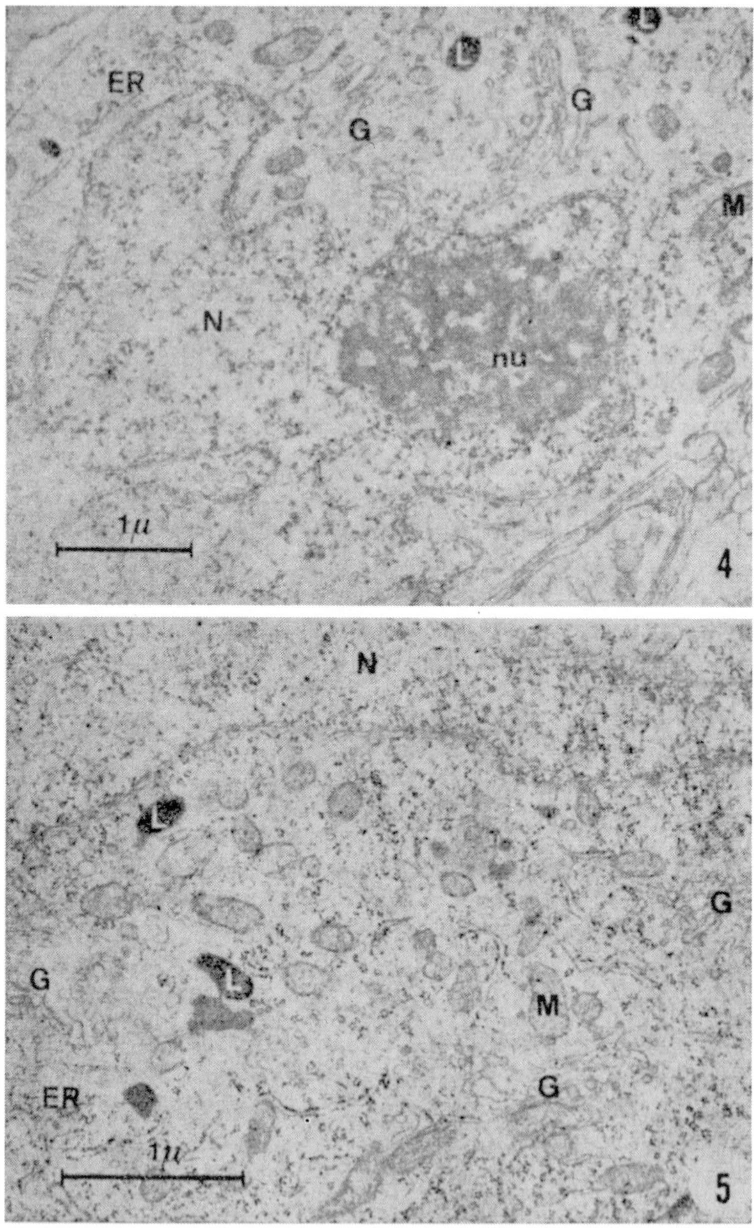

FIG. 4. Anterior hypothalamic nucleus. The neuron shown in this photograph has a nucleus with undulating limits and profound cytoplasmic infoldings. The cytoplasm has a clear matrix with a small Golgi apparatus, no secretory granules, and few cisternae of endoplasmic reticulum.

FIG. 5. Suprachiasmatic nucleus. Part of the nucleus and a portion of the cytoplasm of a cell which contains a relatively small Golgi apparatus, few lysosomes, and scattered elements of endoplasmic reticulum are shown. For abbreviations see Fig. 1.

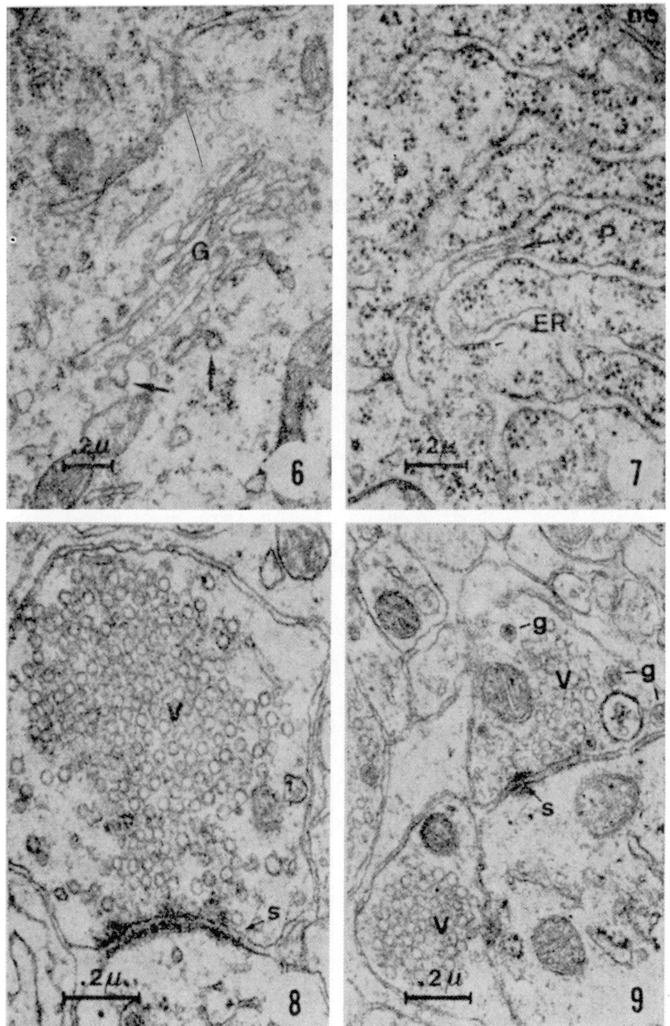

FIG. 6. Anterior hypothalamic nucleus. A Golgi apparatus (G) is shown in detail; few vesicles are present around the flattened cisternae; some coated vesicles are seen near to the Golgi apparatus and are budding or fusing with Golgi cisternae (—▶).

FIG. 7. Anterior hypothalamic nucleus. A detail of the rough endoplasmic reticulum (ER) is shown. Numerous polysomes (P), both free and membrane-bound, are present. The arrow indicates a point of fusion between two cisternae of the endoplasmic reticulum, with a loss of ribosomes at the fusion point. A nerve ending (ne) is present in the upper part of the picture.

FIG. 8. Nerve ending present in the anterior hypothalamic nucleus. The presynaptic bouton contains clear vesicles (v) arranged in crystal-like form. A thickening of the pre- and post-synaptic membrane is present in the synaptic contact (S).

FIG. 9. Other two typical nerve endings in the anterior nucleus. These two nerve endings contain a different population of vesicles (v) and dense granules (g), and are making contacts with the same postsynaptic profile.

are not so closely associated with the blood capillaries. The large nucleus (located centrally with a pale, evenly distributed chromatin) has a prominent nucleolus and numerous profound cytoplasmic infoldings (Fig. 4).

The cytoplasm has a pale appearance with a clear cytoplasmic matrix: the mitochondria are small, elongated, and randomly scattered (Fig. 5). The Golgi apparatus is often located near the nucleus and is small, comprising few elongated sacs and few vesicles. These vesicles contain almost no dense material or elementary granules (Figs. 4–6). In the same area we found many microtubules that do not seem to have any particular relationship with the Golgi elements (Fig. 5). The endoplasmic reticulum is scarce and is formed mainly by a few, small, elongated cisternae scattered within the cytoplasm (Figs. 4, 5). These seldom assume the typical stacked appearance (Fig. 7), but when they do so, the parallel lamellae are not numerous. In some cases the cisternae of the rough endoplasmic reticulum seem to fuse and thereby loose their ribosomes at the zone of fusion (Fig. 7). This aspect of the rough endoplasmic reticulum has already been described in the cells of arcuate nucleus (Mazzucca, 1968) and in prolactin cells of the anterior pituitary (Clementi and De Virgiliis, 1967). The polysomes are found in rosette form or in spirals and are abundant on the membranes of the endoplasmic reticulum and free in the cytoplasm (Figs. 6, 7). Few elements of smooth endoplasmic reticulum are present and seem to derive from the cisternae of the rough endoplasmic reticulum (Figs. 6, 7). Many large granules containing an amorphous, dense substance are present; they are very similar in appearance to those described in the neurosecretory cells as lysosomes (Figs. 4, 5).

These nuclei have great importance from an endocrinological point of view. It is probable that estrogen-sensitive neurons are localized in this area (Köves and Halász, 1969). Exogenous estradiol is taken up and retained by cells of the anterior and suprachiasmatic nuclei (Kato and Villee, 1967; Stumpf, 1968). Furthermore, the cells that control the cyclic release of Luteinizing Hormone (LH) from the anterior pituitary also seem to be localized in these nuclei (Antunes-Rodriguez and McCann, 1967; Mess and Martini, 1968); small but significant quantities of Luteinizing Hormone Releasing Factor (LHRF) are also present in this area (McCann, 1968). These nuclei also seem important in the regulation of the secretion of thyrotropin.

Flament-Durand and Desclin (1968) postulated that a large area comprising the anterior nucleus and part of the ventromedial nucleus is responsible for the synthesis of Thyrotropin Releasing Factor. It would be interesting herefore to verify whether these neurons are able to synthesize releasing factors. However, the morphological evidence does not appear to support this possibility. The cells present in these nuclei have a small Golgi complex which apparently does not elaborate secretory granules. The endoplasmic reticulum is scarce and polysomes are more often free in the cytoplasm than attached on the membranes. It might be that these neurons control the synthesis and the release of releasing factors in other neurons (e.g., those of the arcuate nucleus) rather than actually synthesize them.

Very little is known about the nervous afferences of these nuclei. The suprachiasmatic nucleus contains a large number of serotoninergic nerve endings, while the anterior nuclei contain mainly adrenergic terminals (Fuxe and Hökfelt, 1969). Cholinesterase is also present in moderate amount (Abrahams *et al.*, 1957). The afferent fibers make contact with both the body and the dendrites of the neurons. We were able to distinguish two types of nerve terminals: one type contains only clear vesicles (Figs. 8, 9), and the other type in addition contains some dense granules larger than synaptic vesicles (Fig. 9). Frequently we have seen two nerve endings, one very close to the other, making contact with the same postsynaptic structure. Sometimes only one of these endings contains synaptic vesicles, and the other contains both dense-core granules and synaptic vesicles (Fig. 9).

V. Ventromedial Nucleus

The ventromedial nucleus is located in the medial hypothalamus and contains two main cell types, as already reported by Szentágothai *et al.* (1968) and Marsala (1968). One cell type has a dense cytoplasm and is large; the other has paler cytoplasm and is smaller. The smaller cell type seems to have a fine structure similar to that described in the anterior nuclei. The fine structure of the larger cell type is slightly different in that they have a dense cytoplasmic matrix; the rough endoplasmic reticulum is abundant and is formed by numerous staks of parallel membranes (Fig. 10). Mitochondria, lysosomes, and Golgi apparatus are nearly identical to those described in the

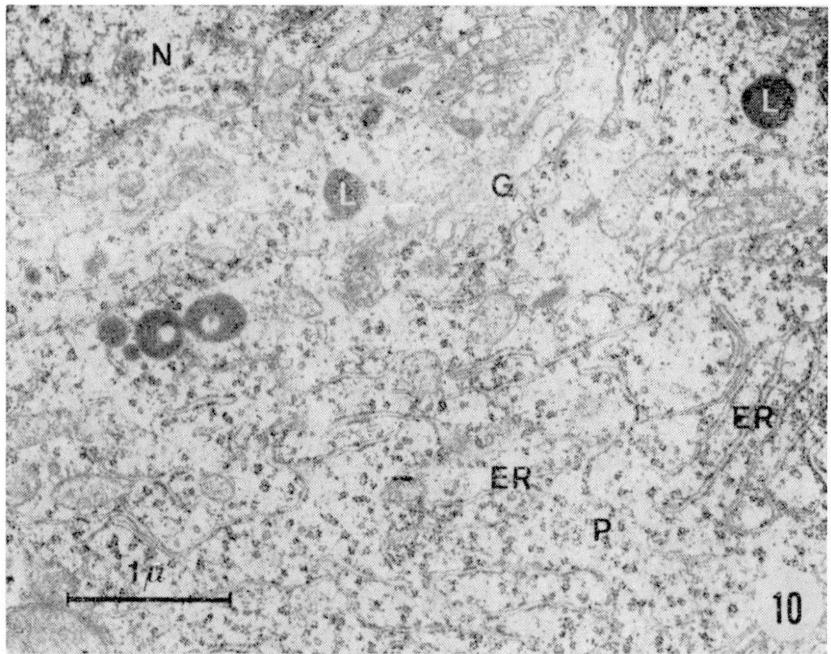

FIG. 10. Ventromedial nucleus. The cell shown here is similar to those of the anterior nucleus, but is richer in endoplasmic reticulum (ER) and polysomes (P). The Golgi apparatus (G) is still quite small and no secretory granules are present. For additional abbreviations see Fig. 1.

smaller cell of this nucleus. The significance and the physiological role of the two cell types, one of which is larger and richer in endoplasmic reticulum, is still unclear. The endocrinological function of this nucleus is also poorly understood. Part of the nucleus seems to control the release of Follicle-Stimulating Hormone (FSH) from the anterior pituitary (Flament-Durand, 1965). Recently, Frohman and co-workers (1968) presented strong evidence favoring the view that the ventromedial nucleus might be the hypothalamic site responsible for the control of Growth Hormone (GH) secretion. The possible localization of the neurons sensitive to glucose in this nucleus (Oomura *et al.*, 1969) is a stimulating idea if one keeps in mind the intimate relationships between GH and glucose metabolism.

VI. Arcuate Nucleus

The arcuate nucleus is located in the midposterior portion of the hypothalamus, in the periventricular region beneath the ventral zone of the infundibular recess, and below the ependymal cells lining the third ventricle. It contains round, clear cells. In the rat, we observed only one cell type, which is similar to those described in guinea pig studies (Mazzucca, 1968). However, some small differences were evident among these cells, apparently due to their functional status at the moment of fixation. We feel that the two types of cells described by Zambrano and De Robertis (1968) in the rat could probably be interpreted as different functional stages of the same type of cell. The cell nucleus is central and round and does not show infoldings of the nuclear membrane. The matrix of the cytoplasm is clear and the Golgi apparatus is the most highly developed structure; it occupies a large area and comprises numerous sacs and vacuoles (Figs. 11–13). Vesicles containing a dense substance and numerous microtubules are also present. The rough endoplasmic reticulum is formed by parallel elements at the periphery of the cell and by a few flattened cisternae scattered in the cytoplasm (Figs. 11, 12); polysomes are also numerous. The mitochondria are small and elongated. The lysosome-like bodies contain a dense, finely granular substance. These cells seem clearly to be of the secretory type.

The arcuate nucleus is probably the site where production of several releasing factors occurs. In fact it seems that the basal secretion of LH is controlled by this nucleus (Szentágothai *et al.*, 1968); data reported by other authors indicate that FSHRF is also present in this area (Watanabe and McCann, 1968). The arcuate nucleus is a part of the hypophysiotropic area that Hálasz *et al.* (1962), Flament-Durand (1965), and Szentágothai and co-workers (1968) postulated to control the anterior pituitary. The fine structure of the cells of the arcuate nucleus support this possibility; in fact our observations indicate that these cells could very well have a secretory activity. However, since we do not know the precise chemical nature of the releasing factors nor their site of synthesis and storage in the cells, it is rather difficult to draw final conclusions based only on morphological data.

Some recent results obtained with fluorescence techniques have

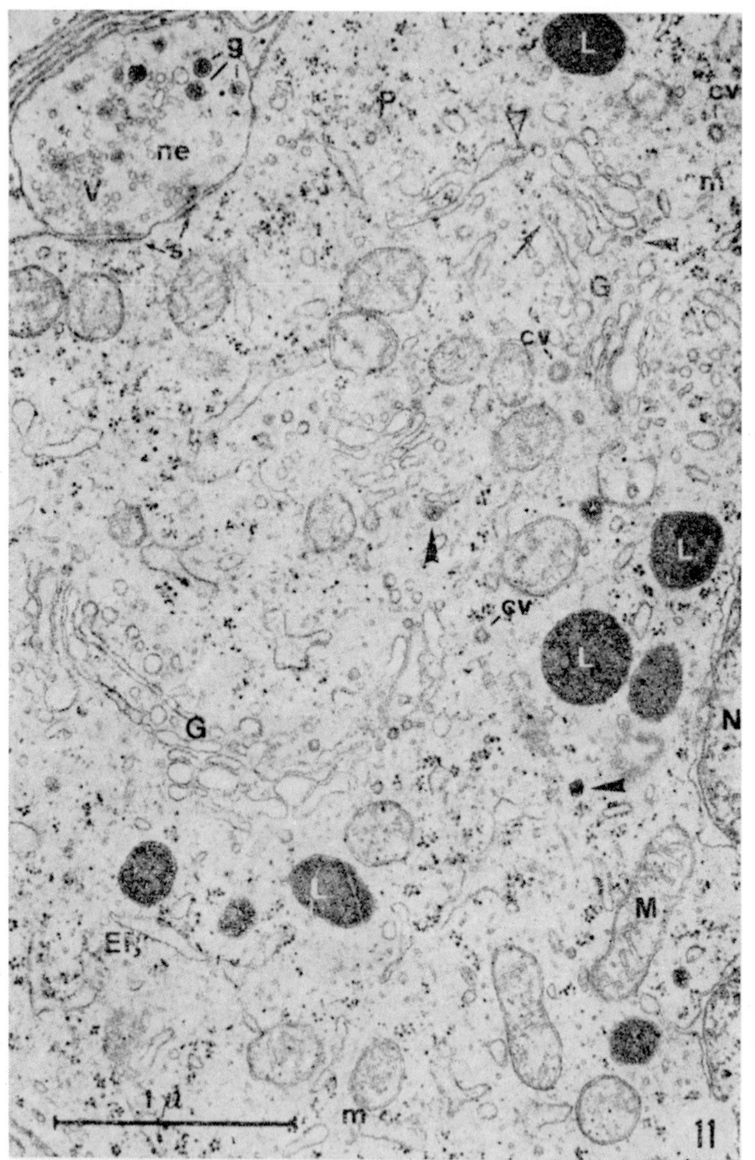

FIG. 11. Neuron of the arcuate nucleus. The large Golgi apparatus characteristic of these cells is formed by numerous stacks and vesicles which sometimes contain a dense substance (—▶). Secretory granules are also present (▶). In the Golgi area (G) it is possible to see quite a number of coated vesicles probably derived by budding from the endoplasmic reticulum (▷); microtubules (m) and lysosomes (L) are also present in the cytoplasm. The endoplasmic reticulum (ER) is formed by few elements containing polysomes (P); the latter are also numerous and free in the cytoplasm. In the left upper corner a nerve ending (ne) is present; this makes contact with the cell body at two points (S); it contains clear synaptic vesicles (V) and dense-core granules (g), which seem to have close relation with the membrane at the point of the synaptic junctions. For additional abbreviations see Fig. 1.

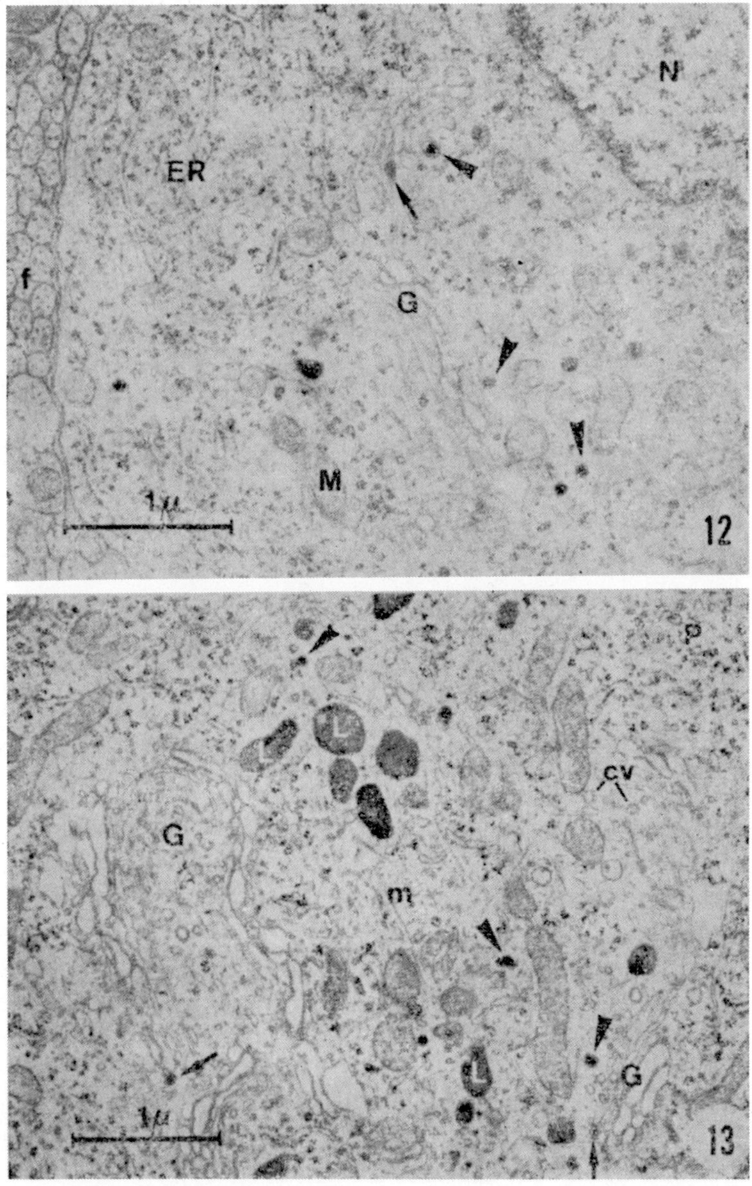

FIG. 12. Arcuate nucleus. This neuron is very rich in secretory granules (▶) and in granules being formed in the Golgi cisternae (—▶); the rough endoplasmic reticulum is also very abundant. Numerous axons profiles (f) are observed outside the neurons.
FIG. 13. Arcuate nucleus. A portion of the cytoplasm of a neuron containing numerous lysosomes and a rather large Golgi apparatus (G) is shown. A dense substance is also present in some elongated cisternae of the Golgi apparatus (▶) and numerous secretory granules (—▶) are also found. Polysomes (P) are particularly numerous in both free and bound forms. Microtubules (m) are also seen. For additional abbreviations see Fig. 1.

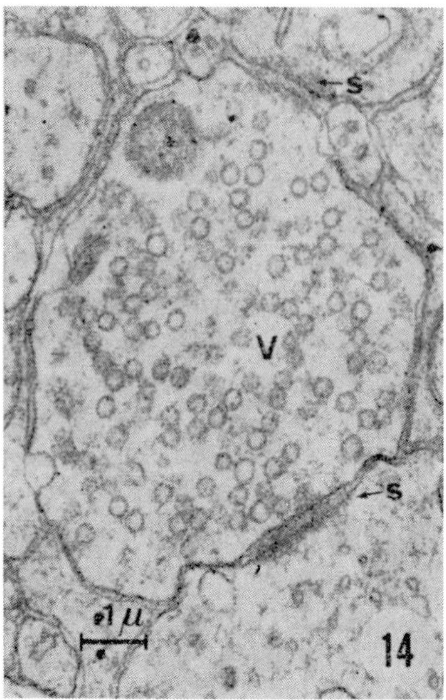

FIG. 14. Arcuate nucleus. A nerve ending making contact with a cell dendrite is shown. The presynaptic enlargement contains only synaptic vesicles evenly distributed in the presynaptic area. For abbreviations see previous Figures.

shown that about 10–15% of the cells of the arcuate nucleus contain a large amount of dopamine (Fuxe and Hökfelt, 1967, 1969). In fact these cells give origin to the dopaminergic fibers of the tubero-infundibular system ending in the median eminence. However, since we could not find different types of neurons in the arcuate nucleus, it is very difficult to assess which cells produce the releasing factors and which produce dopamine. The afferences to the arcuate cells are numerous and have different morphological aspects. The axosomatic and axodendritic synapses are of two types: one contains clear synaptic vesicles (Fig. 14) and the other contains synaptic vesicles and dense-core granules as well (Fig. 11).

VII. Median Eminence

By means of optical microscopy the median eminence of the rat can be divided into three regions: the ependymal layer, the inner zone, and the external zone or palisade (Röhlich *et al.*, 1965; Kobayashi and Matsui, 1969). These regions can now be more clearly defined by morphological differences at the ultrastructural level (Rinne, 1966; Monroe, 1967; Rodriguez, 1969; Akmayev, 1969). The ependymal cells bordering the third ventricle are long cuboidal cells; their surface facing the ventricle has many microvilli, large bulbous expansions (Leonhardt, 1968) and cilia of the usual structure, and few pinocytotic vesicles. The junctions between the ependymal cells are formed by numerous zonulae occludens and desmosomes, and by numerous infoldings of the cell membranes. A large Golgi apparatus is present in the cytoplasm, formed by packed cisternae and vesicles. The endoplasmic reticulum is limited to a few elements containing polysomes, the latter being mainly free in the cytoplasm. The mitochondria are small and few. Lysosome-like bodies are abundant. Some of the ependymal cells are very long and cross the entire median eminence, reaching with their processes the external zone near the capillaries of the portal system (Figs. 15, 16). The ependymal cells do not have a fine structure to support the idea that they have a secretory activity. The fact that ependymal cells are bordering the third ventricle and ending with foot processes on the neurohemal part of the median eminence near the terminals of the monoaminergic and neurosecretory systems suggests that these cells probably have an important function in the physiology of the median eminence.

Beneath the layer of ependymal cells we have found many axons containing large, dense granules of about 1300 Å diameter. They represent the fibers forming the hypothalamo-neurohypophyseal tract which transports the classical neurosecretory granules. Sometimes these fibers show sizeable enlargements filled with secretory granules and with other cytoplasmic organelles that have been described as "Herring bodies."

The palisade or external zone of the median eminence is characterized by a complex relationship between the capillaries of the portal system and the fiber terminals of the tubero-hypophyseal system. At low magnification, micrographs of this area (Fig. 15) show that numerous nerve fibers end in close proximity to the fenestrated capillaries of the portal system and are separated from them only by a few collagen fibers and by two thin basement membranes. Foot processes of the ependymal cells can be often seen in this region (Figs. 15, 16). The nerve endings also can be divided into two main

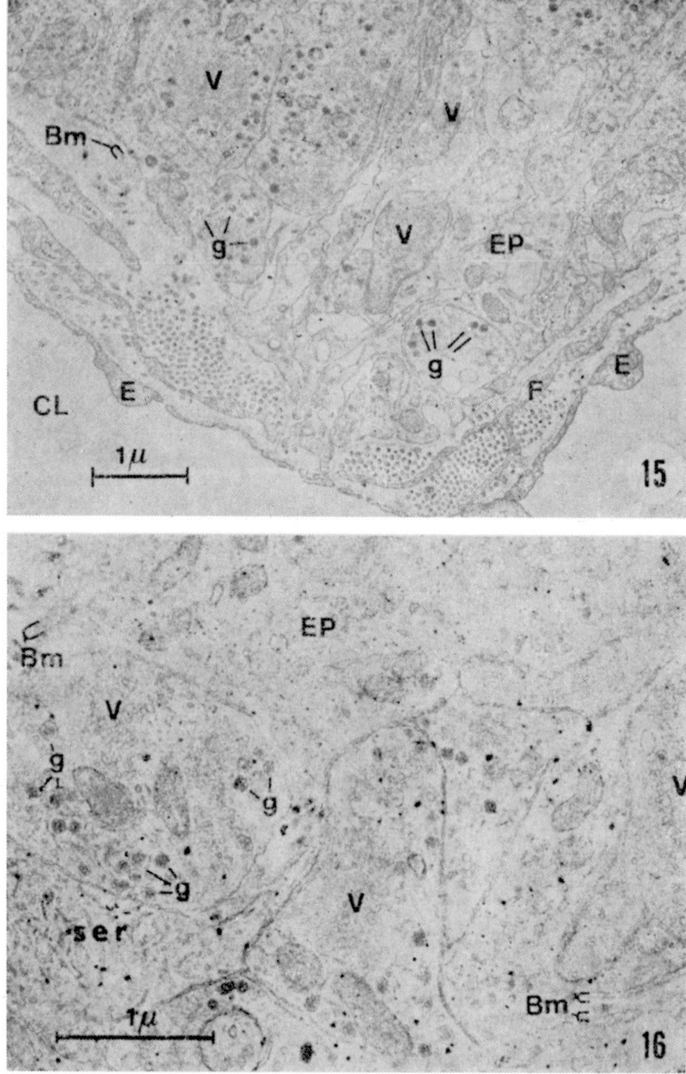

FIG. 15. Neurohemal or external part of the median eminence. This low-magnification picture shows the general structure of this part of the median eminence that is characterized by the ending of numerous nerve terminals near the fenestrated capillaries (E) of the pituitary portal system. Some ependymal cells (EP) are also present among the nerve endings. The capillary is separated from the nerve endings by two thin basement membranes (Bm) and sometimes by fibroblasts (F).

FIG. 16. External part of the median eminence. At higher magnification it is possible to see in more detail the structure of nerve endings. The synaptic vesicles (V) and the granules (g) are evenly distributed in the nerve endings and no clumps of vesicles are visible near the cell membranes. In one nerve terminal a large number of sacculi or profiles resembling smooth endoplasmic reticulum (ser) are present and intermingled with synaptic vesicles, suggesting a probable connection of the latter with the cisternae of endoplasmic reticulum. For additional abbreviations see previous Figures.

types. Some of them contain only clear synaptic vesicles of about 500 Å diameter (Fig. 17) and larger clear vesicles that are variable in diameter and often elongated. With good fixation methods these vesicles appear evenly distributed in the nerve terminal and not clumped or aggregated. The second type of nerve endings which are more numerous, contain these synaptic vesicles as well as numerous dense-core granules (Figs. 15–18). They are surrounded by a typical three-layered membrane similar to that of synaptic vesicles (Fig. 18). The dense-core granules are often at a distance from the plasma membrane facing the basement membrane, but can also be found very near to it. Another type of cell process is present near the basement membrane and contains a few synaptic vesicles which are clumped and retained in a small limited area near the plasma membrane. The remainder of the process is filled with filament-like substance (Fig. 17). These processes are infrequent and quite different from the previously described synapses which are completely filled with synaptic vesicles. It is not clear at present whether these processes are of nervous or ependymal origin.

The tubero-infundibular neurosecretory system and the dopaminergic tubero-infundibular system terminate in this neurohemal zone. Biochemical and endocrinological experiments show that this region contains all the releasing factors (Szentágothai *et al.*, 1968) and also a large quantity of dopamine and noradrenaline (Fuxe and Hökfelt, 1969). It would be therefore very important to know which terminals contain the releasing factors and which contain dopamine. At present no answer to this question can be given. There is evidence indicating that dopamine (like other amines in peripheral and central nervous tissue) is stored in nerve endings and particularly in dense-core granules (De Robertis *et al.*, 1965). The number of the granules in the nerve endings of the median eminence is roughly proportional to the content of catecholamines present in the region, both in normal and experimental situations (Rinne and Arstila, 1966; Mazzuca, 1966; Pellegrino de Iraldi and Etcheverry, 1967; Matsui, 1967; Rinne *et al.*, 1967; Pfeifer *et al.*, 1968). It is interesting that a similar correlation exists also in castrated animals in which both the number of dense granules and the hypothalamic content of catecholamine (Donoso and Stefano, 1967) and especially of dopamine (Fuxe and Hökfelt, 1969) increase.

The releasing factors are also present in nerve endings of the basal hypothalamus (Clementi *et al.*, 1970) and there is positive evidence in support of their storage in dense granules. After castration the number of granules and the number of clear vesicles increase in the nerve endings of the median eminence. According to Kobayashi and Matsui (1969) this indicates a relation between granules and gonadotropin releasing factors. Moreover, dense granules increase in size after adrenalectomy and many of them lose their central core, wich suggests (Akmayev *et al.*, 1967) that they are correlated with the storage of the Corticotropin Releasing Factor (CRF). Even stronger evidence is given by the experiments of Ishii and his co-workers (1970); using gradient centrifugation, they separated from horse median eminence a fraction rich in dense granules and containing CRF activity. Unfortunately there was a high content of vasopressin in these fractions; this could have been responsible for the CRF activity. The nature of the substance contained in the clear synaptic vesicles of this area is unknown. Some authors have postulated that they might contain acetylcholine (Kobayashi and Matsui, 1969). However, the cholinesterase content of the median eminence seems too low to support this assumption (Shute and Lewis, 1966). These clear vesicles could also be the storage sites of dopamine and may have lost their dense aspect as a result of aldehyde fixation. With permanganate fixation it is in fact possible to show some dense material inside the usually clear vesicles (Hökfelt, 1967).

VIII. Pars Tuberalis

In the rat the cells of the pars tuberalis are numerous: they are located very close to the median eminence and to the infundibulum. These are large cells with a central nucleus; in this region there are wide intercellular spaces where microvilli and cilia are projected. A prominent Golgi apparatus is present in the cytoplasm and is formed by numerous cisternae and vesicles; however, vesicles containing dense material are rare. The endoplasmic reticulum is formed by large flattened sacs, with some bound polysomes. Free polysomes are more numerous. Numerous, large, dense, secretory granules limited by a membrane are also present in the cytoplasm. Lysosome-like bodies and many glycogen granules are scattered throughout the cytoplasm (Figs. 19, 20). Because of their typical fine structure,

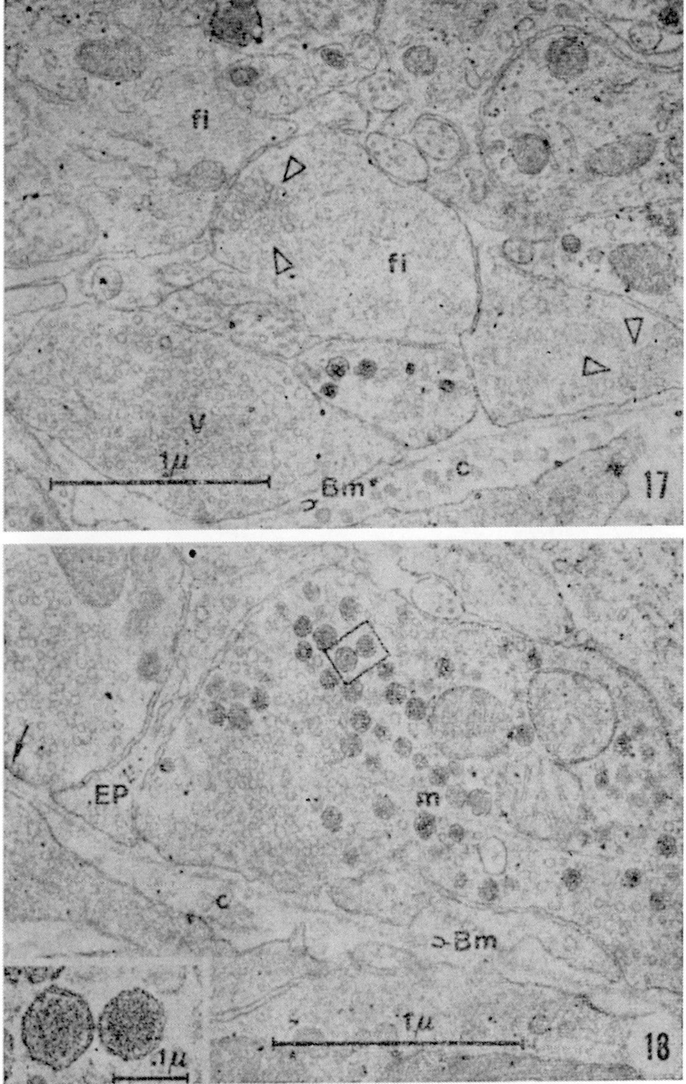

FIG. 17. External part of the median eminence. A nerve ending is shown which contain only clear vesicles (V) that have different shape and size and fill the entire nerve profile. Serial sections have shown that dense granules are not contained in this type of nerve ending. Profiles containing a small number of vesicles clumped and placed eccentrically near the cell membrane (▷) are also present. These also contain a large amount of filamentous-like material (fi). It is not yet clear if these processes are of nervous or ependymal origin.

FIG. 18. External part of the median eminence. A detail of two nerve endings is shown. The dense-core granules and the synaptic vesicles are scattered in the cytoplasm without any specific localization; microtubules (m) are also present. A thickening of the part of the membrane facing the basement membranes is frequently observed (—▶). In the inset two dense-core granules are shown at higher magnification. For additional abbreviations see previous Figures.

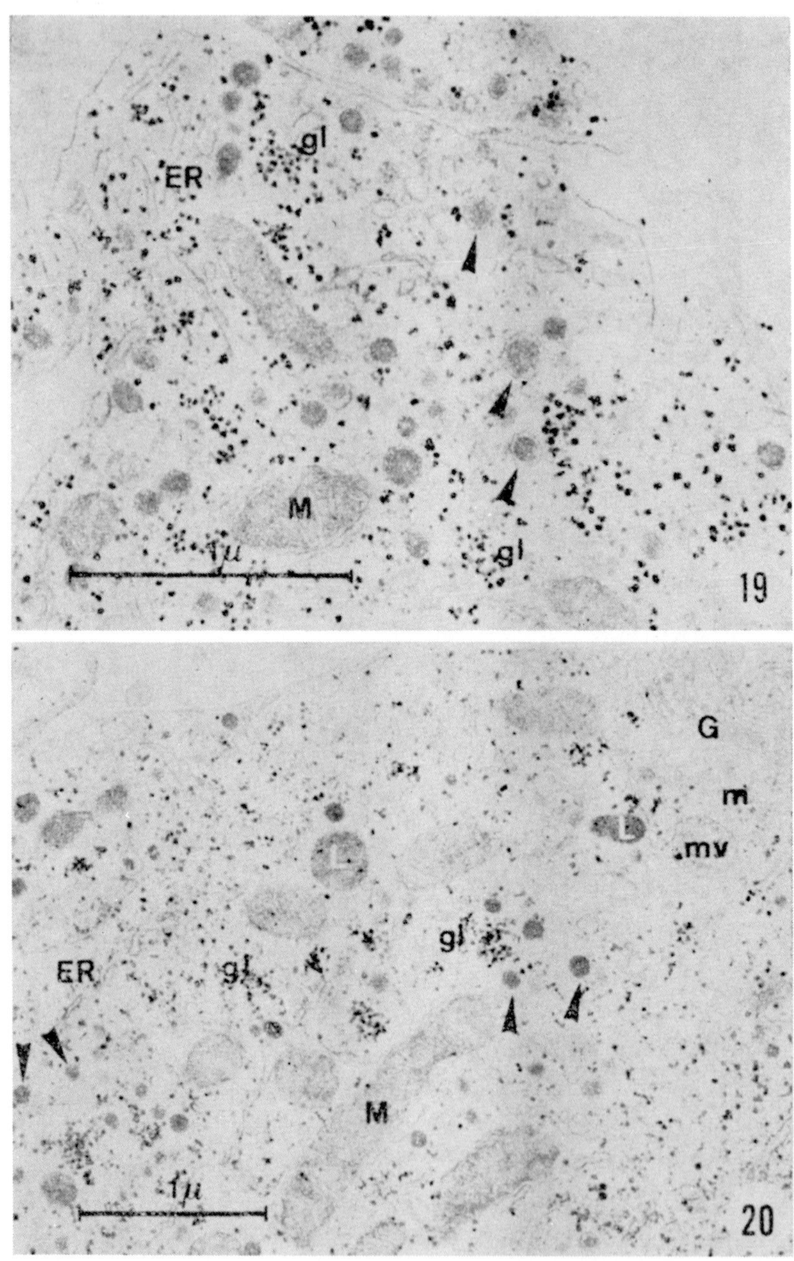

FIG. 19. Secretory cell of the pars tuberalis. The cell is characterized by numerous dense granules (▶) similar to secretory granules of endocrine cells, few cisternae of endoplasmic reticulum (ER), and a large number of glycogen granules (gl).

FIG. 20. Secretory cell of the pars tuberalis showing in detail the Golgi apparatus (G) and the relationships of the cisternae of the endoplasmic reticulum (ER) with Golgi apparatus, mitochondria (M), and glycogen granules (gl). A multivesicular body (mv) is also present. For additional abbreviations see previous Figures.

these cells can be classified as secretory.

The few reports concerning the fine structure of the cells present in this zone are not in agreement. Kobayashi *et al.* (1963) reported no secretory granules in the cytoplasm, and Stutinski *et al.* (1964) and Rinne (1966) described two types of secretory cells very similar to the gonadotropic and corticotropic cells of the anterior pituitary. Our findings confirm the presence of secretory cells; however, they do not appear similar to those found in the pars distalis. The main differences are: a high content of glycogen granules, the form of the endoplasmic reticulum, and the presence of numerous cilia and microvilli. The importance of the pars tuberalis for the function of the hypothalamo-pituitary system becomes evident if one considers that the blood supply from the median eminence to the anterior pituitary may pass through the pars tuberalis. It is also possible that part of the blood in contact with the cells of the pars tuberalis might reach the hypothalamus through a reverse flow (Szentágothai *et al.*, 1968). Therefore it seems reasonable that the cells of the pars tuberalis, in view of their strategic position, could play an important role in the " short " feedback mechanisms controlling the anterior pituitary (see the article by Motta and associates in this volume).

IX. Blood Capillaries

The blood capillaries of all hypothalamic nuclei have a structure similar to that of brain capillaries (Fig. 21). They have a continuous endothelium which forms a rather thick wall; the cell junctions are of the type described as " tight junctions " with fusion between the two opponent cell surface membranes. Plasmalemmal vesicles are not numerous and it is difficult to see any opening at the blood or tissue fronts. There is a thick basement membrane beneath the endothelium which probably results from the fusion of the basement membrane of the endothelium with that of the underlying pericytes or neurons. There is practically no connective tissue between the capillaries and the nervous tissue.

The pericytes lie like a second layer below the endothelium. In the arcuate nucleus, however, the pericytes are less numerous and very often the cells of the nucleus border the capillaries (Fig. 21). The permeability of this type of capillary has been recently investi-

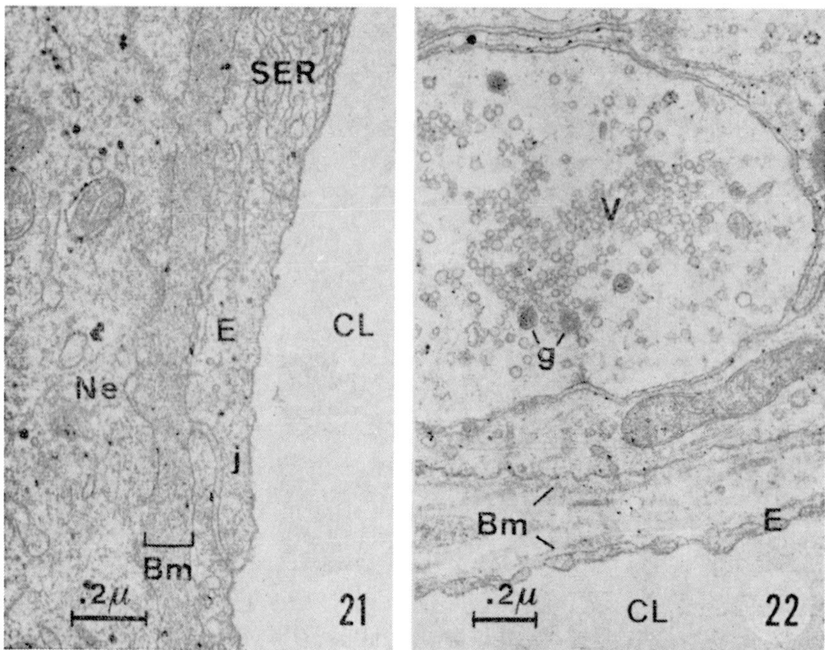

FIG. 21. Structure of endothelium of blood capillaries in the arcuate nucleus. The endothelium is rather thick, with few plasmalemmal vesicles opening in blood or tissue front and with a tight junction (j) between two endothelial cells. The basement membrane between endothelial cells and neurons (Ne) is very thick and probably results from the fusion of the basement membranes of capillary endothelium and neuron; CL, capillary lumen; SER, smooth endoplasmic reticulum.

FIG. 22. Wall of a capillary in the external part of the median eminence. The endothelium is fenestrated, the basement membrane is very thin, and the space between nerve endings and capillary is occupied by few collagen fibers. For additional abbreviations see previous Figures.

gated (Reese and Karnovsky, 1967). It has been shown that they are relatively impermeable to both small and large molecules. It is also known that monoamines and other substances (e.g., Adrenocorticotropic Hormone) (Ferrari, 1958) are not able to cross the blood/brain barrier at the hypothalamic level.

The capillaries of the median eminence have a completely different appearance. They are formed by a highly fenestrated, flattened endothelium; they are separated from the nervous tissue by a large pericapillary space in which we have found a few pericytes and a loose network of collagen fibers. The basement membrane of the endothelium is very thin and never fuses with the tissue basement membrane (Fig. 22). This type of capillary has a high permeability to small and large molecules (Clementi and Palade, 1969) and is present in all endocrine glands and other tissues where tissue/blood exchanges are rapid and intense. These findings suggest that an exchange of large molecules (such as releasing factors, dopamine, and even hormones from the anterior pituitary) between blood and nervous tissue could take place only in the external zone of the median eminence and in the pars tuberalis. These molecules could not cross the capillary walls very easily in other regions of the hypothalamus.

X. Conclusions

From the data reported here on the structure of some hypothalamic nuclei it seems useful to divide the hypothalamic cells into three types: (1) neurons present in the supraoptic and paraventricular nuclei and which have a typical neurosecretory structure; (2) neurons present in the anterior, suprachiasmatic, and ventromedial nuclei which have a fine structure similar to the neurons present in other areas of the central nervous system; and (3) cells present in the arcuate nucleus and which are neurosecretory in nature but which have a structure different from that of supraoptic neurons. These findings are relevant in the light of the possibility that releasing and inhibiting factors which regulate the activity of the anterior pituitary are synthesized in some of these nuclei. However, only the cells in the arcuate nucleus have a morphological structure that suggest that they could produce such factors and export them in granules.

The releasing factors in the median eminence are accumulated in the nerve endings and probably in granules. We do not know at present if the releasing factors and dopamine (which seems to be the mediator that controls the release of such substances) (Schneider and McCann, 1969) are stored in the same nerve endings or in different ones. Combined morphological and biochemical research will undoubtedly throw some light on the problem of the precise localization of the structures that synthesize, store, and release the releasing and inhibiting factors.

ACKNOWLEDGMENT

This work was supported in part by a grant from C.N.R.

The skilful assistance of Mrss. F. Crippa, B. Jezzi, P. Oldani. and P. Tinelli was highly appreciated.

REFERENCES

Abrahams, V. C., Koelle, G. B., and Smart, P. (1957). *J. Physiol.* (London) **139**, 137.

Acher, R., Chauvet, J., and Olivry, G. (1956). *Biochim. Biophys. Acta* **22**, 428.

Akmayev, I. G. (1969). *Z. Zellforsch.* **96**, 609.

Akmayev, I. G., Réthelyi, R., and Majorossy, K. (1967). *Acta Biol. Acad. Sci. Hung.* **18**, 187.

Antunes-Rodriguez, J., and McCann, S. M. (1967). *Endocrinology* **81**, 666.

Bargmann, W. (1968). *In* "Handbook of Experimental Pharmacology, Neurohypophyseal Hormones and Similar Polypeptides" (B. Berde, ed.). Vol. XXIII, pp. 1–39. Springer, New York.

Bargmann, W., and Sharrer, E. (1951). *Am. J. Sci.* **39**, 255.

Bern, H. A., and Knowles, F. G. (1966). *In* "Neuroendocrinology" (L. Martini and W. F. Ganong, eds.). Vol. I, pp. 137–189. Academic Press, New York.

Bindler, E., La Bella, F. S., and Sanwall, M. (1967). *J. Cell Biol.* **34**, 185.

Brooks, C. McC., Ishikawa, T., Koizumi, K., and Lu, H. H. (1966). *J. Physiol.* (London) **182**, 217.

Ceccarelli, B., and Pensa, P. (1968). *Lo Sperimentale* **118**, 197.

Clementi, F., and De Virgiliis, G. (1967). *Path. Biol.* **15**, 119.

Clementi, F., and Palade, E. G. (1969). *J. Cell Biol.* **41**, 33.

Clementi, F., Ceccarelli, B., Cerati, E., De monte, M. L., Felici, M., Motta, M., and Pecile, A. (1970). Proc. Intern. Symposium on Cell Biology and Cytopharmacology. (In press).

Davidson, J. M. (1966). *In* "Neuroendocrinology" (L. Martini and W. F. Ganong, eds.). Vol. I, pp. 565–611. Academic Press, New York.

De Robertis, E., Pellegrino de Iraldi, A., Rodriguez de Lores Arnaiz, G., and Zieher, L. M. (1965). *Life Sciences* **4**, 193.

Donoso, A. O., and Stefano, F. J. E. (1967). *Experientia* **23**, 665.

Everett, J. W. (1964). *Physiol. Rev.* **44**, 373.

Ferrari, W. (1958). *Arch. Ital. Sci. Farmacol.* **8**, 131.

Flament-Durand, J. (1965). *Endocrinology* **77**, 446.

Flament-Durand, J., and Desclin, L. (1968). *J. Endocrinol.* **41**. 531.

Fraschini, F., and Motta, M. (1967). *Program 49th Meeting Endocrine Soc.*, p. 128.

Frohman, L. A., Bernardis, L. L., and Kant, K. J. (1968). *Science* **162**, 580.

Fuxe, K., and Hökfelt, T. (1967). *In* "Neurosecretion" (F. Stutinsky, ed.), pp. 173–177. Springer-Verlag, Berlin.

Fuxe, K., and Hökfelt, T. (1969). *In* "Frontiers in Neuroendocrinology 1969" (W. F. Ganong and L. Martini, eds.), pp. 47–97. Oxford Univ. Press, New York.

Ginsburg, M. (1968). *In* "Handbook of Experimental Pharmacology". (B. Berde, ed.). Vol. XXIII, pp. 286–371. Springer-Verlag, Berlin.

Ginsburg, M., Jayasena, K., and Thomas, P. J. (1966). *J. Physiol.* (London) **183**, 145.

Halász, B., Pupp, L., and Uhlarik, S. (1962). *J. Endocrinol.* **25**, 147.

Herlant, M. (1967). *In* "Neurosecretion" (F. Stutinski, ed.), pp. 20–35. Springer-Verlag, Berlin.

Hökfelt, T. (1967). *Brain Res.* **5**, 121.
Hökfelt, T. (1968). *Z. Zellforsch.* **91**, 1.
Ishii, S., Iwata, T., and Kobayashi, H. (1969). *Endocrinol. Jap.* **16**, 171.
Karnovsky, M. J. (1965). *J. Cell Biol.* **27**, 137.
Kato, J., and Villee, A. C. (1967). *Endocrinology* **80**, 567.
Kobayashi, H., and Matsui, T. (1969). *In* "Frontiers in Neuroendocrinology 1969" (W. F. Ganong and L. Martini, eds.). pp. 3–46. Oxford Univ. Press, New York.
Kobayashi, T., Kobayashi, T., Yamamoto, K., and Inatomi, M. (1963). *Endocrinol. Jap.* **10**, 69.
König, J. F. R., and Klippel, R. A. (1963). "The Rat Brain." Williams and Wilkins, Baltimore.
Konstantinova, M. (1967). *Z. Zellforsch.* **83**, 549.
Köves, K., and Halász, B. (1969). *Neuroendocrinology* **4**, 1.
Lederis, K. (1961). *Gen. Comp. Endocrinol.* **1**, 80.
Lederis, K. (1969). *In* "Advances in the Biosciences" (G. Raspé, ed.). Vol. I, pp. 155–166. Pergamon Press-Vieweg, Braunschweig.
Leonhardt, H. (1968). *Z. Zellforsch.* **88**, 297.
Marsala, J. (1968). *Folia Morphol.* **16**, 348.
Matsui, T. (1967). *Neuroendocrinology* **2**, 99.
Mazzucca, M. (1966). *J. Microscop.* (*Paris*) **5**, 63a.
Mazzucca, M. (1968). *J. Microscop.* (*Paris*) **7**, 135.
McCann, S. M. (1968). *Am. J. Physiol.* **202**, 395.
McCann, S. M., and Dhariwal, A. P. S. (1966). *In* "Neuroendocrinology" (L. Martini and W. F. Ganong, eds.). Vol. I, pp. 261–296. Academic Press, New York.
Mess, B., and Martini, L. (1968). *In* "Recent Advances in Endocrinology" (V. H. T. James, ed.), pp. 1–49. J. and A. Churchill, London.
Mess, B., Fraschini, F., Motta, M., and Martini L. (1967). *In* "Hormonal Steroids" (L. Martini, F. Fraschini and M. Motta, eds.), pp. 1004–13. Excerpta Medica, Amsterdam.
Monroe, B. G. (1967). *Z. Zellforsch.* **76**, 405.
Nemetschek-Gansler, H. (1965). *Z. Zellforsch.* **67**, 844.
Nibelink, D. W. (1961). *Am. J. Physiol.* **200**, 1229.
Odake, G. (1967). *Z. Zellforsch.* **82**, 46.
Olivecrona, H. (1957). *Acta Physiol. Scand.* **40**, *Suppl.* 136.
Oomura, Y., Ono, T., Ooyama, H., and Wayner, M. J. (1969). *Nature* **222**, 282.
Osinchak, J. (1964). *J. Cell Biol.* **21**, 35.
Palay, S. Z. (1960). *Anat. Record* **138**, 417.
Pecile, A., and Müller, E. (1966). *In* "Neuroendocrinology" (L. Martini and W. F. Ganong, eds.). Vol. I, pp. 537–564. Academic Press, New York.
Pfeifer, A. K., Szabo, D., Palkovits, M., and Okrös, I. (1968). *Exp. Brain Res.* **5**, 79.
Pellegrino de Iraldi, A., and Etcheverry, G. J. (1967). *Brain Res.* **6**, 614.
Pepler, W. J., and Pearse, A. G. E. (1957). *J. Neurochem.* **1**, 193.
Ramchandran, L. K., and Winnick, K. (1957). *Biochim. Biophys. Acta* **23**, 533.
Reese, T. S., and Karnovsky, M. J. (1967). *J. Cell Biol.* **34**, 207.
Rinne, U. K. (1966). *Z. Zellforsch.* **74**, 98.
Rinne, U. K., and Arstila, A. U. (1966). *Med. Pharmacol. Exp.* **15**, 357.
Rinne, U. K., Sonninen, V., and Helminen, H. (1967). *Med. Pharmacol. Exp.* **17**, 108.
Rodriguez, E. M. (1969). *Z. Zellforsch.* **93**, 182.
Röhlich, P., Vigh, B., Teichmann, I., and Aros, B. (1965). *Acta Biol. Acad. Sci. Hung.* **15**, 431.
Sachs, H., and Takabatake, Y. (1964). *Endocrinology* **75**, 943.

Sachs, H., Portanova, R., Haller, E. W., and Share, L. (1967). *In* " Neurosecretion " (F. Stutinsky, ed.), pp. 146–154. Springer Verlag, Berlin.
Schneider, H. P. G., and McCann, S. M. (1969). *Endocrinology* **89**, 121.
Shute, C. C. D., and Lewis, P. R. (1966). *Brit. Med. Bull.* **22**, 221.
Sokol, H. W., and Valtin, H. (1967). *Nature* **214**, 314.
Stumpf, W. E. (1968). *Science* **162**, 1001.
Stutinsky, F., Porte, A., and Stoeckel, M. E. (1964). *C .R. Acad. Sci. (Paris)* **259,** 1765.
Szentágothai, J., Flerkó, B., Mess, B., and Halász, B. (1968). " Hypothalamic Control of the Anterior Pituitary." Akadémiai Kiadó, Budapest.
Van Dyke, H. B., Chow, B. F., Greep, R. O., and Rothen, A. (1941). *J. Pharmacol. (Kyoto)* **74**, 190.
Watanabe, S., and McCann, S. M. (1968). *Endocrinology* **82**, 664.
Zambrano, D., and De Robertis, E. (1966a). *Anat. Record* **54**, 446.
Zambrano, D., and De Robertis, E. (1966b). *Z. Zellforsch.* **73**, 414.
Zambrano, D., and De Robertis, E. (1967a). *Z. Zellforsch.* **81**, 264.
Zambrano, D., and De Robertis, E. (1967b). *Z. Zellforsch.* **76**, 458.
Zambrano, D., and De Robertis, E. (1968). *Z. Zellforsch.* **87**, 409.

Hypothalamic Neurosecretion

F. S. Stutinsky

I. Definition

The concept of neurosecretion is everywhere accepted today by all biologists. It signifies, as stated recently by Bargmann (1966), " the production and the release of hormones by nerve cells that have the cytological characteristics of secretory cells." In their monograph " Neuroendocrinology " Scharrer and Scharrer (1963) wrote: " The term *neurosecretion* should not be applied to the humoral transmission of nervous impulses by, for example, acetylcholine, and should not be extended to nerve cells simply because they contain electron dense granules." But a good definition of neurosecretion is still under discussion: the cytological character of glandular cells can be observed in certain nerve cells (arcuate nucleus) that elaborate neurotransmitters, and it is assumed that their axons terminate on the basal membrane of the same capillaries as the neurosecretory fibers. If we admit that, besides a possible local effect, the neurotransmitter acts on the gland by the way of a vascular link, as it is postulated by Fuxe and Hökfelt (1967), then we must consider the neurotransmitter as a hormone and the cells of the arcuate nucleus as neurosecretory cells. On the other hand, substances that exhibit real hormonal characteristics (i.e., the releasing factors produced in the hypothalamus and reaching the anterior lobe through the portal vessels) presumably originate in nervous cells, the identity of which is still unknown.

All neurosecretory cells are able to conduct nerve impulses and possess the same membrane properties as other nerve cells (Kandel, 1962; Yagi and Bern, 1963) and thus can be called " neuroglandular cells."

II. Technical Remarks

The problem of the selective stainability of neurosecretory elements has been reviewed recently by Bargmann (1966). The different stains (chrome alum hematoxylin, CAH; paraldehyde fuchsin, AF; and pseudoisocyanine) can be used on slides (Bargmann, 1949; Halmi, 1952; Gabe, 1953; Sterba, 1964), or in total brain preparations (Braak, 1962; Oksche *et al.*, 1964; Mautner, 1964). In spite of the great value of these technics for identification and study of neurosecretory cells, the results must be considered very critically under experimental conditions. It has been shown (Stutinsky *et al.*, 1963) that the usually employed AF method remarkably stains lysosomes in the cells of the supraoptic nucleus of the rat, an observation confirmed in electron microscopy by Osinchak (1964), who observed dense bodies of large size with acid phosphatase activity. Unpublished observations by Stutinsky and associates also indicate that when granules of a certain number of nerve terminals in the median eminence of hypophysectomized rats undergo degeneration with coalescence of elementary granules and membrane transformations of the granules, the resulting inclusions become stainable by Periodic Acid-Schiff stain (PAS) and probably also with AF. These figures may then suggest the existence of more neurosecretory material (NSM) than is really present.

It must also be emphasized that these degenerations occur from time to time in normal terminals and that autophagic granules, as well as lysosomes, exhibit a weak yellow-green fluorescence.

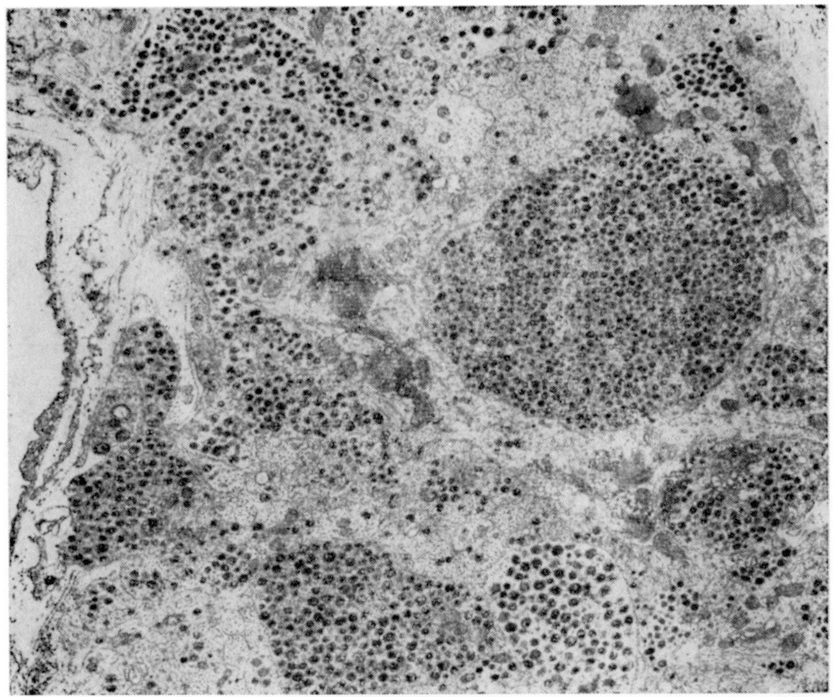

FIG. 1. Neurohypophysis of the rat. Glutaraldehyde-osmium fixation; uranyl acetate and lead citrate stain; numerous dense-core granules (8,500 ×).

Elementary granules that are enveloped by a definite membrane have been identified in the diencephalic neurosecretory nerve fibers by Palay (1955, 1957), Bargmann *et al.* (1957), and many others. The size of the granules ranges from 1000 Å to 3000 Å according to the species. These elementary granules make up the stainable NSM, but it must be kept in mind that isolated granules of this size cannot be seen with the optic microscope. At least three elementary granules must be closely packed together to become visible (Follenius, 1965).

A number of papers have been published concerning different aspects and variations of these granules under different experimental conditions, but fixations formerly used for electron microscopy are unable to preserve correctly the dense core of the elementary granule, as clearly demonstrated by Porte *et al.* (1968). Accordingly, in this article we shall consider only the results obtained after glutaraldehyde-

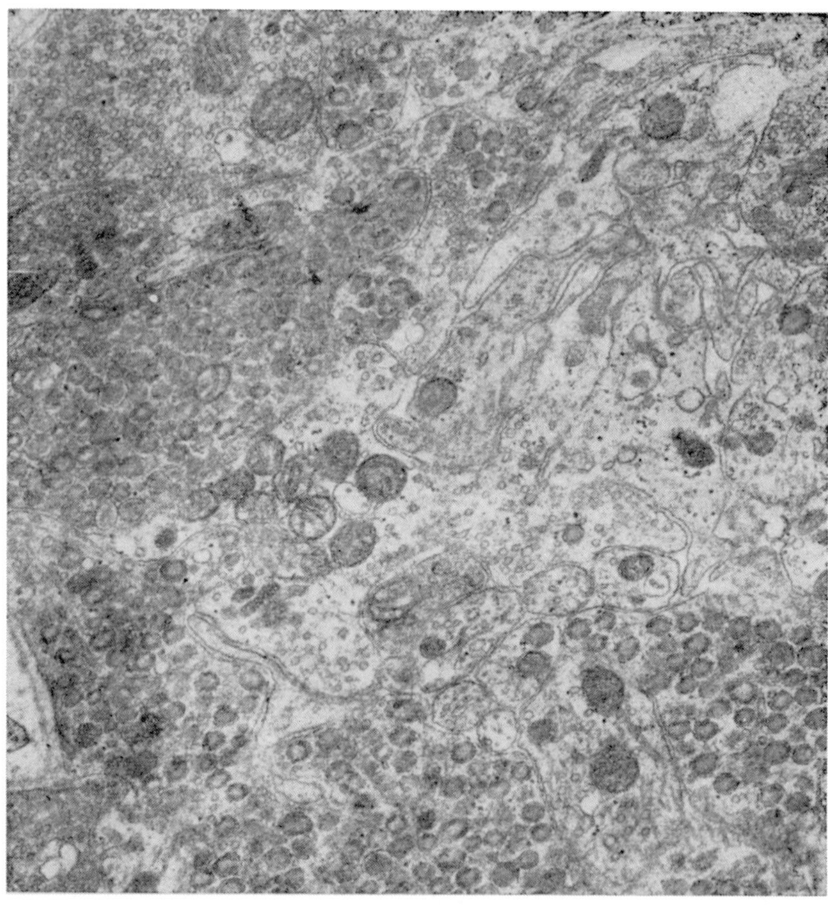

FIG. 2. Neurohypophysis of the rat. Glutaraldehyde-osmium fixation; lead citrate, 20'; translucent granules (23,000 ×).

osmium fixation. After osmium fixation alone the dense-core elementary granules become translucent and many artefacts have thus been described as experimentally induced modifications (Figs. 1, 2).

III. Nervous Properties of Neuroglandular Cells

Neuroglandular cells are first of all also nerve cells. They are stainable by silver methods, are bi-or pluri-polar, and contain Nissl bodies and neurofibrillae. They receive adrenergic afferences (Fuxe and Hökfelt, 1967) and react upon electrical stimulation by secretion of the " posterior lobe " hormones, Antidiuretic Hormone (ADH) and oxytocin (Harris, 1948; Anderson and McCann, 1956; Stutinsky and Guerné, 1965). The nerve cells of the preoptic nucleus in a teleost-fish (*Carassius auratus*) exhibit a membrane potential of 50 mV and an action potential of more than 117 mV, followed by a slow diphasic wave of a hyperpolarizing postpotential (Kandel, 1964).

The neuroglandular cells of the supraoptic nucleus and the paraventricular nucleus can also be stimulated by afferences originating from a large number of nuclei, and the excretion of hormone parallels the decrease of the NSM (Shimazu *et al.*, 1954). The decrease of NSM has also been reported by Jasinski *et al.*, (1967) after an intranasal injection of hypertonic saline, which elicits a marked increase in the firing rate of isolated nerve cells, in a fish (*Carassius auratus*). The decrease of NSM begins in the perikaryon, continues into the axon, and finally occurs in the nerve terminals. Only 60 to 90 min are necessary for an almost complete repletion of the whole neuron.

These recent observations confirm earlier data on the action of hypertonic saline upon the electrical activity of these nerve cells in the rabbit (Cross and Green, 1959) and in the cat (Brooks *et al.*, 1962). Finally, the axons originating from the cells of the supraoptic and paraventricular nuclei form the final common pathway of well-known neuroendocrine reflexes, e.g., the milk-ejection reflex. These neurons, like other nerve cells, also react to pharmacological substances such as acetylcholine (Pickford, 1959).

In summary, these cells receive cholinergic and adrenergic afferences and exhibit all the morphological, electrophysiological, and pharmacological characteristics of nerve cells.

IV. Glandular Properties of Neurosecretory Cells

The neuroglandular cells also show all the structures that have been described in glandular cells. Ergastoplasmic formations and Golgi complexes are well developed and elaborate the characteristic elementary granules that first appear in the Golgi cisternae following the same process and which have been described in other glandular cells (Fig. 3) and as shown by Droz (1965). The immature granules are smaller and posses a more opaque core which only incompletely fills the isolated Golgi vesicle. Their maturation (this is very important for further discussion, see below) occurs on the way from the Golgi complex to the periphery of the cell and the beginning of the axon. Axons contain a solely homogenous population of granules (from about 1500 Å–2000 Å in size). A mixture of immature and adult granules is always seen in the normal cell. Numerous electron-dense bodies of larger size are present; these are the equivalent of multivesicular bodies or lysosomes.

The structures of cells of the supraoptic and paraventricular nuclei are very similar. The glandular character of these neurons is also accentuated by the close relationship that exists between the presence of abundant NSM and the presence of posterior lobe hormones. The distribution of ADH nicely follows that of the Gomori-positive material. Like NSM, ADH is found in the neural lobe and in the hypothalamus. Even three months after hypophysectomy, when the hypothalamus and the median eminence are filled up with Gomo-

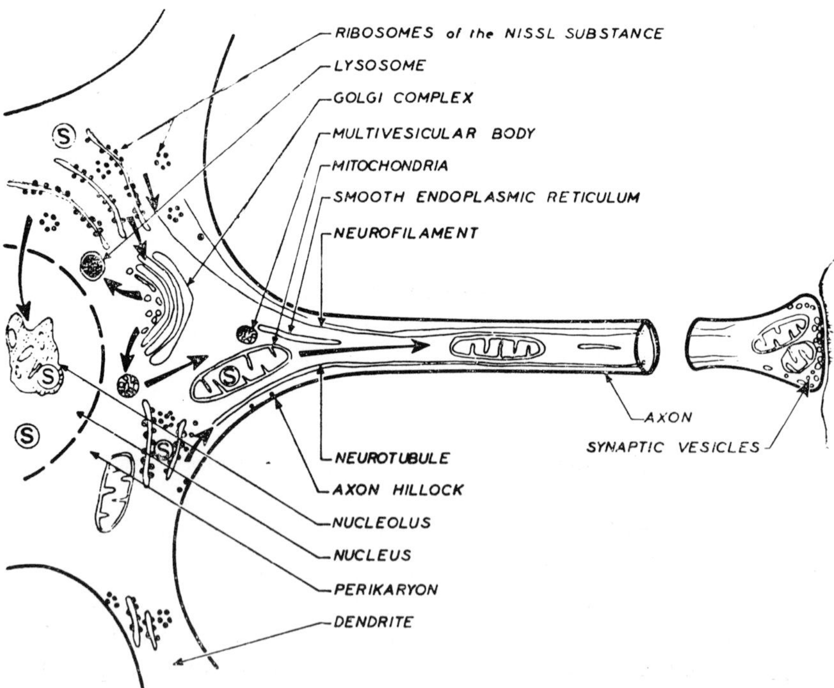

FIG. 3. Diagrammatic drawing of loci of protein synthesis (S) in the neuron (arrows) after injection of tritiated leucine (Courtesy of Prof. Droz, Paris).

ri stained material, the hormone is still present (Diamond, 1952; Stutinsky, 1951). NSM decreases in the posterior lobe after osmotic stimulation, as does the hormone content (Ortmann, 1951). In rats that are stimulated in this way, definite morphological modifications appear in the neurons of the supraoptic and paraventricular nuclei. In the optical microscope the neurons are enlarged, as are the basophilic zone, the Golgi complex, the nuclei, and the nucleoli.

The electron microscope reveals the proliferation of ergastoplasm which accumulates at the periphery of the neuron. The Golgi complexes show hyperactivity and proliferation of saccules in which many immature elementary granules are produced, although the mature granules are very few.

The lysosomes are reduced in number. At the end of the first week the axons are almost empty, and this aspect becomes even more pronounced after three weeks. Nevertheless the decrease of NSM is never complete in the paraventricular nucleus.

Under these conditions the uptake of radioactive cystine is more intense in the neurosecretory neurons, and its transport to the neural lobe as well as its excretion is accelerated. This increased turnover also occurs during lactation and exposure to continuous light (Flament-Durand, 1967). Neuroglandular cells can also be implicated in feedback mechanisms.

V. Neuroglandular Cells and Feedback Mechanisms

A. Gonads

Estrogens act on the neurosecretion, as shown by Gastaldi (1952), Stutinsky (1953a), Legait (1955), Bugnon (1957), and Flament-Durand (1966). The increase is particularly obvious in the neurons, as much in the cells of the supraoptic nucleus as in those of the paraventricular nucleus. This increase is even more prominent after intracerebral implantation of crystals of estradiol (Lisk, 1966; Stutinsky, 1968). This has been confirmed (Fig. 4) by electron microscopy (Stutinsky and associates, unpublished observations). This effect of estrogens can also be noted after early postnatal injections (Kawashima, 1965) and is seen among estradiol-concentrating neurons studied by autoradiography; those of the paraventricular nucleus are clearly involved (Stumpf, 1968).

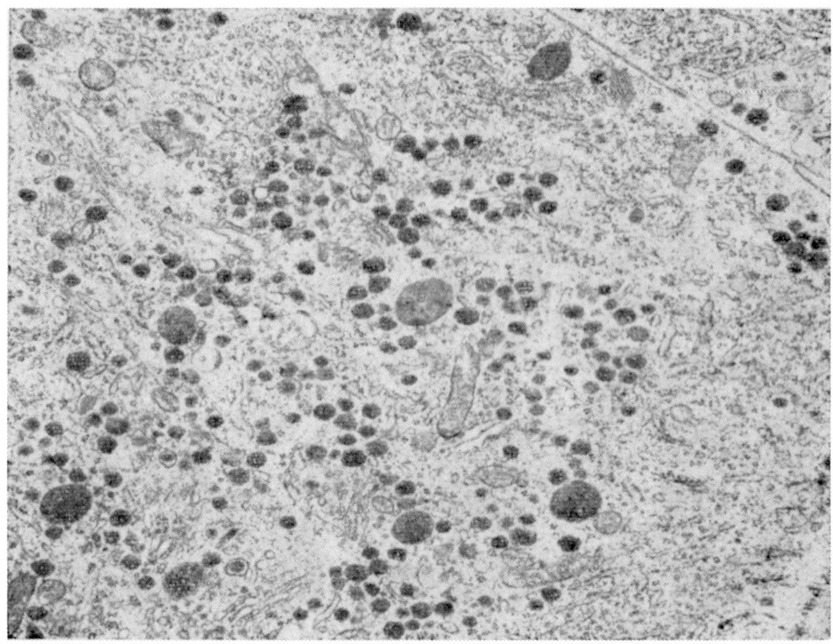

FIG. 4. Storage of neurosecretory granules in a neuron of the paraventricular nucleus of the rat after treatment with estrogens (18,500 ×).

Castration also exerts an effect upon the ultrastructure of the supraoptic and paraventricular nuclei. One month after castration both nuclei show signs of hyperactivity characterized by dilated cisternae of the endoplasmic reticulum and an increase in the number of free ribosomes. After six months some neurosecretory neurons show larger lysosomes. These cells have few neurosecretory granules (Zambrano and de Robertis, 1968a).

After castration the arcuate complex also shows hyperactivity after one month and even after six months. A considerable rise of granulated vesicles is noted (Zambrano and de Robertis, 1968b). In lactating rats, which normally exhibit signs of hyperactivity in their hypothalamic neurosecretory nuclei, castration elicits decrease of this situation in the supraoptic nucleus but not in the paraventricular nucleus, which presumably remains stimulated by the milk-ejection reflex (Flament-Durand and Desclin, 1968).

An effect upon the neurosecretory nuclei is also postulated in permanent estrus caused by continuous exposure to light (Fiske and Greep, 1959). This is confirmed by the increase of the radioactive cysteine incorporation and an accelerated migration to the posterior lobe of the radioactive material (Flament-Durand and Desclin, 1967).

B. Adrenals

A very curious aspect has been described by Akmayev *et al.* (1967) after bilateral adrenalectomy. It is similar to that obtained after hypophysectomy. Dense osmophilic vesicles occurring normally in moderate number in the nerve endings of the rat's median eminence (zona palissadica), increase in number and size during the second week after bilateral adrenalectomy. This observation is in good agreement with the optical microscope observation of Bock and Mühlen (1968a) in the median eminence of the mouse after the same experimental procedure. An obvious increase in stainable NSM appears in the outer zone in contact with the pars tuberalis. But the gradual decrease of the dense osmiophilic cores (Bock and Mühlen, 1968b), a process said to begin in the center of the granule, may be a good example of the frequent fixation or staining artefacts (see above).

C. Thyroid

Thiouracil has generally caused a reduction in the volume of the ganglion cell nuclei, while treatment with thyroxine has led to an enlargement of the nuclei. The neurosecretory material diminished under the influence of both thiouracil and excess thyroxine; the same results have been obtained by thyroidectomy (Talanti, 1967a, b). Similar effects on the NSM after thyroxine were obtained by Courrier and Colonge (1957).

D. Pancreas

In rats, pancreatectomy elicits a decrease of neurosecretion when diabetes mellitus ensues (Stutinsky and Mialhe, 1962). Degenerative modifications in the neurosecretory nuclei have also been described after pancreatectomy in the dog (Hagen, 1955; Goebels, 1957).

VI. The Transport Theory

Since Bargmann's first publications (1949) it has been suggested that the neurosecretory material synthesized in the perikaryon was transported to the nerve terminals by axonal flow. This hypothesis has been supported by powerful experimental evidence. Every surgical interruption of the neurosecretory pathway (Bargmann, 1949; Bargmann *et al.*, 1950) elicits the storage of the stainable material in the proximal part of the sectioned stalk (Hild, 1951; Hild and Zetler, 1951; and many others). The same observation was made in different species of hypophysectomized animals: anguilla, frog, rat, dog (Stutinsky, 1951, 1957; and many others), and even in man (Sloper and Adams, 1956; Sloper, 1962). Such a storage is not the privilege of neurosecretory fibers only, and also occurs proximal to a constriction in sympathetic axons after 24 hr (Kapeller and Mayor, 1966).

That this storage is not passive is shown by the increased speed of the phenomenon after osmotic stimulation in the hypophysectomized anguilla (Arvy *et al.*, 1954). Active transport is also suggested by the use of radioactive aminoacids; 30 min after the administration of radioactive cysteine there is an active uptake in the supraoptic and paraventricular nuclei. Ten hours later, a progressive accumulation of the isotope appears at the stump of the stalk in hypophysectomized rats. This proximo-distal migration can be accelerated by different physiological conditions (Flament-Durand, 1967).

Electron microscopic studies of hypophysectomized rats reveal modifications in the infundibular stump as well as in the neurosecretory neurons of the supraoptic and paraventricular nuclei. Immediately after the operation there appears a simultaneous storage of neurosecretory material in the endings and an obvious increase in the quantity of glycogen. This modification is still present after two months, but only true neurosecretory endings are involved. There is no modification in catecholamine-containing endings or in the second type of elementary granules (see above). The degenerative aspects also concern only the first type of neurosecretory material (see above). After the first 24 hr the hypothalamic neurosecretory cells show signs of glandular activity and the appearance of multivesicular bodies and autophagic figures of secretory granules; after the fourth day an enormous increase of ergastoplasm can be

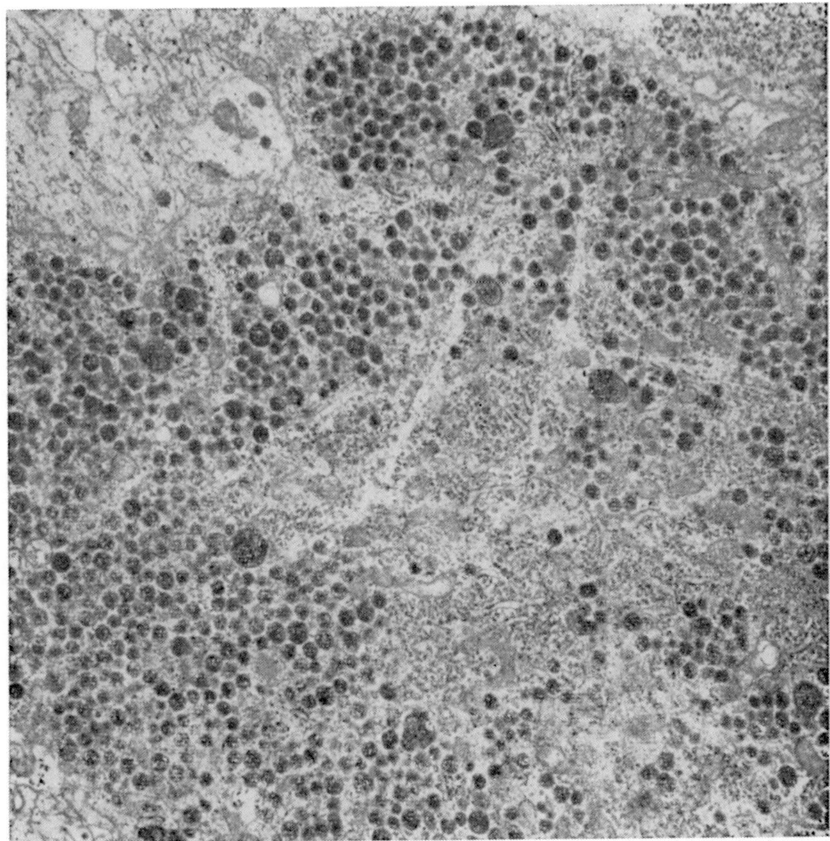

FIG. 5. Storage of neurosecretory granules in a neuron of the supraoptic nucleus in a four-day hypophysectomized rat; only the peripheral zone of the cell is shown (13,000 ×)

observed (Fig. 5), as well as glycogen, lysosomes, and very large amounts of stored granules (Stutinsky *et al.*, unpublished observations).

These observations mean that we are dealing in the same cell at the same moment with both secretion and catabolic involution of the synthesized granules. Thus the electron microscopic observations confirm all earlier findings, but it appears that the electron microscope more rapidly demonstrates the transformation of the remaining stump to an equivalent of the neural lobe. A similar regeneration has been reported after the removal of the urophysis in teleost fishes (Fridberg *et al.*, 1966).

A number of arguments, especially by Spatz (1958) and his group, have been put forth in opposition to the transport theory. These have recently been reinforced by the experimental work of Christ (1962). After electrocoagulation he observed a storage of the secretory material not only in the proximal stump but also in the distal part. His interpretation of these structures as aspects of axon degeneration is a return to one of the earliest interpretations (Tello, 1912). It must be conceded that the silver staining methods for the Herring bodies and the existence of numerous enlarged terminal rings simulate obvious axon degeneration, but the interpretation of these aspects may be quite different. Even in normal animals, a great number of rings and other complicated terminal-like structures can be seen close to the Herring bodies. In addition in electron microscopic observations in the rat, axoaxonal synapses have been observed on neurosecretory fibers (Klein *et al.*, 1968). This observation confirms those of Kobayashi *et al.* (1965) in teleosts and of Bargmann *et al.* (1967) in the rat.

A more serious argument against the transport theory is given by Green and Maxwell (1959), who estimate that the high speed necessary for the movement of the NSM could not be reached. More recent observations by Jasinski *et al.* (1967) suggest that this speed is higher than formerly supposed.

Pure biochemical observations by Sachs *et al.* (1967) suggest that there may be a synthesis of vasopressin along the nerve fiber from the perikaryon to the posterior lobe; such a maturation has also been observed by Dean *et al.* (1968). This interpretation is reinforced by the observation of Gerschenfeld *et al.* (1960), who found a maturation of the elementary granules during their migration. Similar observations are reported by Monroe (1965).

As far as the rat is concerned, we cannot agree with this formulation. In the perikaryon, the diameters of elementary granules vary with their age, but after they have left the Golgi apparatus and have penetrated the axon, they show a high degree of constancy; this is at least true for granules originating from the two neurosecretory nuclei. Such a constancy in diameter has also been asserted by Follenius (1963) in the fish, where two origins for neurosecretory granules exist (the preoptic nucleus and the nucleus lateralis tuberi).

The problem may be not only one of a progressive maturation; the possibility that two different kinds of neurons synthesize the two posterior lobe hormones must also be considered.

Bindler *et al.* (1967) suggest that vasopressin and oxytocin are each stored in a different type of nerve ending. Evidence is given by Dean *et al.* (1968) for the storage of oxytocin and vasopressin in different protein carriers (neurophysin I and II) in separate neurosecretory granules.

A separate origin is also suggested by observations on the fate of the neurosecretion in the posterior lobe of mice with genetic diabetes insipidus (Weyl-Sokol and Valtin, 1967). This is in agreement with histochemical work on the neurohypophysis in the rat, where only certain masses of NSM showed a positive reaction for arginine, an amino acid present only in vasopressin (Howe, 1962).

VII. Targets for Neurosecretory Fibers

A. Posterior Lobe

The major portion of the hypothalamic neurosecretory fibers end in the posterior lobe, around the capillaries. Ancient observations made by silver stains concerning possible innervation of pituicytes and of other glial (ependymal) cells (Knowles, 1967) seem to have been confirmed by electron microscopy in teleost fishes (Follenius, 1967).

B. Median Eminence

Numerous neurosecretory fibers terminate in the median eminence on the portal vessels, but their presence may be overlooked in paraffin preparations because of the inability of chrome hematoxylin or AF to show minute aggregates of neurosecretory granules. Experimental conditions, such as hypophysectomy or mechanical stimulation of the bottom of the third ventricle, may enhance the quantity of NSM around the portal vessels (Fig. 6).

These fibers seem closely intermingled with fluorescent fibers that contain catecholamines. The electron microscopic observations of Barry and Cotte (1960), Monroe (1967), Kobayashi *et al.* (1965), Oota (1963), and Mazzuca (1965) in mammals, of Oota and Kobayashi (1962) and Oksche (1967) in birds, and of Oota and Kobayashi (1963) in the bullfrog show that in the median eminence there are

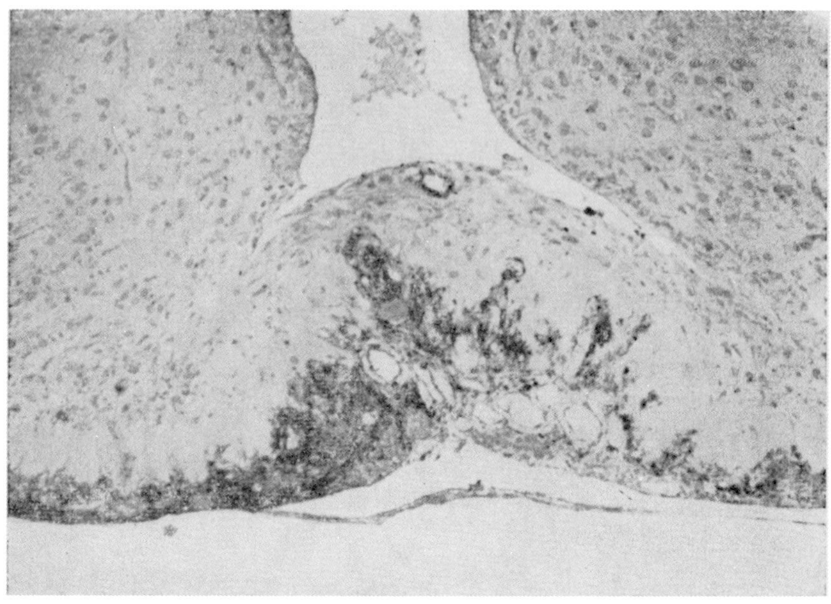

FIG. 6. Median eminence of the rat. Bouin-AF stain; after mechanical stimulation of the bottom of the third ventricle storage of NSM around the portal vessels (130 ×).

different kinds of nerves. Oota (1963) demonstrated in electron microscopical investigations that the zona externa of the mouse is supplied with densely packed neurosecretory and non neurosecretory fibers. At least part of the former may belong to the adrenergic fibers described by Björklund *et al.* (1968).

The zona externa of the median eminence in birds and mammals is richly provided with nerves, demonstrable by impregnation preparations (Wingstrand, 1951; Nowakowski, 1951; Stutinsky, 1955a, 1958) in the mouse, ox, sheep, dog, etc. They resemble mostly the " Nodulus Fasern " described by Knoche (1953, 1958). A part of these fibers can be identified as neurosecretory by AF staining.

At least three types of vesicles (Fig. 7) can be described in the rat (Monroe, 1965; Stutinsky, unpublished observations) as well as in the guinea pig (Mazzuca, 1966):

1. Neurosecretory endings containing synaptic vesicles and elementary granules with a diameter of 2000 Å. The dense cores completely fill up the vesicles and are in close contact with the membrane.

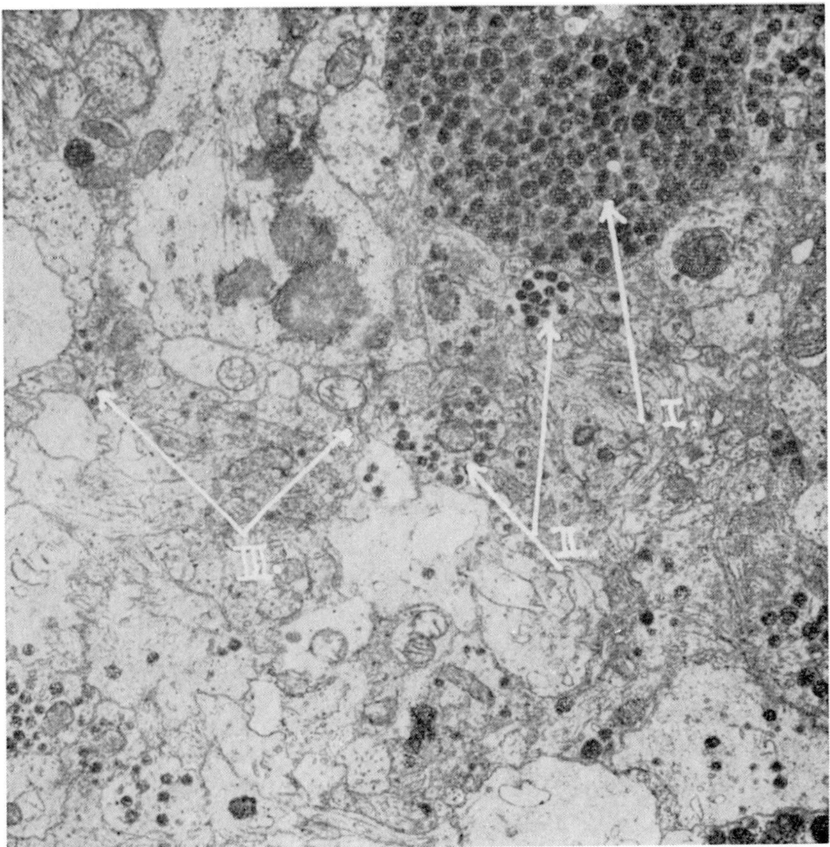

FIG. 7. Median eminence of the rat containing three types of neurosecretory granules (14,000 ×).
I. Neurosecretory endings containing synaptic vesicles and elementary granules with a diameter of 2000 Å. The dense cores completely fill up the vesicles and are in close contact with the membrane
II. Nerve endings with synaptic vesicles and a few, small, dense-core granules in vesicles from about 600 Å to 900 Å with a definite external membrane
III. Nerve endings with synaptic vesicles very similar to the neurosecretory elements but of smaller size, varying from 800 Å to 1200 Å.

2. Nerve endings with synaptic vesicles and a few, small, dense-core granules in vesicles from about 600 Å to 900 Å with a definite external membrane.
3. Nerve endings with synaptic vesicles very similar to the neurosecretory elements but of a smaller size, varying from 800 Å to 1200 Å.

These large terminals are the most numerous. Curiously enough, Mazzuca (1966) shows that reserpine empties only the second type of nerve endings. It is obvious that the absolute diameter and appearance of the terminal inclusions have no significance; they can vary from one species to another. Only the comparison of the endings in the same species permits a classification. In the rat a fourth type of ending cannot be excluded. This nerve terminal could contain synaptic vesicles only (but the existence of such terminals is questionable and may be due to technical reasons such as orientation of the piece).

C. Neurosecretory Fibers and the Pars Distalis

It has been shown by Da Lage (1958), Stutinsky (1958), Follenius (1963), and others that many neurosecretory fibers penetrate into all lobes of the teleostean hypophysis. In electron microscope preparations it appears that the nerve terminals are in contact with the vessels and the basal membranes, and only rarely with glandular cells.

D. Neurosecretory Fibers and the Pars Tuberalis

In normal animals the neurosecretory fibers can very closely approach the pars tuberalis, and in hypophysectomized rats silver-stained fibers can be traced through this part of the gland (Vasquez-Lopez and Williams, 1952); at least a part are neurosecretory fibers (Stutinsky, 1951). Such aspects can also be seen in other experimental conditions.

E. Neurosecretory Fibers and the Pars Intermedia

The pars intermedia is well provided with nerve fibers, as it has been shown by silver impregnations (for references see Dellmann, 1962). Neurosecretory fibers have been described in many species.

As early as 1953, Stutinsky (1953b) emphasized that the fibers stained by chrome hematoxylin are much more numerous in cold-blooded vertebrates than in mammals. In these latter animals (especially the ewe and ox), silver impregnates very rich nervous plexuses which cover the major part of the pars intermedia. The general aspect of these fibers and their endings are very similar to those that exist in the posterior lobe, and the continuity of these two systems cannot be denied. The nerve fibers penetrate in the pars intermedia along the vascular spaces and great numbers of fibers seem to terminate along the vessels in bulbs, reticular formations, and rings.

Beside these complex endings (which do not appear to be in direct connection with secretory cells of the pars intermedia), there are also terminal rings on the external borderline of cell groups, close to certain glandular or connective cells. In the best impregnations it does not seem that all intermediate lobe cells are in contact with nerve endings, and only certain territories posses a dense nervous plexus. It is also doubtless that the stains for the optical microscope are valuable for the cell groups close to the neural lobe and that the affinity for CAH and AF vanishes in the deeper entering fibers.

Finally, the visualization of neurosecretory fibers by these staining procedures may be variable and intermittent as we have shown in the rat. Under normal conditions there are only a few neurosecretory fibers seen, but under certain experimental conditions they may become very numerous. Whatever the explanation for this observation may be, it must be kept in mind that several kinds of nerve fibers may be present. Fluorescent fibers have been detected in the frog pars intermedia (Enemar and Falck, 1965; Cohen, 1967). Using the electron microscope, Kobayashi and Oota (1964), Follenius and Porte (1962), Knowles (1965), and recently Bargmann *et al.* (1967) and Vincent and Anand-Kumar (1968) have shown neurosecretory terminals as well as synaptic vesicles presumably containing only neurotransmitter hormone.

Bargmann and co-workers (1967) distinguished three kinds of synaptic contacts in the pars intermedia of the cat: (1) nerve endings containing empty synaptic vesicles, (2) nerve endings containing empty synaptic vesicles and granules with dense core (looking like catecholamine granules), and (3) nerve endings containing synaptic vesicles and elementary neurosecretory electron-dense granules. They propose that the different kinds of endings be termed cholinergic, adrenergic, and peptidergic synapses, respectively.

F. Neurosecretion and the Third Ventricle

Neurosecretory fibers and NSM are also in close relationship with the ependymal layer of the third ventricle and with the ventricle itself. Gomori-positive material staining with chrome hematoxylin has been seen in the lumen of the ventricle of the dog (Bargmann *et al.*, 1950; Hild, 1951) and rat (Stutinsky, 1955b). The electron microscope has shown neurosecretory fibers in the ventricle of the rabbit (Leonhardt and Lindner, 1967), of the mouse (Wittkowski, 1968) and the rat (Stutinsky, unpublished observations) (Fig. 8). Dendrites and nerve cells are visible between ependymal cells in the preoptic nucleus of anguilla (Stutinsky, 1953c; Knowles, 1967), of rana (Dierickx, 1962), and of bufo (Collin and Barry, 1954).

This close relationship may explain the presence of posterior lobe hormones in the cerebrospinal fluid, as was claimed a long time ago and recently confirmed with modern techniques by Heller *et al.* (1968), in spite of the low concentrations found in comparison with the blood. There are reports of evidence that the special diffe-

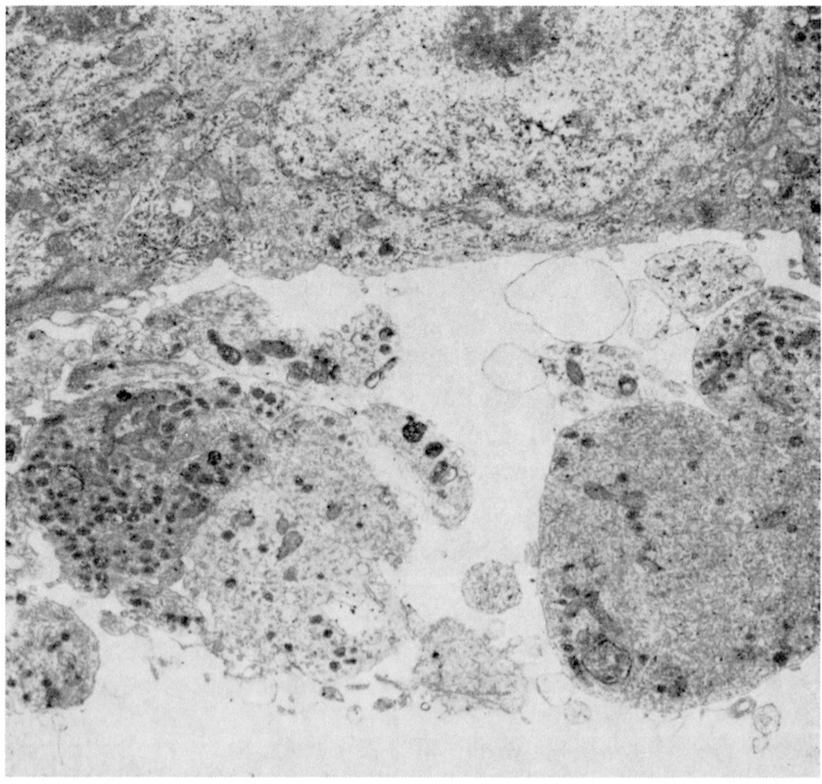

FIG. 8. Presence of neurosecretory fibers in the third ventricle of the rat (11,000 ×).

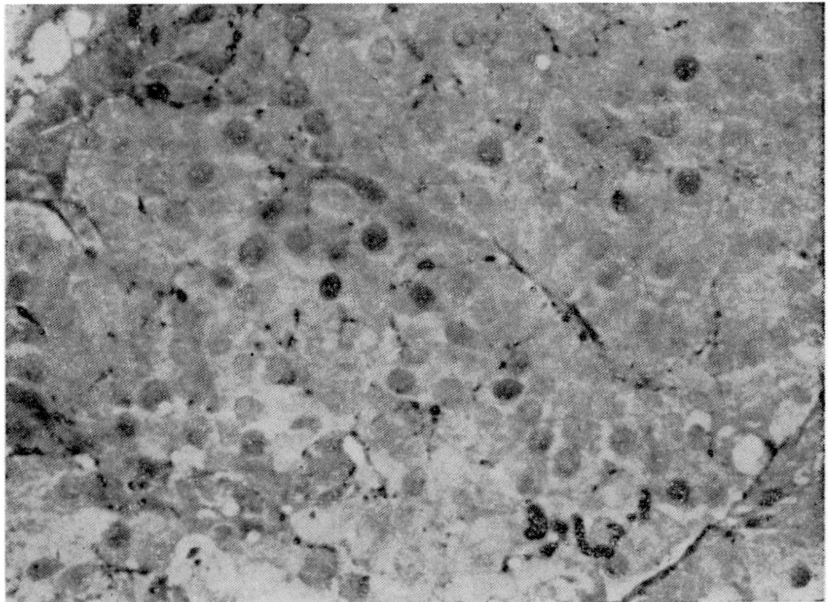

FIG. 9. Pars intermedia of the rat. Bouin-AF stain; rich nervous plexus formed by neurosecretory fibers in the pars intermedia after mechanical stimulation of the third ventricle (550 ×).

rentiated ependymal cells show glandular character and that they possibly can assume a physiological link between the third ventricle and the portal vessels by means of their vascular feet (Löfgren, 1960; Levêque *et al.*, 1966; Knowles 1967; Vigh *et al.*, 1963).

It may also be significant that mechanical stimulation of the third ventricle can influence the amount of NSM present in the pars intermedia (Stutinsky, unpublished observations), as found in the rat (Fig. 9).

VIII. Conclusions

The concept of neurosecretion substantially explains the origin and release of the posterior lobe hormones. The vascular relationship (portal vessels) of the hypothalamus to the anterior lobe of the pituitary provides the explanation for nervous regulation of the gland. But the origin of the releasing factors is still completely unknown.

No cells of the hypothalamus except the neurons of the supraoptic and the paraventricular nuclei (and perhaps neurons of the arcuate nucleus and certain ependymal cells) show true glandular characters. This constitutes a serious problem because the chemical properties of the relasing factors are very close to those of the substances transported in elementary granules from the peptidergic neurons into the neurohypophysis and the median eminence. One would expect the releasing factors to be transported in similar granules. It seems very unlikely that the aminergic nerve endings of the median eminence can account for the different releasing factors, and the " permissive " role of these nerve endings remains under discussion.

So it appears that the concept of neurosecretion, although supporting data have increased considerably since the first observations, has not been sufficiently developed to give a correct interpretation of all known facts. Our knowledge regarding the critical relationship between neurosecretion and releasing factors must await coming work for completion.

Acknowledgment

This work was achieved by the Research Team for Comparative Neuroendocrinology No. 178 of the C.N.R.S. I am greatly indebted to Dr. Porte, chief of the Department of Electron microscopy, and his co-workers Misses Klein and Stoeckel, and for technical assistance to Misses Gertner, Haller, and Steible.

References

Akmayev, I. G., Rethelyi, M., and Majorossy, K. (1967). *Acta Biol. Acad. Sci. Hung.* **18**, 187.

Anderson, B., and McCann, S. M. (1956). *Acta Physiol. Scand.* **35**, 312.

Arvy, L., Fontaine, M., and Gabe, M. (1954). *C. R. Soc. Biol.* (Paris) **148**, 1759.

Bargmann, W. (1949). *Z. Zellforsch. Mikroskop. Anat.* **34**, 610.

Bargmann, W. (1966). *Intern. Rev. Cytol.* **19**, 183.

Bargmann, W., Hild, W., Ortmann, R., and Schiebler, Th. H. (1950). *Acta Neuroveget.* (Wien) **1**, 233.

Bargmann, W., Knoop, A., and Thiel, A. (1957). *Z. Zellforsch. Mikroskop. Anat.* **47**, 114.

Bargmann, W., Lindner, E., and Andres, K. H. (1967). *Z. Zellforsch. Mikroskop. Anat.* **77**, 282.

Barry, J., and Cotte, G. (1960). *C. R. Soc. Biol.* (Paris) **154**, 2054.

Bindler, E., Labella, F. S., and Sauval, (1967). *J. Cell Biol.* **34**, 185.

Björklund, A., Enemar, A., and Falck, B. (1968). *Z. Zellforsch. Mikroskop. Anat.* **89**, 590.

Bock, R., and Aus der Mühlen, K. (1968a). *Z. Zellforsch. Mikroskop. Anat.* **92**, 130.
Bock, R., and Aus der Mühlen, K. (1968b). *Z. Zellforsch. Mikroskop. Anat.* **94**, 434.
Braak, H. (1962). *Z. Zellforsch. Mikroskop. Anat.* **58**, 265.
Brooks, Ch. McC., Ushiyama, J., and Lange, G. (1962). *Am. J. Physiol.* **202**, 487.
Bugnon, C. (1957). M. D. Thesis, University of Paris.
Christ, J. (1962). *Mem. Soc. Endocrinol.* **12**, 125.
Cohen, A. G. (1967). *Nature* **215**, 55.
Collin, R., and Barry, J. (1954). *Ann. Endocrinol.* (Paris) **15**, 533.
Courrier, R., and Colonge, A. (1957). *C. R. Acad. Sci.* (Paris) **245**, 3883.
Cross, B. A., and Green, J. D. (1959). *J. Physiol.* (London) **148**, 554.
Da Lage, C. (1958). *In* "Pathophysiologia Diencephalica" (S. B. Curri and L. Martini, eds.). pp. 118–121. Springer-Verlag, Wien.
Dean, C. R., Hope, D. B., and Kázik, T. (1968). *Brit. J. Pharmacol.* **34**, 192.
Dellmann, H. D. (1962). *Z. Hirnforsch.* **5**, 249.
Diamond, H. C. (1952). *Am. J. Physiol.* **171**, 171.
Dierickx, K. (1962). *Arch. Intern. Pharmacodyn.* **140**, 708.
Droz, B. (1965). *C. R. Acad. Sci.* (Paris) **260**, 320.
Enemar, A., and Falck, B. (1965). *Gen. Comp. Endocrinol.* **5**, 377.
Fiske, V. M., and Greep, R. O. (1959). *Endocrinology* **64**, 175.
Flament-Durand, J. (1966). *Ann. Soc. Roy. Soc. Med. Nat.* **19**, 1.
Flament-Durand, J. (1967). *In* "Neurosecretion" (F. S. Stutinsky, ed.). pp. 60–76. Springer-Verlag, Berlin.
Flament-Durand, J., and Desclin, L. (1967). *Ann. Endocrinol.* (Paris) **28**, 240.
Flament-Durand, J., and Desclin, L. (1968). *C. R. Acad. Sci.* (Paris) **267**, 1205.
Follenius, E. (1963). *Gen. Comp. Endocrinol.* **3**, 66.
Follenius, E. (1965). *Ann. Sci. Nat. Zool.* **7**, 32.
Follenius, E. (1967). *In* "Neurosecretion" (F. S. Stutinsky, ed.). pp. 42–55. Springer-Verlag, Berlin.
Follenius, E., and Porte, A. (1962). *Mem. Soc. Endocrinol.* **12**, 51.
Fridberg, G., Bern, H. A., and Fleming, W. R. (1966). *J. Exp. Zool.* **34**, 164.
Fuxe, K., and Hökfelt, T. (1967). *In* "Neurosecretion" (F. S. Stutinsky, ed.). pp. 165–177. Springer-Verlag, Berlin.
Gabe, M. (1953). *Bull. Microscop. Appl.* **3**, 153.
Gastaldi, E. (1952). *Boll. Soc. Ital. Biol. Sper.* **28**, 1094.
Gerschenfeld, H. M., Tramezzani, J., and De Robertis, E. (1960). *Endocrinology* **66**, 741.
Goebels, H. (1957). *Acta Anat.* **30**, 307.
Green, J. D., and Maxwell, D. S. (1959). *In* "Comparative Endocrinology" (A. Gorbman, ed.). pp. 368–392. J. Wiley and Sons, New York.
Hagen, E. (1955). *Acta Anat.* **25**, 1.
Halmi, N. S. (1952). *Stain Technol.* **27**, 61.
Harris, G. W. (1948). *J. Physiol.* **107**, 418.
Heller, H., Hasan, S. H., and Sarfi, A. Q. (1968). *J. Endocrinol.* **41**, 273.
Hild, W. (1951). *Z. Zellforsch. Mikroskop. Anat.* **35**, 33.
Hild, W., and Zetler, G. (1951). *Arch. Exp. Pathol. Pharmacol.* **213**, 139.
Howe, A. (1962). *Mem. Soc. Endocrinol.* **12**, 241.
Jasinski, A., Gorbman, A., and Hara, T. (1967). *In* "Neurosecretion" (F. S. Stutinsky, ed.). pp. 106–123. Springer-Verlag, Berlin.
Kandel, E. R. (1962). *Fed. Proc.* **21**, 361.
Kandel, E. R. (1964). *J. Gen. Physiol.* **47**, 691.
Kapeller, K., and Mayor, D. (1966). *J. Anat.* **100**, 439.
Kawashima, S. (1965). *J. Fac. Sci. Univ. Tokyo, Sect. IV*, **10**, 497.
Klein, M. J., Porte, A., and Stutinsky, F. S. (1968). *C. R. Acad. Sci.* (Paris) **267**, 1007.

Knoche, H. (1953). *Acta Anat.* **18**, 208.
Knoche, H. (1958). *Z. Zellforsch. Mikroskop. Anat.* **48**, 602.
Knowles, F. (1965). *Nature*, **206**, 1168.
Knowles, F. (1967). *In* "Neurosecretion" (F. S. Stutinsky, ed.). pp. 8–19. Springer-Verlag, Berlin.
Kobayashi, H., and Oota, Y. (1964). *Gunma Symp. Endocrinol.* **1**, 63.
Kobayashi, H., Hirano, T., and Oota, Y. (1965). *Arch. Anat. Microscop. Morphol. Exp.* **54**, 277.
Legait, H. (1955). *Arch. Anat. Microscop. Morphol. Export.* **44**, 323.
Leonhardt, H., and Lindner, E. (1967). *Z. Zellforsch. Mikroskop. Anat.* **78**, 1.
Levêque, F. T., Stutinsky, F. S., Porte, A., and Stoeckel, M. E. (1966). *Z. Zellforsch. Mikroskop. Anat.* **69**, 381.
Lisk, R. D. (1966). *Neuroendocrinology* **1**, 83.
Löfgren, F. (1960). *Acta Morphol. Neerl. Scand.* **3**, 55.
Mautner, W. (1964). *Z. Zellforsch. Mikroskop. Anat.* **64**, 813.
Mazzuca, M. (1965). *J. Microscopie* **4**, 225.
Mazzuca, M. (1966). *Colloque Annuel S.F.M.E.*, (Bordeaux) pp. 639–669.
Monroe, B. G. (1965). *Anat. Record* **151**, 389.
Monroe, B. G. (1967). *Z. Zellforsch. Mikroskop. Anat.* **76**, 405.
Nowakowski, H. (1951). *Z. Nervenheilk.* **165**, 261.
Oksche, A. (1967). *In* "Neurosecretion" (F. S. Stutinsky, ed.). pp. 77–88. Springer-Verlag, Berlin.
Oksche, A., Mautner, W., and Farner, D. S. (1964). *Z. Zellforsch. Mikroskop. Anat.* **64**, 83.
Oota, Y. (1963). *J. Fac. Sci. Univ. Tokyo, Sect. IV*, **10**, 155.
Oota, Y., and Kobayashi, H. (1962). *Ann. Zool.* (Tokyo) **35**, 128.
Oota, Y., and Kobayashi, H. (1963). *Z. Zellforsch. Mikroskop. Anat.* **60**, 667.
Ortmann, R. (1951). *Z. Zellforsch. Mikroskop. Anat.* **36**, 92.
Osinchak, J. (1964). *J. Cell Biol.* **21**, 35.
Palay, S. L. (1955). *Anat. Record* **121**, 348.
Palay, S. L. (1957). *In* "Ultrastructure and Cellular Chemistry of Neural Tissue" (H. Waelsch, ed.). pp. 31–49. Hoerber-Harper, New York.
Pickford, M. (1959). *J. Physiol.* (London) **149**, 41.
Porte, A., Klein, M. J., Stoeckel, M. E., and Stutinsky, F. S. (1968). *C. R. Acad. Sci.* (Paris) **267**, 1051.
Sachs, H., Portanova, R., Haller, E. W., and Share, L. (1967). *In* "Neurosecretion" (F. S. Stutinsky, ed.). pp. 146–154. Springer-Verlag, Berlin.
Scharrer, E., and Scharrer, B. (1963). "Neuroendocrinology." p. 21. Columbia Univ. Press, New York.
Shimazu, K., Okada, M., Ban, T., and Kurotsu, T. (1954). *Med. J. Osaka Univ.* **5**, 701.
Sloper, J. C. (1962). *Symp. Zool. Soc.* **9**, 29.
Sloper, J. C., and Adams, C. V. M. (1956). *J. Pathol. Bacteriol.* **72**, 587.
Spatz, H. (1958). *In* "Pathophysiologia Diencephalica" (S. B. Curri and L. Martini, eds.). pp. 53–77. Springer-Verlag, Berlin.
Sterba, G. (1964). *Acta Histochem.* **17**, 268.
Stumpf, W. E. (1968). *Science*, **162**, 1001.
Stutinski, F. S. (1951). *C. R. Soc. Biol.* **145**, 367.
Stutinsky, F. S. (1953a). *Ann. Endocrinol.* (Paris) **14**, 101.
Stutinsky, F. S. (1953b). *Ann. Biol.* **29**, 487.
Stutinsky, F. S. (1953c). *Z. Zellforsch. Mikroskop. Anat.* **39**, 276.
Stutinsky, F. S. (1955a). Sci. Thesis. Masson et Cie. Paris.
Stutinsky, F. S. (1955b). *C. R. Ass. Anat.* **92**, 1256.

Stutinsky, F. S. (1957). *Arch. Anat. Microscop.* **46**, 95.
Stutinsky, F. S. (1958). *In* "Pathophysiologia Diencephalica" (S. B. Curri and L. Martini, eds.). pp. 78–103. Springer-Verlag, Wien.
Stutinsky, F. S. (1968). *Arch. Anat. Histol. Embryol. Exp.* **51**, 683.
Stutinsky, F. S., and Guerné, Y. (1965). *C. R. Soc. Biol.* (Paris) **159**, 1420.
Stutinsky, F. S., and Mialhe, P. (1962). *Bull. Ass. Anat.* (Toulouse) **98**, 1249.
Stutinsky, F. S., Porte, A., Tranzer, J. P., and Terminn, Y. (1963). *C. R. Soc. Biol.* (Paris) **157**, 2294.
Talanti, T. (1967a). *Z. Zellforsch. Mikroskop. Anat.* **79**, 92.
Talanti, T. (1967b). *Acta Physiol. Scand.* **70**, 80.
Tello, J. F. (1912). *Trab. Inst. Invest. Cajal.* (Madrid) **10**, 145.
Vasquez-Lopez, E., and Williams, P. C. (1952). *Ciba Foundation Colloquia on Endocrinol.* **4**, 154.
Vigh, B., Aros, B., Wenger, T., Koritsansky, S., and Cegledi, G. (1963). *Acta Biol. Acad. Sci. Hung.* **13**, 407.
Vincent, D. S., and Anand-Kumar, T. C. (1968). *J. Endocrinol.* **41**, XVIII.
Weyl-Sokol, H., and Valtin, H. (1967). *Nature* **214**, 314.
Wingstrand, K. G. (1951). "Structure and Development of the Avian Hypophysis." Thesis, C. W. K. Glerup, Lund.
Wittkowski, W. (1968). *Z. Zellforsch. Mikroskop. Anat.* **92**, 207.
Yagi, K., and Bern, H. A. (1963). *Science* **142**, 491.
Zambrano, D., and de Robertis, E. (1968a). *Z. Zellforsch. Mikroskop. Anat.* **86**, 14.
Zambrano, D., and de Robertis, E. (1968b). *Z. Zellforsch. Mikroskop. Anat.* **86**, 409.

A System Analysis of Some Hypothalamic Functions

J. P. Schadé

I. Introduction

Cerebral tissue organization is characterized by a tricompartmental arrangement. The neuronal compartment consists of an *n*-dimensional network of neurons of variable size; the glial compartment encompasses the different types of glial elements and the vascular compartment comprises the circulatory system. Between these compartments is the extracellular space which, according to some authors, consists of a hydrated net of large molecules. The exchange of substances between the compartments, and the intra-and inter-compartmental relationships, are dealt with in other articles in this volume.

By treating the brain and its subdivisions (namely, the hypothalamus) as a tissue, the neurophysiologist can list all passive and active electrical properties of neurons and neuronal aggregates, and in this way draw a picture of the function of individual neurons and groups of neurons. However, one can also treat the brain as a system, and in doing so one has to decide directly what sort of data he should collect as a structural basis of the model. Since this book deals with the hypothalamus, an attempt will be made to analyze some of the functional properties of the hypothalamic neurons from the view point of a system analyst.

Neurons in the hypothalamus are known to respond to thermal-osmotic, humoral, and sensory stimuli. We shall define the hypothal, amic integrating neurons as specific sensors with the following properties:

1. The receptive part of the neuron consists of a perikaryon and a set of tapering dendrites that receive synaptic endings of other neurons. The membrane surface of the receptive pole will respond to synaptic activity of other neurons and will also convert environmental stimuli into local response-generating activities. The receptive pole is sensitive to at least a particular type of environmental stimulus.
2. The conductive part of the neuron is differentiated to conduct action potentials away from the receptive pole.
3. The transmitting part of the unit shows membrane and cytoplasmic differentiation related to synaptic transmission or neurosecretory activity. This part either releases hormones (releasing or inhibiting factors) or is part of a neuronal circuit that either produces one of these substances or stimulates other cerebral structures to a specific behavioral response; e.g., glucose-sensitive neurons will probably not be involved in the secretion of a hormone, but may be involved in the stimulation of neuronal circuits that ultimately elicit a hunger response.

It is difficult to distinguish direct environmental effects upon the neuron under study from secondary changes in synaptic input due to environmental conditions (temperature, glucose, hormones, etc.) at other sites. Therefore the analysis of the data has been restricted to some studies employing microelectrophoresis techniques.

II. Properties of Hypothalamic Units

Hypothalamic neurons discharge uniformly with a low firing rate. The majority of neurons in various species under different anesthesia conditions show firing rates of less than 10 per second. Cross and Silver (1966) made a critical review of the literature on spontaneous unit discharges and the response to various stimuli of hypothalamic neurons. It is clear from their own results and from data in the literature that those neurons in the hypothalamus that fire at less than one per second are more abundant than faster firing units. In the hypothalamic island (see the article by Cross and Dyer in this volume) the average firing rate tends to be much higher.

In normal unanesthetized preparations the firing rate rarely exceeds 50 per second. Even when excited by various stimuli, values up to 90 spikes per second are seldom observed. This would indicate that the receptive pole of the neuron is influenced by slow-acting substances, whereby the neuronal membrane slowly changes its resting potential. A likely explanation would be that circulating substances such as glucose and hormones, or factors such as temperature, act upon the neuronal membrane by mimicking synaptic mechanisms. The conversion rate of these changes may be slow, as is well known for similar phenomena in sense organs.

The hypothalamic neurons conduct action potentials at a low speed; values of 0.2–1.1 m/sec have been found. The slow conduction velocity may be related in some instances to the transport and excretion of releasing factors or hormones.

Few reports have been published on the statistical analysis of the interspike intervals and the distribution of firing frequency under unanesthetized conditions.

Oomura and co-workers (1964) reported on these parameters for neurons in the lateral region and ventromedial nucleus in the cat. As far as the interspike interval histogram is concerned, the data on neurons in the lateral region nicely fitted into an exponential distribution. The analysis of the number of spikes per unit of time revealed that the distribution resembled the Poisson curve.

For neurons in the ventromedial nucleus, a Gaussian curve described the discharge frequency. Similar observations for the interspike interval histogram have been made in the rabbit hypothalamus (Findlay and Hayward, 1969), but only under waking conditions because slow sleep and paradoxical sleep markedly changed this characteristic firing pattern. Lincoln (1967), in a careful study, also reported that neurons in many areas of the hypothalamus show an exponential distribution of resting period firing frequencies.

These data indicate that either a random or a systematic pattern of discharges is produced by hypothalamic neurons. In this respect the hypothalamic units resemble motor neurons in the spinal cord (where similar observations have been made) in contrast to, e.g., neurons in the cerebral cortex, whose discharges do not fit any particular distribution.

A. Hypothalamic Thermostat

Hellon (1967) performed a beautiful series of experiments on single hypothalamic neurons during controlled changes of cerebral temperature. Of 227 neurons in the anterior hypothalamus, 23 elements showed a temperature sensitivity. The majority (17) exhibited an increase in firing rate as temperature was rising, while the remaining six neurons increased their firing rate in response to a fall in temperature.

From a cybernetic point of view the behavior of a group of neurons possessing a set-point was most interesting. These units were only effected by temperature when this had gone above or below a certain level (Figs. 1, 2). The threshold temperature or set-point of the measuring device was within the range of 38.5 to 39.5 °C (hypothalamic temperature).

These experiments clearly point out that a small percentage of neurons lying within 2 mm of the midline of the anterior hypothalamus of the rabbit have a high degree of temperature sensitiviy. Both positive and negative sensors could be distinguished, since they were diffusely scattered throughout the area. Eisenman (1965) showed the presence of temperature sensors in the preoptic region and septum of the cat. These neurons also exhibited a threshold or set-point in the relationship of firing frequency and temperature.

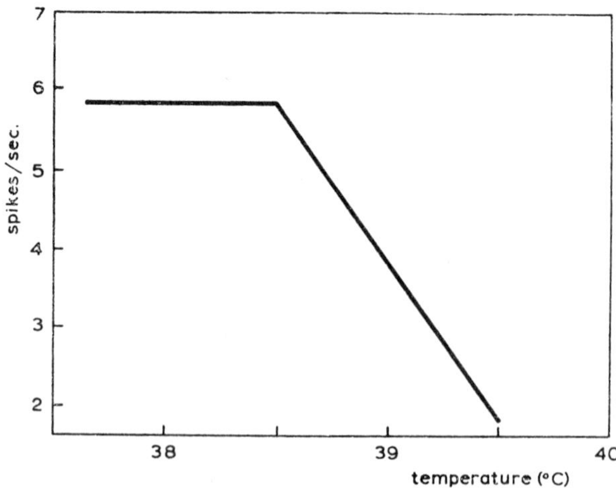

Fig. 1. Average number of spikes per second of a hypothalamic neuron that shows a decrease in firing rate by cooling (From Hellon, 1967).

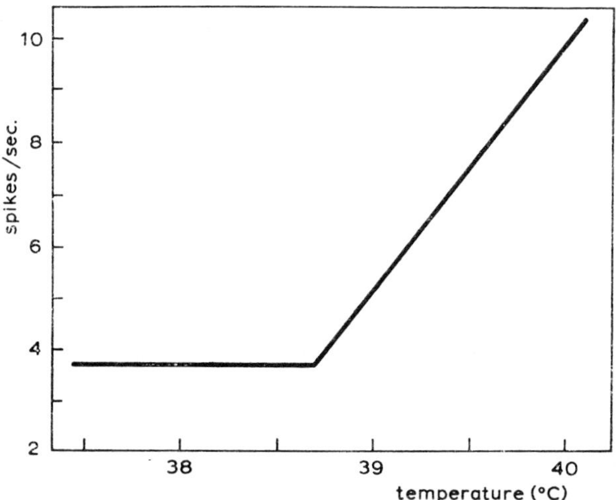

FIG. 2. Average number of spikes per second of a hypothalamic neuron that shows an increase in firing rate by warming. (From Hellon, 1967).

The T^+ sensors and T^- sensors form part of a hypothalamic integrating system, which constitutes the cerebral portion of the thermostat. The sensors act as monitor devices for changes of temperature in the circulatory system. It seems safe to assume that this physical factor is mediated through the extracellular space between blood vessels, neuroglia, and neurons.

How the sensors gain and maintain knowledge about the norm or set-point is entirely unknown at the present time. Analog type of information about temperature shift is converted by the receptive pole of the sensors via changes of the neuronal membrane in a hyperpolarizing or depolarizing direction. It effects in this way the conducivity of the membrane and the firing rate of the spike-generating area of the sensor.

The set-point of hypothalamic portions of the body thermostat may be adjusted by neuronal stimuli that influence the sensors via excitatory or inhibitory pathways. Monoaminergic transmitters play a significant role in modulating hypothalamic functions. The enzymes for biosynthesis and inactivation of the monoamines are abundantly present in the anterior hypothalamus. When applied intraventricularly or locally, serotonin induces shivering, whereas noradrenaline abolishes shivering in the cat (see the article by Feldberg in this volume).

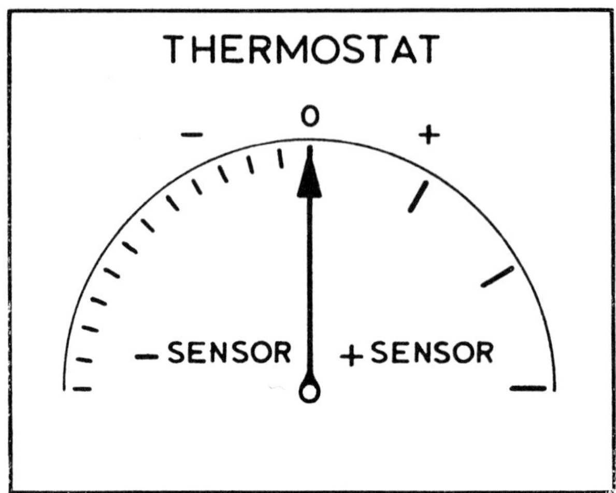

FIG. 3. Schematic representation of the hypothalamic thermostat. Note the difference in scale for the negative and positive part of the sensing device.

Serotonin apparently mimics the function of synaptic mechanisms, leading to a rise in temperature; noradrenaline mimics the function of synaptic pathways leading to a drop in temperature. Acetylcholine, in general an excitatory transmitter for hypothalamic neurons (see Shute and Lewis, 1967; Dyball and Koizumi, 1969), is also present in various areas, and is known to play a role in temperature regulation.

The interpretation of the observations on the influence of neurotransmitters upon temperature-regulatory mechanisms is considerably hampered by the fact that significant differences exist among mammalian species.

B. Hypothalamic Glucostat

The neuronal regulation of food and water intake has already been emphasized by a number of authors (see Anand, 1967). The lateral and ventromedial hypothalamic region are referred to as feeding and satiety centers, respectively (Anand, 1967). There is sufficient evidence in the literature to support the concept that glucose-sensitive neurons (hereafter referred to as glucosensors) are present in the hypothalamus. Both the systemic injection of glucose and the electro-osmotic application of glucose through a micropipette alter the discharge rate of a large proportion of neurons in the lateral hypothalamic region and the ventromedial hypothalamic nucleus of the rat brain.

Oomura and colleagues (1969) performed an elegant series of experiments by applying glucose through a micropipette to the membrane surface of neurons in various areas of the hypothalamus. Of 57 neurons tested in the ventromedial hypothalamus, almost half (24) were plus-glucosensors, the remainder showing no effect. Quite a different population of neurons was found in the lateral hypothalamic region; here, 21 of 64 were plus-glucosensors and 15 were minus-glucosensors, with 28 neurons showing no effect (Fig. 4). In both areas a relatively high percentage of neurons was observed with glucose-sensitive properties. A large number of neurons were investigated in other cerebral structures as far as glucose sensitivity was concerned, but in neither the cerebral cortex nor in the thalamus could any glucosensors be detected.

These experiments support the concept of the existence of a neuronal glucostat (Fig. 5). The basic regulatory system consists of two sets of neurons that are distributed in unequal proportion over two hypothalamic areas. It is interesting to note that although the areas are generally referred to as feeding and satiety centers, both contain plus-glucosensors. This observation points to the presence in the hypothalamus of an integrating system characterized by two measuring devices and a set-point. The unequal number of plus-and minus-sensors may indicate a difference in the scale of the measuring devices.

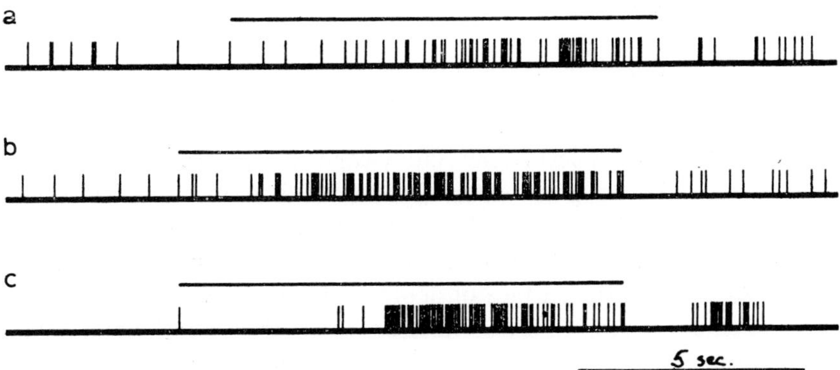

FIG. 4. Response of a hypothalamic neuron to glucose application: (a) current 4nA; (b) current 6nA; (c) current 15nA (From Oomura et al., 1969).

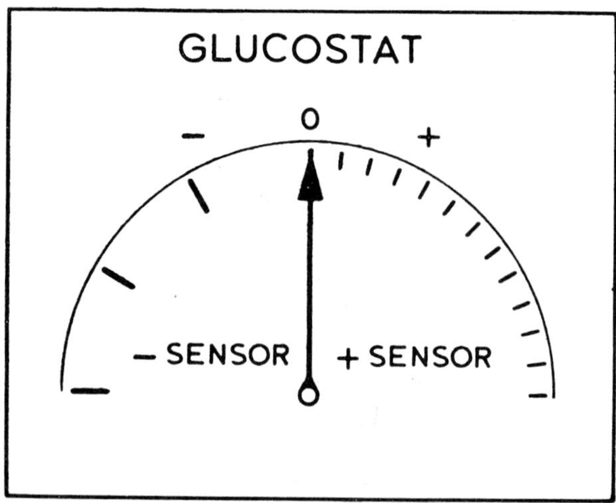

FIG. 5. Schematic representation of the glucostat. Note the difference in scale for the negative and positive part of the sensing device.

An increase of circulating glucose above the normal is apparently measured in smaller steps than is a decrease in glucose content of the blood (which is monitored by a few sensors on a rough scale). Temporary adjustment of the set-point will be achieved by changes in the neuronal input of the sensors. It is known that stimulation of other cerebral structures either enhances or diminishes the intake of food. Long-lasting adjustment of the set-point is demonstrated by lesions in the hypothalamic areas, since loss of sensors will then result in aphagia or hyperphagia.

C. Hypothalamic Steroidstat

It has been a mystery so far how circulating hormones affect the brain at the cellular level. Our knowledge in this field has recently been advanced to a great extent by Ruf and Steiner (1967a,b) and by Steiner and co-workers (1968, 1969) who iontophoretically applied hormonal substances, dexamethasone and Adrenocorticotropic Hormone (ACTH), to the surface of neurons. They showed very elegantly the existence of steroid-sensitive neurons in the hypothalamus and midbrain of the rat. They report that of 337 individual hypothalamic and mesencephalic neurons, only 57 showed a decrease or an increase in firing rate upon direct application of minute amounts of dexame-

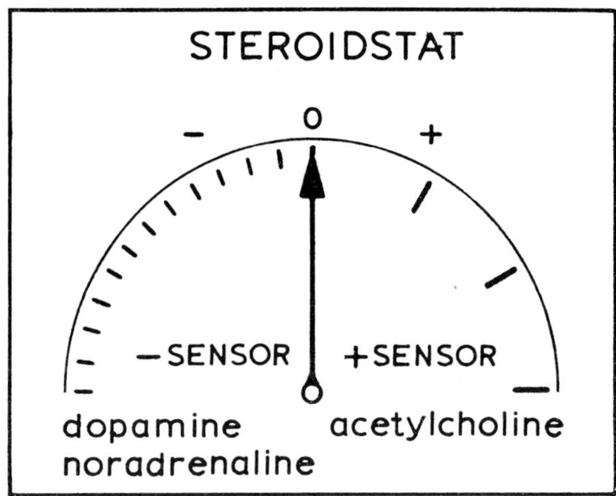

FIG. 6. Schematic representation of the steroidstat. Note the difference in scale for the negative and positive part of the sensing device.

thasone to the neuronal membrane. The steroid minus-sensors far exceed in number the plus-sensors. The steroid-sensitive neurons are scattered over a wide area in the hypothalamus. As far as other localizations of steroid sensors are concerned, no neurons in the cerebral cortex, dorsal hippocampus, or thalamus were found to respond to dexamethasone. The same phenomenon of an unequal distribution of plus and minus sensors also seems to exist for the steroidstat, as was already noted for other measuring devices in the hypothalamus.

Steroid-sensitive neurons were also exposed to the action of neurotransmitters in order to mimic the activity of excitatory and inhibitory presynaptic terminals upon the membrane of the sensors. In general, acetylcholine increased the firing rate, while noradrenaline and dopamine decreased the firing rate of the steroid sensors. Dopamine exhibited a more potent effect than did noradrenaline. As could be expected from the difference in action upon the neuronal membrane of hormones and neurotransmitters, the former acted slower but the action lasted longer. Here again a dual sensitivity of hypothalamic neurons was shown, both to circulating substances and to regular neurotransmitters acting via presynaptic endings. It seems likely that the set-point of the steroidstat (Fig. 6) is determined by the circulating steroid; deviations from the norm are regulated by a change in firing rate of the steroid sensors. The activity of synaptic pathways may provide a neuronal basis for adjustment of the set-point. This latter mechanism is able to act in a fast way by influencing the membrane conductivity and thus making it the threshold of the spike-generating area of the sensor.

III. Conclusions

The hypothalamic integrating neuron (HIN) is a specialized unit that shows dual properties:

1. Neurotransmitter sensitivity: both excitatory and inhibitory synaptic mechanisms have been demonstrated.
2. Sensitivity to environmental factors such as temperature, glucose, and hormones.

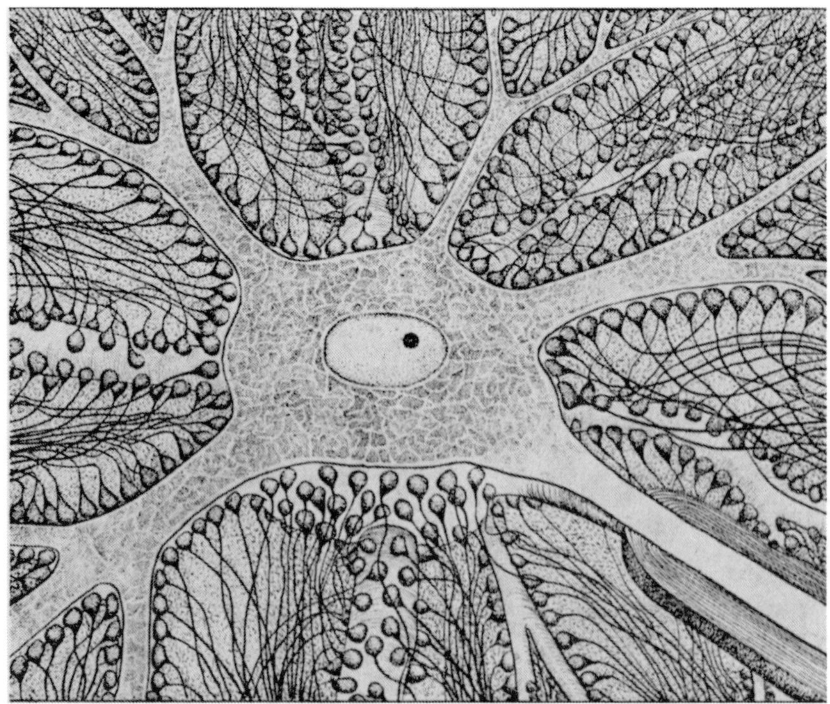

FIG. 7. Model of a hypothalamic integrating neuron with excitatory and inhibitory input.

A hypothalamic integrating system consists of two sets of these neurons showing sensitivity to the same factors. However, as far as the environmental component is concerned, these integrating neurons have either a plus or minus sensor property. One of the most striking findings is the fact that the plus-and minus-sensors are present in unequal numbers. This indicates that the measuring system has a different scale either below or above the norm. We are apparently dealing with a stochastic set-point over a group of neurons that may be adjusted by neuronal synaptic mechanisms.

The characteristic properties of the HIN may be found in neurons and neurosecretory elements of the hypothalamus. The phenomenon of neurosecretion means, in this respect, that cells characterized by morphological and physiological features (action potentials and synaptic mechanisms) as true neurons are able to produce substances that

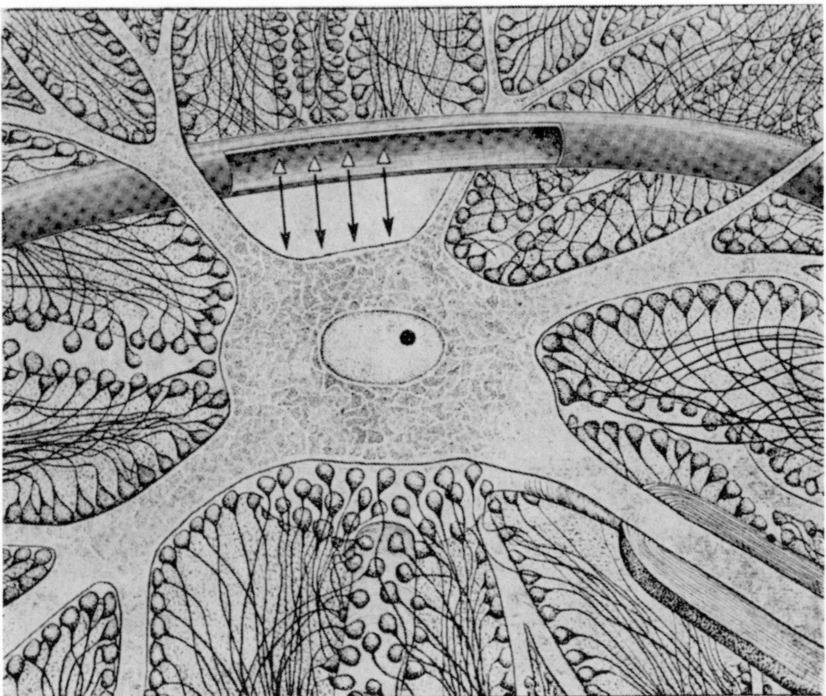

FIG. 8. Model of a hypothalamic neuron. A part of the receptor surface will respond in a quantitative manner to certain factors or substances in the circulation.

act as hormones or releasing or inhibiting factors. These substances are probably released in a quantitative manner by action potentials arriving at the transmitting part of the cell.

The HIN receives information by parallel input, in contrast to computing units that operate mainly by serial inputs (Fig. 7). The receptive pole of the HIN is not only the seat of the interplay between the hundreds and thousands of excitatory and inhibitory synapses, but is also the receptor site of environmental stimuli.

The HIN possesses the qualities of a hybrid computing unit. It handles both digital and analog types of information. In general, sophisticated digital/analog (D/A) and analog/digital (A/D) converters are integrated parts of a hybrid computing unit. In the neuron

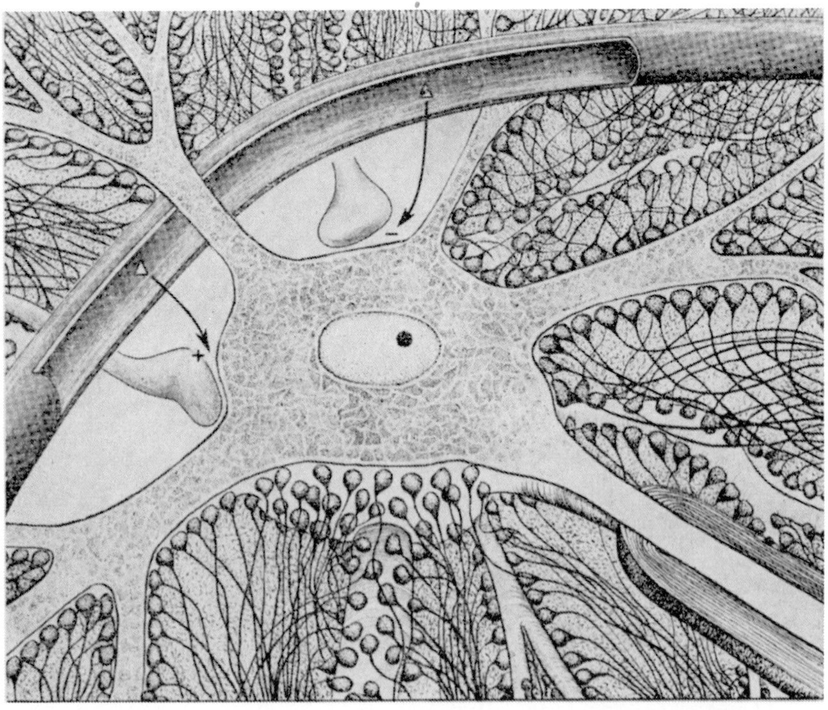

FIG. 9. Model of a hypothalamic neuron. The circulatory factor may influence the synaptic transmission in either a positive (increasing firing rate) or negative way (decreasing firing rate). The two possibilities, as shown in the figure, actually do not occur in the same neuron, since positive or negative effects have been shown to occur in separate units.

the synapse acts as a D/A converter. It may be regarded as a transistor switch that can open a microcircuit according to whether it receives a digital 1 or 0 control command. The microcircuit releases quantal amounts of neurotransmitter. As in the input of an amplifier network, these switches act as relay contacts. The binary signals of the presynaptic terminals are supplied to sets of D/A switches (the branch terminals of the presynaptic fiber), each supplying a given voltage to an analog amplifier (the postsynaptic neuron, or HIN), the sum of their voltages giving the required voltage.

Environmental factors either directly affect the postsynaptic neuronal surface (the analog signals are then directly added or subtracted by the analog amplifier) or indirectly via the synapse (Fig. 8, 9). Control signals are usually logic signals and the digital part of the unit is always given the job of controlling both its own actions and those of the analog part. It can be assumed that the RNA molecules in the neuron provide these control signals. The control signals determine the threshold properties of the HIN; however, their exact nature is still unknown.

A major question is the adjustment of the set-point of the system. It seems most likely that the set-point is adjusted by neuronal and not by environmental factors. The same situation is encountered in the communication between the digital (neuronal) and the analog (environmental) parts of a hybrid computing unit. The digital part must be able to vary the information pattern in the analog part during the course of a run, according to its own calculation. Similarly, the analog part must be able to change the digital calculations when certain conditions are satisfied in its own computing. Lack of knowledge about the information-processing properties of the HIN still hampers the construction of a suitable model.

References

Anand, B. (1967). *In* " Handbook of Physiology " Williams & Wilkins, Baltimore.

Cross, B. A., and Silver, I. A. (1966). *Brit. Med. Bull.* **22**, 254.

Dyball, R. E. J., and Koizumi, K. (1969). *J. Physiol.* **201**, 711.

Eisenman, J. (1965). *Physiologist* **8**, 158.

Findlay, A. L. R., and Hayward, J. N. (1969). *J. Physiol.* **201**, 237.

Hellon, R. F. (1967). *J. Physiol.* **193**, 381.

Lincoln, D. W. (1967). *J. Endocrinol.* **37**, 177.

Oomura, Y., Kimura, K., Ooyama, H., Maeno, T., Iki, M., and Kuniyoshi, M. (1964). *Science* **143**, 484.

Oomura, Y., Ono, T., Ooyama, H., and Wayner, M. J. (1969). *Nature* **222**, 282.
Ruf, K., and Steiner, F. A. (1967a). *Science* **156**, 667.
Ruf, K., and Steiner, F. A. (1967b). *Acta Endocrinol.* **119**, 38.
Schadé, J. P. (1969). "Die Funktion des Nervensystems." Fischer, Stuttgart.
Schadé, J. P., and Ford, D. B. (1967). "Basic Neurology." Elsevier Publishing Co., Amsterdam.
Shute, C. C. D., and Lewis, P. R. (1967). *Brain* **90**, 497.
Steiner, F. A., Pieri, L., and Kaufmann, L. (1968). *Experientia* **24**, 1133.
Steiner, F. A., Ruf, K., and Akert, K. (1969). *Brain Res.* **12**, 74.

Some Endocrine Applications of Electrophysiology

C. H. SAWYER

I. Introduction

Electrophysiological recording methods have been employed in hypothalamic research for less than 20 years, and the first unit studies on hypothalamic neurons were reported just 10 years ago. Pioneers in this area of neuroendocrinology include von Euler (1953) who recorded slow DC " osmopotentials " from the anterior hypothalamus of the cat following the intracarotid injection of hypertonic saline, Porter (1953, 1954) who described electroencephalographic (EEG) changes in the cat and monkey hypothalami relative to pituitary-adrenal function, and Cross and Green (1959) who observed changes in unit firing patterns in the rabbit hypothalamus induced by osmotic changes in the carotid blood.

Although every electrophysiological study on hypothalamic neurons has potential neuroendocrine significance, only those investigations bearing directly on hypothalamo-hypophyseal function or the feedback action of hormones on the brain will be discussed here. For a consideration of afferent pathways to the hypothalamus as traced by electrophysiological methods the reader is referred to excellent papers by Sutin (1966), Dreifuss and associates (1968), and to the article by Feldman and Dafny in this volume. More details on hypothalamic unit activity relative to neuroendocrine function may be found in a recent review by Beyer and Sawyer (1969).

Each of the electrical recording techniques has its own virtues and limitations. The DC method, employing nonpolarizable electrodes, reveals slow potential changes (Aladjalova, 1964) but their meaning and localization are difficult to assess (Kawamura *et al.*, 1967). EEG methods, employing macroelectrodes with tips more than 100 μ in diameter, record moderate to slow wave activity from a wide area. These EEG waves are also difficult to interpret and localize in terms of neuronal function, but the technique has revealed meaningful correlations between endocrine functions and patterns of electrical activity in discrete hypothalamic regions. Generalized changes in sleep-wakefulness and seizure activity are also easily observed in the EEG record.

The unit activity recording technique, with microelectrodes of about 1μ tip, has become the method of choice in hypothalamic neuroendocrine research because the electrical phenomena are easy to interpret as action potentials generated by neurons located near the recording microelectrode (Morrell, 1967). However, the unit method samples only a relatively small population of cells related to a given function and these are usually selected from large neurons generating big action potentials. This bias may adversely influence neuroendocrine studies, where " parvicellular " neurons (Szentágothai, 1964) control adenohypophyseal function. Recently a multiunit background activity technique has been introduced to neuroendocrine studies by Beyer and associates (1967). With a large microelectrode (25 to 50 μ diam) the method permits the integrated simultaneous recording of activity from a large number of neurons in regions such as the median eminence and arcuate nucleus, where it is difficult to register single unit activity.

II. Neurohypophysis

A. Electrical Correlates of Activation

It is well known that neurons of the supraoptic and paraventricular nuclei synthesize vasopressin (ADH) and oxytocin, hormones of the neurohypophysis. These neurons share characteristics of nerve cells and gland cells and their electrophysiological properties are therefore of considerable interest. The suggestion that important osmoreceptors are located in the anterior hypothalamus stems from the brilliant research of Verney (1947) in which he demonstrated the release of ADH in response to increased osmolarity of the carotid blood. The search for osmoreceptors with electrophysiological methods began with the experiments, already cited, of von Euler (1953). His slow DC potential changes suggested that neurons sensitive to alterations in blood tonicity produced a generator receptor potential not unlike that of peripheral receptors.

Recording dramatic EEG changes emanating from the olfactory system in response to intracarotid injections of hypertonic saline in the rabbit, cat, and dog, Sawyer and Gernandt (1956), Sundsten and Sawyer (1959), Holland and associates (1959a, b), and Sawyer and Fuller (1960) proposed that the important osmoreceptors might reside in the olfactory bulb or tubercle. Sundsten and Sawyer (1961) disproved this interesting hypothesis by demonstrating the osmotically induced activation of the neurohypophysis in rabbits with " hypothalamic islands." However, osmotic stimuli trigger unit responses in the isolated olfactory bulb (Freedman, 1963), and Moyano and Brooks (1968) have recently shown that electrical stimulation of the olfactory bulb increases the firing rate of neurons in or near the ipsilateral supraoptic nucleus. Thus there appear to be secondary osmoreceptor sites outside the hypothalamus which may possibly modulate the activity of the primary osmoreceptor cells.

Meanwhile Cross and Green (1959) had initiated the pioneer unit recording studies which were eventually to demonstrate the electrical properties of osmosensitive neurons in the isolated hypothalamus. These authors described the osmoreceptors as highly specific, since they generally failed to respond to tactile, auditory, or visual stimulation. Specificity of the osmoreceptors was also claimed by Joynt (1964) in his unit studies in the cat hypothalamus. However, Brooks (1962, 1966) and Koizumi (1964) and their co-workers, who confirmed that intracarotid injections of hypertonic solutions altered the firing rate of supraoptic neurons, reported that the discharge rate of these neurons was also strongly influenced by somatic stimuli. Research in Brooks' laboratory by Suda and associates (1963) and Ishikawa and co-workers (1966a) demonstrated that neurons in the isolated hypothalamus fired at a faster resting level but that they retained their responsiveness to osmotic stimuli and ability to release ADH. Recently osmosensitive cells have been investigated in the opossum hypothalamus (Zeballos *et al.*, 1967; Wang, 1969). In all species so far studied these cells have also proved to be very sensitive to cholinergic drugs (Brooks *et al.*, 1962, 1966; Koizumi *et al.*, 1964; Joynt, 1966; Ishikawa *et al.*, 1966a; Zeballos *et al.*, 1967; Wang, 1969).

Oxytocin is released from the neurohypophysis in response to suckling or milking stimulation of the mammary glands or distension of the uterus (Cross, 1966). Whereas electrical stimulation of the nipples did not alter the firing rate of supraoptic or paraventricular neurons (Cross and Green, 1959), either the application of suction to the nipples with a vacuum pump or the distension of the uterine cervix increased the firing rate of paraventricular neurons and activated the milk ejection reflex (Brooks *et al.*, 1966). Similar stimuli increased electrical activity in the pituitary stalk and elevated pressure in the mammary ducts (Ishikawa *et al.*, 1966b). This work establishes an important correlation between electrical activity in the paraventricular nucleus and oxytocin release.

As mentioned above, the neurosecretory cells of the hypothalamo-hypophyseal system have properties of both gland cells and nerve cells. Intracellular electrical recordings in fishes have established the neural properties of these cells (Kandel, 1964; Bennett *et al.*, 1968) and demonstrated their slow conduction rates of about 0.5 m/sec. Conduction speeds along the stalk in the cat were estimated at 0.6 to 1.4 m/sec (Ishikawa *et al.*, 1966b). Yagi *et al.* (1966) have reemphasized and extended the original observation of Cross and Green (1959) that mammalian neurosecretory cells of the supraoptic nucleus have neuronal properties of generation and conduction of action potentials.

B. Effects of Neurohypophyseal Hormones on Hypothalamic Electrical Activity

Vasopressin has been reported to induce both sleep-like EEG activity (Faure, 1957; Kawakami and Sawyer, 1959a) and EEG arousal (Capon, 1960) in the rabbit. In the rat under urethane anesthesia, Beyer and his colleagues (1967) found that vasopressin depressed multiunit background activity in the diencephalon secondarily to activation of carotid baroreceptors by the elevation in blood pressure. In the rabbit, vasopressin usually increased hypothalamic background activity and activated the EEG.

In the rabbit, Kawakami and Sawyer (1959a) found that oxytocin could evoke a sleep-like " EEG afterreaction " (see below). In the anestrous cat, oxytocin activated anterior and lateral hypothalamic cells but decreased the firing rate of neurons of the ventromedial hypothalamus (VMH) (Kawakami and Saito, 1967). In the estrous cat, oxytocin failed to influence the VMH or anterior hypothalamus, but it clearly depressed unit firing in lateral hypothalamic cells, which were unresponsive to Luteinizing Hormone (LH) or osmotic stimuli. There are suggestions that a feedback action of oxytocin may influence oxytocin release (Folley, 1959) or synthesis (Fendler and Telegdy, 1962).

III. Adenohypophysis and Target Organs

A. Electroencephalographic Recordings

1. Changes in Hypothalamic Electroencephalographic Patterns Related to Hormonal Action

Porter (1953, 1954) described changes in amplitude and frequency in hypothalamic EEG records of cats and monkeys following injection of epinephrine adequate to trigger Adrenocorticotropic Hormone (ACTH) release; the responses were altered by adrenalectomy or treatment with cortisone. Important contributions on the effects of sex hormones on EEG changes in the rabbit brain have been made by Faure (1954, 1957) who described an Olfacto-Bucco-Ano-Genital Syndrome (OBAGS) following hormone treatment or low-frequency stimulation of rhinencephalic structures. In the estrous rabbit, Sawyer (1955) observed a histamine-pentobarbital synergism which activated hypothalamic EEG changes and stimulated the release of pituitary ovulating hormone(s) in this reflexly ovulating species. The same sequence could be induced by intraventricular norepinephrine without pentobarbital (Sawyer, 1963). In the estrous cat, Porter and associates (1957) reported lateral hypothalamic EEG changes in response to vaginal stimulation. A somewhat different EEG afterreaction to coitus or vaginal stimulation occurs in the estrous rabbit (Sawyer and Kawakami, 1959); its EEG patterns include phases of slow wave sleep and paradoxical sleep (PS) followed by Faure's OBAGS.

2. *Excitability Changes Due to Hormones*

The rabbit's EEG afterreaction could also be elicited with exogenous pituitary hormones or low-frequency stimulation of the rhinencephalon (Fig. 1) or hypothalamus (Kawakami and Sawyer, 1959a). The threshold to electrical stimulation was markedly influenced by exogenous and endogenous hormones (Kawakami and Sawyer, 1959b; Sawyer and Kawakami, 1961; Kawakami and Sawyer, 1964; Sawyer *et al.*, 1966; Kawakami *et al.*, 1966; Sawyer, 1967; Kawakami and Sawyer, 1967; Endröczi, 1967). The release of pituitary ovulating hormone(s) is inhibited when the threshold is elevated, and it has been proposed that the antifertility steroids exert their critical blocking effects at both hypothalamic (Sawyer and Kawakami, 1961; Kawakami and Sawyer, 1967; Exley *et al.*, 1968; Schally *et al.*, 1968) and pituitary (Hilliard *et al.*, 1966; Spies *et al.*, 1969) levels of action. Changes in EEG afterreaction threshold postcoitum and throughout pregnancy are seen in Fig. 2.

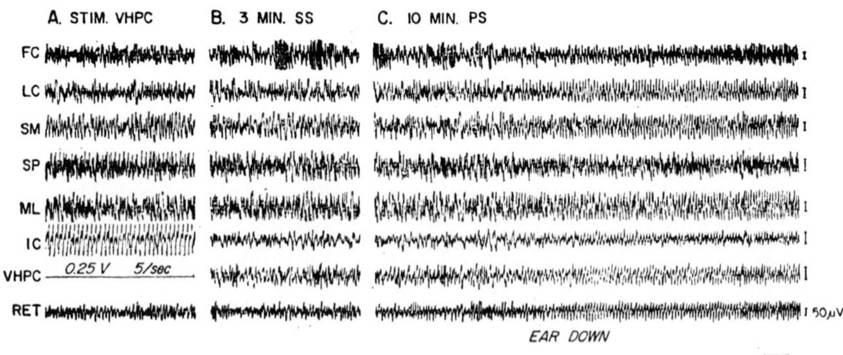

FIG. 1. Phases of the rabbit's EEG afterreaction including episodes of slow wave sleep (SS) and paradoxical sleep (PS) at intervals after stimulating the ventral hippocampus (VHPC) for 30 sec at 0.25 V and a frequency of 5 cps. The record is alert or aroused in A. Drooping of the ear indicates the onset of PS. Abbreviations: FC, frontal cortex; IC, internal capsule; LC, limbic cortex; ML, lateral mammillary nucleus; RET, reticular formation; SM, stria medullaris; SP, septum (From Kawakami and Sawyer, 1967).

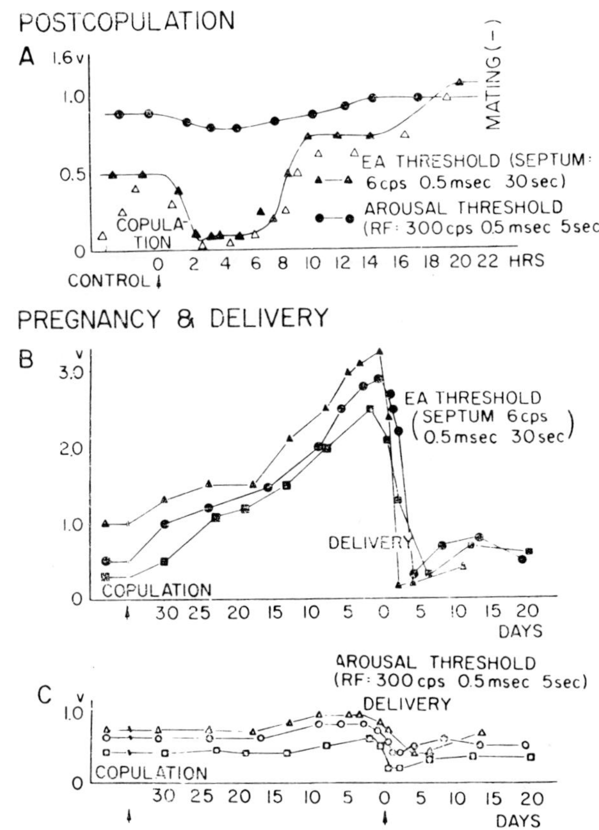

FIG. 2. A: EEG afterreaction (EA) and arousal threshold changes in voltage following copulation. Note the sharp drop in EA threshold a few hours after coitus and the abrupt elevation afterward. B and C: Changes of EEG afterreaction threshold (B), and EEG arousal (C) during pregnancy and after delivery. Note sharp drop in arousal threshold prior to onset of delivery (From Kawakami *et al.*, 1966).

The threshold of EEG arousal, evoked by stimulating the posterior hypothalamus or midbrain reticular formation at high frequency, is also influenced by sex hormones but not so dramatically as the afterreaction threshold (Kawakami and Sawyer, 1959b). The arousal threshold is low at natural or hormonally-induced estrus, and its changes appear to be more closely related to estrous behavior than to other aspects of pituitary-gonadal function (Sawyer, 1962). It drops sharply at the end of pregnancy (Fig. 2), a period of heightened receptivity in the rabbit doe. Feldman and Davidson (1966) reported that hydrocortisone had no effect on the EEG arousal threshold in rabbits.

Adrenocortical steroids, however, do influence seizure thresholds. Woodbury and Vernadakis (1966) reviewed the evidence for the lowering of seizure thresholds by cortisol and cortisone and their elevation by desoxycorticosterone. Endröczi and Lissák (1962) observed that the EEG seizure threshold of electrical stimulation of the hippocampus was markedly lowered by cortisone in cats. In rabbits, Feldman and Davidson (1966) reported localized electrical seizures in VMH and septum following the intravenous injection of cortisol; in cats, intraventricular cortisol induced electrical seizures starting in the anterolateral hypothalamus and hippocampus (Feldman, 1966). In rats, seizure thresholds were lowered by estrogen and raised by testosterone (Werboff and Corcoran, 1961; Woolley and Timiras, 1962; other references in Sawyer, 1967). The threshold of localized seizures in the medial amygdala of female rats is lowered at puberty (Terasawa and Timiras, 1968a) and following treatment with estrogens (Terasawa and Timiras, 1968b).

3. *Sleep Studies*

The phase of " hippocampal hyperactivity " in the female rabbit's EEG afterreaction to coitus (Sawyer and Kawakami, 1959) was recognized as paradoxical sleep (reviewed by Jouvet, 1967) almost simultaneously by Kawakami and Sawyer (1962) and Faure and colleagues (1963). Its characteristics have been thoroughly studied by Faure (1965a, 1965b, 1968) and Kawakami (1966). Although rabbits of either sex, fully habituated to their environment, show cyclic episodes of PS, the threshold of the induced response is lowered by female sex hormones and elevated by androgen (Kawakami and Sawyer, 1959b). Both Kawakami (1966) and Faure (1968) have described increases in spontaneous PS relative to amounts of slow wave sleep (SS) in natural or induced estrus; the elevated PS/SS ratio does not develop in rabbits with small lesions in the basal tuberal-premammillary region of the hypothalamus (Kawakami, 1966).

In confirmation of the suggestion (Sawyer and Kawakami, 1959) that the postcoital reponse might be induced by a feedback action of released pituitary hormones, Spies and his associates (1970) have noted that PS fails to occur postcoitally in the estrogen-primed hypophysectomized doe. Moreover, they were unable to induce episodes of PS in the hypophysectomized rabbit with exogenous LH, prolactin, or oxytocin, which are effective in the intact female. Changes in the distribution of PS relative to the estrous cycle in the rat have been reported by Yokoyama (1966) and Colvin (1968) and their co-workers. Although the functional significance of PS is not understood, its importance to endocrinology has been emphasized by the observations of Mandell and Mandell (1965) that ADH and ACTH are released during episodes of PS in man, and by the report of Honda and colleagues (1969) that the release of Growth Hormone (GH), which is activated during SS, is inhibited during PS.

B. Evoked Potentials

The evoked potential technique (Landau, 1967) in which a population of neurons is induced to discharge in response to a stimulus applied some distance away, has been useful not only in tracing afferent pathways to the hypothalamus but also in studying the effects of hormones on transmission, i.e., latency, amplitude, wave form, and recovery time of the electrical reponse. Although with macroelectrodes the precise source of the evoked potential is generally not easy to establish, Feldman and colleagues (1961) were able to show that there were both oligosynaptic and polysynaptic routes between sciatic nerve and hypothalamus in cats, and that intravenous cortisol increased the amplitude of the long latency potentials. Adrenalectomy increased the conduction time and delayed neuronal recovery (Feldman, 1962). These findings in acute experiments were confirmed in chronic unanesthetized cats by Chambers and his co-workers (1963) and more recently by Feldman and Davidson (1966) in unanesthetized rabbits.

The effects of female sex hormones and oxytocin on evoked potentials in the arcuate nucleus have been studied in chronic ovariectomized cats and rabbits by Kawakami and associates (1966) and Kawakami and Terasawa (1967). In hormonally-induced estrus the potentials evoked by stimulating the sciatic nerve or the amygdala were facilitated; those resulting from stimulation of the hippocampus or dorsal sacral spinal root were inhibited. Under " pro1esterone dominance " the opposite effects were observed. Oxytocin reversed the effects of both estrogen-progesterone and progesterone alone. Estrogen enhanced somatosensory (sciatic) transmission to the hypothalamus and progesterone viscerosensory (dorsal sacral root) transmission, both by way of the reticular formation which appeared to act as a highgain amplifier. Endröczi and colleagues (1968) have reported that ovariectomy in rats elevated the threshold of evoked potentials in the preoptic region stimulated in the midbrain reticular formation; estrogen restored the threshold to normal. Thus hormonal effects on evoked potentials are interpreted as lowering or raising thresholds of synaptic transmission with the consequent involvement of more or fewer units, respectively, in the responses.

C. Unit Studies

Compared with the sharply localized, large-celled supraoptic and paraventricular nuclei responsible for neurohypophyseal secretion, the locations of parvicellular neurons controlling adenohypophyseal function are less well defined. Their secretions, the releasing factors, are part of a complex humoral circuit which includes tropins of the anterior pituitary gland and hormones of the target organs, both of which can act back on the brain in internal and external feedback circuits. Electrophysiological evidence for both long and short feedback loops will be given below.

1. Steroids

a. Adrenal cortex. The sensitivity of hypothalamic cells to adrenal steroids has been investigated by Slusher and associates (1966) with a multiple-microelectrode technique. In unanesthetized cats the intravenous or direct intracerebral injection of cortisol quickly affected the firing patterns of the hypothalamus and midbrain. An even more direct approach was the microelectrophoretic infusion of dexamethasone into hypothalamic cells by Ruf and Steiner (1967); the neurons were inhibited almost immediately. Steiner and co-workers (1969) have recently extended the study with the local application of ACTH, which was generally stimulatory to hypothalamic neurons. According to Feldman and Dafny (1966) intravenous cortisol increases the firing rate of anterior hypothalamic cells and alters their responsiveness to incoming peripheral impulses. In a study of the effects of ACTH on multiple unit activity in the rat diencephalon (to be reported below), Sawyer and colleagues (1968) noted that dexamethasone depresses multiunit levels in the hypothalamus.

b. Ovary. Cross and his colleagues introduced the unit technique to the study of hypothalamo-adenohypophyseal function. In the rat under urethane anesthesia, Barraclough and Cross (1963) studied the responsiveness of lateral hypothalamic neurons to various stimuli during the different phases of the estrous cycle. Urethane was a happy choice of anesthetic, since light dosages do not block ovulation (Haller and Barraclough, 1968; Terasawa *et al.*, 1969) and "sleep-wakefulness" changes can be monitored in the EEG (Ramirez *et al.*, 1967). During estrus more units were inhibited and fewer facilitated in response to peripheral stimulation than during diestrus. In spayed rats, estrogen increased the inhibitory and decreased the excitatory responses to pain, cold, and cervical stimuli (Cross 1964, 1965; Lincoln, 1967; Lincoln and Cross, 1967). Intravenous progesterone in propylene glycol at a dosage of 40 to 400 μg blocked the excitatory response of lateral hypothalamic neurons to cervical probing (Barraclough and Cross, 1963). Since pain and cold were still effective in accelerating these neurons, it was inferred that progesterone selectively depressed the response of lateral hypothalamic neurons to genital stimulation.

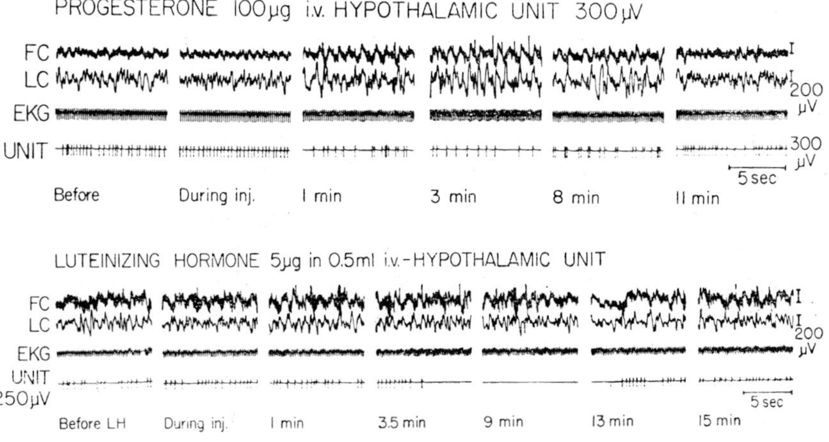

FIG. 3. Effects of progesterone and luteinizing hormone (LH) on the cortical EEG and unit firing of neurons in the ventromedial hypothalamus of the rat under light urethane anesthesia. The changes induced by progesterone here appear to be related to a generalized sleep-like response, whereas the LH effect appears to be exerted more specifically on the hypothalamic neuron. For abbreviations see Fig. 1 (From Sawyer, 1967, after the work of Ramirez *et al.*, 1967).

Ramirez and co-workers (1967) examined the effects of progesterone on the spontaneous discharge of thalamic and hypothalamic units in the urethanized rat. They found that most units were depressed in their firing rate within a matter of seconds but that simultaneously the cortical EEG became synchronized (Fig. 3). Since all units affected had been previously observed to decelerate their discharges during EEG synchronization, it was suggested that the progesterone influence was indirect and nonspecific. Subsequently Beyer and colleagues (1967) recorded hypothalamic multiple unit activity and found that a drop in background activity accompanied EEG spindling. Intravenous progesterone induced EEG synchrony, apparently by raising the blood pressure and triggering the carotid sinus reflex with resultant inhibition in neuronal activity. More recently Komisaruk and associates (1967) studied the actions of progesterone on both hypothalamic multiunit and single unit activity simultaneously. The firing rates of 80% of the units were suppressed with the onset of EEG synchrony, and responses to peripheral stimuli, including vaginal probing, were depressed.

These findings might explain the lack of responsiveness of lateral hypothalamic units in the progestin-treated rats as a generalized action of progesterone on brain excitability associated with spindle sleep. However, Cross and his colleagues (Cross and Silver, 1966) maintain that progesterone selectively suppresses the response of hypothalamic cells to vaginal stimulation. A possible explanation for these disagreements lies in differences in recording site; the English group recorded from the lateral hypothalamus and the California investigators from the ventromedial region. Such an explanation receives support from the findings of Kawakami and Saito (1967) in the unanesthetized cat and rabbit, and Faure and colleagues (1967) in the estrous rabbit, namely, that vaginal stimulation simultaneously excites neurons of the arcuate nucleus and inhibits the firing pattern of ventromedial and dorsomedial hypothalamic neurons. These changes were independent of the state of vigilance of the animal. The lateral hypothalamus was not tested by these investigators.

Consistent with these findings are the recent observations of Terasawa and Sawyer (1970) on the effects of progesterone (1 to 2 mg subcutaneously in oil in the estrogen-primed ovariectomized rat) on multiunit background activity in the arcuate nucleus and me-

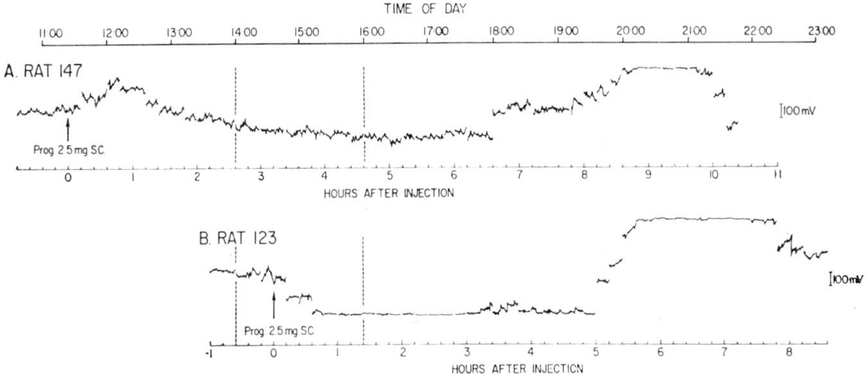

FIG. 4. Records of multiunit activity in the arcuate nucleus showing diurnal change in responsiveness of its neurons to progesterone (2.5 mg injected subcutaneously). The ovariectomized rats were in artificial proestrus from estrogen priming. The elevation in multiunit activity following progesterone injection around 11:00 may be correlated with advancement of the "critical period" (14:00–16:00) during which the release of pituitary ovulating hormone is triggered at proestrus. A sleep-like EEG (not shown) follows progesterone treatment at either hour (Terasawa and Sawyer, 1970).

dian eminence. The initial response was activation if the progesterone was injected in the morning and depression if it was injected after the onset of the "critical period" (14:00); there appears to be a diurnal variation in responsiveness of the neurons to the steroid (Fig. 4). The overall response to progesterone injected in the morning was triphasic, activation followed by depression and secondary activation. The afternoon response involved only the last two phases. However, the effect on the EEG in both cases was synchronization and sleep. Thus there is evidence that progesterone exerts both specific and generalized effects on hypothalamic neurons, and it is essential to monitor the EEG to differentiate between them.

2. *Pituitary Hormones*

The suggestion of Sawyer and Kawakami (1959, 1961) and others that the tropic hormones of the adenohypophysis may exert internal or short-loop feedback effects on the brain was mentioned earlier. A few studies have been made of the effects of the tropins on hypothalamic activity in the absence of the target organ.

Reference has been made above to a study of the effects of ACTH and dexamethasone on multiunit background activity of the hypothalamus. In the adrenalectomized rat under light urethane anesthesia, ACTH injections (2 units) induced a short latency (10 min) rise in multiunit activity in the arcuate nucleus and a fall in activity in the basolateral thalamus, zona incerta, and entopeduncular nucleus (Sawyer *et al.*, 1968). The excitatory response in the arcuate nucleus suggests that ACTH may exert an initially positive feedback action on the neurons controlling its secretion. As compared with the short latency effect of ACTH, Follicle-Stimulating Hormone (FSH) administration to ovariectomized rats elicited a long latency (2 hr) rise in multiunit activity in the arcuate nucleus and ventral part of the ventromedial nucleus (Kawakami and Sawyer, unpublished observations).

The effects of LH on hypothalamic unit activity have been studied more extensively than those of the other pituitary tropins. Ramirez and co-workers (1967) found specific units at the base of the ventromedial nucleus in which the LH effects were totally independent of EEG activity (Fig. 3, lower tracing). The characteristic response to LH (5 to 10 μg intravenously) was depression in firing rate lasting 3 to 30 min; this response could not be duplicated with saline, vasopressin, boiled LH, or peripheral stimuli. In the cat, Kawakami and Saito (1967) also noted inhibition of firing of VMH neurons following intravenous injection of LH, but there was simultaneously a clear increase in frequency of discharge in the arcuate nucleus; the latter was independent of EEG changes, whereas the VMH effects paralleled sleep-like changes in the cortical EEG. Ramirez and associates (1967) did not record from the arcuate nucleus.

Recently Terasawa and co-workers (1969) have investigated the effects of LH on multiple unit activity of the arcuate nucleus and medial preoptic area in the estrogen-primed ovariectomized rat under urethane. Intravenous injections of LH (20 to 40 μg)gradually increased multiunit activity of the arcuate nucleus to a peak approximately an hour and a half after injection. With a similar time course the multiunit activity in the medial preoptic area was depressed by LH (Fig. 5). These responses to LH appeared to be specific, since no clear changes were observed in other diencephalic areas or in the EEG, and boiled LH was without effect. To test the independence of the arcuate nucleus a semicircular cut with a small stereotaxic knife was made to separate the two areas. This is the frontal cut of Halász and Gorski (1967) which blocks cyclic ovulation. The an-

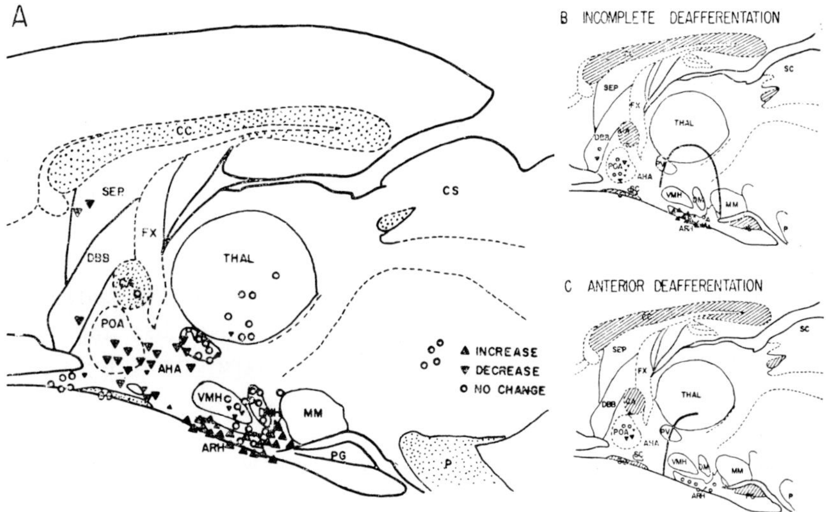

FIG. 5. A: Diagram of sagittal section of rat brain showing areas responding to Luteinizing Hormone (LH) injection by increased and decreased multiple unit activity. Abbreviations: AHA, anterior hypothalamic area; ARH, arcuate nucleus; CA, anterior commissure; CO, optic chiasm; CS, superior colliculus; DBB, diagonal band (Broca); DMH, dorsomedial hypothalamic nucleus; FX, fornix; MM, mammillary body; P, pons; PG, pituitary gland; POA, preoptic area; PV, paraventricular nucleus; SEP, septum; THAL, thalamus; VMH, ventromedial hypothalamic nucleus. The smaller triangles represent lesser changes in multiunit activity. B: Posterior or incomplete deafferentation does not alter the response to LH. C: Anterior deafferentation completely eliminates the reponse of ARH neurons to LH (From Terasawa *et al.*, 1969).

terior deafferentation (Fig. 5 C) prevented the excitatory influence of LH on arcuate neurons, suggesting that the effect was synaptically mediated and dependent on connections with the anterior hypothalamus or medial preoptic region, which controls the cyclic release of LH in the rat (Everett, 1961). The rise in activity in the arcuate nucleus was interpreted as a positive feedback influence, since it could be duplicated by ovulation-inducing electrochemical stimulation of the medial preoptic region (Terasawa and Sawyer, 1969).

IV. Conclusions

With electrophysiological recording methods it has been possible to correlate patterns of hypothalamic neural activity with control of pituitary secretion and to observe the effects of hormones, including pituitary tropins, on activity levels and responsiveness of hypothalamic neurons. This is merely a starting point for a deeper analysis of hypothalamic functions. EEG, evoked potential and multiple unit methods have provided useful information as to *when* and *where* neuronal activity is occurring relative to hormonal secretion. However, greater use of the single unit technique in conjunction with the other methods will be required to gain a deeper understanding of underlying mechanisms and relationships.

A start has been made in the use of computers for the statistical analysis of hypothalamic neuronal firing patterns (Findlay and Hayward, 1969; Kawakami and Saito, 1969). This type of scrutiny holds much greater promise than the mean frequency discharge analysis methods employed heretofore (Burns, 1968). Techniques of unit recording from freely moving conscious animals must be perfected, and intracellular recordings must be made from mammalian hypothalamic neurons if insight is to be gained into the synaptic events preceding neuronal firing. It will still be essential to monitor the EEG and desirable to have other polygraphic information such as electrocardiogram, respiratory rate, and blood pressure. Only then will it be possible to judge the specificity of discharge of neurons relative to changes in their internal environment.

Acknowledgments

Original work presented in this paper was supported by grants from the National Institutes of Health (NB 01162 and AM 06468) and the Ford Foundation. The assistance of the UCLA Brain Information Service is gratefully acknowledged (NINDB contract 43-66-59) and special thanks are due to Mrs. S. Elise Bozzo for bibliographic and secretarial help.

References

Aladjalova, N. A. (1964). "Progress in Brain Research." Vol. VII. Elsevier Publishing Co., New York.

Barraclough, C. A., and Cross, B. A. (1963). *J. Endocrinol.* **26**, 339.

Bennett, M. V. L., Gimenez, M., and Ravitz, M. J. (1968). *Anat. Record* **160**, 313.

Beyer, C., and Sawyer, C. H. (1969). *In* "Frontiers in Neuroendocrinology 1969" (L. Martini and W. F. Ganong, eds.). Oxford Univ. Press, New York, pp. 255–287.

Beyer, C., Ramirez, V. D., Whitmoyer, D. I., and Sawyer, C. H. (1967). *Exp. Neurol.* **18**, 313.

Brooks, C. McC., Ushiyama, J., and Lange, G. (1962). *Am. J. Physiol.* **202**, 487.

Brooks, C. McC., Ishikawa, T., Koizumi, K., and Lu,H. H. (1966). *J. Physiol.* (London) **182**, 217.

Burns, B. D. (1968). "The Uncertain Nervous System." Arnold, London.

Capon, A. (1960). *Arch. Int. Pharmacodyn.* **127**, 141.

Chambers, W. F., Friedman, S. L., and Sawyer, C. H. (1963). *Exp. Neurol.* **8**, 458.

Colvin, G. B., Whitmoyer, D. I., Lisk, R. D., Walter, D. O., and Sawyer, C. H. (1968). *Brain Res.* **7**, 173.

Cross B. A. (1964). *Symp. Soc. Exp. Biol.* **18**, 157.

Cross, B. A. (1965). *In* "Proc. II Intern. Congr. Endocrinol." (S. Taylor, ed.). *Excerpta Med. Int. Congr. Ser.* **83**, 513–516.

Cross, B. A. (1966). *In* "Neuroendocrinology" (L. Martini and W. F. Ganong, eds.). Vol. I. Academic Press, New York, pp. 217–259.

Cross, B. A., and Green, J. D. (1959). *J. Physiol.* (London) **148**, 554.

Cross, B. A., and Silver, I. A. (1966). *Brit. Med. Bull.* **22**, 254.

Dreifuss, J. J., Murphy, J. T., and Gloor, P. (1968). *J. Neurophysiol.* **31**, 237.

Endröczi, E. (1967). *In* "Symposium on Reproduction" (K. Lissák, ed.). Akadémiai Kiadó, Budapest, pp. 39–63.

Endröczi, E., and Lissák, K. (1962). *Acta Physiol. Acad. Sci. Hung.* **21**, 257.

Endröczi, E., Babichev, V., Hartmann, G., and Koranyi, L. (1968). *Endocrinol. Exp.* **2**, 1.

Everett, J. W. (1961). *In* "Sex and Internal Secretions" (W. C. Young, ed.). Vol. I. Williams and Wilkins, Baltimore, pp. 497–555.

Exley, D., Gellert, R. J., Harris, G. W., and Nadler, R. D. (1968). *J. Physiol.* (London) **195**, 697.

Faure, J. (1954). *Rev. Neurol.* **90**, 338.

Faure, J. (1957). *Rev. Pathol. Gen. Physiol. Clin.* **57**; No. 690, 1029; No. 691, 1263; No. 692, 1445.

Faure, J. (1965a). *In* "Proc. II Intern. Congr. Endocrinol." (S. Taylor, ed.). *Excerpta Med., Int. Congr.* **83**, 608–611.

Faure, J. (1965b). *Colloq. Int. C.N.R.S.* (Paris) *No.* 127, 241.

Faure J. (1968). *Actualités Neurophysiologiques*, 8th Ser., p. 251.

Faure, J. Vincent, J. D., LeNouëne, J., and Geissmann, A. (1963). *C.R. Soc. Biol.* **157**, 799.

Faure, J., Vincent, J. D., Bensche, Cl., Favarel-Garrigues, B., and Dufy, B. (1967). *J. Physiol.* (Paris) **59**, 405.

Feldman, S. (1962). *Arch. Neurol.* **7**, 460.

Feldman, S. (1966). *Epilepsia* **7**, 271.

Feldman, S., and Dafny, M. (1966). *Israel J. Med. Sci.* **2**, 621.

Feldman, S., and Davidson, J. M. (1966). *J. Neurol. Sci.* **3**, 462.

Feldman, S., Todt, J. C., and Porter, R. W. (1961). *Neurology* **11**, 109.

Fendler, K., and Telegdy, G. (1962). *Acta Physiol. Acad. Sci. Hung.* **22**, 59.

Findlay, A. L. R., and Hayward, J. N. (1969). *J. Physiol.* (London) **201**, 237.

Folley, J. S. (1959). *In* "Comparative Endocrinology" (A. Gorbman, ed.). Wiley, New York, p. 144.

Freedman, S. (1963). *Anat. Record* **145**, 229.

Halász, B., and Gorski, R. A. (1967). *Endocrinology* **80**, 608.

Haller, E. W., and Barraclough, C. A. (1968). *Proc. Soc. Exp. Biol. Med.* **129**, 291.

Hilliard, J., Croxatto, H. B., Hayward, J. N., and Sawyer, C. H. (1966). *Endocrinology* **79**, 411.

Holland, R. C., Cross, B. A., and Sawyer, C. H. (1959a). *Am. J. Physiol.* **196**, 796.

Holland, R. C., Sundsten, J. W., and Sawyer, C. H. (1959b). *Circ. Res.* **7**, 712.

Honda, Y., Takahashi, K., Takahashi, S., Azumi, K., Irie, M., Sakuma, M., Tsushima, T., and Shizumi, K. (1969). *J. Clin. Endocrinol. Metab.* **29**, 20.

Ishikawa, T., Koizumi, K., and Brooks, C. McC. (1966a). *Neurology* **16**, 101.

Ishikawa, T., Koizumi, K., and Brooks, C. McC. (1966b). *Am. J. Physiol.* **210**, 427.

Jouvet, M. (1967). *Physiol. Rev.* **47**, 117.

Joynt, R. J. (1964). *Neurology* **14**, 584.

Joynt, R. J. (1966). *Arch. Neurol.* **14**, 331.

Kandel, E. R. (1964). *J. Gen. Physiol.* **47**, 691.

Kawakami, M. (1966). *Progr. Brain Res.* **21 B**, 90.

Kawakami, M., and Saito, H. (1967). *Japan. J. Physiol.* **17**, 466.

Kawakami, M., and Saito, H. (1969). *Japan. J. Physiol.* **19**, 243.

Kawakami, M., and Sawyer, C. H. (1959a). *Endocrinology* **65**, 631.

Kawakami, M., and Sawyer, C. H. (1959b). *Endocrinology* **65**, 652.

Kawakami, M., and Sawyer, C. H. (1962). *Fed. Proc.* **21**, 356.

Kawakami, M., and Sawyer, C. H. (1964). *Exp. Neurol.* **9**, 470.

Kawakami, M., and Sawyer, C. H. (1967). *Endocrinology* **80**, 857.

Kawakami, M., and Terasawa, E. (1967). *Japan. J. Physiol.* **17**, 65.

Kawakami, M., Terasawa, E., Tsuchihashi, T., and Yamanaka, K. (1966). *In* "Steroid Dynamics" (G. Pincus, J. Tait and T. Nakao, eds.). Academic Press, New York, pp. 237–302.

Kawamura, H., Whitmoyer, D. I., and Sawyer, C. H. (1967). *Electroencephalog. Clin. Neurophysiol.* **22**, 337.

Koizumi, K., Ishikawa, T., and Brooks, C. McC. (1964). *J. Neurophysiol.* **27**, 878.

Komisaruk, B. R., McDonald, P. G., Whitmoyer, D. I., and Sawyer, C. H. (1967). *Exp. Neurol.* **19**, 494.

Landau, W. M. (1967). *In* "The Neurosciences" (G. C. Quarton, T. Melnechuk and F. O. Schmitt, eds.). Rockefeller Univ. Press, New York, pp. 469–482.

Lincoln, D. W. (1967). *J. Endocrinol.* **37**, 177.

Lincoln, D. W., and Cross, B. A. (1967). *J. Endocrinol.* **37**, 191.

Mandell, A. J., and Mandell, M. P. (1965). *Am. J. Psychiat.* **122**, 391.

Morrell, F. (1967). *In* "The Neurosciences" (G. C. Quarton, T. Melnechuk and F. O. Schmitt, eds.). Rockefeller Univ. Press, New York, pp. 452–469.

Moyano, H. F., and Brooks, C. McC. (1968). *Fed. Proc.* **27**, 320.

Porter, R. W. (1953). *Am. J. Physiol.* **172**, 515.

Porter, R. W. (1954). *Recent Progr. Hormone Res.* **10**, 1.

Porter, R. W., Cavanaugh, E. B., Critchlow, B. V., and Sawyer, C. H. (1957). *Am. J. Physiol.* **189**, 145.

Ramirez, V. D., Komisaruk, B. R., Whitmoyer, D. I., and Sawyer, C. H. (1967). *Am. J. Physiol.* **212**, 1376.

Ruf, K., and Steiner, F. A. (1967). *Science* **156**, 667.

Sawyer, C. H. (1955). *Am. J. Physiol.* **180**, 37.

Sawyer, C. H. (1962). *Excerpta Med. Int. Congr. Ser.* **47**, 642.

Sawyer, C. H. (1963). *In* "Proc. I Int. Pharmacol. Meeting" (R. Guillemin, ed.). Vol. I. Pergamon Press, New York, pp. 27–46.

Sawyer, C. H. (1967). *In* "Hormonal Steroids" (L. Martini, F. Fraschini, and M. Motta, eds.). *Excerpta Med. Int. Congr. Ser.*, **132**, 123-135.

Sawyer, C. H., and Fuller, G. R. (1960). *Electroencephalog. Clin. Neurophysiol.* **12**, 83.

Sawyer, C. H., and Gernandt, B. E. (1956). *Am. J. Physiol.* **185**, 209.

Sawyer, C. H., and Kawakami, M. (1959). *Endocrinology* **65**, 622.

Sawyer, C. H., and Kawakami, M. (1961). *In* "Control of Ovulation" (C. A. Villee, ed.). Pergamon Press, New York, pp. 79–100.

Sawyer, C. H., Kawakami, M., and Kanematsu, S. (1966). *Res. Publ. Assoc. Res. Nervous Mental Disease* **43**, 59.
Sawyer, C. H., Kawakami, M., Meyerson, B., Whitmoyer, D. I., and Lilley, J. J. (1968). *Brain Res.* **10**, 213.
Schally, A. V., Carter, W. H., Saito, M., Arimura, A., and Bowers, C. Y. (1968). *J. Clin. Endocrinol. Metab.* **28**, 1747.
Slusher, M. A., Hyde, J. E., and Laufer, M. (1966). *J. Neurophysiol.* **29**, 157.
Spies, H. G., Stevens, K. R., Hilliard, J., and Sawyer, C. H. (1969). *Endocrinology* **84**, 277.
Spies, H. G., Whitmoyer, D. I., and Sawyer, C. H. (1970). *Brain Res.* (In press).
Steiner, F. A., Ruf, K., and Akert, K. (1969). *Brain Res.* **12**, 74.
Suda, I., Koizumi, K., and Brooks, C. McC. (1963). *Japan. J. Physiol.* **13**, 374.
Sundsten, J. W., and Sawyer, C. H. (1959). *Proc. Soc. Exp. Biol. Med.* **101**, 524.
Sundsten, J. W., and Sawyer, C. H. (1961). *Exp. Neurol.* **4**, 548.
Sutin, J. (1966). *Int. Rev. Neurobiol.* **9**, 263.
Szentágothai, J. (1964). *Progr. Brain Res.* **5**, 135.
Terasawa, E., and Sawyer, C. H. (1969). *Endocrinology.* **85**, 143.
Terasawa, E., and Sawyer, C. H. (1970). *Exp. Neurol.* (In press).
Terasawa, E., and Timiras, P. S. (1968a). *Am. J. Physiol.* **215**, 1462.
Terasawa, E., and Timiras, P. S. (1968b). *Endocrinology* **83**, 207.
Terasawa, E., Whitmoyer, D. I., and Sawyer, C. H. (1969). *Am. J. Physiol.* **217**, 1119.
Verney, E. B. (1947). *Proc. Roy. Soc. Ser. B* **135**, 25.
Von Euler, C. (1953). *Acta Physiol. Scand.* **29**, 133.
Wang, M. B. (1969). *Neuroendocrinology* **4**, 51.
Werboff, J., and Corcoran, J. B. (1961). *Am. J. Physiol.* **201**, 830.
Woodbury, D. M., and Vernadakis, A. (1966). *In* "Methods in Hormone Research" (R. I. Dorfman, ed.). Vol. V. Academic Press, New York, pp. 1–57.
Woolley, D. E., and Timiras, P. S. (1962). *Endocrinology* **70**, 196.
Yagi, K., Azuma, T., and Matsuda, K. (1966). *Science* **154**, 778.
Yokoyama, A., Ramirez, V. D., and Sawyer, C. H. (1966). *Gen. Comp. Endocrinol.* **7**, 10.
Zeballos, G. A., Wang, M. B., Koizumi, K., and Brooks, C. McC. (1967). *Neuroendocrinology* **2**, 88.

Effects of Extrahypothalamic Structures on Sensory Projections to the Hypothalamus

S. FELDMAN and N. DAFNY

I. Introduction

The various autonomic, behavioral, and endocrine regulatory functions of the hypothalamus are influenced to a considerable degree by peripheral stimuli as well as by other brain regions, particularly those regions anatomically close to the hypothalamus (Brady, 1960; Ingram, 1960; Mangili *et al.*, 1966). It is the purpose of this article to review our electrophysiological experiments on the effects of extrahypothalamic structures upon the sensory projections to the hypothalamus.

Experiments were perfomed on cats. In the evoked potentials studies, locally anesthetized, immobilized, and artificially respirated animals were used, whereas the microelectrode recordings were perfomed on cats under pentobarbital anesthesia. Concentric stainless-steel bipolar electrodes were introduced stereotaxically for the recording of the evoked potentials, and either platinum-iridium or stainless-steel microelectrodes were used for single-cell recording. While the evoked potentials were studied in the whole hypothalamus, the single-cell recordings were perfomed mainly in the anterior (F12-13; L0.5-1.5; H-3 — -5) and posterior (F8.5-9; L0.5-1; H-2 — -3) hypothalamic regions (Jasper and Ajmone-Marsan, 1954). The sensory stimuli consisted of single photic, acoustic, and sciatic nerve stimulation.

The responses were amplified, recorded on an oscilloscope, and photographed. To test the effects of the various brain regions on the responses in the hypothalamus, single conditioning or high-frequency stimulation were applied to these structures.

The effects of acute and chronic brain lesions on the sensory projections were also investigated. The spontaneous activity of the single cells and the changes following stimulation were evaluated from at least ten sweeps (2000 msec each) for every experimental condition. In many experiments one hundred sweeps were photographed. The total number of spikes was counted for each unit in all sweeps, and the data were evaluated statistically for changes in the rate and pattern of firing of the cells. Changes in the rate of firing were determined by using the Critical Ratio (CR) Test: [$CR = (E - S)/(E \div S)^{1/2}$, where E and S signify evoked and spontaneous activity, respectively]. The change of pattern of firing was determined by counting ten periods of 200 msec each following the stimulus and by applying the Chi-square (X^2) Test calculated at various levels of significance. A unit showing either a significant change in the rate and/or pattern of firing was defined as a responsive unit. The position of electrodes and brain lesions were determined histologically at the end of the experiments.

II. Results

A. Effects of Midbrain Reticular Formation

Following the stimulation of the sciatic nerve, both short- (7–10 msec) and long- (20–40 msec) latency evoked potentials were recorded in the hypothalamus (Feldman *et al.*, 1959). Results of brainstem lesion and stimulation experiments (Feldman, 1963) suggest an oligosynaptic pathway through the medial lemniscus to the posterolateral hypothalamus as well as a polysynaptic pathway extending through the midbrain reticular formation to the ventromedial hypothalamus and preoptic area. Both short- and long-latency responses were reduced by high-frequency (100—200 cps) stimulation applied to the midbrain reticular formation. Furthermore, single conditioning reticular stimuli inhibited the response, and recovery was only complete at a stimulus separation of 300 msec. The stimulation of numerous points in the hypothalamus (where the sciatic responses were recorded) evoked short-latency (1–3 msec) potentials in the dorsal hippocampus and suggested the existence of a unidirectional oligosynaptic pathway from the hypothalamus to the hippocampus. In contrast to its inhibitory effect on the afferent projections to the hypothalamus, the reticular formation facilitated conduction from the hypothalamus to the dorsal hippocampus. Thus, 200-cps reticular stimulation as well as conditioning reticular stimuli at stimulus separations up to 200 msec, increased the hippocampal responses and caused the appearance of additional long-latency waves which did not exist in the test response (Feldman, 1962).

The neuronal recovery in the hypothalamus was studied by delivering paired shocks to the sciatic nerve at varying time intervals. The amount of recovery was determined by expressing the amplitude of the response to the second of a pair of shocks as a percentage of that of the first of the same pair. The evoked potentials in the posterior hypothalamus, in spite of their short latency, have shown in intact animals a prolonged recovery, which was not complete even at a stimulus separation of 1000 msec. In view of the inhibitory effect of the reticular formation on the somatosensory-evoked potentials in the hypothalamus the effects of anterior pontine and midbrain reticular formation lesions on the recovery cycle of the short-latency responses were studied. Such lesions very considerably shortened the recovery cycle, with complete recovery of the test response at stimulus separations of 100–200 msec (Feldman, 1963).

At the single-cell level in the posterior hypothalamus, single conditioning reticular stimuli preceding the sensory stimulus by 25 msec considerably modified the responses of about 60% of the units to photic, acoustic, and sciatic stimulation. In the majority of units the sensory stimulation alone caused an increase in the rate of firing, while the reticular-sensory interaction produced a decrease in firing-rate, when compared with the effects of the sensory modality alone. The units that were inhibited by sensory stimulation showed mainly further inhibition as a result of interaction with the reticular formation. In addition, the pattern of firing, which during the 2000 msec following sensory stimulation had shown a triple peak response, was modified. It now showed either early inhibition or an early increase in firing in the first 200 msec after stimulation, with subsequent inhibition. Thus there were hypothalamic units that, as a result of reticular-sensory interaction, showed significant changes in the

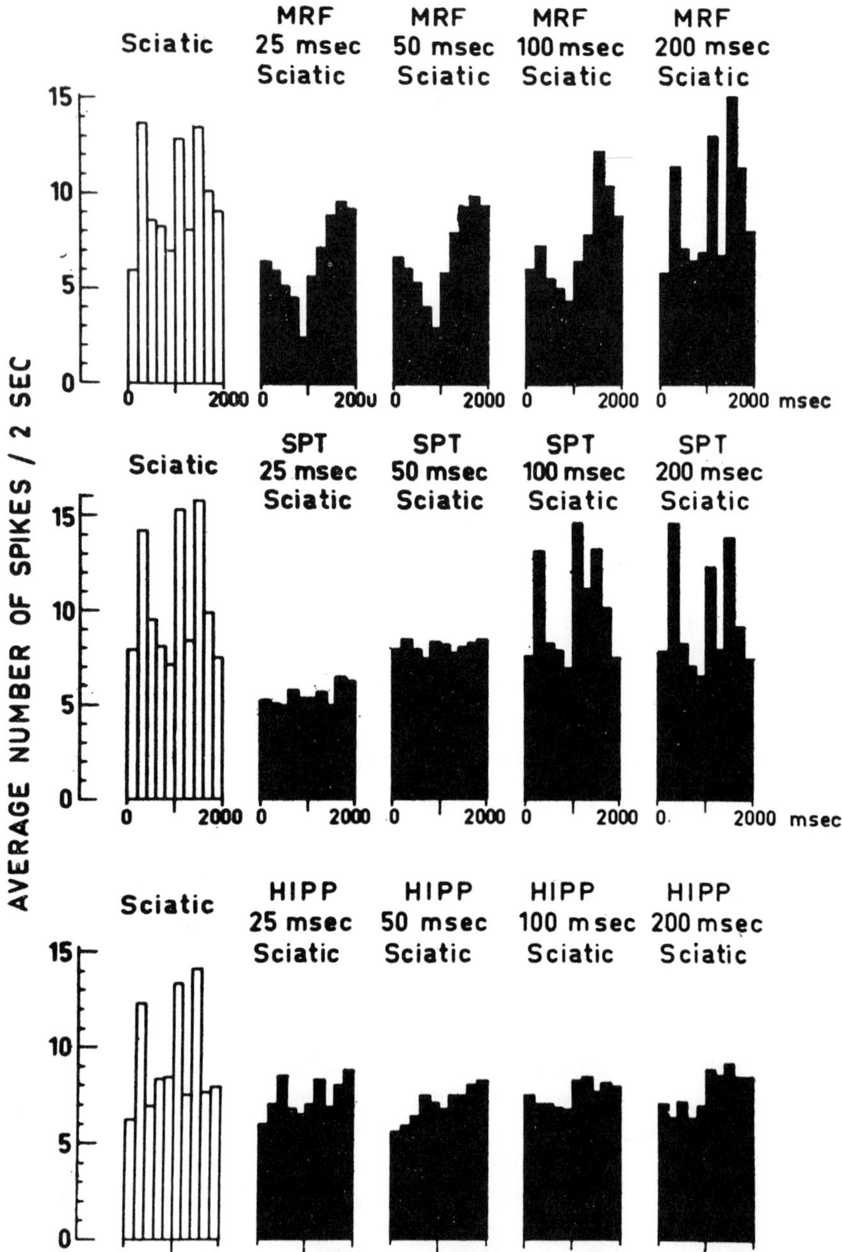

FIG. 1. Compound histograms during a period of 2000 msec after stimulation, showing the effects of sensory stimuli on the spontaneous activity (white bars) and the effects of interaction, at different time intervals (25–200 msec), between a conditioning central and a sciatic stimulus (black bars), in groups of 25 units each. Abbreviations: MRF, midbrain reticular formation; SPT, septum; Hipp, hippocampus (From Dafny and Feldman, 1969).

pattern of firing but no modification in the overall rate of discharge. At an interval of 200 msec between the conditioning reticular and test sensory stimuli, the compound histogram returned to its control pattern (Fig. 1). The early increase in unit firing following reticular-photic interaction may possibly be related to our previous finding, that single reticular conditioning stimuli at short stimulus separations facilitate the short-latency (10 msec), photic-evoked potentials in the anterior hypothalamus and preoptic area (Feldman, 1964). Midbrain reticular formation lesions caused a considerable reduction in the spontaneous activity of the posterior hypothalamic units as well as in their sensory responsiveness. However, the units that were facilitated by sensory stimuli in such animals showed a considerable increase in the firing when compared with the rates in intact cats. Thus a fivefold increase in firing was observed in many units following photic stimulation (Dafny and Feldman, 1969).

B. Effects of Hippocampus and Septum

In contrast to the uniformly inhibitory effects of the reticular formation, the dorsal hippocampal inhibition on the sciatic-evoked potentials varied in accordance with the position of the hypothalamic electrode and was obtained in about half of the recordings, particularly in the rostral hypothalamus. High-frequency (100–200 cps) and single conditioning hippocampal stimuli abolished sciatic responses, and this inhibition was even more prolonged in the case of the potentials recorded in the preoptic area. The ventral hippocampus and the amygdala also had inhibitory effects on the hypothalamic-evoked potentials, but not necessarily at the same points at which the hippocampal effects were exerted (Feldman, 1962).

The hippocampus and septum have shown predominantly inhibitory effects on posterior hypothalamic units, affecting the responsiveness to sensory stimulation of 40–50% and 50–60% of the units, respectively. Conditioning hippocampal or septal stimuli preceding the sensory stimulus by 25 msec reduced the rate of firing and abolished the pattern of the triple peak response in the compound histograms. While the septal-sensory interaction at varying intervals produced complete recovery of the histogram at a stimulus separation of 100 msec, the hippocampal-sensory interaction caused a much more prolonged inhibition, and recovery was obtained only at an interval of 350 msec (see Fig. 1) (Feldman and Dafny, 1968a; Dafny and Feldman, 1969).

C. Effects of Corpus Striatum

Single conditioning stimuli applied to the head of the caudate nucleus caused a variable degree of inhibition of both short- and long-latency evoked potentials in the hypothalamus. At a 50-msec separation almost complete inhibition was observed, with recovery at 200 msec. For the purpose of further clarifying the role of the caudate nucleus in afferent somatosensory projections to the hypothalamus, bilateral electrolytic lesions were placed in the head of both caudate nuclei. In such animals, which were alert and had normal behavior, neuronal recovery of the sciatic-evoked potentials was determined 3–10 days after the operation. While there was no appreciable change in the prolonged recovery cycle of the long-latency responses, which propagate through multisynaptic systems to the anterior hypothalamus, there was a considerable shortening of the neuronal recovery of the short-latency evoked potentials in the posterior hypothalamus. Thus, in cats with 10 days lesions, complete recovery was attained at a stimulus separation of 100 msec and persisted throughout the recovery cycle of 1000 msec (Feldman, 1966).

In intact cats the stimulation of the caudate nucleus caused a significant reduction (CR Test) in both spontaneous and evoked single-cell activity when a conditioning caudate stimulus was applied 25 msec before a photic, acoustic, or a sciatic stimulus (Dafny and Feldman, 1967; Feldman and Dafny, 1968b). The mean firing rate of units in the posterior hypothalamus following sensory stimuli (5.5–6.3 spikes/sec) was reduced to 2.8–3.3 spikes/sec following the caudate-sensory interaction. The caudate stimulation significantly changed the rate and the pattern of firing (X^2 Test). The pattern of the peaks was changed as the result of the interaction. At intervals of 25 and 50 msec between the ipsilateral caudate nucleus and the sensory stimulus it became an ascending histogram. At 100 msec the three peaks became evident and complete recovery of the pattern was obtained at an interval of 200 msec. In contrast to these findings, a single conditioning pallidal stimulus at 25, 50, 100, and 200 msec has very considerably reduced the mean firing rate (to 1.7–2.5 spikes/

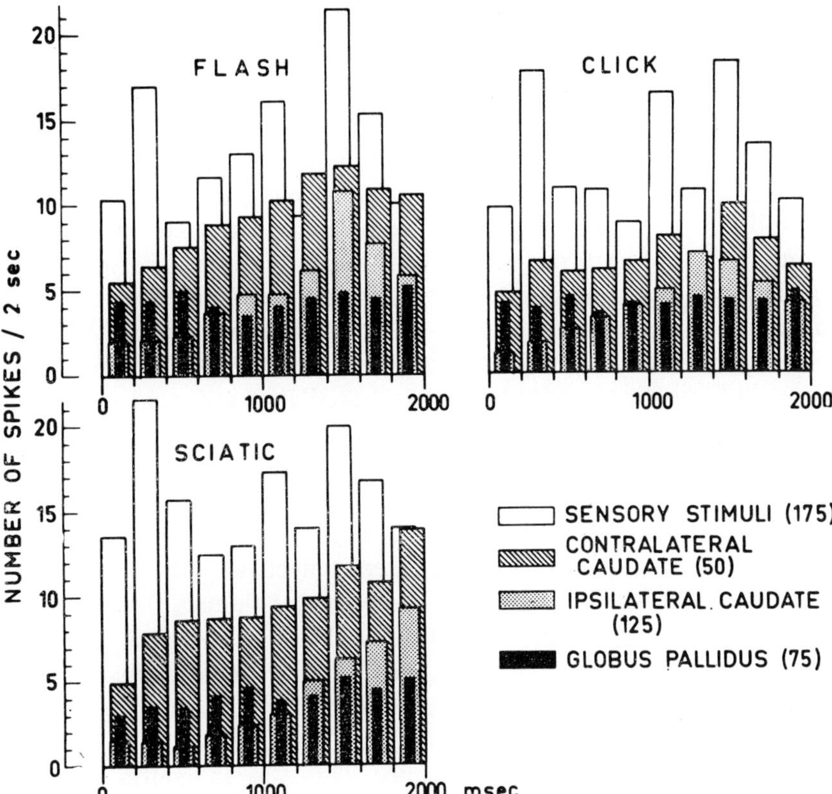

FIG. 2. Compound, superimposed histograms during a period of 2000 msec after stimulation, showing the effects of the three sensory modalities on the spontaneous activity of units in the posterior hypothalamus and comparing the effects of different conditioning striatal stimuli preceding the sensory stimulus by 25 msec (From Feldman and Dafny, 1968b).

sec) and has abolished the pattern of the compound histogram (Fig. 2). In some of these units there was no significant change in the rate of firing as a result of caudate stimulation; however, the same units showed a very significant ($p < 0.001$) change in the pattern of firing, due to complete early inhibition. This demonstrates the importance of determining responsiveness and of examining each cell as to changes in both the rate and pattern of firing.

Further statistical analysis of the results of interaction between caudate and sensory stimuli has demonstrated that 70–80% of the cells respond significantly to an ipsilateral caudate nucleus stimulation when there is a change in the rate of firing, more than 90 percent showing inibition. Following contralateral caudate or globus pallidus stimulation the ratio between cells that were facilitated or inhibited by the central stimulus was essentially the same as after ipsilateral caudate stimulation; however, the total number of responsive units was reduced. At 2000 msec after stimulation, conditioning striatal stimuli caused various degrees of complete cessation in the spontaneous and evoked firing of the hypothalamic units. The ipsilateral caudate was much more effective in this respect than either the contralateral caudate or the globus pallidus. The analysis of the relationship between the effects of striatal stimuli and sensory responsiveness of the single cells has demostrated that the caudate mainly affected units that were facilitated by peripheral stimuli and responded uniformly to all three sensory modalities. No such effect was observed in units influenced by the globus pallidus. These and other differences in responsiveness suggest that the caudate and globus pallidus may, to a certain extent, affect two different populations of hypothalamic neurons (Dafny and Feldman, 1968).

Chronic striatal lesions caused a considerable change in the responsiveness of hypothalamic neurons (Dafny and Feldman, 1967; Feldman and Dafny, 1968b). Ipsilateral or bilateral caudate nucleus lesions produced no change in the overall percentage of responsive units in the posterior hypothalamus. This was in contrast to the finding in the anterior hypothalamus in which the responsiveness was reduced. However, more responsive cells were facilitated by the three sensory modalities. Contralateral caudate lesions were not effective. While bilateral pallidal lesions reduced the percentage of posterior hypothalamic units responsive to sensory stimulation, many more facilitatory cells were found in such preparations, when compared to intact animals. As a result of these changes there was a considerable increase in the overall rate of firing of the units after sensory stimulation. Thus, while in intact animals the rate of firing of anterior hypothalamic units was 4.3–4.5 cps, sensory stimulation after ipsilateral caudate nucleus lesions resulted in rates that were increased to 7.2–8.8 cps. This effect was even more pronounced in animals with globus pallidus lesions. Effective striatal lesions also caused changes in the pattern of the compound histograms of unit firing (Fig. 3). In animals with chronic lesions of globus pallidus or midbrain reticular formation, or in *cerveau isolé* preparations,

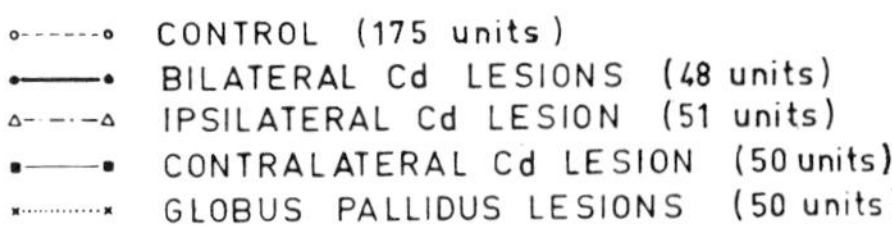

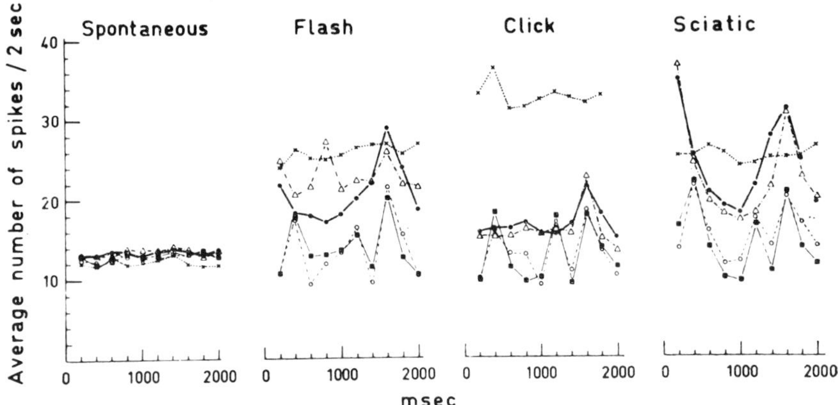

FIG. 3. Average curves showing the spontaneous activity of single cells in the posterior hypothalamus during a period of 2000 msec in intact cats (control) as well as in animals with various striatal lesions. The figure also demonstrates the effects of photic, acoustic, and sciatic stimulation on the pattern of firing of units in intact animals and in those with chronic caudatal or pallidal lesions (From Feldman and Dafny, 1968b).

the interaction of caudate and sensory stimuli caused a considerable reduction in the responsiveness of the units when compared with that of intact cats, indicating that caudatal inhibitory effects on the hypothalamus are mediated both directly and through the globus pallidus and the midbrain. Furthermore, in the animals with pallidal and reticular lesions the percentage of facilitatory units was considerably increased; caudate stimulation has also caused an increase in the rate of the spontaneous firing of hypothalamic units.

D. EFFECTS OF BRAIN LESIONS ON SENSORY CONVERGENCE

When convergence of sensory stimuli on a unit was defined as responsiveness to a change in the rate and/or a pattern of firing after stimulation of two or three sensory modalities, it was found that sensory convergence occurred in 82.1% of 223 units and in 68.2% of 715 units in the anterior and posterior hypothalamus, respectively. Further analysis of changes in the rate of firing of the

TABLE I

EFFECTS OF BRAIN LESIONS ON SENSORY CONVERGENCE

Brain Lesions	+ +/+ + +, %	Mixed, %	— —/— — —, %
Control	44.6	24.1	31.3
Caudate nucleus	68.3	12.7	9.5
Globus pallidus	70.0	—	30.0
Reticular formation	75.0	—	25.0

(Significance brackets in table: Control vs. Caudate nucleus $p<0.001$; $p<0.06$; $p<0.01$.)

[1] The control group consists of 115 units; 218 units were studied in animals with brain lesions. The headings + +/+ + + and — —/— — — indicate uniform responsiveness of units to two or three sensory modalities by facilitation or inhibition, respectively. " Mixed " indicates heterogenous responsiveness of the units.

posterior hypothalamic units, basing the test on the normal approximation to hypergeometric distribution, has shown that a high correlation exists between the responsiveness of single cells to the three sensory modalities ($p < 0.0001$). Bilateral caudate nucleus lesions reduced the percentage of convergent cells in the anterior hypothalamus to 37.5% ($p < 0.0001$). There was no change in convergence in the posterior hypothalamus following caudate nucleus lesions; however, globus pallidus and midbrain reticular formation lesions reduced the percentage of convergence to 29.2% ($p < 0.001$) and 38.0% ($p < 0.02$), respectively. The p values relate to intact animals.

In animals with brain lesions the correlation between the responsiveness to the three sensory modalities became less significant or nonexistent. While in the intact animals the three sensory stimuli produced (in the same cell) uniform facilitatory, inhibitory, or mixed responses, the brain lesions changed this proportion and most units reacted by facilitation; relatively few cells were inhibited by the three sensory modalities or showed a mixed response. These differences between intact animals and those with brain lesions are presented in Table I; the respective p values are calculated by the Chi-square Test.

III. Conclusions

The present experiments demonstrate that the reticular formation, hippocampus, septum, and corpus striatum modulate to a very considerable degree the sensory projections to the hypothalamus. The effects of the aforementioned structures on somatosensory-evoked potentials are inhibitory. In addition, our studies have demonstrated in the case of the reticular formation and the caudate nucleus, that electrolytic lesions have very considerably shortened the neuronal recovery of the short-latency evoked potentials in the posterior hypothalamus (which have all the properties of responses propagating in oligosynaptic pathways). As the animals with the caudate lesions were fully alert, it may be assumed that the shortening in the neuronal recovery was not related to an altered state of consciousness. The effect produced by the lesions can most probably be attributed to the removal of phasic inhibitory influences exerted normally by these structures on the hypothalamus. The extrahypothalamic effects on single-cell responses following the stimulation of three sensory modalities were also predominantly inhibitory (as demonstrated by interaction of central and peripheral stimuli and by finding that lesions in the reticular formation, caudate nucleus, and the globus pallidus have produced facilitatory phenomena in unit firing which were not observed in intact animals).

The hypothalamus has close anatomical connection with various brain regions, and the influences demonstrated in the present experiments are probably mediated by these pathways. Prominent in this respect is the "limbic system-midbrain circuit" (Nauta, 1958), which consists of reciprocal connections of the hippocampus, septum, and amygdala with a medial region of the midbrain tegmentum. The hypothalamus has major relays to both regions of this circuit and therefore occupies a central position within it. As is evident from our results, the hypothalamus is greatly influenced by the structures within the circuit, and because it receives sensory projections it serves as a nodal region in which central and peripheral inputs are interacted. The reticular inhibitory effects on the conduction from the periphery to the hypothalamus, and of facilitation from the hypothalamus to the hippocampus, correspond to its reported influences on distal and proximal segments, respectively, of classical sensory pathways (Bremer and Stoupel, 1959). This may support the proposal that the hypothalamo-hippocampal pathways normally mediate the transmission of sensory impulses to the hippocampus.

The demonstrated striatal effects on the hypothalamus are probably mediated by an oligosynaptic pathway from the caudate nucleus to the globus pallidus and thence to the hypothalamus. The existence of such a caudatal-pallidal-hypothalamic inhibitory pathway is supported by our observations, by recording short-latency hypothalamic responses upon stimulation of the caudate nucleus and medial globus pallidus (Feldman, 1966), and by the presence of anatomical connections between the caudate and the globus pallidus and between the globus pallidus and the hypothalamus (Voneida, 1960; Johnson and Clemente, 1959). However, from the different patterns of inhibition exerted by the caudate and globus pallidus on the units in the hypothalamus, it can be concluded that the caudate affects the hypothalamus through the globus pallidus as well as through other neural pathways.

The modification of sensory convergence on single cells caused by the extrahypothalamic lesions is another indication that different brain structures affect the sensory responsiveness of the hypothalamus and subsequently its function.

Acknowledgment

These investigations have been aided by Agreement No. 4X5108 with the National Institutes of Health, Bethesda, Md., U.S.A. The technical assistance of Mr. N. Conforti is gratefully acknowledged.

References

Brady, J. V. (1960). Neurophysiology, *In* "Handbook of Physiology" Sect. 1 (J. Field, ed.). Vol. III, pp. 1529–1552. American Physiological Society, Washington, D.C.
Bremer, F., and Stoupel, N. (1959). *Arch. Int. Physiol. Biochem.* **67**, 240.
Dafny, N., and Feldman, S. (1967). *Electroencephalog. Clin. Neurophysiol.* **23**, 546.
Dafny, N., and Feldman, S. (1968). *Exp. Neurol.* **21**, 397.
Dafny, N., and Feldman, S. (1969). *Electroencephalog. Clin. Neurophysiol.* **26**, 578.
Feldman, S. (1962). *Exp. Neurol.* **5**, 269.
Feldman, S. (1963). *Electroencephalog. Clin. Neurophysiol.* **15**, 672.
Feldman, S. (1964). *Ann. N.Y. Acad. Sci.* **117**, 1953.
Feldman, S. (1966). *Electroencephalog. Clin. Neurophysiol.* **21**, 249.
Feldman, S., and Dafny, N. (1968a). *Electroencephalog. Clin. Neurophysiol.* **25**, 150.
Feldman, S., and Dafny, N. (1968b). *Brain Res.* **10**, 402.
Feldman, S., Van Der Heide, C., and Porter, R. W. (1959). *Am. J. Physiol.* **196**, 1163.
Ingram, W. R. (1960). Neurophysiology, *In* "Handbook of Physiology" Sect. 1 (J. Field, ed.). Vol. II, pp. 951–978. American Physiological Society. Washington, D.C.
Jasper, H. H., and Ajmone-Marsan, C. (1954). "A Stereotaxic Atlas of the Diencephalon of the Cat." National Research Council of Canada, Ottawa.
Johnson, T. N., and Clemente, C. D. (1959). *J. Comp. Neurol.* **113**, 83.
Mangili, G., Motta, M., and Martini, L. (1966). *In* "Neuroendocrinology" (L. Martini and W. F. Ganong, eds.). Vol. I, pp. 297–370. Academic Press, New York.
Nauta, W. J. H. (1958). *Brain* **81**, 319.
Voneida, T. J. (1960). *J. Comp. Neurol.* **115**, 75.

Characterization of Unit Activity in Hypothalamic Islands with Special Reference to Hormone Effects

B. A. CROSS and R. G. DYER

I. Introduction

The idea that circulating hormones modulate the patterning of activity in hypothalamic neurons is implicit in most of the schema that have been proposed to account for behavioral and neuroendocrine responses. The idea is supported by a mass of evidence from many different sources, but most of this is highly circumstantial and it has yet to be proved that any designated pattern of hypothalamic neuronal activity is indispensable for the performance of a particular behavioral or neuroendocrine reaction.

In the laboratory of the senior author, interest has for some years focused on the effect of manipulating ovarian hormone levels on unit activity recorded with steel microelectrodes in various hypothalamic regions of the adult female rat under urethane anesthesia. Changes have been detected in the resting discharge rate of neurons and in their responsiveness to test stimuli, e.g., probing the cervix *per vaginam* with a tapered glass rod and pain or cold delivered to the tail. In general, estrogens whether exogenous (Cross, 1964, 1965) or endogenous (Cross and Silver, 1966; Lincoln and Cross, 1967) enhanced the inhibitory responses of neurons in both anterior and lateral regions to these stimuli. They also reduced the firing rate of neurons in the anterior and preoptic areas (Lincoln, 1967). The main effect of luteal hormone was to reduce the excitatory responses of neurons to cervical probing, while inhibitory responses were unaffected (Barraclough and Cross, 1963; Cross and Silver, 1965). None of these alterations in hypothalamic activity induced by ovarian hormones were seen in neurons of the unspecific nuclei of the thalamus (Cross and Silver, 1965, 1966).

Interpretation of these experiments is complicated by at least two problems. Does urethane anesthesia itself alter the activity of hypothalamic neurons? Also, to what extent are hormonal effects mediated by nervous inputs from other regions of the brain, e.g. the limbic system or brain stem (see the articles by Sawyer and by Feldman and Dafny in this volume). To overcome these difficulties a technique was developed (Cross and Kitay, 1967) to enable unit recording from diencephalic islands in unanesthetized decerebrate rats.

II. Preparation and Properties of the Islands

The islands are prepared by means of a stereotaxically oriented cylindrical cutting device in rats given a short-acting barbiturate (sodium methohexitone, Brietal, Eli Lilly Ltd.) via a cannulated tail vein. The cylinder encloses and protects the hypothalamus and its blood vessels while the forebrain rostral to the midcollicular level is aspirated. Bleeding is controlled with thrombin and the evacuated part of the cranial cavity filled with agar. Finally the residual cortex and hippocampus are gently aspirated from the cylinder to reveal the third ventricle and dorsal thalamus. Surgery is completed within half an hour, by which time the effect of the Brietal has worn off with the reappearance of vigorous spinal reflexes.

The anatomical contents of the islands are strikingly constant. The entire hypothalamus (but not the anterior preoptic area) is included together with portions of the septum, and anterior and medial thalamic nuclei.

Actively firing neurons can be detected in the islands within minutes after completing the operation and are routinely recorded for periods of up to 6 hr, after which the animal is killed with nembutal given intravenously and the islands fixed by perfusion for subsequent histological examination. The islands can, however, survive for much longer periods; in one case units were recorded 22 hr after preparation. Calculations of the number of active units recorded per millimeter of electrode track in whole brain preparations and in diencephalic islands show that there is no diminution of active cells in the latter.

In comparison to the intact hypothalamus of rats under urethane, the islands show the following distinctive features:

1. The histogram of cell firing frequency is shifted to the right; i.e., there are more faster firing hypothalamic cells than in intact brains. One probable explanation for this is the exclusion of inhibitory nervous inputs from other brain regions (Koizumi *et al.*, 1964; see also the article by Sawyer in this volume).
2. Units are completely unresponsive to peripheral nervous stimuli; e.g., pain or cold on the tail, which in the intact hypothalamus either excites or inhibits about 70% of cells.
3. Spontaneous changes in firing rate, which are seen in about half the hypothalamic units recorded in intact brains, are absent in the island preparations. In intact brains these changes are associated with fluctuations of the electroencephalogram (EEG) between synchronized and desynchronized patterns.

III. Use of Islands for Studying Direct Humoral Influences

The foregoing characteristics of the islands make this preparation very suitable for studying neuronal responsiveness to humoral substances conveyed in the circulating blood. The regular discharge rates of hypothalamic units in the islands for periods of an hour or more simplify assessment of rate changes in either direction, and afferent nervous influences can be completely excluded.

Two distinct procedures can be adopted:

1. The single neuron response. Here a cell is recorded continuously for 20 to 60 min and the agent under test (and control solutions) injected intravenously one or more times during this period. It is essential that the response be reversible and this method is therefore suitable only for substances having a brief duration of action.

2. Total cell activity. Here the resting rates of many neurons are determined and quantitative comparisons made between the firing characteristics of cell populations from islands prepared in rats of different physiological states.

Both these approaches have been used to study the action of urethane and of pituitary hormones.

A. Effect of Urethane

Since all the earlier unit recording in rats has been done under urethane, it was important to see if this anesthetic had any discernible direct action upon hypothalamic neurons. The following evidence was obtained:

1. With the single neuron method intravenous injection of a dose of urethane sufficient to block spinal reflexes (125 mg as 25% solution) did not change cell firing rate, whereas subsequent injection of a subanesthetic dose of Brietal (1.0 mg) abruptly arrested cell firing for up to 10 min, after which recovery to the preinjection rate occurred.

2. Firing-frequency histograms were prepared for 200 island cells without anesthesia and 147 island cells under urethane anesthesia. They did not differ significantly from each other, though both distributions were significantly ($p < 0.01$) faster than a population of 618 cells from similar hypothalamic regions in intact brains under urethane.

3. The similarity of the cell populations in the urethanized and nonurethanized islands was also indicated by the close correspondence in mean number of stable units detected per millimeter of electrode tract, viz., 3.7 and 4.3, respectively.

B. Effect of Intravenous Injection of Oxytocin

It has been reported that intracarotid injection of oxytocin induces large changes in firing rate in a considerable number of hypothalamic neurones in immobilized cats (Kawakami and Saito, 1967). Earlier findings by Findlay (personal communication) in the senior author's laboratory in Cambridge indicated that intravenous oxytocin was without effect on unit firing in rabbits under urethane anesthesia unless uterine contractions or milk ejection resulting from the oxytocin elicited EEG arousal responses. This sort of response implies an indirect route of excitation via afferent pathways to the hypothalamus.

As the half-life of oxytocin in the blood is less than 2 min in the rat (Ginsburg and Smith, 1959) and the hormone is water-soluble the single cell response method is admirably suited for testing for direct effects on island neurones.

A total of 65 neurons from 18 female rats were tested in this way. The distribution was as follows: anterior hypothalamic area, 31; paraventricular nucleus, 9; dorsomedial and ventromedial nuclei, 16; lateral hypothalamic area, 9. Three dose levels of oxytocin (0.1, 0.5, and 1.0 unit) were used and altogether 155 injections were given. In addition, 29 control injections of normal saline were made. With one exception all tests were uniformly negative. Since in 11 cases subsequent intravenous injection of subanesthetic doses of Brietal caused immediate reduction of firing rate for periods of 1 to 10 min, it was clear that the hormone was circulating through the island. The one positive cell, situated in the anterior hypothalamic area, showed a short-lived 30% increase in frequency on three tests with 0.5 unit oxytocin. The small change in rate following this high intravenous dose was in marked contrast to the large effects reported by Kawakami and Saito (1967).

C. Effect of Hypophysectomy

At the time of writing, results were available on the hypothalamic activity in islands prepared from seven animals hypophysectomized by the transauricular approach one month previously. All animals failed to gain weight after operation and showed persistent anestrus. Though the sample of cells was small, results were very uniform from animal to animal. The mean firing frequency (1.6/sec) was similar to that observed in intact brains, but the frequency histograms were markedly different because very few cells were firing at more than three per second. Despite this slowing of neuron firing there was no apparent reduction in the number of stable units recorded, viz., 4.1 per millimeter.

It seems likely that the diminished activity of hypothalamic cells was at least in part secondary to the metabolic deficiencies resulting from the loss of anterior pituitary hormones, but further experiments are needed to establish the mechanism of this effect.

D. Effect of Stage of the Estrous Cycle

Suggestive evidence of a cyclic variation in hypothalamic unit responses in female rats was reported by Barraclough and Cross (1963). Recent experiments with islands prepared from cyclic rats in proestrus, estrus, and metestrus or diestrus have confirmed this (Fig. 1). The study involved 307 cells mainly in the anterior area in 33 rats. The most striking new observation was that the highest firing rate occurred in animals prepared on the day of proestrus (mean rate 5.8/sec) and the lowest 24 hr later in the cycle, i.e., during

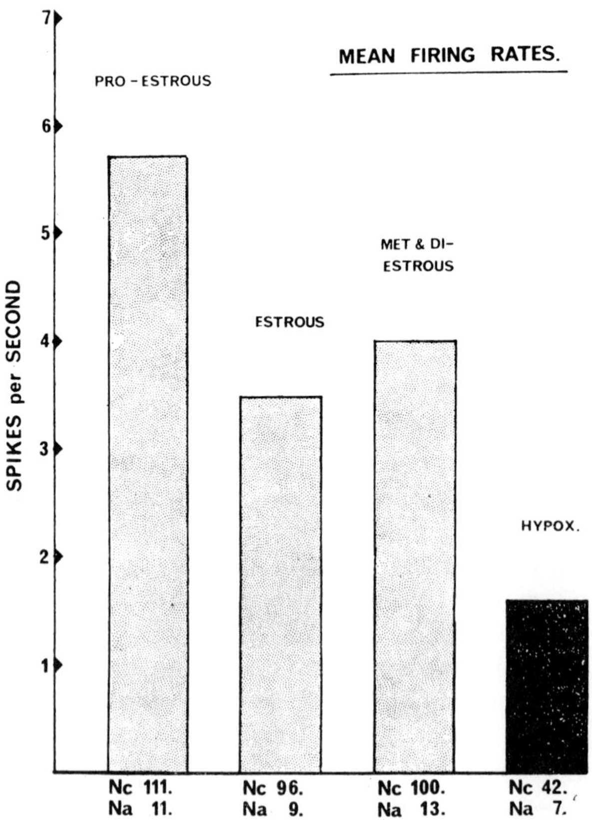

Fig. 1. Histogram of mean firing frequencies in cyclic and hypophysectomized (hypox) female rats: Nc, No of cells; Na, No of animals.

estrus (mean rate 3.5/sec). The mean firing rates in these two groups differed at the 1% level of significance.[1] The firing rate of cells in the mixed group of metestrous and diestrous rats (mean rate 4.0/sec) was intermediate between those of the proestrous and estrous groups and significantly lower than the former ($p < 0.01$). If the number of cells firing faster than three per second was taken as the criterion the estrous and metestrous/diestrous groups also differed significantly by the X^2 Test at the 5% level. There can be little doubt therefore that the various phases of the rat estrous cycle are associated with measurable differences of hypothalamic unit activity. Since in these experiments the recording continued for up to 6 hr it is most likely that the effects were due to the direct action of hormones from the animals' ovaries. If so, the high activity in the anterior hypothalamic region in proestrous rats could be ascribed to the hormonal output of the preovulatory ovary, and the much reduced activity 24 hr later in the estrous group could be due to the hormones issuing from the freshly ovulated ovary. The analysis of these events is being continued.

IV. Conclusions

1. The unit activity of acute diencephalic islands prepared by the technique of Cross and Kitay (1967) differs from that of intact brains in ways that make the preparation especially suitable for testing direct humoral effects on the hypothalamus.
2. Urethane has no detectable direct effect on hypothalamic unit activity, in sharp contrast to the short-acting barbiturate sodium methohexitone (Brietal), which causes immediate depression of cell firing.
3. Intravenous injections of large doses of oxytocin cause no significant change in hypothalamic unit activity.
4. Hypophysectomy performed one month before preparation of the island results in a drastic reduction in cell-firing activity.
5. Hypothalamic units fire faster in islands prepared from proestrous rats than from estrous rats, and the rate during metestrus-diestrus is intermediate.

[1] Because of the exponential distribution of unit firing rates (Cross and Silver, 1966; Lincoln, 1967) significance levels are calculated after log transformation of the data.

Acknowledgments

We are much indebted to Mrs. Wendy Musgrave for skilled technical assistance and Mr. Roger Francis for his care of the animals. The work was supported by grants from the Science Research Council and the Agricultural Research Council.

References

Barraclough, C. A., and Cross, B. A. (1963). *J. Endocrinol.* **26**, 339.

Cross, B. A. (1964). *In* Symp. Soc. Exptl. Biol. (G. M. Hughes, ed.). Vol. XVIII. Cambridge Univ. Press, Cambridge, pp. 157–193.

Cross, B. A. (1965). *In* "Proc. II Inter. Congr. Endocrinology" (S. Taylor, ed.). *Excerpta Med. Inter. Congress. Ser.* **83**, 513–516.

Cross, B. A., and Kitay, J. I. (1967). *Exp. Neurol.* **19**, 316.

Cross, B. A., and Silver, I. A. (1965). *J. Endocrinol.* **31**, 251.

Cross, B. A., and Silver, I. A. (1966). *Brit. Med. Bull.* **22**, 254.

Ginsburg, M., and Smith, M. W. (1959). *Brit. J. Pharmacol.* **14**, 327.

Kawakami, M., and Saito, H. (1967). *Japan. J. Physiol.* **17**, 466.

Koizumi, K., Ishikawa, T., and Brooks, C. M. (1964). *J. Neurophysiol.* **27**, 878.

Lincoln, D. W. (1967). *J. Endocrinol.* **37**, 177.

Lincoln, D. W., and Cross, B. A. (1967). *J. Endocrinol.* **37**, 191.

Central Monoaminergic Systems and Hypothalamic Function

K. Fuxe and T. Hökfelt

I. Introduction

It is well known that there exist networks of noradrenaline (NA) and 5-hydroxytryptamine (5-HT) nerve terminals in the hypothalamus and the preoptic area (Fuxe, 1965; Fuxe and Hökfelt, 1969a,b; Fuxe *et al.*, 1969a). Furthermore there is a tubero-infundibular dopamine (DA) neuron system, which terminates in the external layer of the median eminence (Fuxe, 1963, 1964; Fuxe and Hökfelt, 1966; Lichtensteiger and Langemann, 1966). All three monoamine neuron systems (DA, NA, and 5-HT) in the hypothalamus and the preoptic area have important neuroendocrine functions. Experiments made in our laboratory to elucidate these functions will be discussed.

II. Central Noradrenaline Neurons

A. Anatomy

The huge networks of NA nerve terminals present in the hypothalamus and the preoptic area probably all derive from fibers originating from NA cell bodies in the pons and the medulla oblongata (Andén *et al.*, 1966a). The evidence for this is as follows:

1. After a complete deafferentation of the hypothalamus practically all the NA nerve terminals have disappeared in this area by 10 to 30 days after the operation. Only the DA nerve terminals in the median eminence remain (Fuxe *et al.*, unpublished observations).

2. After a large lesion in the middle part of the caudal reticular tegmentum of the mesencephalon there is a marked reduction of the demonstrable NA nerve terminals in the hypothalamus (Andén *et al.*, 1966a,b).

3. The ascending NA fibers have been found to be separated into a dorsal and ventral bundle in the reticular tegmentum of the mesencephalon. Recent evidence suggests that the ventral NA bundle mainly innervates the hypothalamus, the preoptic area, and the ventral telencephalon, whereas the dorsal NA bundle, which originates mainly from NA cell bodies of the locus coeruleus, primarily innervates the cortical areas. The ventral NA bundle turns medially along the lemniscus medialis in the cranial mesencephalon to enter the lateral hypothalamic area via the area tegmenti ventralis. These results have been obtained in part from stereotaxic lesions of the two bundles (Ungerstedt, personal communication), from studies on large brain slices after incubation with α-methyl-NA and 6-hydroxytryptamine (6-HT) (Hökfelt and Fuxe, unpublished observations), and from studies on outgrowing NA axons in very young rats (Olson and Fuxe, unpublished observations). The *in vitro* experiments have also given further support to the view that one single NA neuron gives rise to a huge collateral network. Thus, a NA cell body in the nucleus reticularis lateralis may innervate, e.g., the spinal cord, the diencephalon, and the cerebellum (Andén *et al.*, 1966c, 1967a). It was found in the *in vitro* experiments that the NA cell bodies of, e.g., the locus coeruleus, besides giving rise to the large number of ascending NA fibers, probably give rise to a number of fibers innervating certain areas in the medulla oblongata and pons (Hökfelt and Fuxe, unpublished observations). In view of this it is likely that NA cell bodies in the pons and the medulla oblongata which innervate the hypothalamus and the preoptic area, also have collaterals that innervate other parts of the brain and the spinal cord.

The distribution of the NA nerve terminals in the hypothalamus and the preoptic area has been described in detail in previous papers (Fuxe, 1965; Fuxe and Hökfelt, 1969a).

B. Possible Neuroendocrine Function

So far these studies have mainly been performed by following the amine turnover in the central NA neurons under various endocrinological conditions. Two amine-synthesis inhibitors have been used to study the NA turnover: the tyrosine hydroxylase inhibitor, α-methyl-p-tyrosine methylester (H44/68; Andén *et al.*, 1966d); and the dopamine-β-oxidase inhibitor, *bis* (1-methyl-hexahydro-1, 4-diazepinyl-4-tiokarbonyl) disulfide (FLA 63; Corrodi *et al.*, 1970). The rate of depletion of neuronal amine stores after amine-synthesis inhibition is a measure of the overall amine turnover, which in turn is highly dependent on the nervous impulse flow (Andén *et al.*, 1966b; Andén *et al.*, 1969). (For other ways of measuring amine turnover see the article by Glowinski in this volume and the book edited by Hooper, 1969). In the experiments with synthesis inhibitors the doses used have been in excess of those necessary to cause a practically complete inhibition of the enzymes; the excess was used to assure that under various endocrinological conditions no variations in the degree of inhibition of tyrosine hydroxylase and DA-β-oxidase would occur.

1. Gonadotropin Secretion

In particular after hypophysectomy (2–4 weeks) but also after castration (1–7 months) it has been found that there is a decreased rate of amine depletion in some NA nerve terminals of the brain after injection of H 44/68. This has been found both histologically and biochemically (Fuxe *et al.*, unpublished observations). These data suggest a decreased amine turnover in the NA nerve terminals of the brain under these endocrinological conditions. When castrated male rats are treated with testosterone for three days with a dose of 10 mg per rat, the NA turnover is again increased to normal. On the other hand, treatment of castrated female rats with estrogen, even in relatively high doses of 10 μg per rat, did not increase the amine turnover to normal in the hypothalamic NA nerve terminals (Fuxe *et al.*, unpublished observations). This dose of estrogen markedly activated the tubero-infundibular DA neurons (see below).

In contrast to this, other workers using intraventricular injections of labeled tyrosine and noradrenaline (Anton-Tay and Wurtman 1968; Donoso *et al.*, 1969) have obtained results indicating an increased NA turnover after castration. The explanation for this discrepancy remains to be given.

After hypophysectomy no increases in brain NA levels were obtained, whereas after castration a small increase in brain NA levels could be seen (from 0.25 to 0.30 μg/g) (Jonsson *et al.*, 1970a). Increase in NA levels of the hypothalamus after castration has previously been reported (Donoso *et al.*, 1967). The fact that increases in NA levels are present after castration may favour the view that the NA turnover is decreased, since in states with increased NA turnover, such as stress, the NA levels in the nerve terminals tend to decrease. This is probably due to the fact that synthesis of NA is not sufficiently increased (see Corrodi *et al.*, 1968).

Using the tool H 44/68, it could not be demonstrated that the NA turnover in the various NA nerve terminals of the brain was obviously changed in pregnancy or during lactation (Jonsson *et al.*, 1970a). These are endocrinological states with low Follicle-Stimulating Hormone (FSH) and Luteinizing Hormone (LH) secretion and high prolactin secretion. A possible increase in whole brain NA levels were found only in lactation (Jonsson *et al.*, 1970a).

A large number of studies with neuroleptic drugs such as reserpine, chlorpromazine, perphenazine, and fluphenazine, all of which probably act by blocking catecholamine (CA) neurotransmission (Andén *et al.*, 1967b), have shown that these drugs, when given systemically or after implantation in the hypothalamus, induce pseudopregnancy and artificial lactation (see Meites, 1957; Coppola *et al.*, 1965; Mishkinsky *et al.*, 1966; Ben-David, 1968; Van Maanen and Smelik, 1968). It has been postulated that this effect is due to decreased secretion of Prolactin Inhibitory Factor (PIF), which is caused by the CA receptor blockade elicited by these drugs. It is not known, however, if the endocrine effects are caused by blockade of NA or DA receptors. Furthermore, a direct action on the pituitary gland is quite possible, since neuroleptic drugs such as haloperidol and pimocide are strongly accumulated in the pituitary gland. Thus, the effects obtained with these drugs have to be interpreted with caution.

In view of the results presented above it seems possible that the NA neurons belonging to the ventral pathways participate in control of gonadotropin secretion. They may directly control some of the preoptic-tubero tracts, causing the peak release of Luteinizing Hormone Releasing Factor (LHRF) necessary for ovulation by way of, e.g., the dense plexus of NA nerve terminals present in the periventricular preoptic area, where estradiol-concentrating neurons exist (Stumpf, 1968). The NA neurons may also be involved in regulation of prolactin secretion (see Section IV, B, 1).

Recently the effect of an antifertility steroid, norethisterone (17α-ethynyl-19-nortestosterone), on central NA neurons has been studied. Preliminary results indicate that under the influence of this progestational steroid, there was a decrease in the amine turnover of the NA nerve terminals of castrated female rats, as revealed with the help of amine-synthesis inhibitors (H 44/68) (Fuxe and Hökfelt, unpublished observations). Since the NA neurons seem to participate in the regulation of gonadotropin secretion (see above), it may be that progestational compounds of this type exert part of their effects on the ovarian cycle via decreasing activity in the central NA neurons. The effect of other progestational compounds is under study (see Section IV, B, 1).

Daily intravenous injections of prolactin (1 mg/rat for three days) have not been found to have marked influence on the turnover in the NA nerve terminals of the hypothalamus. The tubero-infundibular DA neurons are markedly activated by this dose, however (see below). Furthermore, biochemical and histochemical data suggest that melatonin, which is known to change serum gonadotropin levels (Adams *et al.*, 1965; Motta *et al.*, 1967), in relatively high doses (total dose 0.5–5 mg/rat), does not influence the activity in the NA neurons (Hökfelt and Fuxe, unpublished observations). The NA levels were not changed, confirming the work of Anton-Tay *et al.* (1968).

2. *Possible Effect on Adrenocorticotropic Hormone Secretion*

It is known that in sham rage and during various types of stress there is a markedly increased activity in the central NA neurons (Fuxe and Gunne, 1964; Corrodi *et al.*, 1968; Reis and Fuxe, 1968). Under these states there is also an increase in Adrenocorticotropic Hormone (ACTH) secretion from the anterior pituitary (see Mangili *et al.*, 1966). It has not as yet been shown if these two phenomena are related. It has recently been found that stress-induced activation of the NA neurons is not blocked by dexamethasone (300 μg, intramuscularly) given immediately before the stress (Fuxe *et al.*, unpublished observations). Furthermore, data obtained from intraventricular amine injections and studies with amine synthesis inhibitors indicate that the NA turnover is increased 6 days after adrenalectomy (Javoy *et al.*, 1968; Fuxe *et al.*, unpublished observations). The NA turnover is restored to normal again by treatment with cortisol (Fuxe *et al.*, unpublished observations).

3. *Possible Effect on Antidiuretic Hormone and Oxytocin Secretion*

The supraoptic nucleus (Antidiuretic Hormone, ADH, production) and the paraventricular nucleus (oxytocin production) are both densely innervated by NA nerve terminals, and these probably participate in the neural control of ADH and oxytocin secretion. Thus, during both hypo- (0.2% NaCl) and hyper-osmotic (1.8% NaCl) stimulation involving intravenous infusions, there was an activation of practically all the NA nerve terminals in the brain *inter alia* of the terminals in the supraoptic and paraventricular nuclei (Andén *et al.*, personal communication).

III. Central 5-Hydroxytryptamine Neurons

A. Anatomy

The 5-HT nerve terminals in the hypothalamus and the preoptic area mainly arise from 5-HT fibers originating from the 5-HT cell bodies in the mesencephalon (Dahlström and Fuxe, 1964; Fuxe, 1965). Each 5-HT neuron probably gives rise to a large number of collaterals. The 5-HT fibers ascend medially in the mesencephalon laterally to the midline and enter the medial forebrain bundle by turning somewhat laterally when reaching the border between the mesencephalon and the diencephalon where most of the 5-HT fibers are densely aggregated, lying ventral and immediately medial to the fasciculus retroflexus (Dahlström and Fuxe, 1964). In the medial forebrain bundle the 5-HT fibers are spread out in various bundles, some of which lie immediately lateral to the fornix and some of which occupy a ventrolateral position along the dorsal surface of the optic tract. Some also lie immediately ventral to the ventral part of the crus cerebri and the retrolenticular part of the capsula interna. The distribution of the 5-HT nerve terminals is not known to the same extent as that of the CA nerve terminals, owing to technical difficulties (Fuxe, 1965; Fuxe and Jonsson, 1967). However, new possibilities have been opened by the introduction of 6-HT into fluorescence histochemistry (Jonsson *et al.*, 1969), which gives a fluorescence yield much larger than that obtained from 5-HT (Jonsson and Sandler, 1969). The fluorescent product has an emission peak around 500 mμ and does not fade as rapidly as that formed by 5-HT.

So far, high densities of 5-HT nerve terminals in the hypothalamus have been found primarily in the nucleus suprachiasmaticus (Fuxe 1965) and in the middle third of the retrochiasmatic area and of the corresponding area at the level of the anterior median eminence (Hökfelt and Fuxe, unpublished observations).

B. Possible Neuroendocrine Function

Melatonin has been found to cause rapid but brief increases in 5-HT concentrations in the brain (Anton-Tay *et al.*, 1968). These effects on central 5-HT neurons may be related to the influence on gonadotropin secretion caused by melatonin. This is supported by the fact that increases in brain 5-HT probably cause a blockade of superovulation in the immature rat (Kordon *et al.*, 1968). Furthermore, an increase in central 5-HT neurotransmission may in certain endocrinological states inhibit LH secretion from the anterior pituitary and thereby hinder ovulation. There also exists strong support for the view that estrous behavior is under the control of inhibitory 5-HT pathways, since 5-HT membrane-pump blocking agents such as imipramine (see Carlsson *et al.*, 1968) were the most potent of the drugs that inhibit estrous behavior (Meyerson, 1966). The 5-HT nerve terminal systems in the suprachiasmatic area and the anterior ventral hypothalamus are probably the main ones responsible for these effects. In accordance with this scheme, the mechanism for triggering the release of LHRF for ovulation is lost in rats with suprachiasmatic lesions (Antunes-Rodrigues and McCann, 1967).

It should be pointed out that α-ethyl-tryptamine, which has strong inhibitory actions on ACTH secretion (Gold and Ganong, 1967), probably can stimulate 5-HT receptors in the central nervous system. (Meek and Fuxe, unpublished observations). Central 5-HT receptor stimulation is also found after treatment with lysergic acid diethylamide (Andén *et al.*, 1968). Furthermore, it has been found that after adrenalectomy the 5-HT turnover is decreased. The 5-HT turnover is restored at least to normal after treatment with cortisol or dexamethasone (Fuxe, unpublished observations). Thus it is probable that central 5-HT neurons are affected by the degree of activity in the pituitary-adrenal axis and that central 5-HT neurons in turn can participate in the control of ACTH regulation.

IV. The Tubero-Infundibular Dopamine Neurons

A. Anatomy

The DA cell bodies of this system are mainly situated in the nucleus arcuatus and the anterior periventricular hypothalamic nucleus. The short axons give rise to a very densely packed plexus of DA nerve terminals in the external layer of the median eminence, close to the primary capillary plexus of the hypophyseal portal system (Fuxe, 1964; Fuxe and Hökfelt, 1966). It is probable that the DA nerve terminals establish (at least in part) an axoaxonic influence on other nerve terminals (Hökfelt, 1967, 1968), which may store one of the peptidergic factors regulating gonadotropin secretion (see Fuxe and Hökfelt, 1969a,b).

B. Possible Neuroendocrine Function

1. Possible Role in Gonadotropin Secretion

In pregnancy, pseudopregnancy, and lactation it has been found that DA cell bodies belonging to the tubero-infundibular system exhibit a marked increase in number and intensity (Fuxe *et al.*, 1967a,b). Furthermore, as shown with the help of amine-synthesis inhibitors (Fuxe *et al.*, 1967a,b, 1969b,c) and by studies on the decline of labeled DA in the median eminence (Jonsson *et al.*, 1970b), the amine turnover in the DA nerve terminals of the median eminence is strongly increased in these endocrinological states when compared to normal cycling rats.

It must be pointed out that the activity of the nigro-neostriatal DA neurons did not seem to be affected during pregnancy, pseudopregnancy, and lactation. This indicates that there is a selective and marked increase in the activity of the tubero-infundibular DA neurons under these circumstances.

We have postulated that the increased activity found in this system is related to the low FSH–LH secretion, high prolactin secretion, and the blockade of ovulation found in these rats. One mechanism for such a relation, which could explain the present findings, is that the DA nerve terminals establish axoaxonic contacts with LHRF and/or Follicle-Stimulating Hormone Releasing Factor (FSHRF) containing terminals. The DA could then act by way of presynaptic inhibition to decrease the release and/or synthesis of FSHRF and/or particularly of LHRF, the release of which causes ovulation in normal cycling rats. In support of this hypothesis we have found in preliminary experiments, using the lactic acid production test (Hamberger, 1968) for measuring LH, that systemic reserpine treatment or treatment with selective DA receptor blocking agents will increase LH secretion from the anterior pituitary. This is particularly true for states in which the activity of the tubero-infundibular DA system is increased (e.g., after estrogen treatment of castrated rats; Hamberger *et al.*, personal communication).

A direct action of DA on the anterior pituitary seems to be excluded, since repeated experiments have shown that incubation of pituitaries with concentrations of DA up to 10 μg/ml does not influence the LH secretion from the anterior pituitary (Hamberger *et al.*, personal communication). Furthermore, in endocrinological conditions with marked changes in ACTH and growth hormone secretion (such as adrenalectomy with and without hydrocortisone treatment, immobilization stress, insulin-induced hypoglycemia with or without bone injury), there are no demonstrable changes in the activity of the tubero-infundibular DA neurons when the methods described above are applied, which, strongly links this neuron system with the regulation of gonadotropin secretion (Fuxe and Hökfelt, 1967).

In agreement with this result it has been found that, after castration (when FSH–LH secretion is high), the activity in the system compared to that found in most stages (metestrus-diestrus) of the ovarian cycle will decrease (Fuxe *et al.*, 1970a). Furthermore, the system is rapidly activated again above normal levels by sex hormones such as testosterone (in doses down to 0.3 mg) and estradiol (in doses down to 0.15 μg) but much less so by progesterone and practically not at all by hydrocortisone, as revealed when using amine-synthesis inhibitors (Fuxe *et al.*, 1969b) and isotope measurements of labeled DA (Jonsson *et al.*, 1970b). These findings strongly indicate that the tubero-infundibular DA neurons at least partially mediate the negative feedback action that testosterone and estrogen exert on FSH–LH secretion by way of inhibiting the release of FSHRF and/or LHRF into the portal system, as has been discussed above. Furthermore, there is support for the view that in this endocrinological situation the sex hormones act centrally by decreasing FSHRF and LHRF release (McCann *et al.*, 1968). Estradiol and testosterone have not been found to have the same strong actions on the turnover of the NA neurons (Jonsson *et al.*, personal communication).

During the ovarian cycle there are clear-cut variations in the activity of the tubero-infundibular DA neurons, with a somewhat lower activity in stages of proestrus-estrus and somewhat higher activity in stages of metestrus-diestrus (Fuxe *et al.*, 1970a). During the ovarian cycle, variations in the number and intensity of DA cell bodies in the arcuate nucleus have also been observed semiquantitatively (Fuxe *et al.*, 1967a) and quantitatively (Lichtensteiger, 1969). As pointed out above, these changes are lost in castrated rats, in which the activity remains low. These cyclic variations are also lost in androgen-sterilized rats and in rats exposed to constant light (Fuxe *et al.*, 1970b); however, in these two states of blocked ovulation and constant estrus the activity in the DA system remains relatively high, i.e., it is similar to that found in diestrus. Estrogen is probably produced in high amounts in these rats (Barraclough, 1967) and this may in part explain the relatively high activity observed in the DA neurons. Since, obviously, FSHRF and LHRF could be secreted in sufficiently high amounts to produce estrogen, the main deficiency in gonadotropin secretion in these rats is probably due to the lack of sufficient LHRF secretion for ovulation. It is possible that the tubero-infundibular DA systems may contribute to this effect, since the activity is constantly relatively high and never low as in certain periods of proestrus-estrus. These results further tend to support the view that an inhibitory action on LHRF secretion may be the main effect of the DA released from the tubero-infundibular neurons.

Castration of the androgen-sterilized rats markedly reduced the activity in the tubero-infundibular DA neurons. This was probably due to the discontinuations of the high serum levels of estrogen. Furthermore high doses of estrogen and testosterone cause activation of the neurons. Thus the sensitivity of the hypothalamic nerve cells to sex hormones is probably not markedly reduced after androgen sterilization.

Since estrogen and testosterone still cause an activation of tubero-infundibular DA neurons after hypothalamic islands have been made (Fuxe *et al.*, unpublished observations), and since estrogen is accumulated in the arcuate nucleus (Stumpf, 1968), it seems possible that the sex hormones exert their action on the DA systems by directly influencing the activity of the DA cell bodies.

It should be pointed out that neural stimuli ending on the soma dendrite of the DA system are also of importance for maintaining the activity in the system, since the activity is reduced in male and female rats after cutting off the neural connections of the hypothalamus, which is the case in hypothalamic islands (Fuxe *et al.*, unpublished observations).

The effect of hypophyseal hormones on the tubero-infundibular DA neurons of hypophysectomized and castrated rats has recently been studied with amine-synthesis inhibitors. It was found that intravenous injections of prolactin in doses of 0.5–2 mg/rat given daily for 3 days caused marked activation of the tubero-infundibular DA neurons, whereas FSH (up to 300 μg/100 g), LH (up to 300 μg/100 g), ACTH (up to 125 μg/rat), and vasopressin (up to 800 mU/rat) were without any effects when given intravenously daily for 3 days (Hökfelt and Fuxe, unpublished observations). These results definitely show that prolactin strongly acts directly on the brain (see also Clemens and Meites, 1968). Thus, hypophyseal hormones can directly influence neuron systems in the brain. This action of prolactin on the brain, causing direct or indirect activation of the tubero-infundibular DA neurons, is probably of physiological importance. After hypophysectomy during lactation the marked activation of the tubero-infundibular DA neurons will rapidly subside (Hökfelt and Fuxe, unpublished observations). Thus, prolactin probably is of great importance for the activation of the tubero-infundibular DA neurons in pregnancy and lactation. For the maintenance of the activation, neural stimuli from mammals and the female genital organs are probably of some importance (Fuxe *et al.*, 1969c).

The fact that hypophysectomy decreases the activity in the system more than does castration also indicates that in the normal animal not only sex hormones but also prolactin may be of importance (perhaps of the greatest importance) for regulating the activity in the tubero-infundibular DA neurons. This is supported by the fact that the activity is relatively low in proestrus-estrus, at which stage estradiol secretion is maximal.

The important role of prolactin is further illustrated by the fact that if the adenohypophysis is transplanted onto the iris of an hypophysectomized-castrated rat, there is a marked activation of the tubero-infundibular DA system (Olson *et al.*, personal communication).

In view of the potent effects of prolactin on the tubero-infundibular DA neurons, one may suspect that a neuron system in the hypothalamus probably participates in regulating the secretion of prolactin (see above). In support of the existence of a noradrenergic link in such a mechanism it was found that a NA-receptor blocking agent, phenoxybenzamine, and a dopamine-β-oxidase inhibitor, FLA63, cause an activation of the tubero-infundibular DA neurons. This activation was not seen if the hypophysis had been removed previously (Fuxe and Hökfelt, unpublished observations).

Thus the speculation can be made that the adrenergic blocking agent, phenoxybenzamine, which blocks NA-receptor sites in the central nervous system (Andén *et al.*, 1967b), could block ovulation by primarily decreasing Prolactin Inhibiting Factor (PIF) secretion. This would in turn lead to increased prolactin secretion, activation of the DA neurons, and finally blockade of LHRF secretion. Work is in progress to obtain further support for this view. This hypothesis would in part explain the complex data in the literature on the role of CA neurons in the regulation of gonadotropin secretion (see Lippmann *et al.*, 1967).

In trying to interpret the results existing in the literature with regard to the effect of drugs that interfere with monoamine neurotransmission upon gonadotropin secretion (Gold and Ganong, 1967), it has to be remembered that a drug which blocks DA-receptors in the neostriatum does not necessarily cause a blockade of the DA-receptors existing in the median eminence. The DA-receptors in the median eminence may very well be different from those present in other part of the brain, particularly in view of the fact that the peptidergic LHRF-secreting neurons containing the hypothetical DA-receptor sites also have glandular characteristics and that the DA neurons may regulate the LHRF neurons via an axo-axonic influence. Furthermore, the drugs may act directly on the hypophysis.

Recently, evidence has been obtained that ergot alcaloids (ergocornine and 2-Br-α-ergokryptine) in doses of 3 mg/kg cause a marked inactivation of the tubero-infundibular DA neurons in lactation down to normal activity (Fuxe and Hökfelt, unpublished observations). These potent actions on the tubero-infundibular DA neurons are probably related to the marked antiprogestational effects of these drugs in the same doses (Weidmann and Flückiger, personal communication).

The effects of some progestational steroids (ethynodiol diacetate and norethisterone) and the estrogen mestranol (ethinyl-estradiol-3-methylether) on the tubero-infundibular DA neurons have recently been studied, using synthesis inhibitors. These steroids are used as antifertility compounds. Preliminary findings indicate that ethynodiol diacetate alone, in a low dose and mestranol alone cause a marked activation of the tubero-infundibular DA neurons of castrated female rats. Norethisterone proved ineffective in causing increased turnover in the DA neurons; if anything, a decreased activity seemed to occur. These results, taken together with the findings on the NA neurons with norethisterone, indicate that antifertility steroids may influence either DA or NA neurons, or both, and therefore may have different sites of action in the brain. Thus the antifertility effects of these steroids may be at least partly mediated via a direct or indirect effect on central monoamine neurons (Fuxe and Hökfelt, unpublished observations).

V. Conclusions

Evidence is rapidly accumulating that the dense networks of DA, NA, and 5-HT nerve terminals in the hypothalamic and preoptic regions are of great importance for the regulation of gonadotropin secretion. Previous studies and the results given in the present article indicate that there may be a link by which central NA neurons may influence the activity in the tubero-infundibular DA neurons.

The tubero infundibular DA neurons may inhibit secretion of LHRF and/or possibly FSHRF via an axoaxonic influence in the median eminence. These neurons are activated by estrogen and testosterone, particularly in the castrated state. They are also markedly activated by prolactin. They are inactivated by ergot alcaloids.

In view of the present results, a coupling between central NA neurons and the tubero-indundibular DA neurons may exist, since prolactin secretion may be controlled by the NA neurons. Thus the activity of the DA neurons and consequently also LHRF and/or possibly FSHRF secretion from the median eminence can be partly controlled by the activity in the central NA neurons by way of neural and hormonal links involving, e.g., PIF and prolactin. The latter, when injected, causes marked activation of the tubero-infundibular DA neurons (see above).

Drugs, such as phenoxybenzamine that selectively block NA neurotransmission, produce pseudopregnancy, and cause blockade of ovulation (Sawyer, 1959; Everett, 1964). In agreement with the present hypothesis, the tubero-infundibular DA neurons are activated by phenoxybenzamine. This activation is blocked by hypophysectomy. On the other hand, a drug that selectively stimulates NA receptors, such as catapresan, decrease the activity of the tubero-infundibular DA neurons.

In view of the high density of NA terminals in the periventricular-preoptic area and the turnover changes and changes in levels of NA found following castration, noradrenergic mechanisms may also be directly involved in control of FSH–LH secretion.

Moreover, in view of the high density of 5-HT nerve terminals in the supra-and retro-chiasmatic area and the data discussed in the present article, the 5-HT neurons may participate in regulating estrous behavior and the neuronal mechanisms responsible for the peak secretion of LHRF necessary for ovulation.

Acknowledgment

This study has been supported by a grant (1-RO3-NH16825-01) from the National Institute of Health, by the Swedish Medical Research Council (Project nr 14X-715-04A and B70-14X-2887-01), and by grants from Magnus Bergwalls Stiftelse and O. & E. Erikssons Stiftelse.

References

Adams, J. C., Wan, L., and Sohler, A. (1965). *J. Endocrinol.* **31**, 295.

Andén, N. E., Dahlström, A., Fuxe, K., Olson, L., and Ungerstedt, U. (1966a). *Experientia* **22**, 44.

Andén, N. E., Dahlström, A., Fuxe, K., Larsson, K., Olson, L., and Ungerstedt, U. (1966b). *Acta Physiol. Scand.* **68**, 313.

Andén, N. E., Fuxe, K., and Larsson, K. (1966c). *Experientia* **22**, 842.

Andén, N. E., Corrodi, H., Dahlström, A., Fuxe, K., and Hökfelt, T. (1966d). *Life Sci.* **5**, 561.

Andén, N. E., Fuxe, K., and Ungerstedt, U. (1967a). *Experientia* **23**, 838.

Andén, N. E., Corrodi, H., Fuxe, K., and Hökfelt, T. (1967b). *Europ. J. Pharmacol.* **2**, 59.

Andén, N. E., Corrodi, H., Fuxe, K., and Hökfelt, T. (1968). *Brit. J. Pharmacol.* **34**, 1.

Andén, N. E., Corrodi, H., and Fuxe, K. (1969). *In* " Metabolism of Amines in the Brain " (G. Hooper, ed.). pp. 38–47. Macmillan, London.

Anton-Tay, F., and Wurtman, R. J. (1968). *Science* **159**, 1245.

Anton-Tay, F., Chou, C., Anton, S., and Wurtman, R. J. (1968). *Science* **162**, 277.

Antunes-Rodrigues, J., and McCann, S. M. (1967). *Endocrinology* **81**, 666.

Barraclough, C. A. (1967). *In* " Neuroendocrinology " (L. Martini and W. F. Ganong, eds.). Vol. II, pp. 61–99. Academic Press, New York.

Ben-David, M. (1968). *Neuroendocrinology* **3**, 65.

Carlsson, A., Fuxe, K., and Ungerstedt, U. (1968). *J. Pharm. Pharmacol.* **20**, 150.

Clemens, J. A., and Meites, J. (1968). *Endocrinology* **82**, 878.

Coppola, J. A., Leonardi, R. G., Lippman, W., Perrine, J. W., and Ringers, I. (1965). *Endocrinology* **77**, 485.

Corrodi, H., Fuxe, K., and Hökfelt, T., (1968). *Life Sci.* **7**, 107.

Corrodi, H., Fuxe, K., and Hamberger, B. (1970). *Europ. J. Pharmacol.* (To be published).

Dahlström, A., and Fuxe, K. (1964). *Acta Physiol. Scand.* **62**, *Suppl. 235*, 1.

Donoso, A. O., Stefano, F. J. E., Biscardi, A. M., and Cukier, J. (1967). *Experientia* **23**, 665.

Donoso, A. O., de Gutierrez Moyano, M. B., and Santolaya, R. C. (1969). *Neuroendocrinology* **4**, 12.

Everett, J. W. (1964). *Physiol. Rev.* **44**, 373.

Fuxe, K. (1963). *Acta Physiol. Scand.* **58**, 383.

Fuxe, K. (1964). *Z. Zellforsch.* **61**, 710.

Fuxe, K. (1965). *Acta Physiol. Scand.* **64**, *Suppl. 247*, 39.

Fuxe, K., and Gunne, L.-M. (1964). *Acta Physiol. Scand.* **62**, 493.

Fuxe, K., and Hökfelt, T. (1966). *Acta Physiol. Scand.* **66**, 245.

Fuxe, K., and Hökfelt, T. (1967). *In* " Neurosecretion " (F. Stutinsky, ed.). pp. 165–175. Springer-Verlag, Berlin.

Fuxe, K., and Hökfelt, T. (1969a). *In* " Frontiers in Neuroendocrinology 1969 " (W. F. Ganong and L. Martini, eds.). pp. 47–96. Oxford University Press, New York.

Fuxe, K., and Hökfelt, T. (1969b). *In* " Progress in Endocrinology " (C. Gual, ed.). *Excerpta Med. Intern. Congr. Ser.* **184**, 495.

Fuxe, K., and Jonsson, G. (1967). *Histochemie* **11**, 161.

Fuxe, K., Hökfelt, T., and Nilsson, O. (1967a). *Life Sci.* **6**, 2057.

Fuxe, K., Hökfelt, T., and Nilsson, O. (1967b). *Excerpta Med. Intern. Congr. Ser.* **133**, 541.

Fuxe, K., Hökfelt, T., and Ungerstedt, U. (1969a). *In* " Metabolism of Amines in the Brain " (G. Hooper, ed.). pp. 10–22. MacMillan, London.

Fuxe, K., Hökfelt, T., and Nilsson, O. (1969b). *Neuroendocrinology* **5**, 107.

Fuxe, K., Hökfelt, T., and Nilsson, O. (1969c). *Neuroendocrinology* **5**, 257.

Fuxe, K., Hökfelt, T., and Nilsson, O. (1970a). *Neuroendocrinology* (To be published).

Fuxe, K., Hökfelt, T., and Nilsson, O. (1970b). *Neuroendocrinology* (To be published).

Gold, E. M., and Ganong, W. F. (1967). *In* " Neuroendocrinology " (L. Martini and W. F. Ganong, eds.). Vol. II, pp. 377–437. Academic Press, New York.

Hamberger, L. (1968). M.D. Thesis, Gothenburg, Sweden.

Hökfelt, T. (1967). *Brain Res.* **5**, 121.
Hökfelt, T. (1968). M.D. Thesis, Stockholm, Sweden.
Hooper, G. (1969). "Metabolism of Amines in the Brain." pp. 1–74. MacMillan, London.
Javoy, F., Glowinski, J., and Kordon, C. (1968). *Europ. J. Pharmacol.* **4**, 103.
Jonsson, G., and Sandler, M. (1969). *Histochemie* **17**, 207.
Jonsson, G., Fuxe, K., Hamberger, B., and Hökfelt, T. (1969). *Brain Res.* **13**, 190.
Jonsson, G., Fuxe, K., Hökfelt, T., and Andén, N. E. (1970a). *Neuroendocrinology.* (To be published).
Jonsson, G., Fuxe, K., and Hökfelt, T. (1970b). *Neuroendocrinology.* (To be published).
Kordon, C., Javoy, F., Vassent, G., and Glowinski, J. (1968). *Europ. J. Pharmacol.*, **4**, 169.
Lichtensteiger, W. (1969). *J. Pharmacol. Exp. Ther.* **165**, 204.
Lichtensteiger, W., and Langemann, H. (1966). *J. Pharmacol. Exp. Therap.* **151**, 400.
Lippmann, W., Leonardi, R., Bell, J., and Coppola, J. A. (1967). *J. Pharmacol. Exp. Therap.* **156**, 258.
Mangili, G., Motta, M., and Martini, L. (1966). *In* "Neuroendocrinology" (L. Martini and W. F. Ganong, eds.). Vol. I, pp. 297–370. Academic Press, New York.
McCann, S. M., Dhariwal, A. P. S., and Porter, J. C. (1968). *Ann. Rev. Physiol.* **30**, 589.
Meites, J. (1957). *Proc. Soc. Exp. Biol. Med.* **96**, 728.
Meyerson, B. (1966). *Acta Physiol. Scand.* **67**, 411.
Mishkinsky, J., Lajtos, Z. K., and Sulman, F. G. (1966). *Endocrinology* **78**, 919.
Motta, M., Fraschini, F., and Martini, L. (1967). *Proc. Soc. Exp. Biol. Med.* **126**, 431.
Reis, D., and Fuxe, K. (1968). *Brain Res.* **7**, 448.
Sawyer, C. H. (1959). *In* "Recent Progress in the Endocrinology of Reproduction" (C. W. Lloyd, ed.). pp. 1–20. Academic Press, New York.
Stumpf, W. E. (1968). *Science* **162**, 1001.
Van Maanen, J. M., and Smelik, P. G. (1968). *Neuroendocrinology* **3**, 177.

Metabolism of Catecholamines in the Central Nervous System and Correlation with Hypothalamic Functions

J. GLOWINSKI

I. Introduction

In order to be considered as neurotransmitters, central monoamines (noradrenaline, NA; dopamine, DA; 5-hydroxytryptamine, 5-HT) must fulfill precise conditions. That they do partly cope with those theoretical requirements has been demonstrated during the past ten years.

These substances are located intraneuronally. They are highly concentrated in the axon terminals of special neuron systems which have been extensively studied, particularly in the rat (Hillarp *et al.*, 1966). These amines are synthesized intraneuronally from their original precursors: tyrosine for catecholamines (CA) and tryptophane for 5-HT. They are stored mainly in neuron vesicles, where they are protected from inactivation by enzymes. Recent *in vivo* or *in vitro* studies have revealed that these central amines can be released from nerve terminals. Local extraneuronal or intraneuronal inactivating systems, including enzymes or reuptake processes, have also been described. Less is known of the amine effects on postsynaptic receptor cells in various structures of the central nervous system. Data about functional characteristics of some systems of monoaminergic neurons have been recently obtained, particularly in sleep and extrapyramidal mechanisms. Information concerning the role of aminergic neurons in endocrine and non endocrine mechanisms of the hypothalamus is also emerging. The highly complex and intricate organization of hypothalamic mechanisms of control over autonomic and endocrine functions makes it particularly difficult to conduct experimentation and interprete data in this field.

Some functions of hypothalamic amines in the integration of particular hypothalamic mechanisms are extensively reviewed in this volume. The localization of hypothalamic aminergic neurons are also described on the basis of histochemical data (see the article by Fuxe and Hökfelt in this volume). Therefore, it appears of particular interest to summarize characteristics of monoamine metabolism in these neurons and to review the experimental approaches by which modifications can be detected. Furthermore, it seems necessary to discuss the pharmacological means available to modify the activity of these neurons and consequently to alter their functions. These two complementary lines of research used to study the role of NA- and DA-containing neurons will be examined from a general point of view. Then some examples pertaining to the relationships of monoamines with hypothalamic mechanisms will be given to illustrate the notions developed in the first part of this section.

Although both CA and 5-HT neurons are often associated with the regulation of hypothalamic mechanisms, we shall focus our attention on CA. Many of the experimental approaches or concepts used for the study of CA neurons can be applied to 5-HT neurons as well (see the article by Wurtman in this volume).

II. Notions on the Metabolism of Central Neurons Containing Catecholamines

A. Schematic Model of a Noradrenergic Nerve Ending

To appreciate the various means of analyzing the changes in CA neuron activities, it seems first necessary to schematically describe the dynamic events occurring in these neurons and particularly in

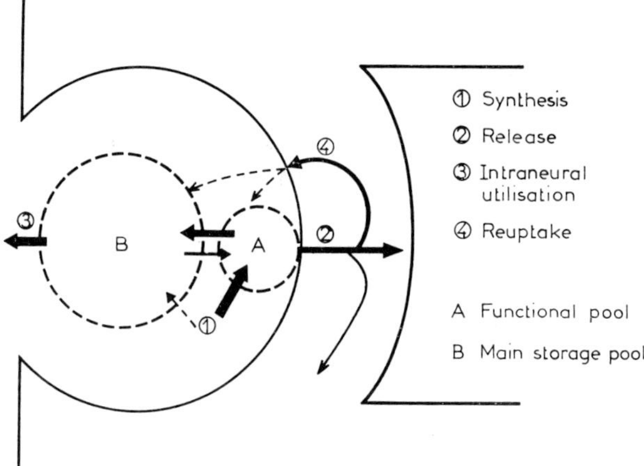

FIG. 1. Schematic representation of events occurring in nerve endings and synapses of NA-containing neurons.

nerve endings. From the data obtained by different experimental approaches and also from studies made on peripheral sympathetic neurons and central NA neurons, a general simplified model (Fig. 1) for a NA nerve ending can be proposed. Four main processes can be distinguished: synthesis, storage, utilization, and finally inactivation of the neurotransmitter (Glowinski and Baldessarini, 1966; Hornykiewicz, 1966; Iversen, 1967).

1. Synthesis

CA synthesis occurs in the numerous nerve endings as well as in the cell body and axon of the neuron. Hydroxylation of tyrosine, decarboxylation of dihydroxyphenylalanine (DOPA) and β-hydroxylation of DA are the three steps of NA biosynthesis. The fact of particular interest is that the three enzymes involved are not all contained in the synaptic vesicles which are the main storage sites of NA as well as DA. Tyrosine hydroxylase and DOPA decarboxylase are soluble enzymes; on the other hand, dopamine-β-hydroxylase is found after differential centrifugation of sucrose brain homogenates in particles and likely within synaptic vesicles. The main rate-limiting step of CA biosynthesis is the transformation of tyrosine to DOPA. Another process, the intraneuronal uptake of dopamine in vesicles where the β-hydroxylation takes place, may also regulate NA synthesis.

2. Storage

As revealed by histochemical data and subcellular distribution studies, NA as well as DA are highly concentrated in nerve endings. Most of the amines are localized in granules or vesicles of 400–500 Å diameter. A few larger granulated vesicles (1000 Å) are generally observed in NA-containing endings, particularly in the hypothalamus. It seems that small quantities of amine are also present in a very loosely bound form outside the vesicles. The mechanism by which the extravesicular amine is protected against inactivation is still unknown (Glowinski, 1970). Beside this morphological information, the concept of a heterogeneous distribution of CA in neurons, or even in nerve endings, has been developed on the basis of pharmacological and biochemical data. It appears clear that NA as well as DA are localized in many pools, which can be visualized as compartments. At least three compartments can be distinguished: a functional (C-I), a main storage (C-II), and another smaller storage compartment (C-III), the features of which are less known. The characteristics of these compartments are not only different in NA and DA systems of central neurons, but also between various systems of NA or DA neurons. However, for simplification it can be assumed that the functional compartment (C-I) corresponds to a small pool of amine (NA or DA) in which the turnover is very fast (half-life of less than 30 min). In the main storage compartment corresponding to a much larger pool of the amine the half-life of the amine is about 2–5 hr (for example, about 2 hr for striatal dopamine and 3–4 hr for global hypothalamic NA). In the third compartment the half-life of the amine is more than 10 hr. Our present knowledge does not yet allow to give an exact anatomical basis to these various pools.

3. Utilization

The amine synthesized and stored is utilized intra- and extraneuronally. Although extraneuronal utilization, corresponding to the process of release in the synaptic cleft, is of major importance from a physiological view point, its quantitative estimation is particularly difficult in the central nervous system. Recently, Kopin and associates (1968) have demonstrated that in peripheral noradrenergic neurons the newly synthesized amine was preferentially released during electrical stimulation. In our laboratory we found that newly synthesized striatal dopamine was preferentially released (Besson *et al.*, 1969a). Similar results were obtained with NA and 5-HT (Besson and associates, personal communication). Under normal conditions the functional pool (C-I) likely contains mainly the newly synthesized amine.

4. Inactivation

The two enzymes monoamine oxidase (MAO) and catechol-*o*-methyltransferase (COMT) inactivate catecholamines in the central nervous system as they do in the peripheral sympathetic system. The oxydative deamination mainly occurs intraneuronally and the *o*-methylation extraneuronally. Therefore an increased extraneuronal utilization of CA is associated with an increase of the *o*-methylated metabolites normetoxyadrenaline or methoxydopamine. The reuptake process is another very important inactivating mechanism. Most of the released amines are recaptured and restored in the neurons. As shown by pharmacological and kinetic studies, the uptake processes in NA and DA neurons are different, but DA and NA can compete for NA or DA uptake processes (Fuxe and Ungerstedt, 1968; Snyder and Coyle, 1969). Exogenous amines introduced in the peripheral circulation can thus be taken up and stored particularly in some hypothalamic CA terminals localized outside the blood-brain barrier (Lichtensteiger and Langemann, 1966). A specific accumulation of exogenous CA is also observed in the various central CA neurons after their direct introduction in the cerebrospinal fluid (Glowinski and Axelrod, 1966).

The dynamic events occurring in DA nerve endings are generally similar to those described for NA endings. However, some differences of metabolism can be observed in the two kinds of neurons. For example, the overall turnover of DA in the nigrostriatal system is faster than that of NA in most of the NA neurons (Costa and Neff, 1966). As already mentioned, these two types of neurons do not react similarly to some psychotropic drugs. Differences in turnover or in sensitivity to drugs are also found among various systems of NA neurons (Iversen and Glowinski, 1966).

B. Estimation of Metabolic Changes in Central Catecholaminergic Neurons

Three main approaches have been used successfully to detect changes in synthesis or utilization of CA in the central nervous system.

1. Measurement of Turnover

Modifications of NA or DA turnover in the whole brain or in some particular structures of the central nervous system have been detected by: (1) following the decline of endogenous amine levels after the selective inhibition of catecholamine synthesis with α-methyl-*p*-tyrosine or its soluble methylester; or (2) following the changes in NA or DA specific activities between 1 and 6 hr after the labeling of central CA neurons. This labeling can be obtained by injecting the radioactive amines into the cerebrospinal fluid or by injecting their labeled precursors (tyrosine or DOPA) into either the blood or cerebrospinal fluid.

The second approach has an important advantage: the physiological or pharmocological effects studied are not modified. However, in both cases the changes observed are likely to be reflecting turnover modifications of the amine in its main storage pool (C-II). This pool may act as a " reservoir " for the small functional pool (C-I). Therefore changes are only likely to be seen when the activity of the neuron is markedly affected. These approaches are inadequate to appreciate changes occurring only in the functional pool.

2. Measurement of Synthesis

If estimations of changes in amine synthesis rate are based on previous turnover studies, underestimated values will result because the amines are not localized in a single compartment (Sedvall *et al.*, 1968). Direct estimations of modification in CA synthesis in brain tissues can be made by measuring the specific activity of CA a short time after the intravenous (Spector *et al.*, 1965), intracisternal, or local injection of labeled tyrosine (Javoy *et al.*, 1970). In some cases the amines newly synthesized are immediately utilized; therefore a better estimate of increase in CA synthesis can be obtained by measuring the specific activity of CA at the end of a short, continuous injection of ^{3}H tyrosine (Sedvall *et al.*, 1968). Both techniques (and particularly the estimation of turnover) can reveal changes in overall global utilization of the amines, but neither can distinguish effects on intraneuronal (depletion) and extraneuronal (release) utilization. However, an increased turnover associated with an increase in *o*-methylated metabolite formation is a good indication of an enhancement of extraneuronal utilization.

3. Direct Measurement of Release

Changes in amounts of transmitters liberated from the neurons are difficult to estimate but their occurrences can be directly correlated with apparent physiological or pharmacological effects. In rare cases the labeled amines previously taken up (Stein and Wise, 1969) or synthesized from precursors (Besson *et al.*, 1969 b) have been measured in collected fractions of brain-structure superfusates. In these studies local superfusion was used to collect the released amines. Similar experiments have also been carried out *in vitro* on brain slices (Baldessarini and Kopin, 1966) or isolated structures (Besson *et al.*, 1969 a, b). In our laboratory, changes in the amount of CA released have been recently demonstrated under various physiological or pharmacological experimental conditions and have been simultaneously correlated with the estimation of synthesis. This was obtained by short-term incubation of isolated brain structures of animals treated with ^{3}H tyrosine and by measuring ^{3}H catecholamines in tissues and in the medium at the end of incubation. This approach makes it possible to appreciate changes in metabolism occurring mainly in the functional pool according to the model previously described.

C. Effects of Some Drugs on Central Catecholamine Metabolism

Some psychotropic drugs are generally used to study functions supposedly related to aminergic neurons. They can be arbitrarily classified in three groups: drugs that increase transmitter quantities in synaptic clefts; drugs that decrease transmitter quantities, mainly because they interfere with storage or synthesis processes; and drugs that compete with the neurotransmitter at the receptor site (Glowinski and Baldessarini, 1966). These types are discussed below.

1. An extraneuronal increase of DA or NA can be obtained by precursors of amine synthesis, such as DOPA or dihydroxyphenylserine, which increase respectively DA and NA. Desmethylimipramine has a similar effect on NA-containing neurons by inhibiting the reuptake process. Amphetamine affects NA as well as DA neurons by enhancing the amount of amines released. In NA neurons these effects are associated with an increased transmitter synthesis, possibly by a feedback mechanism.

2. A decrease of extraneuronal CA is obtained with inhibitors of synthesizing enzymes (α-methyl-*p*-tyrosine for tyrosine hydroxylase; α-methyl-DOPA for DOPA decarboxylase, and disulfiram for dopamine-β-hydroxylase). Compounds competing for storage sites (such as metabolites formed from α-methyl-DOPA or α-methyl-*m*-tyrosine) may act as false transmitters and thus decrease the amount of amine available for extraneuronal release. A depletion of CA can be also seen with drugs that affect the storage capacity of vesicles. The best example is given by reserpine, which inhibits the uptake-storage process in granules or vesicles and leads to an enhanced intraneuronal utilization of the amine.

3. Recent experiments suggest that different types of drugs may react at the postsynaptic sites of CA synapses. The compounds usually involved in these reactions are neuroleptics such as chlorpromazine and derivatives, which are very potent in blocking DA receptors in the striatum. These drugs may indirectly activate DA neurons by complex feedback process. Some may also activate NA neurons in a similar way.

III. Involvement of Catecholaminergic Neurons in the Integration of Endocrine and Non Endocrine Hypothalamic Mechanisms

Using the complementary experimental approaches described above, various authors have shown that noradrenergic and/or dopaminergic neurons are involved in some hypothalamic functions. Only a few studies will be examined. As examples, we shall discuss some problems related to non endocrine and endocrine hypothalamic functions. Observations obtained with histochemical fluorescence techniques have largely contributed to our knowledge in this field. These reports are reviewed in the article by Fuxe and Hökfelt in this volume.

A. Catecholamines and Non Endocrine Functions

1. Thermoregulation

It is well known that the hypothalamus is concerned with temperature regulation. Feldberg and Myers (1965) have postulated that the body temperature of the cat is maintained at its normal level by a balance between the NA and 5-HT systems in the hypothalamus. Intraventricular injection of NA in rats produced a fall of rectal temperature (Feldberg and Lotti, 1967). The shivering induced by injection of 5-HT in the anterior hypothalamus could be relieved by injection of NA into the same area. Since external temperature changes are without effect on the endogenous hypothalamic level of NA, Simmonds and Iversen (1969) attempted to detect a change of the activity of NA-containing neurons by estimating the turnover of the amine in the hypothalamus. These authors followed the temporal changes of NA specific activity in rats exposed to heat or to cold (see Section II, B, 1.). NA-containing neurons were previously labeled by intracisternal injection of ^{3}H noradrenaline. In an environmental temperature of 32 °C, which leads to an increase in rectal temperature of 2.9 °C, the turnover of NA was selectively accelerated in the hypothalamus. No changes in the activity of NA neurons were detected with this experimental approach in rats exposed to cold. The localization of the effect in the hypothalamus indicates that generalized heat stress is not involved and suggests that hypothalamic NA may regulate heat loss in the rat (Simmonds and Iversen, 1970). Although Corrodi *et al.* (1967) observed more generalized effects after biochemical and histochemical estimation of monoamine disappearance rates following synthesis inhibition, which impairs normal thermoregulation, they have drawn similar conclusions. They also suggested that 5-HT neurons were involved in thermoregulation, since the 5-HT terminals of the suprachiasmatic nucleus were differently affected by exposure of the animals to heat or cold (Corrodi *et al.*, 1967).

2. *Stress*

As already mentioned, inhibition of monoamine synthesis suppresses thermoregulation. Under these conditions the exposure of animals to high environmental temperature is a stressful situation and an increased activity of the ascending neurons of the rhombencephalon is observed. Other stresses such as muscular exercise (Spector *et al.*, 1965) or application of mild electric shocks to the paws (Thierry *et al.*, 1968a) increase the turnover of brain NA. There is little doubt that in both cases the stress was associated with an increased temperature of the animals. In our experiments in which aminergic neurons were previously labeled (see Section II, B), the electric shocks induced an acceleration of NA turnover, particularly in the brain stem, but a marked effect was also seen in the hypothalamus. Similarly, we observed an activation of central 5-HT neurons under these conditions (Thierry *et al.*, 1968b). These results support the hypothesis of an involvement of hypothalamic NA and 5-HT neurons in thermoregulation. Moreover, they raise the question of a similar role in Adrenocorticotropic Hormone (ACTH) control.

3. Self-Stimulation

Of particular interest are the experiments of Stein (1962) on self-stimulation, which strongly suggest that NA released from terminals of the ascending noradrenergic fibres running in the medial forebrain bundle mediates rewarding or positively reinforcing effects on behavior. This is supported by various complementary results. Drugs that increased the amounts of NA in synapses (amphetamine imipramine, MAO inhibitors) facilated self-stimulation. Conversely, drugs with the opposite effect on NA (α-methyl-*p*-tyrosine, reserpine) or those that may block " NA-receptors " (chlorpromazine) decreased self-stimulation (Stein, 1962). Furthermore, it was recently shown that rewarding electrical stimulation of the medial forebrain bundle released ^{14}C noradrenaline (that had been previously taken up in NA neurons) into solutions perfused through the hypothalamus and amygdala (Stein and Wise, 1969). In a final series of experiments, Wise and Stein (1969) showed that the inhibition of dopamine-β-hydroxylase by disulfiram sharply reduced the rate of self-stimulation. The effect of the inhibitory drug was antagonized by intraventricular injection of small doses of L-noradrenaline. Intraventricular administration of d-noradrenaline, DA, or 5-HT was ineffective. These results gave further support to the hypothesis of an involvement of NA in rewarding mechanisms.

B. Cathecholamines and Endocrine Functions

Among the considerable amount of recent data concerned with the role of hypothalamic CA in the control of adenohypophyseal secretions, the interrelationships between CA neuron activity and pituitary gonadotropin release are particularly interesting. The hypothalamus is very rich in NA nerve terminals. They are localized mainly in the supraoptic and in the paraventricular nuclei. An important tubero-infundibular DA neuron system has been described. The cell bodies of these DA neurons are mainly found in the arcuate and periventricular nuclei, with their terminals in the median eminence (Fuxe *et al.*, 1967; Barry and Leonardelli, 1967). There is good evidence of coincidental changes of the metabolism of NA and DA in the anterior part of the hypothalamus and hypothalamic Luteinizing Hormone Releasing Factor (LHRF) or pituitary Luteinizing Hormone (LH) content. In the rat the NA levels in this hypothalamic area are maximal during proestrus or after gonadectomy, and are minimal at estrus (Stefano and Donoso, 1967). They are reduced by a simultaneous administration of estradiol and progesterone (Donoso and Stefano, 1967). Changes in accumulation and release of CA synthesized from ^{3}H tyrosine have also been noticed between proestrus and estrus in these structures (Javoy and associates, personal communication). Indications of an increased synthesis of ^{3}H noradrenaline from ^{3}H tyrosine, occurring many days after castration, have been recently obtained (Donoso *et al.*, 1969). The modifications of DA levels in the anterior hypothalamus are concomitant with those of NA, but seem to be mostly in opposite direction. Donoso and his co-workers observed that DA levels were: (1) lower in proestrus than in estrus, (2) reduced by gonadectomy (both in female and male rats), and (3) increased to normal levels by the simultaneous injection of high doses of estradiol and progesterone.

Lichtensteiger (1969) observed significant changes in the fluorescence intensity of DA cells of the tubero-infundibular neuron system during the estrous cycle, with a maximal intensity at estrus. Fuxe *et al.* (1967) detected a marked increase of the fluorescence of DA cells of the arcuate nucleus only during pregnancy. With the help of synthesis inhibitors they observed a diminution of the rate of amine depletion in the DA nerve terminals of the median eminence as compared to that of normally cycling rats. This effect could be counteracted by estrogen treatment. Drugs interfering with the metabolism of monoamines, such as reserpine and MAO inhibitors, can block the ovulation in the rat (Meyerson and Sawyer, 1968). These drugs probably act by different mechanisms. A similar effect on ovulation process can be obtained by adrenergic blocking agents (Everett, 1964) or by neuroleptic drugs such as chlorpromazine (Barraclough and Sawyer, 1957), which may block DA or NA receptor cells.

More recent pharmacological evidence has provided complementary arguments for the hypothesis of a role of hypothalamic DA-containing neurons in gonadotropin regulation and particularly in the ovulation process. Ovulation induced in immature rats by seric and chorionic gonadotropin was inhibited by the peripheral administration of α-methyl-*p*-tyrosine or α-methyl-DOPA. The inhibiting effect was observed only when the drugs were administered during a fraction of the critical period that takes place on the day of proestrus. They probably induced a temporary blockade of DA transmission by diminishing the availability of the newly synthesized amine in the functional pool (C-I as described previously). The action of α-methyl-*p*-tyrosine could be reversed by DOPA but not by dihydroxyphenylserine (Kordon and Glowinski, 1969). Furthermore, similar inhibitory effects on ovulation were obtained by direct intrahypothalamic infusion of α-methyl-DOPA throughout the median eminence region but not elsewhere in the hypothalamus (Kordon, 1970). Possibly the DA activates release of adenohypophysiotropic hormones from the median eminence into the primary plexus of the hypophyseal portal system. NA-containing neurons appear also to be involved in the control of LHRF release, but perhaps this happens at an earlier stage of the complex regulatory mechanism of ovulation process.

IV. Conclusions

The rapid development of various techniques and concepts in the field of CA now permits the detection and study of changes in the metabolism and regulation of CA neurons in very discrete areas of the central nervous system. Moreover, as it has been described in this article, slight and specific modifications of CA-containing neuron activities can be estimated during events occurring in normal physiological situations. It is also possible to modify the metabolism and consequently the functions of some CA neurons by microinjections of suitable drugs into specific nuclei. However, the effects observed after local modifications of amine quantities in some synaptic clefts probably represent summations of various and perhaps opposite actions of these amines on postsynaptic membranes of the receptor cells.

Further research is needed to obtain information about the effects of CA on receptor sites. They should facilitate the elucidation of the specific role of these amines in various endocrine and non endocrine hypothalamic functions.

Data reported with respect to the functional characteristics of aminergic neurons must be interpreted with caution. Indeed, NA and DA neurons represent only a small fraction of the hypothalamic neuronal population.

ACKNOWLEDGMENTS

I am very grateful to my colleagues Dr. Thierry, Dr. Javoy, and Dr. Kordon, and to Mrs. Rotillon for their collaboration in the preparation of this manuscript.

REFERENCES

Baldessarini, R. J., and Kopin, I. J. (1966). *Science* **152**, 1630.

Barraclough, C. A., and Sawyer, C. H. (1957). *Endocrinology* **61**, 341.

Barry, J., and Leonardelli, J. (1967). *C. R. Acad. Sci.* (Paris) **265**, 557.

Besson, M. J., Cheramy, A., Feltz, P., and Glowinski, J. (1969a). *Proc. Nat. Acad. Sci.* **62**, 745.

Besson, M. J., Cheramy, A., and Glowinski, J. (1969b). *Europ. J. Pharmacol.* **7**, 111.

Corrodi, H., Fuxe, K., and Hökfelt, T. (1967). *Acta Physiol. Scand.* **71**, 224.

Costa, E., and Neff, N. H. (1966). *In* " Proc. 2nd Symposium of the Parkinson's Disease Information and Research Center " (E. Costa, L. J. Coete and M. D. Yahr, eds.). pp. 141–156. Raven Press, New York.

Donoso, A. O., and Stefano, F. J. E. (1967). *Experientia* **23**, 665.

Donoso, A. O., De Gutierrez Moyano, M. B., and Santolaya, R. C. (1969). *Neuroendocrinology* **4**, 12.

Everett, J. W. (1964). *Physiol. Rev.* **44**, 373.

Feldberg, W., and Lotti, V. J. (1967). *Brit. J. Pharmacol.* **31**, 152.

Feldberg, W., and Myers, R. D. (1965). *J. Physiol.* **177**, 239.

Fuxe, K., and Hökfelt, T. (1967). *In* " Neurosecretion " (F. Stutinsky, ed.). pp. 165–177. Springer Verlag, Berlin.

Fuxe, K., and Ungerstedt, V. (1968). *Europ. J. Pharmacol.* **4**, 135.

Fuxe, K., Hökfelt, T., and Nilsson, O. (1967). *Life Sci.* (Oxford) **6**, 2057.

Glowinski, J. (1970). *In* " Handbook of Neurochemistry " (A. Lajtha, ed.). Plenum Publishing Company, New York. (In press).

Glowinski, J., and Axelrod, J. (1966). *Pharmacol. Rev.* **18**, 775.

Glowinski, J., and Baldessarini, R. (1966). *Pharmacol. Rev.* **18**, 1201.

Hillarp, N. A., Fuxe, K., and Dahlström, A. (1966). *In* " Mechanisms of Release of Biogenic Amines " (U. S. Von Euler, S. Rosell and B. Uvnäs, eds.). pp. 31–57. Pergamon Press, New York.

Hornykiewicz, O. (1966). *Pharmacol. Rev.* **18**, 925.

Iversen, L. L. (1967). *In* " The Uptake and Storage of Noradrenaline " Cambridge University Press, Cambridge, pp. 108–132.

Iversen, L. L., and Glowinski, J. (1966). *J. Neurochem.* **13**, 671.
Javoy, F., Hamon, M., and Glowinski, J. (1970). *Europ. J. Pharmacol.* (In press).
Kopin, I. J., Breese, G. R., Krauss, K. R., and Weise, U. R. (1968). *J. Pharmacol. Exp. Ther.* **161**, 271.
Kordon, C. (1970). *J. Physiol.* (In press).
Kordon, C., and Glowinski, J. (1969). *Endocrinology* **58**, 924.
Lichtensteiger, W. (1969). *J. Pharmacol. Exp. Ther.* **165**, 204.
Lichtensteiger, W., and Langemann, H. (1966). *J. Pharmacol. Exp. Ther.* **151**, 400.
Meyerson, B. J., and Sawyer, C. H. (1968). *Endocrinology* **83**, 170.
Sedvall, G. C., Weise, V. K., and Kopin, I. J. (1968). *J. Pharmacol. Exp. Ther.* **159**, 274.
Simmonds, M. A., and Iversen, L. L. (1969). *Science* **163**, 473.
Snyder, S. H., and Coyle, J. T. (1969). *J. Pharmacol. Exp. Ther.* **165**, 78.
Spector, S., Sjoerdsma, A., and Udenfriend, S. (1965). *J. Pharmacol. Exp. Ther.* **147**, 86.
Stefano, F. J. E., and Donoso, A. O. (1967). *Endocrinology* **81**, 1405.
Stein, L. (1962). *In* " Recent Advances in Biological Psychiatry " (J. Wortis, ed.). pp. 288–308. Plenum Press, New York.
Stein, L., and Wise, D. (1969). *J. Comp. Physiol. Psychol.* **67**, 189.
Thierry, A. M., Javoy, F., Glowinski, J., and Kety, S. S. (1968a). *J. Pharmacol. Exp. Ther.* **163**. 163.
Thierry, A. M., Fekete, M., and Glowinski, J. (1968b). *Europ. J. Pharmacol.* **4**, 384.
Wise, D., and Stein, L. (1969). *Science* **163**, 299.

The Role of Brain and Pineal Indoles in Neuroendocrine Mechanisms

R. J. WURTMAN

I. Introduction

Serotonin (5-hydroxytryptamine) and its derivatives are synthesized or stored in a variety of mammalian cells and appear to undergo distinct fates in each. This article considers serotonin and related physiologically active indoles in three tissues of especial interest to students of the hypothalamus, i.e., the brain, the pituitary, and the pineal. It should be recognized that these tissues contain only a small fraction of the serotonin in the body, most of the amine being present in enterochromaffin cells, mast cells, circulating platelets, and platelets sequestered in the lungs, spleen, and elsewhere. For a general discussion of the distribution and possible functions of serotonin, the reader is referred to recent reviews by Garattini and Valzelli (1965), Garattini and associates (1968), Page (1968), and Wurtman and co-workers (1968a).

II. Biosynthesis of Serotonin

A. Tryptophan

The circulating precursor for all physiologically active indoles is the essential amino acid L-tryptophan. The quantities of this substance present in most dietary proteins tend to be extremely low; moreover, most of the tryptophan molecules which enter the body are unavailable for serotonin synthesis inasmuch as they are incorporated into peptide chains or destroyed by the enzyme tryptophan pyrrolase. Hence it is possible that the amount of tryptophan available to a given cell and the rate at which that cell metabolizes the amino acid by alternate pathways might influence its ability to synthesize serotonin.

The concentrations of tryptophan in human plasma (Wurtman *et al.*, 1968b) and in certain rat tissues (Wurtman *et al.*, 1968c) undergo regular diurnal fluctuations which result in part from meal eating [i.e., the overflow of dietary tryptophan beyond the portal circulation; the stimulation of tryptophan uptake into tissues by insulin secreted in response to dietary carbohydrates (Wurtman, 1970)]. Although the concentrations of serotonin in the rat pineal (Quay, 1963) and the brain of the turtle (Quay, 1967) are known to vary as a function of time of day, no data are yet available on the relationships between these rhythms and the rhythms in plasma or tissue tryptophan content. The mechanism by which neurons and other cells take up tryptophan from the plasma is also largely unexplored. Perhaps because only a fraction of the tryptophan that enters the body is converted to serotonin and related indoles, it has generally been assumed that this relationship also holds for the tryptophan in any tissue. That this assumption need not be correct has recently been demonstrated by *in vitro* studies on the rat pineal: this organ was found to convert at least 40 times as much ^{14}C-tryptophan to ^{14}C-serotonin, ^{14}C-5-hydroxyindole acetic acid, and ^{14}C-melatonin as to ^{14}C-protein (Wurtman *et al.*, 1969).

Before tryptophan can be converted to serotonin, it must first (1) be present in the plasma, (2) be taken up into cells which contain the necessary enzymes and cofactors, and (3) not be incorporated into peptides or proteins or be otherwise metabolized. It is safe to predict that processes which modify one or more of these steps will also be shown to influence serotonin synthesis.

B. Enzymes

The conversion of tryptophan to serotonin is catalyzed by two enzymes: tryptophan hydroxylase and aromatic L-amino acid decarboxylase (AAAD) (see Fig. 1). Both enzymes have been identified in the brain (Grahame-Smith, 1964; Snyder and Axelrod, 1964) and the pineal (Lovenberg *et al.*, 1967; Snyder and Axelrod, 1964; Shein *et al.*, 1967). Tryptophan hydroxylase demonstrates considerably more substrate specificity than AAAD, which can also catalyze the decarboxylation of tyrosine, 1-dihydroxyphenylalanine, and histidine (Lovenberg *et al.*, 1962). Tryptophan hydroxylase activity can be inhibited by the synthetic amino acid parachlorophenylalanine (Koe and Weissman, 1966). Animals treated with this drug show a decrease in brain serotonin content (Jequier *et al.*, 1967); brain catecholamine levels are unaffected.

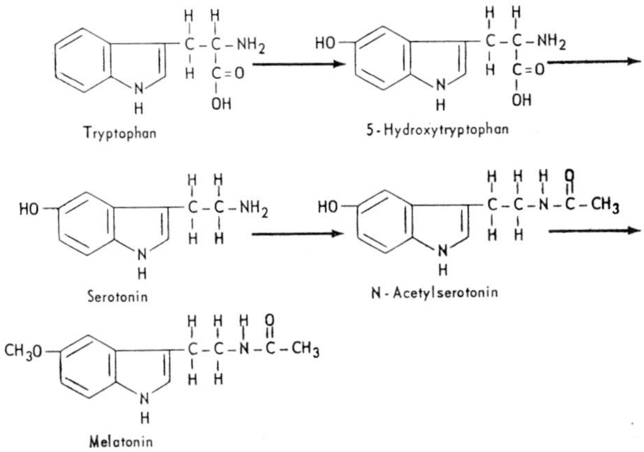

Tryptophan → 5-Hydroxytryptophan → Serotonin → N-Acetylserotonin → Melatonin

Fig. 1. Synthesis of serotonin and melatonin from tryptophan. Tryptophan is oxidized to 5-hydroxytryptophan through the action of the enzyme tryptophan hydroxylase. This amino acid is converted to the amine serotonin by aromatic L-amino acid decarboxylase (AAAD). Most of the amine in the pineal and essentially all of it elsewhere is metabolized by oxidative deamination. This process, catalyzed by the enzyme monoamine oxidase, yields an aldehyde which can be oxidized to 5-hydroxyindole acetic acid (5-HIAA) or reduced to 5-hydroxytryptophol. In the pineal, serotonin can be acetylated to form N-acetylserotonin, which is converted to melatonin through the action of hydroxyindole-*o*-methyl transferase (HIOMT).

III. Brain Serotonin

A. Distribution

Serotonin was first demonstrated in mammalian brain by Twarog and Page (1953). It is unevenly distributed in this organ; in most species highest concentrations are found in the hypothalamus and midbrain, while lowest levels are present in the cerebrum and cerebellum (Garattini and Valzelli, 1965). The elegant histochemical fluorescence analyses of Hillarp, Dahlström, Fuxe, and their collaborators have provided much information about the distribution of serotonin-containing neurons in the brain (Hillarp *et al.*, 1966; Fuxe *et al.*, 1968). In general, the cell bodies of most of these neurons lie within the midbrain, especially in the raphe nuclei; elsewhere in the brain the serotonin is present almost exclusively within nerve terminals (Andén *et al.*, 1965). Most of the axons of the serotonin-containing cells in rat brain ascend via the medial forebrain bundle (Dahlström and Fuxe, 1965); lesions of this tract cause a marked decline in the concentration of serotonin within the telencephalon (Moore *et al.*, 1965).

B. Physiological Disposition

Although the general outlines of serotonin metabolism in the brain are known, precise information on the fate of the serotonin stored in or released from nerve endings is generally unavailable, probably because of the difficulties involved in introducing a radioactive tracer into these serotonin stores. If radioactive serotonin is injected into the peripheral circulation only small amounts of the amine cross into the brain (Page, 1968); this probably reflects the action of a physiological blood-brain barrier. Alternative methods of labeling brain serotonin, described below, also have serious drawbacks:

1. The circulating precursor tryptophan can be labeled by administering systemically the tritiated or carbon-labeled amino acid. Any radioactive serotonin synthesized from this material will constitute an authentic tracer for brain serotonin. However, such synthesis continues until long after the brain has started to release the newly formed amine (i.e., for as long as the ^{3}H-

or ^{14}C-tryptophan is still present in neurons); hence it is extremely difficult to attribute minute-to-minute changes in the radioactive serotonin content of the brain to either synthesis or turnover. Moreover, so little of the administered labeled amino acid is converted to serotonin that it becomes prohibitively expensive to use this method for labeling brain serotonin stores in large numbers of animals.

2. Isotopically labeled 5-hydroxytryptophan can be administered systemically. This amino acid crosses readily into the brain and little, if any, of it is incorporated into brain protein; hence a considerably larger fraction of the injected radioactive molecules are actually converted to serotonin. However, inasmuch as circulating 5-hydroxytryptophan is not the physiological precursor for brain serotonin, the investigator has no guarantee that the amine formed from this amino acid will undergo the same fate as endogenous serotonin.

3. Labeled serotonin can be injected into the lateral cerebral ventricle using the stereotaxic method originally described by Glowinski and associates (1965) for labeled catecholamines, or using a nonstereotaxic modification (Noble *et al.*, 1967). While there is substantial evidence that tracer doses of ^{3}H-catecholamines administered in this manner do mix with endogenous brain dopamine and/or norepinephrine, the evidence that labeled serotonin taken up from the cerebrospinal fluid provides an adequate tracer for brain serotonin is suggestive but certainly not compelling (Aghajanian and Bloom, 1967).

Brain serotonin appears to exist within several metabolic compartments which turn over at different rates; the average turnover time for the amine is 4 to 5 hr in the rat (Aghajanian *et al.*, 1966), but the half-life of disappearance for the most active pool may be in the order of minutes (Udenfriend and Weissbach, 1958). The disappearance of serotonin from central neurons can result, as does that of catecholamines, from either of two processes:

1. The amine can be released from nerve endings, possibly into synapses, where it may function as a neurotransmitter. Thus radioactive serotonin taken up by brain slices *in vitro* is liberated

when the tissues are stimulated electrically (Chase *et al.*, 1967); stimulation of the midbrain raphe nuclei in the rat causes forebrain serotonin levels to decline (Aghajanian *et al.*, 1967); the induction of shivering in cats (by the administration of tranylcypromine) is followed by the release of serotonin into the cerebrospinal fluid (Feldberg and Lotti, 1967). The role of reuptake in the inactivation of released serotonin probably remains to be established.

2. The enzyme monoamine oxidase can metabolize the amine to an aldehyde, which is then either oxidized to 5-hydroxyindole acetic acid (5-HIAA) or reduced to 5-hydroxytryptophol. It appears likely that the monoamine oxidase in the brain actually represents a group of enzymes, of which one acts selectively on serotonin and other tryptamines (see Feldberg, 1968), and another on catecholamines.

C. Functions

1. Serotonin as a Central Neurotransmitter

Much of the experimental evidence which would be necessary to demonstrate that serotonin functions as a central neurotransmitter has already been obtained. Thus:

a. The brain is able to synthesize this amine from its circulating precursor, and contains a powerful enzymatic mechanism for its rapid inactivation.

b. Serotonin is present in nerve endings near the presynaptic membrane (Aghajanian and Bloom, 1967).

c. Serotonin is released from brain neurons following the application of an electrical or a physiological stimulus.

d. The microelectrophoretic application of serotonin to specific central neurons modifies their rates of discharge (Bloom *et al.*, 1964).

Moreover, drugs such as parachlorophenylalanine which modify the synthesis or release of serotonin also interfere with specific functions (described below) which are mediated by brain regions that contain large amounts of the amine.

2. Serotonin and Specific Brain Functions

There is considerable evidence that serotonin placed in the cerebrospinal fluid can elevate body temperature (summarized in Feldberg, 1968). The physiologic significance of these observations is subject to several interpretations. For example, it is possible that the release of serotonin into the cerebrospinal fluid constitutes the first step in the induction of hyperthermy and that the amine acts on a specific receptor zone which is adjacent to the ventricular system. Alternatively, the appearance of serotonin in cerebrospinal fluid during the induction of hyperthermy could simply reflect the heightened activity of serotonin-containing neurons whose terminals lie near the ependyma. Sheard and Aghajanian (1967) have shown that stimulation of the midbrain raphe nuclei in unanesthesized rats raises body temperature at the same time that it lowers the serotonin level and raises the 5-HIAA content of the brain. The systemic administration of tryptophan or of 5-hydroxytryptophan to rabbits causes an elevation in body temperature (Weber and Angell, 1967).

The depletion of brain serotonin which follows the administration of parachlorophenylalanine is associated with insomnia in cats, monkeys, and other mammals (Jouvet, 1967; Koella *et al.*, 1968; Weitzman *et al.*, 1968). This effect appears to be transient and may or may not be associated with a selective suppression of rapid-eye-movement sleep, probably depending on the species studied.

The specific participation of serotonin-containing neurons in the regulation of gonadal function appears likely on the basis of studies by Kordon and his collaborators (Kordon *et al.*, 1968). Rats were given Pregnant Mares' Serum (PMS) and Human Chorionic Gonadotropin (HCG) to cause superovulation, and the number of ova released were counted. The experimental animals also received drugs which modified the metabolism of one or more brain amines. It was observed that treatment regimens which increased brain serotonin levels before or during the critical period suppressed the superovulation; those which elevated brain catecholamines alone did not. For example, nialimide, an inhibitor of monoamine oxidase, decreased the number of ova released; this effect persisted if the animals also received α-methylparatyrosine (which suppresses the rise in brain catecholamine levels), but not if they were treated with parachlorophenylalanine (which blocks the elevation in brain serotonin).

IV. Pituitary Serotonin

The existence of serotonin in the mammalian pituitary was first demonstrated by Björklund and associates (1967) who found small amounts (0.05 to 0.09 μg/g) of the amine in all three lobes of the pig hypophysis. Considerably higher concentrations of serotonin are present in bovine pituitaries and the amine is unevenly distributed in this organ: the pars distalis contains 1.7 to 1.8 μg/g, the pars intermedia contains 5.8 to 6.1 μg/g, and the pars nervosa contains 7.2 to 9.5 μg/g (Piezzi and Wurtman, 1970). The rat pituitary contains about 100 to 150 ng of serotonin; slightly less than half of this amount is present in the pars distalis (Piezzi and Wurtman, 1970). Neither the cellular location of pituitary serotonin in these species nor its possible role in endocrine function has been defined. The administration of melatonin to rats causes a selective increase in the serotonin content of the pars intermedia (Piezzi and Wurtman, unpublished observations).

V. Pineal Serotonin, Melatonin, and Other Indoles

A. Metabolism

Serotonin was first described in mammalian pineal organs by Giarman and his colleagues (Giarman and Day, 1959; Giarman *et al.*, 1960), who noted that the concentration of the amine in human organs was as high as 20 μg/g. Pineal serotonin content appears to vary markedly among species (see Wurtman *et al.*, 1968a); moreover, the concentration of the amine changes several-fold during each 24-hr period in the pineals of rats (Quay, 1963) and monkeys (Quay, 1966).

One function for pineal serotonin is to serve as the precursor for a family of compounds, the methoxyindoles, which in mammals are produced exclusively in this organ. Only the pineal contains the enzyme hydroxyindole-*o*-methyl transferase (HIOMT), which catalyzes the transfer of a methyl group from S-adenosylmethionine to the 5-hydroxyl group of various hydroxyindoles (see Fig. 1) (Axelrod and Weissbach, 1961; Axelrod *et al.*, 1961). The prototype of the pineal methoxyindoles is melatonin, which was first isolated from bovine organs by Lerner and his colleagues (Lerner *et al.*, 1959); other methoxyindoles that have been identified in mammalian pineal organs or can be synthesized by pineal homogenates *in vitro* include 5-methoxytryptophol and 5-methoxyindole acetic acid (Wurtman *et al.*, 1968a). Serotonin is not a very good substrate for the HIOMT present in mammalian pineals; however, the quail epiphysis contains a variant of this enzyme which allows serotonin to be *o*-methylated at least as well as N-acetylserotonin (Axelrod and Lauber, 1968). In evaluating the physiological actions of pineal compounds, it is well to keep in mind that melatonin is only one of several methoxyindoles which this organ can produce and that the pineal also contains biologically active compounds which are totally unrelated to the methoxyindoles (see Wurtman *et al.*, 1968a).

Pineal serotonin is synthesized within the parenchymal cells; about half of the amine present in the rat organ is normally stored within these cells, while half is taken up in pineal sympathetic nerve endings, probably within norepinephrine storage sites (Owman, 1964). The pineals of most species contain large amounts of monoamine oxidase (Wurtman *et al.*, 1963; Wurtman *et al.*, 1964a); at least part of this enzyme is located within sympathetic nerve endings (Snyder *et al.*, 1965).

B. Effect of Environmental Lighting on Melatonin Synthesis

Rats that have been kept in an environment of continuous illumination have pineals which are smaller (Fiske *et al.*, 1960) and demonstrate considerably less HIOMT activity (Wurtman *et al.*, 1963) than those of animals kept in darkness or under light for 12 hr per day. This effect of light is mediated by a specific neural pathway which includes the retinas (Wurtman *et al.*, 1964b), the inferior accessory optic tracts (Moore *et al.*, 1968), and the sympathetic axons to the pineal whose cell bodies reside within the superior cervical ganglia (Wurtman *et al.*, 1964b). It seems likely that light acts by depressing the rate at which norepinephrine is released from the terminal boutons of pineal sympathetic nerves, inasmuch as retinal light exposure decreases the electrical activity measured by electrodes implanted in the pineal (Taylor and Winson, 1969) and norepinephrine added to pineals in organ culture accelerates the synthesis of ^{14}C-melatonin from ^{14}C-tryptophan (Wurtman *et al.*, 1969). The retinal photoreceptor which mediates the pineal response to light has not been identified, nor has its action spectrum been characterized. It is of interest that even though the mammalian pineal has lost the ability, characteristic of organs in certain lower vertebrates, to respond directly to light (Kelly, 1962), its metabolic function continues to be regulated indirectly by environmental illumination. Rats kept in an environment of 12 hr of light and 12 hr of darkness per day show a characteristic daily rhythm in HIOMT activity: enzyme activity is lowest at the end of the daily light period (Axelrod *et al.*, 1965). Since this rhythm is very likely associated with a parallel rhythm in melatonin secretion, it has been suggested that the pineal functions as a "biological clock" which serves to transduce nerve impulses generated by light, or its absence, into hormonal signals (Wurtman and Axelrod, 1965).

C. Physiological Actions of Melatonin

The physiological effects of melatonin and related pineal methoxyindoles have recently been reviewed (Wurtman *et al.*, 1968a) and are described in detail in the article by Fraschini and Martini in this volume. Hence, they are mentioned here only briefly.

The systemic administration of melatonin to rats is associated with evidence of decreased functional activity among several endocrine glands, including the gonads (Chu *et al.*, 1964), the thyroid (De Prospo *et al.*, 1968), and the pituitary (Adams *et al.*, 1965); melatonin also decreases the Melanocyte-Stimulating Hormone (MSH) content of the rat pituitary (Kastin and Schally, 1967). Melatonin implants in the median eminence or the midbrain suppress the rise in pituitary Luteinizing Hormone (LH) levels seen among castrated male rats (Fraschini *et al.*, 1968), suggesting that the indolic hormone might produce some of its endocrine effects by acting on brain centers. Melatonin crosses into the brain from the blood with little difficulty (Wurtman *et al.*, 1964c) and ^{3}H-melatonin injected into the blood or placed in the lateral cerebral ventricles is selectively concentrated within the hypothalamus and the midbrain (Anton-Tay and Wurtman, 1969). Moreover, melatonin implants into the preoptic region of the hypothalamus induce sleep and electroencephalographic changes in cats (Marczynski *et al.*, 1964). Therefore studies were undertaken to determine whether the administration of melatonin produced characteristic changes in the metabolism of possible neurotransmitter substances in the brain.

Rats receiving 150 μg of melatonin intraperitoneally were sacrificed at intervals and the serotonin contents were measured in various brain regions. It was observed that melatonin administration was followed within 20 to 60 min by a decrease in cortical serotonin levels and an increase in the serotonin content of the midbrain (Anton-Tay *et al.*, 1968). This observation suggests that serotonin-containing neurons might be a major site of action of pineal methoxyindoles, and helps to suggest why melatonin can influence sleeping behavior and the electrical activity of the brain as well as endocrine function. It has not yet been determined whether the changes in brain serotonin level reflect alterations in the synthesis, release, or metabolism of the amine.

Acknowledgments

Studies from the author's laboratory described in this report were supported in part by grants from the United States Public Heath Service (AM-11709) and the National Aeronautics and Space Administration (NGR-22-009-272).

References

Adams, W. C., Wan, L., and Sohler, A. (1965). *J. Endocrinol.* **31**, 295.
Aghajanian, G. K., and Bloom, F. E. (1967). *J. Pharmacol. Exp. Therap.* **156**, 23.
Aghajanian, G. K., Bloom, F. E., Lovell, R., Sheard, M., and Freedman, D. X. (1966). *Biochem. Pharmacol.* **15**, 1401.
Aghajanian, G. K., Rosecrans, J. A., and Sheard, M. H. (1967). *Science* **156**, 402.
Andén, N. E., Dahlström, A., Fuxe, K., and Larsson, K. (1965). *Life Sci.* (Oxford) **4**, 1275.
Anton-Tay, F., and Wurtman, R. J. (1969). *Nature.* **221**, 474.
Anton-Tay, F., Chou, C., Anton, S., and Wurtman, R. J. (1968). *Science* **162**, 277.
Axelrod, J., and Lauber, J. (1968). *Biochem. Pharmacol.* **17**, 828.
Axelrod, J., and Weissbach, H. (1961). *J. Biol. Chem.* **236**, 211.
Axelrod, J., MacLean, P. D., Albers, R. W., and Weissbach, H. (1961). *In* "Regional Neurochemistry" (S. S. Kety and J. Elkes, eds.). Pergamon Press, Oxford, pp. 307–311.
Axelrod, J., Wurtman, R. J., and Snyder, S. H. (1965). *J. Biol. Chem.* **240**, 949.
Björklund, A., Falck, B., and Rosengren, E. (1967). *Life Sci.* (Oxford) **6**, 2103.
Bloom, F. E., Costa, E., and Salmoiraghi, G. C. (1964). *J. Pharmacol. Exp. Therap.* **146**, 16.
Chase, T. N., Breese, G. R., and Kopin, I. J. (1967). *Science* **157**, 1461.
Chu, E. W., Wurtman, R. J., and Axelrod, J. (1964). *Endocrinology* **75**, 238.
Dahlström, A., and Fuxe, K. (1965). *Acta Physiol. Scand.* **64**, *Suppl.* 247, 1.
De Prospo, N. D., De Martino, L. J., and McGuinness, E. T. (1968). *Life Sci.* (Oxford) **7**, 183.
Feldberg, W. (1968). *In* "Recent Advances in Pharmacology" (J. M. Robson and R. S. Stacey, eds.). 4th ed. Churchill, London, pp. 349–397.

Feldberg, W., and Lotti, V. J. (1967). *J. Physiol.* **190**, 203.
Fiske, V. M., Bryant, K., and Putman, J. (1960). *Endocrinology* **66**, 489.
Fraschini, F., Mess, B., Piva, F., and Martini, L. (1968). *Science* **159**, 1104.
Fuxe, K., Hökfelt, T., and Ungerstedt, U. (1968). *In* " Advances in Pharmacology " (S. Garattini and P.A. Shore, eds.). Vol. VI. Academic Press, New York, pp. 235–251.
Garattini, S., and Valzelli, L., eds. (1965). "Serotonin." Elsevier Publishing Co., Amsterdam.
Garattini, S., Shore, P. A., Costa, E., and Sandler, M., eds. (1968). " Advances in Pharmacology". Vols. VIA and VIB. Academic Press, New York.
Giarman, N. J., and Day, M. (1959). *Biochem. Pharmacol.* **1**, 235.
Giarman, N. J., Freedman, D. X., and Picard-Ami, L. (1960). *Nature* **186**, 480.
Glowinski, J., Kopin, I. J., and Axelrod, J. (1965). *J. Neurochem.* **12**, 25.
Grahame-Smith, D. G. (1964). *Biochem. J.* **92**, 52.
Hillarp, N. A., Fuxe, K., and Dahlström, A. (1966). *Pharmacol. Rev.* **18**, 727.
Jequier, E., Lovenberg, W., and Sjoerdsma, A. (1967). *Mol. Pharmacol.* **3**, 274.
Jouvet, M. (1967). *Sci. Am.* **216**, 62.
Kastin, A. J., and Schally, A. V. (1967). *Nature* **213**, 1238.
Kelly, D. E. (1962). *Am. Scientist* **50**, 597.
Koe, B. K., and Weissman, A. (1966). *J. Pharmacol. Exp. Therap.* **154**, 499.
Koella, W. P., Feldstein, A., and Czicman, J. S. (1968). *Electroencephalog. Clin. Neurophysiol.* **25**, 481.
Kordon, C., Javoy, F., Vassent, G., and Glowinski, J. (1968). *European J. Pharmacol.* **4**, 169.
Lerner, A. B., Case, J. D., and Heinzelman, R. V. (1959). *J. Am. Chem. Soc.* **81**, 6084.
Lovenberg, W., Weissbach, H., and Udenfriend, S. (1962). *J. Biol. Chem.* **237**, 89.
Lovenberg, W., Jequier, E., and Sjoerdsma, A. (1967). *Science* **155**, 217.
Marczynski, T. J., Yamaguchi, N., Ling, G. M., and Grodzinska, L. (1964). *Experientia* **20**, 435.
Moore, R. Y., Wong, S. R., and Heller, A. (1965). *Arch. Neurol.* **13**, 346.
Moore, R. Y., Heller, A., Bhatnagar, R. K., Wurtman, R. J., and Axelrod, J. (1968). *Arch. Neurol.* **18**, 208.
Noble, E., Wurtman, R. J., and Axelrod, J. (1967). *Life Sci.* (Oxford) **6**, 281.
Owman, C. (1964). *Int. J. Neuropharmacol.* **2**, 105.
Page, I. H. (1968). "Serotonin." Year Book Medical Publishers, Chicago.
Piezzi, R., and Wurtman, R. J. (1970). *Endocrinology* (In press).
Quay, W. B. (1963). *Gen. Comp. Endocrinol.* **3**, 473.
Quay, W. B. (1966). *Proc. Soc. Exp. Biol. Med.* **121**, 946.
Quay, W. B. (1967). *Comp. Biochem. Physiol.* **20**, 217.
Sheard, M. H., and Aghajanian, G. K. (1967). *Nature* **216**, 495.
Shein, H., Wurtman, R. J., and Axelrod, J. (1967). *Nature* **213**, 730.
Snyder, S. H., and Axelrod, J. (1964). *Biochem. Pharmacol.* **13**, 805.
Snyder, S. H., Fischer, J. E., and Axelrod, J. (1965). *Biochem. Pharmacol.* **14**, 363.
Taylor, A. N., and Winson, R. W. (1969). *Anat. Record* **163**, 327.
Twarog, B. M., and Page, I. H. (1953). *Am. J. Physiol.* **175**, 157.
Udenfriend, S., and Weissbach, H. (1958). *Proc. Soc. Exp. Biol. Med.* **97**, 748.
Weber, L. J., and Angell, L. A. (1967). *Biochem. Pharmacol.* **16**, 2451.
Weitzman, E. D., Rapport, M. M., McGregor, P., and Jacoby, J. (1968). *Science* **160**, 1361.
Wurtman, R. J. (1970). *In* " Mammalian Protein Metabolism " (H. N. Munro, ed.). Vol. IV. Academic Press, New York. (In press).
Wurtman, R. J., and Axelrod, J. (1965). *Sci. Am.* **213**, 50.

Wurtman, R. J., Axelrod, J., and Phillips, L. S. (1963). *Science* **142**, 1071.
Wurtman, R. J., Axelrod, J., and Barchas, J. D. (1964a). *J. Clin. Endocrinol. Metab.* **24**, 299.
Wurtman, R. J., Axelrod, J., and Fischer, J. E. (1964b). *Science* **143**, 1329.
Wurtman, R. J., Axelrod, J., and Potter, L. T. (1964c). *J. Pharmacol. Exp. Therap.* **143**, 314.
Wurtman, R. J., Axelrod, J., and Kelly, D. E. (1968a). "The Pineal." Academic Press, New York.
Wurtman, R. J., Rose, C. M., Chou, C., and Larin, F. F. (1968b). *New Engl. J. Med.* **279**, 171.
Wurtman, R. J., Shoemaker, W. J., and Larin, F. (1968c). *Proc. Nat. Acad. Sci. U.S.* **59**, 800.
Wurtman, R. J., Shein, H. M., Axelrod, J., and Larin, F. (1969). *Proc. Nat. Acad. Sci. U.S.* **62**, 749.

Distribution of Cholinesterase and Cholinergic Pathways

C. C. D. SHUTE

I. Evidence of Cholinergic Mechanisms

It would not be unreasonable to suppose that cholinergic effects in the hypothalamus are relatively unimportant when compared with aminergic effects or with cholinergic mechanisms in other regions of the brain. It has been shown, for instance, that levels of acetylcholine (ACh), choline acetyltransferase (ChA) and acetylcholinesterase (AChE), unlike those of monoamines, are appreciably lower in the hypothalamus than in the thalamus (Macintosh, 1941; Hebb and Silver, 1956; Burgen and Chipman, 1951). Nevertheless these three substances are all present in significant amounts, suggesting that cholinergic mechanisms are operative; this conclusion is supported by many observations on the effects of cholinomimetic (ACh-like) drugs.

There is some evidence that not all parts of the hypothalamus are equally responsive to cholinergic influences. Studies on unit activity have tended to show active responses in neurons located dorsally or caudally. Thus, in the cat, Abrahams and Edery (1964) found that activity in the ventral hypothalamus was not effected by eserine perfusion. Diagrams by Bloom and associates (1963) of responsive areas show cells facilitated by ACh in either the dorsal or the caudal hypothalamus, especially the mammillary and supramammillary regions. Inhibitory responses to ACh occurred mainly in the ventral and rostral hypothalamus, but the significance of such responses may need to be reconsidered since the report (Adams, 1968) that many ventral hypothalamic cells are inhibited nonspecifically by manipulations of the animal's body.

Bloom and colleagues (1963) point out that the chemical sensitivity of neurons may or may not reflect the type of synaptic transmission; e.g., in certain *Aplysia* neurons nonsynaptic areas of the soma respond to electrophoretically applied ACh. If the ACh effects are indeed physiological, one would expect a degree of correlation between responses to ACh-like drugs and to electrical stimulation. To some extent this is true. For instance, similar effects produced on respiration and on visceral activity in cats were ascribed by Emmelin and Jacobsohn (1945) to a hypothalamic sympathetic center subject to cholinergic control. On the other hand, Baxter (1967) reported opposite behavioral effects resulting from carbachol and electrical stimulation of the perifornical regions, the former producing retreat behavior and the latter causing attack. Since the carbachol response had a long latency, Baxter concluded that the drug might be diffusing via the ventricle to the periaqueductal gray, which on the evidence of unit recording is active in defense reactions.

The distribution pattern of hypothalamic cells found by Bloom and co-workers (1963) to be influenced by ACh did not differ much from that of cells responding to monoamines. This is remarkable because AChE and the monoamines are not similarly distributed in the hypothalamus (Shute and Lewis, 1966a, b), and their results may reflect a sampling bias. Responses were not obtained from the lateral hypothalamic (ventral perifornical) region where both ACh-like and adrenomimetic drugs produce behavioral effects in rats, the former causing drinking and the latter eating (Grossman, 1960). A possible explanation of these responses is that cells in this region are relatively dispersed. Increased water intake is also produced by ACh-like drugs acting on various extrahypothalamic structures, including the hippocampus (Fisher and Coury, 1962; Grossman, 1964). Primary hippocampal efferents are noncholinergic, but since they are cholinoceptive (Shute and Lewis, 1966a; Lewis and Shute, 1967) they might produce drinking by direct projections onto the hypothalamus or by interneurons which are also cholinoceptive.

The effects of topically applied ACh and noradrenaline on feeding are different in kind. Some other behavioral effects of these drugs (e.g., on temperature regulation; Bligh and Maskrey, 1970) may be qualitatively similar but in opposite directions. It is tempting to postulate, in the hypothalamus and elsewhere, a mode of neuronal control in which ACh is in most instances facilitatory and one or other monoamine is inhibitory. Although such a mechanism may operate in some structures such as the striatum (where there is evidence that in Parkinsonism the normal balance between cholinergic and dopaminergic innervation is disturbed, with depression of the latter), results accumulating from unit recording experiments in different regions of the brain do not support so simple a model.

II. Distribution of Acetylcholinesterase in Cells and Fiber Tracts

A. Light Microscopy

The presence of AChE in a neuron can, of course, provide only indirect evidence of its cholinergic or cholinoceptive function, but in view of the difficulty in relating neuropharmacological findings to normal physiology, it becomes necessary to fall back on histochemical methods to determine patterns of cholinergic innervation. All peripheral cholinergic neurons have high concentrations of AChE in the cell cytoplasm, along the axon, and on axonal terminals, and it seems reasonable to presume that the same is true in the central nervous system. Neurons with these characteristics innervating the hippocampus have been proved to be cholinergic by correlating changes in AChE and ChA levels produced by surgical interruption (Lewis *et al.*, 1967). AChE piles up on the cell body side of the lesion and disappears from the opposite side, and this phenomenon has made it possible to trace AChE-containing, presumed cholinergic pathways in the brain as a whole (Shute and Lewis, 1961, 1963).

The distribution of AChE-containing cells in the hypothalamus and the course taken by AChE-containing fibers as seen by light microscopy has been described and illustrated in some detail in the cat (Shute and Lewis, 1966b). In general, cells rich in AChE, apart from those of the lateral preoptic area (and the paraventricular nucleus which will be considered later), are located mainly in posterior and dorsal regions of the hypothalamus. Aggregations of such cells are found in the supramammillary nucleus, perifornical nucleus and dorsal hypothalamic area, and also in the subthalamic nucleus. Subcortical projections from the lateral preoptic area go to the supraoptic nucleus, to the amygdala via the stria terminalis, and to the nucleus ventralis anterior of the thalamus. Axons of supramammillary neurons travel along the inner aspects of the mammillary peduncles to the mammillary bodies (medial and lateral mammillary nuclei). Other fibers ascend to the nonspecific thalamic nuclei in the ventral part of the massa intermedia. Some fibers from the perifornical nucleus run anteriorly in the descending column of the fornix to the paraventricular thalamic nucleus. Axons from dorsal hypothalamic neurons run forwards by a medial route (see Fig. 1) to the reticular thalamic nucleus and the globus pallidus. Those from cells of the subthalamic nucleus take the lateral route (Fig. 1) across the cerebral peduncle to the entopeduncular nucleus and globus pallidus.

AChE outside cell bodies is also present in much greater quantities in posterior and dorsal regions than elsewhere in the cat hypothalamus. Some of the staining results from enzyme in the dendrites of AChE-containing cells. The rest is in terminals of AChE-containing axons

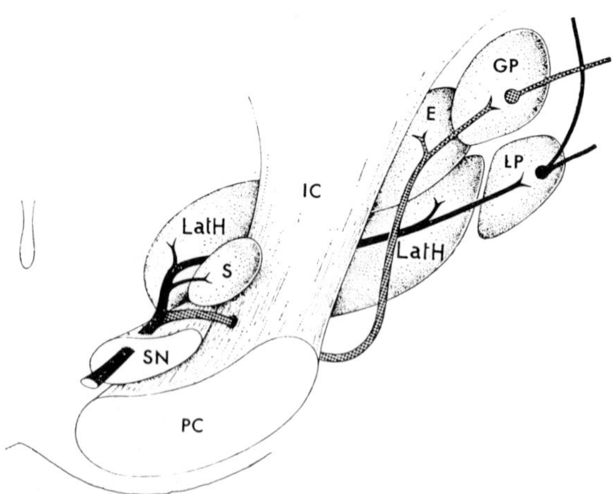

FIG. 1. Diagram illustrating the double route through the ventral diencephalon taken by the ascending cholinergic type of reticular fibers in the cat. Right side of brain seen from behind; lateral pathway stippled. Abbreviations: E, entopeduncular nucleus; GP, globus pallidus; IC, internal capsule; LatH lateral hypothalamus; LP, lateral preoptic area; PC, cerebral peduncle; S, subthalamic nucleus; SN, substantia nigra.

projecting onto these regions or in AChE-containing fibers of passage. The latter include a small projection from the paraventricular thalamic nucleus onto the nucleus reuniens traversing the dorsal part of the periventricular hypothalamic nucleus, and the much more important ventral tegmental pathway of the ascending cholinergic reticular system (Shute and Lewis, 1963, 1967). In cat, the fibers of this system travel by a double route, illustrated diagrammatically in Fig. 1. Fibers taking the more medial route (shown in black) hook over the cerebral peduncle and run forward through the lateral hypothalamus in the medial forebrain bundle to reach the lateral preoptic area. From here corticopetal fibers supply medial and inferolateral cortex and the olfactory bulb. Fibers taking the more lateral course (stippled) pierce the medial part of the cerebral peduncle which then, with the internal capsule, separates them from the hypothalamus. They run via the entopeduncular nucleus to the globus pallidus, which gives rise to corticopetal fibers supplying lateral cortex.

In rat, only a few ascending reticular fibers pierce the cerebral peduncle to take the lateral route. The majority of fibers travel to the globus pallidus and lateral preoptic area by the medial route via the lateral hypothalamus. In the monkey, on the other hand, the medial forebrain bundle is inconspicuous, and the ascending reticular fibers travel by the lateral route to the inner segment of the globus pallidus and to the substantia innominata, which seems to be equivalent to the lateral preoptic area of nonprimates. There is therefore no hypothalamic component of the ascending cholinergic reticular system in the monkey.

In general, however, the dispositions of AChE-containing hypothalamic neurons and tracts in rat, cat, and monkey are more remarkable for their similarities than for their divergences. In all three species, for instance, AChE is present in far larger quantities in cells of the paraventricular nucleus than in those of the supraoptic nucleus, which stain only lightly. The magnocellular portion of the paraventricular nucleus is particularly large and heavily stained in monkey. The axons of the hypothalamo-neurohypophyseal tract are unstained by a standard method for AChE in any of the species, confirming previous observations on dog (Abrahams *et al.*, 1957) and monkey (Holmes, 1961). Koelle and Geesey (1961) reported light to moderate AChE staining of hypothalamo-neurohypophyseal fibers in cat, but experimental conditions were designed to show up very low concentrations of AChE, which may not be diagnostic of cholinergic neurons.

Minor differences between rat, cat, and monkey hypothalami are as follows. In rat, some AChE-containing fibers ascend to the thalamus in the mammillo-thalamic tract but not in the descending columns of the fornix, as in cat. In cat and monkey no mammillothalamic fibers possess AChE, but monkey differs from cat in that only the lateral mammillary nucleus receives a substantial AChE-containing innervation. The perifornical nucleus of monkey is larger than in other species and extends forward into the territory usually ascribed to the dorsomedial nucleus which it populates with AChE-containing cells. The ventral hypothalamus adjacent to the pituitary stalk is virtually devoid of AChE in monkey, whereas in rat a number of cells in the arcuate nucleus contain small or moderate amounts of AChE. Cells rich in AChE are found in the nucleus of the inferior accessory optic root, which is concerned in pineal responses to light and dark.

B. Electron Microscopy

The electron microscope makes it possible to determine the precise localization of AChE and, being more sensitive than the light microscope in detecting small deposits of enzyme, is particularly informative in parts of the hypothalamus where cholinergic innervation is relatively sparse. In cholinergic neurons generally (e.g., those of motor cranial nerves) AChE is located (Lewis and Shute, 1966) on the axonal membrane, in the cisternae of rough endoplasmic reticulum (ER) and in the nuclear envelope with which the cisternae often communicate. Occasionally, as in the medial septal nucleus (the source of the cholinergic supply to the hippocampus), AChE is also present on the plasma membrane of the cell. Cholinoceptive neurons such as the hippocampal pyramidal cells and the granular cells of the dentate gyrus contain AChE on the membrane of dendritic spines, in very small amounts only in rough ER and the nuclear envelope and not at all on the axonal membrane (Shute and Lewis, 1966c). Neurons of doubtful status are the Golgi type II inhibitory cells of the hippocampus and cerebellum. These cells have a characteristic morphology with large amounts of intracytoplasmic AChE, but because of the shortness of their axons it is uncertain whether they should be regarded as cholinergic or cholinoceptive.

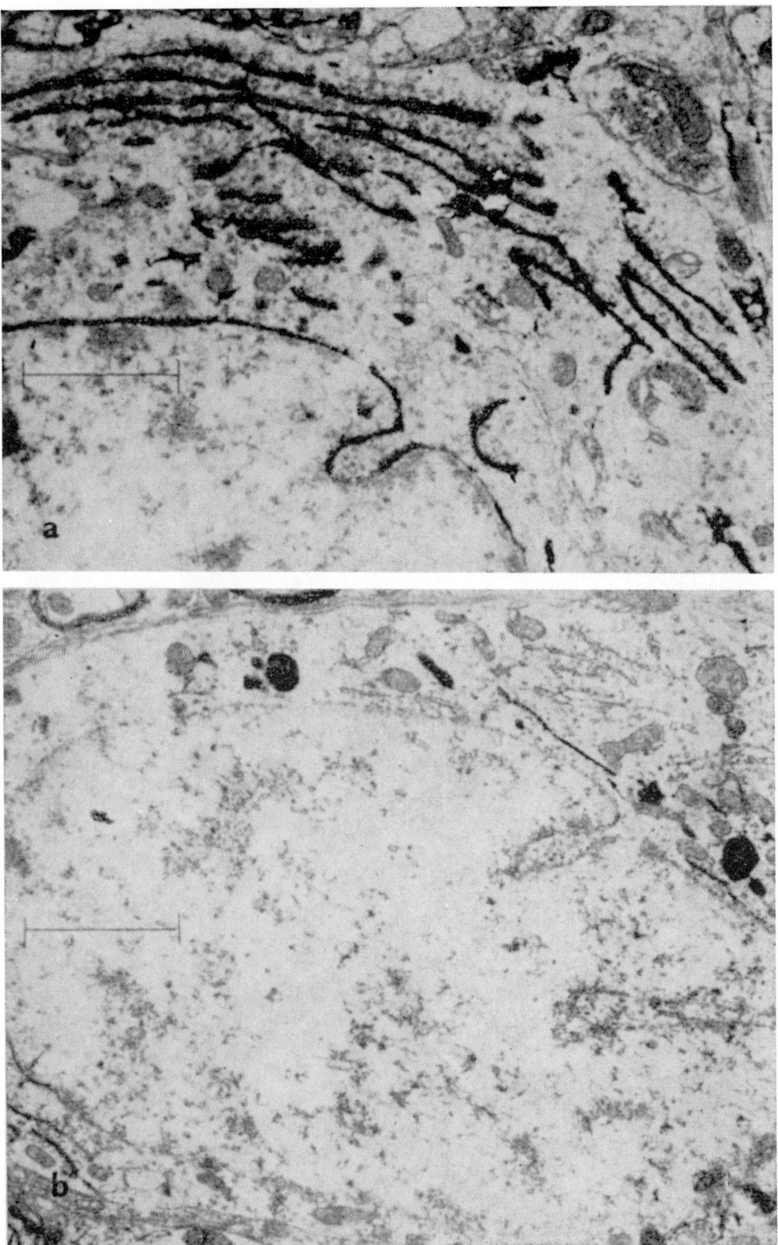

FIG. 2. A: Electron micrograph of neuron from the perifornical nucleus, showing AChE in cisternae of the rough endoplasmic reticulum and in the nuclear envelope. An axosomatic synapse is present (top right). B: Electron micrograph of neuron from the lateral mammillary nucleus. Traces of AChE indicated by arrows. Material illustrated in these and subsequent photographs is from rat brain fixed by perfusion followed by immersion in glutaraldeyde–formalin mixture and subsequently treated by a method for fine localization of AChE similar to that of Lewis and Shute (1966).

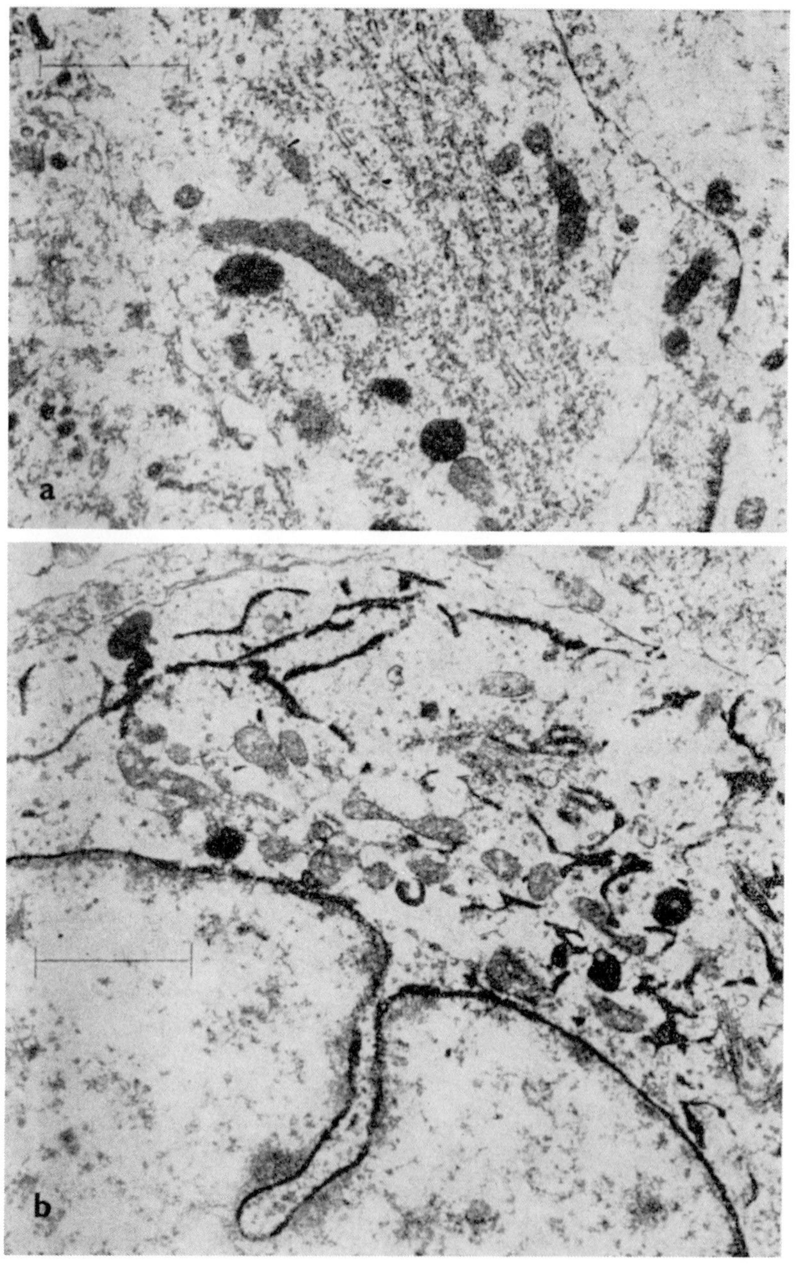

FIG. 3. A: Electron micrograph of supraoptic neuron. The endoplasmic reticulum (ER) is without AChE. B: Electron micrograph of paraventricular neuron showing large quantities of AChE in ER and nuclear envelope.

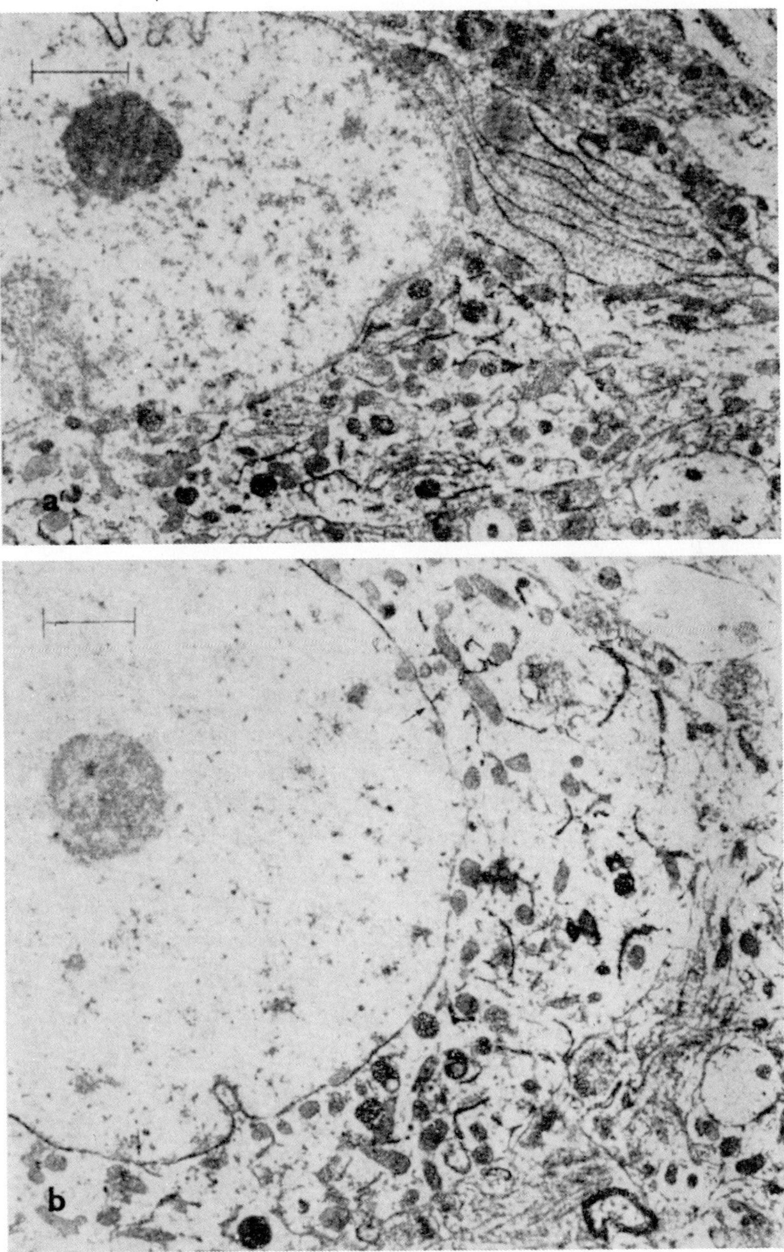

FIG. 4. Electron micrographs of arcuate neurons. In A, AChE occupies much of the retiform endoplasmic reticulum (ER), but less in the stacked cisternae and nuclear envelope. In B, communications between the retiform ER and the nuclear membrane are marked by arrows. A communication with the plasma membrane is present (bottom left).

The fine localization of AChE in rat hypothalamus has been studied in the perifornical nucleus, lateral mammillary nucleus, supraoptic and paraventricular nuclei, and arcuate nucleus (Lewis and Shute, unpublished observations). In perifornical cells AChE is present in large amounts in stacked cisternae of rough ER and in the nuclear envelope as in cholinergic neurons (Fig. 2A). In lateral mammillary cells, on the other hand, AChE is located as in cholinoceptive neurons in short lengths only of ER and nuclear envelope, the greater part of these structures being unstained (Fig. 2B). In supraoptic neurons AChE is either absent from ER and the nuclear envelope (Fig. 3A) or present in minimal amounts. Paraventricular cells have considerably more AChE than those of the supraoptic nucleus. In many of the larger cells the enzyme occupies a high proportion of the available intracisternal and perinuclear space (Fig. 3B). No axons staining for AChE have been found in the supraoptic or paraventricular nucleus.

In the arcuate nucleus many neurons are without AChE, but some contain varying amounts of enzyme. In the latter the AChE-containing ER characteristically has a retiform appearence, although there may be typical stacked cisternae at one pole of the cell (Fig. 4A). This retiform ER opens into the nuclear envelope more frequently than is usual and may even communicate with the plasma menbrane (Fig. 4B). Synapses on the soma and basal dendrites are commonly of the noncholinergic type, i.e., without AChE on the membrane or in the synaptic cleft. Neurons of this type occur elsewhere in the ventral forebrain. In their morphology they are reminiscent of the inhibitory Golgi cells of the hippocampus and cerebellum.

III. Functional Implications

The AChE-containing neurons of cholinergic type which are so characteristic of the perifornical nucleus and other parts of the posterodorsal hypothalamus could be involved in, and perhaps responsible for, the behavioral effects which result from stimulation of these regions. Such effects seem to be mainly rage reactions and attack behavior (Ingram, 1952, 1956). The perifornical rage center may be subject to inhibitory control from the ventromedial nucleus

(Ingram, 1952; Turner *et al.*, 1967), and the ventromedial nucleus, which receives projections from the septum and amygdala, may itself be controlled by inhibitory interneurons (Murphy and Reynaud, 1969). There is evidence (see Feldman, 1963) that various parts of the hypothalamus are subject to inhibitory influences, originating from outside it or from within, which may be responsible for the normally low firing rate of hypothalamic neurons and for the rise in these rates (Cross and Kitay, 1967) which follows surgical isolation. The AChE-containing neurons with retiform ER could be concerned in local inhibitory control. The possibility, however, that chains of inhibitory neurons may be operative, one on another, puts difficulties in the way of relating behavior to specific pathways.

The relative density of AChE-containing neuropil postero-dorsally suggests that cholinergic effects would be most readily obtained in this part of the hypothalamus. On the other hand, the sparsity of such neuropil anteriorly does not mean that cholinergic endings are totally absent there. It has been claimed that the anterior hypothalamus is susceptible to cholinergic influences affecting temperature regulation (Lomax and Jenden, 1966). The low concentration of AChE in supraoptic neurons supports the view (Lederis and Livingston, 1969) that they are cholinoceptive and that ACh acts to produce release of vasopressin (Pickford, 1939, 1945) at this level rather than on the neurohypophysis. Such release has been shown *in vitro* to occur only if the neurohypophysis still has the hypothalamus attached (Daniel and Lederis, 1966). Both supraoptic and paraventricular cells hypertrophy and increase their AChE content in rats treated with saline to overload the antidiuretic release system (Pepler and Pearse, 1957). Oxytocin release is probably also influenced cholinergically, since anticholinesterase injected into the supraoptic nucleus increases spontaneous uterine activity (Abrahams and Pickford, 1956). There does not seem to be a satisfactory explanation at present for the relatively high concentration of AChE in the paraventricular nucleus: possibly it is related to a low spontaneous firing rate.

The ascending AChE-containing fibers in the lateral hypothalamus form part of the extrathalamic portion (Starzl *et al.*, 1951) of the ascending reticular activating system. Cortical arousal is less readily obtained from extrathalamic stimulation of the diencephalon, possibly because the ascending fibers separate into different pathways. On the other hand, bilateral lesions in the lateral part of the posterior hypothalamus which produce sleep (Swett and Hobson, 1968) would interrupt all ascending fibers before they diverge. A quite different basis for sleep production was proposed by Velluti and Hernández-Peón (1963) and by Hernández-Peón and associates (1963), who were able to produce sleep in cats by applying ACh-like drugs anteriorly to the medial preoptic region or posteriorly, and with a shorter latency, to the interpeduncular nucleus. In the former case the production of sleep was blocked by a lesion of the medial forebrain bundle; in the latter case it was not. These workers therefore postulated a descending cholinergic hypnogenic pathway to account for their findings.

There is other evidence of inhibitory centers in both the ventral telencephalon (Sterman and Clemente, 1962; McGinty and Sterman, 1968) and the posteromedial hypothalamus (Swett and Hobson, 1968) which may represent the anterior and posterior sleep-producing areas of Hernández-Peón and associates (1963). However, a descending cholinergic pathway is unlikely because the AChE-containing fibers of the medial forebrain bundle are ascending. It seems more probable that there is a noncholinergic descending inhibitory pathway containing relays in which transmission can be modified by an ascending cholinergic input. It must be remembered too that ACh-like drugs may preferentially elicit inhibitory effects: in the caudate, for instance, carbachol mimics low-frequency stimulation and produces drowsiness (Stevens *et al.*, 1961).

IV. Conclusions

To sum up, therefore, we conclude that cholinergic mechanisms occur in the hypothalamus in the form of a reticular type of input, more important in certain regions than in others, influencing transmission in synaptic relays in a manner which may be facilitatory or which may involve inhibitory interneurons.

References

Abrahams, V. C., and Edery, H. (1964). *Progr. Brain Res.* **6**, 26.
Abrahams, V. C., and Pickford, M. (1956). *J. Physiol.* (London) **133**, 330.
Abrahams, V. C., Koelle, G. B., and Smart, P. (1957). *J. Physiol.* (London) **139**, 137.
Adams, D. B. (1968). *Arch. Ital. Biol.* **106**, 243.
Baxter, B. L. (1967). *Exp. Neurol.* **19**, 412.

Bligh, J., and Maskrey, M. (1970). *J. Physiol.* (London). (In press).
Bloom, F. E., Oliver, A. P., and Salmoiraghi, G. C. (1963). *Int. J. Neuropharmacol.* **2**, 181.
Burgen, A. S. V., and Chipman, L. M. (1951). *J. Physiol.* (London) **114**, 296.
Cross, B. A., and Kitay, J. I. (1967). *Exp. Neurol.* **19**, 316.
Daniel, A. R., and Lederis, K. (1966). *J. Endocrinol.* **34**, x.
Emmelin, N., and Jacobsohn, D. (1945). *Acta Physiol. Scand.* **9**, 97.
Feldman, S. (1963). *Electroenceph. Clin. Neurophysiol.* **15**, 672.
Fisher, A. E., and Coury, J. N. (1962). *Science* **138**, 691.
Grossman, S. P. (1960). *Science* **132**, 301.
Grossman, S. P. (1964). *J. Neuropharmacol.* **3**, 45.
Hebb, C. O., and Silver, A. (1956). *J. Physiol.* (London) **134**, 718.
Hernández-Peón, R., Chávez-Ibarra, G., Morgane, P. J., and Timo-Iaria, C. (1963). *Exp. Neurol.* **8**, 93.
Holmes, R. L. (1961). *J. Endocrinol.* **23**, 63.
Ingram, W. R. (1952). *Electroenceph. Clin. Neurophysiol.* **4**, 395.
Ingram, W. R. (1956). *In* " The Hypothalamus " *Ciba Found. Clinical Symposia* **8**, 117.
Koelle, G. B., and Geesey, C. N. (1961). *Proc. Soc. Exp. Biol. Med.* **106**, 625.
Lederis, K., and Livingston, A. (1969). *J. Physiol.* (London) **201**, 695.
Lewis, P. R., and Shute, C. C. D. (1966). *J. Cell Sci.* **1**, 381.
Lewis, P. R., and Shute, C. C. D. (1967). *Brain* **90**, 521.
Lewis, P. R., Shute, C. C. D., and Silver, A. (1967). *J. Physiol.* (London) **191**, 215.
Lomax, P., and Jenden, D. J. (1966). *J. Neuropharmacol.* **5**, 353.
Macintosh, F. C. (1941). *J. Physiol.* (London) **99**, 436.
McGinty, D. J., and Sterman, M. B. (1968). *Science* **160**, 1253.
Murphy, J. T., and Reynaud, L. P. (1969). *J. Neurophysiol.* **32**, 85.
Pepler, W. J., and Pearse, A. G. E. (1957). *J. Neurochem.* **1**, 193.
Pickford, M. (1939). *J. Physiol.* (London) **95**, 226.
Pickford, M. (1945). *Physiol. Rev.* **25**, 575.
Shute, C. C. D., and Lewis, P. R. (1961). *Bibliotheca Anat.* **2**, 34.
Shute, C. C. D., and Lewis, P. R. (1963). *Nature* **199**, 1160.
Shute, C. C. D., and Lewis, P. R. (1966a). *Nature* **212**, 710.
Shute, C. C. D., and Lewis, P. R. (1966b). *Brit. Med. Bull.* **22**, 221.
Shute, C. C. D., and Lewis, P. R. (1966c). *Z. Zellforsch. Mikroskop. Anat. Abt. Histochem.* **69**, 334.
Shute, C. C. D., and Lewis, P. R. (1967). *Brain*, **90**, 497.
Starzl, T. E., Taylor, C. W., and Magoun, H. W. (1951). *J. Neurophysiol.* **14**, 461.
Sterman, M. B., and Clemente, C. D. (1962). *Exp. Neurol.* **6**, 103.
Stevens, J. R., Kim, C., and MacLean, P. D. (1961). *Arch. Neurol.* **4**, 47.
Swett, C. P., and Hobson, J. A. (1968). *Arch. Ital. Biol.* **106**, 283.
Turner, S. G., Sechzer, J. A., and Liebett, R. A. (1967). *Exp. Neurol.* **19**, 236.
Velluti, R., and Hernández-Peón, R. (1963). *Exp. Neurol.* **8**, 20.

Neurochemical Regulation of Food Intake

S. BALAGURA

I. Introduction

Although the ventral diencephalic area had been suspected to be involved in the regulation of food intake and/or body weight, it was not until Hetherington and Ranson (1939) made their classical experiments that lesions of the ventromedial hypothalamus (VMH) were demonstrated to produce obesity. Later on, Brobeck (1946) emphasized the increase in food intake produced by these lesions. Anand and Brobeck (1951) showed that the destruction of the lateral part of the lateral hypothalamus at the level of the ventromedial nucleus was followed by an abolition of food consumption (i.e., aphagia).

Electrical stimulation of the VMH and the lateral hypothalamus results in effects opposite to those obtained with ablations. Increased food intake has been noticed when animals are stimulated electrically in the lateral hypothalamus (Delgado and Anand, 1953; Anand and Dua, 1955). Stimulation of the VMH decreases food intake (Anand and Dua, 1955) and operant responding for food (Wyrwicka and Dobrzecka, 1960).

In summary, it appears that when the VMH is predominantly active the animal is in a "satiated" state. When the lateral hypothalamus is the active area, however, the animal is in a "hunger" state.

Of course, as are most neural structures, the activity of these areas may depend upon a multitude of incoming stimuli. Great efforts have been made during the past 30 years to determine the possible stimuli that influence the neural control of ingestive behavior, and several theories have been advanced.

Brobeck has proposed an "energy" detecting mechanism (Strominger and Brobeck, 1953; Brobeck, 1957) in which the animal's temperature and also the specific dynamic action of foodstuffs play the central role in the regulation of food intake. There is also a "lipostatic" theory (Kennedy, 1952) in which some lipid or lipoid metabolite would be the substance being measured. Other workers believe that energy intake is in some way controlled by the osmotic state of the organism (Smith, 1966). Mayer (1953, 1955) has proposed a "glucostatic" theory in which glucose plays the leading role, but its utilization rather than its absolute value is what is monitored by the organism. In Mayer's model, the VMH is thought to be the glucoreceptor that would measure glucose utilization.

II. Self-Stimulation of the Lateral Hypothalamus Modified by Insulin and Glucagon

Olds and Milner (1954) reported the first experiments describing what now is known as "self-stimulation." If electrodes are implanted in certain areas of the brain and a lever is made available to the animal in a way such that the occurrence of a lever-press response is followed by electrical stimulation of the brain through the implanted electrodes, the animal will repeat the lever-pressing behavior, thus self-stimulating its brain with electric current; such stimulation is said to be "rewarding." Recently the rate of self-stimulation in the lateral hypothalamus has been found to be related to the state of deprivation of an organism. Self-stimulation rates increase with food deprivation (Margules and Olds, 1962) and decrease following forced-feeding (Hoebel and Teitelbaum, 1962).

To study the possibility that changes in self-stimulation rate are dependent on alterations in glucose metabolism, a series of experiments was carried out in my laboratory. Instead of using food or food deprivation, the two pancreatic hormones insulin and glucagon were used to modify blood glucose level and utilization. Administration of insulin is known to cause hypoglycemia and hyperphagia when administered to rats (Mackay *et al.*, 1940). When glucagon is injected, opposite effects are obtained: hyperglycemia (Unger and Eisentraut, 1964) and hypophagia (Holloway and Stevenson, 1964). Since changes in blood sugar are accompanied by changes in neural activity within the feeding areas of the hypothalamus (Anand *et al.*, 1964; Oomura *et al.*, 1964) and these same areas are involved in hypothalamic reward mechanism, glucose metabolic changes could affect feeding behavior by influencing lateral hypothalamic reward. If so, it was reasoned that the self-stimulation rate, like feeding, should increase with injections of insulin and decrease with injections of glucagon.

The results obtained by Balagura and Hoebel (1967) confirmed this hypothesis. Insulin caused the self-stimulation rate to increase above the preinjection level, whereas glucagon caused it to decrease (Fig. 1). In a control group injected with isotonic saline solution it was possible to demonstrate that the rate of self-stimulation measured during the preinjection period is a good estimate of the rate of self-stimulation that is obtained when neither hormone is injected.

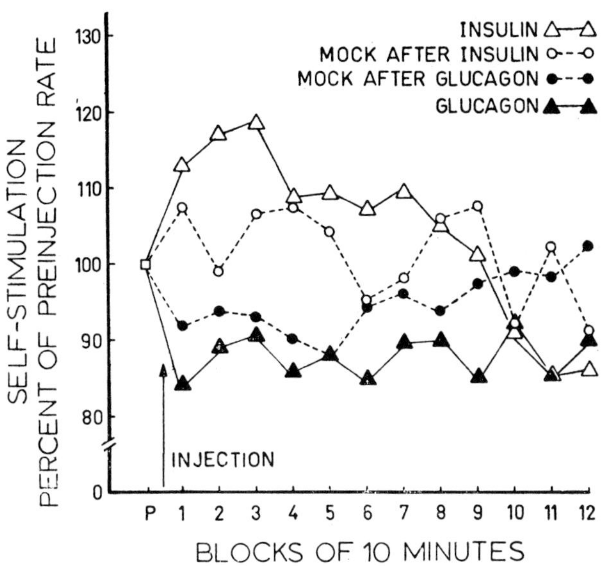

FIG. 1. Mean changes in self-stimulation rate. Explanation in text.

As can be seen in Fig. 1, if the rats were given a mock injection (which under normal conditions had no effect upon self-stimulation rate) after five consecutive sessions with either glucagon or insulin, the rate of self-stimulation was altered in such a way that it increased after insulin sessions and decreased when the mock injection followed glucagon sessions. Thus it appears that the rate of lateral hypothalamic self-stimulation was conditioned to the experimental procedural manipulations.

III. Conditional Glycemic Responses in the Regulation of Food Intake

To test the hypothesis that the conditional changes in self-stimulation rate were in part mediated by changes in glucose levels or glucose utilization, another series of experiments was performed. It was hypothesized that the pancreatic hormones served as the unconditional stimulus and that the injection procedure and experimental manipulation served as the conditional stimulus. In this paradigm, two types of conditional responses were predicted: (1) a conditional change in glycemia, and (2) a conditional change in behavior.

Animals were given a daily intraperitoneal injection of 0.25 mg of glucagon following a 2-hr deprivation period. After five consecutive sessions a series of isotonic saline injections were administered and blood samples were taken 20 min later. Adequate control groups were run to test for the unconditional response to glucagon, pseudoconditioning, and any cumulative effects (Balagura, 1968a). As can be seen in Fig. 2, it was possible to condition glucagon hyperglycemia of a similar magnitude to the hyperglycemia produced by glucagon itself. This hyperglycemic response disappeared after three extinction sessions. The increase in glycemia observed after saline injections during extinction was probably due to the stress produced in the bleeding procedure.

Since conditional insulin hypoglycemia had already been elegantly demonstrated by Alvarez-Buylla and Carrasco-Zanini (1960) and Segura (1962), and since it was possible to condition self-stimulation rate with pancreatic hormones, the next step was to condition food intake, using insulin as the unconditional stimulus.

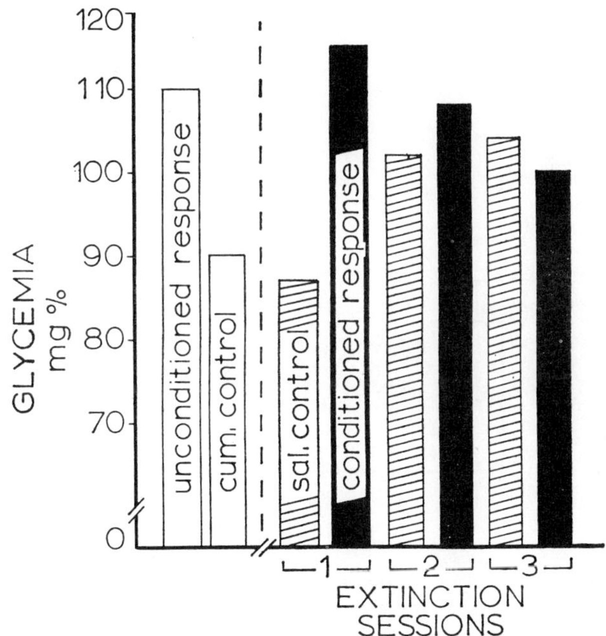

FIG. 2. Conditional glucagon hyperglycemia. Notice that no cumulative effect of glucagon was found. Three sessions were required to extinguish the conditional response.

Rats were trained to lever-press to obtain food (45 mg Noyes pellets). After 3 weeks of adaptation to the test chamber, the experiment was begun. The experimental subjects were given a daily intraperitoneal injection of 0.025 unit of crystalline insulin, and barpressing for food was measured during the following 120 min. As can be seen in Fig. 3, food intake increased with repetition of injections. When saline was administered instead of insulin the animals responded by maintaining an increased food intake. This effect disappeared at the sixth session. Animals given glucagon instead of saline following the insulin conditioning did not maintain a high level of intake but, rather, decreased their lever-pressing to the baseline level.

The results obtained from these experiments indicate that it is possible to condition the hormonal effects of glucagon and insulin. Furthermore, it appears that the conditional stimulus is contained in the manipulation procedure. These conclusions also suggest that

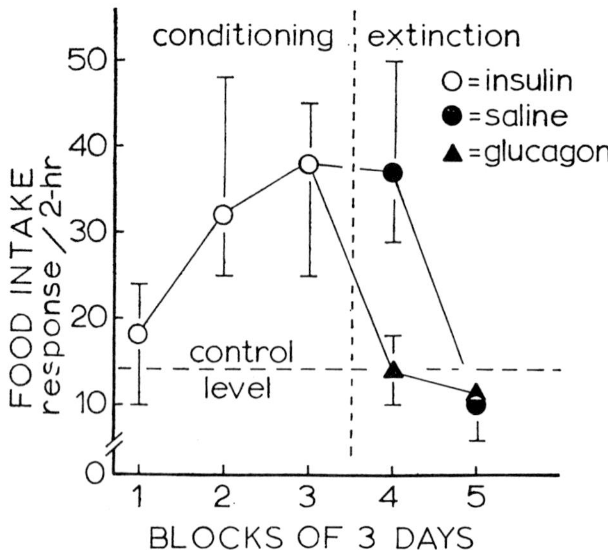

FIG. 3. Conditional insulin hyperphagia. Observe the positive slope during the first nine days of insulin treatment. Subsequently, injections of saline produced eating. This effect was abolished by injecting glucagon rather than saline.

there are two types of conditional responses: one, purely physiological, i.e., changes in blood glucose and, perhaps, tissue glucose utilization, and another, a behavioral response secondary to the physiological changes. These experiments strongly indicate that in many instances behavioral responses of the instrumental type can be modulated or mediated by physiological responses of the classical conditioning type.

IV. Effects of Glucose upon Lateral Hypothalamic Self-Stimulation

It is evident that some correlate of glucose may be playing an important role in the regulation of food intake and self-stimulation of the lateral hypothalamus. Glucose could act to satiate the animal by exerting an osmotic stress on the organism, thus inhibiting food intake (Smith, 1966; Smith and Duffy, 1957). On the other hand, the satiety mechanism that operates after glucose changes could be related to either metabolic or chemoreceptive actions of the hexose (Jacobs, 1964; Mayer, 1955). Both Smith and Jacobs measured the amount of food consumed following a stomach load of either hypertonic saline or glucose solutions.

The following experiment studied the effects of hypertonic glucose and saline solutions on the lateral hypothalamic self-stimulation rate. If both chemicals act solely through their osmotic effects, no qualitative differences should be obtained between the two groups.

Electrodes were implanted in the lateral hypothalamus of female rats. These animals were also prepared with chronic nasogastric tubes. Self-stimulation rate baselines were taken for 30 min prior to stomach loading, after which the subjects were loaded with 3 ml isotonic saline, or 2.0 Osm glucose, or 2.0 Osm saline solutions,

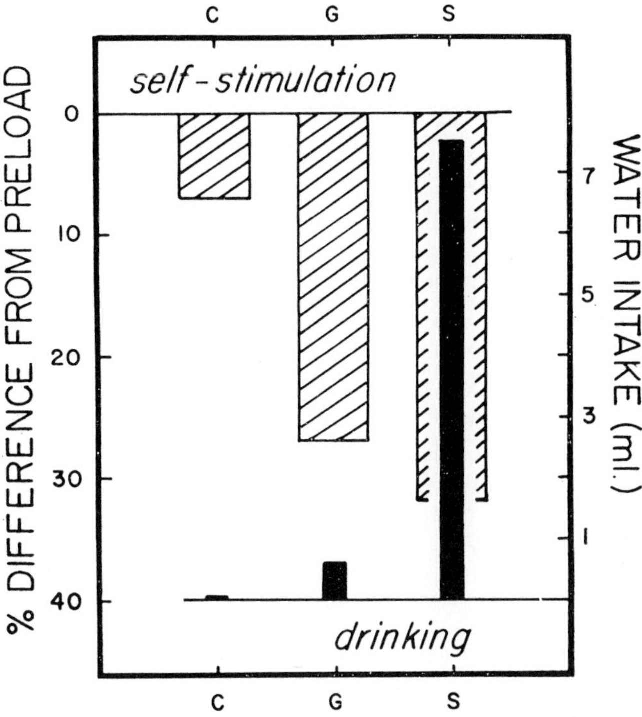

FIG. 4. Hypertonic solutions decreased the rate of self-stimulation, but only hypertonic saline increased drinking. C = 0.3 Osm saline, G = 2.0 Osm glucose, and S = 2.0 Osm saline.

and self-stimulation rates were recorded for the following 120 min. A water bottle was available to the animals at all times during testing.

Figure 4 depicts the results obtained. Both hypertonic glucose and saline depressed self-stimulation activity. Only hypertonic saline produced sufficient osmotic stress to induce drinking behavior. It was concluded that although both hypertonic saline and glucose solutions decrease the rate of self-stimulation, saline acts through changes in osmotic pressure whereas glucose operates primarily through its metabolic effects (Balagura, 1968b).

V. Effects of Intrahypothalamic Administration of Glucose Metabolic Inhibitors

Since 1955 several experimental reports have appeared which suggest the existence of a glucoreceptor system located in the VMH. The initial observations of Waxler and Brecher (1950) of an increased food intake and body weight after injections of gold thioglucose (GTG) were followed by the observations of Marshall *et al.* (1955), and Mayer and Marshall (1956) showing that GTG was selectively taken up by the VMH, producing both VMH lesions and subsequent obesity. Recent electrophysiological evidence (Anand *et al.*, 1964; Oomura *et al.*, 1964) suggests a reciprocal innervation between the VMH and the lateral hypothalamic nuclei. Furthermore, these authors obtained changes in firing rates secondary to blood glucose shifts. Systemic injections of insulin decreased the firing rate of VMH units, whereas hyperglycemia had opposite effects. Epstein and Teitelbaum (1967) have demonstrated that the hyperphagic action of insulin is not lost after destruction of the VMH but disappears after lesions in the lateral hypothalamus. Liebelt and his co-workers (Liebelt and Perry, 1957; Deter and Liebelt, 1964) have repeatedly pointed out that GTG produces lesions not only in the VMH but also in other parts of the brain and perhaps in other tissues outside the central nervous system.

In evaluating the validity of these data, two main points are called immediately to our attention: (1) there have been no published reports on actual biochemical manipulation of the hypothesized hypothalamic glucoreceptores, and (2) it has not been established that the glucoreceptors are located in the VMH.

Although the nature of a glucoreceptive mechanism is not known at the present time, such a mechanism (if it exists) could involve complex biochemical reactions in which some products of glucose oxidation would become available to certain parts of the neuron, producing a series of physicochemical changes that could result in modifications of neural firing rate. Such a possibility suggests the

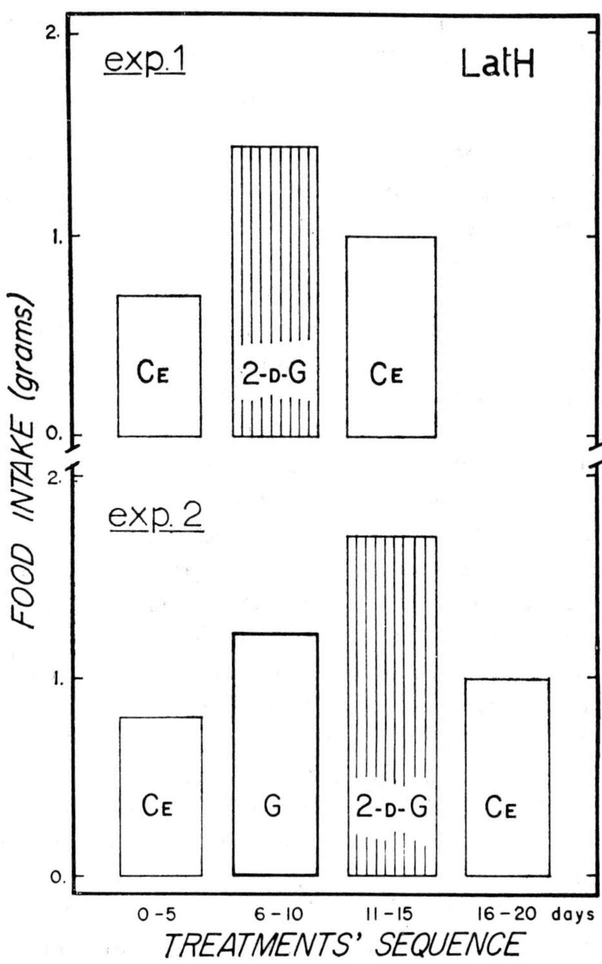

FIG. 5. Effects of D-glucose (G) and 2-deoxy-D-glucose (2DG) administration to the lateral hypothalamus (LatH) on food intake during a 30-min period. CE, control tests with empty cannulae.

importance of a biochemical-experimental approach in attempting a solution to the question of how hunger is metered by the organism. It is now possible to interfere with glucose metabolism at different stages during the oxidation of glucose. The compound 2-deoxy-d-glucose (2DG) is known to inhibit glucose metabolism. This effect may be due to a competition with glucose for the hexokinase substrate (Sols and Crane, 1954; Wick *et al.*, 1957). Furthermore, 2DG may contribute to a progressive failure of the hexokinase, owing to depletion of ATP (Tower, 1958). The antagonistic action of 2DG with regard to glucose metabolism has been behaviorally implicated by Smith and Epstein (1970), who report that 2DG injections are followed by increased feeding in rats and monkeys.

Deposition of 2DG in various areas of the brain by means of cannulae permits the experimenter to produce cellular glucopenia within a restricted area. Our initial experimental group consisted of five female albino rats with cannulae implanted bilaterally in the lateral hypothalamus. In the animals, manipulation of the cannulae without deposition of active chemicals yielded no special effects on food intake. When 2DG was administered (approximately 25 μg), a significant increase in food consumption during the following 30 min was observed. These findings are depicted in Fig. 5. We have replicated these results in another series of experiments with ten additional animals. Implantation of glucose appears to increase eating behavior, but not nearly as much as does 2DG. Currently we are running several experiments implanting glucose and 2DG in the VMH. So far we have only preliminary results suggesting that 2DG depresses food intake when given in the VMH, whereas glucose in the same nucleus increases or potentiates eating caused by food deprivation. It is not clear whether the glucose effects are due to a stimulatory action of the crystals or to a primary metabolic process.

It is hypothesized that, at least in the lateral hypothalamic area, there are two functionally different types of cells. One type pertains to a neural system in charge of controlling eating behavior (effector), and another type is composed of glucosensitive cells (receptor). The receptor cells modulate the activity of the effector cells. From our experiments it seems clear that there are glucoreceptors not only in the VMH but also in the lateral hypothalamus. Thus the glucosensitive area of the brain seems to be more diffuse than originally thought. Figure 6 depicts a theoretical model for glucose-hypothal-

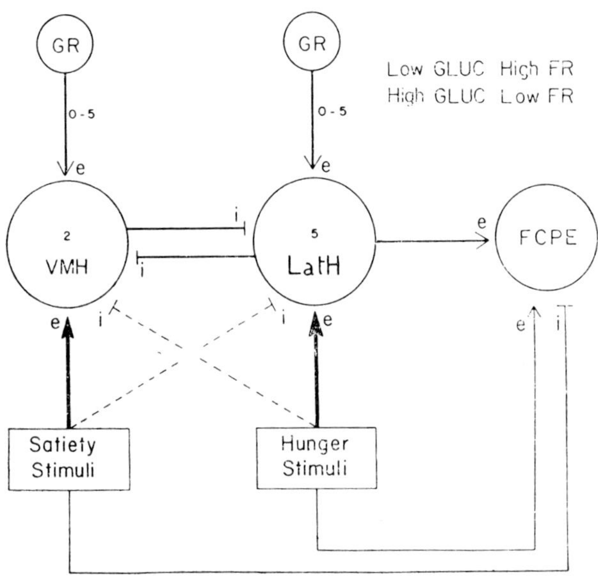

FIG. 6. Hypothetical model for glucose-hypothalamus interaction. It includes glucoreceptors in the lateral hypothalamus (LatH). FCPE, final common path for eating, a construct necessary to explain recuperation of food intake regulation following VMH or LatH lesions.

amic interaction. From our preliminary data we cannot make any predictions as to whether or not a similar functional cell dicotomy (receptor-effector) exists in the VMH. As shown in Fig. 6, the hypothalamic system may receive input from peripheral organs carrying information about physical distention, temperature, etc. It has been hypothesized that there exist peripheral glucoreceptors, mainly in the liver (Russek, 1963). This has been substantiated by the electrophysiological work of Niijima (1969) who found a decreased vagal firing rate following portal injections of glucose. Thus the diencephalon may be integrating information pertaining to both peripheral (hepatic) and central glucose utilization.

At least two other experimental reports provide some evidence for the existence of a glucoreceptor system located in the lateral hypothalamus. Epstein's findings (Epstein and Teitelbaum, 1967) of a loss of the hyperphagic action of insulin in the recovered animal lesioned in the lateral hypothalamus supports the notion of a glucoreceptive function being localized in this area. In addition, the elegant electrophysiological experiments of Ian Silver (personal communication) provide further evidence. Single unit recordings of activity in both VMH and lateral hypothalamic areas were made in rats and rabbits with hypothalamic islands. Immediately following the surgical isolation of the hypothalamus there is a sustained systemic hyperglycemia that can be controlled with injections of insulin or glucose. In general, it was found that the number of spontaneously active units was by far greater in the lateral hypothalamus than in the VMH. Of 300 active units in the lateral area, 180 responded to changes in glucose. Of these, 140 decreased firing rate when glucose increased, 20 increased firing rate when glucose was decreased, and 20 responded ambiguously. On the other hand, of 150 spontaneously active ventromedial units, only 50 responded to glucose changes. Increased glucose decreased firing rate in 12 units, increased firing rate in 20, and produced inconsistent responses in the remaining 10 units.

VI. Conclusions

In summary, a series of observations made in my laboratory seems to indicate that mammals possess a diencephalic glucose-detection system localized in both the ventromedial and the lateral hypothalamus. The neurohumoral system appears to respond not only to direct chemical stimulation but also to psychological factors. Thus the organism may respond with anticipatory neurohumoral changes for the purpose of attuning itself to both internal and external environments.

Acknowledgments

The author wishes to thank his students D. V. Coscina, M. Kanner, L. Devenport, and D. F. Smith for their collaboration and advice at different stages of this study. Part of this research was supported by NIH grants No. MH-14596-02 and MH-14308. A number of the experiments on 2-deoxy-D-glucose were performed in collaboration with Dr. S. P. Grossman.

References

Alvarez-Buylla, R., and Carrasco-Zanini, J. (1960). *Acta Physiol. Latinoam.* **10**, 153.
Anand, B. K., and Brobeck, J. R. (1951). *Yale J. Biol. Med.* **24**, 123.
Anand, B. K., and Dua, S. (1955). *Indian J. Med. Res.* **43**, 113.

Anand, B. K., Chhina, G., Sharma, K., Dua, S., and Singh, B. (1964). *Am. J. Physiol.* **207**, 1146.
Balagura, S. (1968a). *J. Comp. Physiol. Psychol.* **65**, 30.
Balagura, S. (1968b). *J. Comp. Physiol. Psychol.* **66**, 325.
Balagura, S., and Hoebel, B. G. (1967). *Physiol. Behav.* **2**, 337.
Brobeck, J. R. (1946). *Physiol. Rev.* **26**, 541.
Brobeck, J. R. (1957). *Yale J. Biol. Med.* **29**, 565.
Delgado, J. M. R., and Anand, B. K. (1953). *Am. J. Physiol.* **172**, 162.
Deter, R. L., and Liebelt, R. A. (1964). *Texas Rep. Biol. Med.* **22**, 229.
Epstein, A. N., and Teitelbaum, P. (1967). *Am. J. Physiol.* **213**, 1159.
Hetherington, A. W., and Ranson, S. W. (1939). *Proc. Soc. Exp. Biol. Med.* **41**, 465.
Hoebel, B. G., and Teitelbaum, P. (1962). *Science* **135**, 375.
Holloway, S., and Stevenson, J. A. F. (1964). *Can. J. Physiol. Pharmacol.* **42**, 867.
Jacobs, H. L. (1964). *J. Comp. Physiol. Psychol.* **57**, 309.
Kennedy, G. C. (1952). *Proc. Roy. Soc., Ser. B* **140**, 578.
Liebelt, R. A., and Perry, J. H. (1957). *Proc. Soc. Exp. Biol. Med.* **95**, 774.
Mackay, E. M., Callaway, J. W., and Barnes, R. H. (1940). *J. Nutr.* **20**, 59.
Margules, D. L., and Olds, J. (1962). *Science* **135**, 374.
Marshall, N. B., Barrnett, R. J., and Mayer, J. (1955). *Proc. Soc. Exp. Biol. Med.* **90**, 240.
Mayer, J. (1953). *Physiol. Rev.* **33**, 472.
Mayer, J. (1955). *Ann. N. Y. Acad. Sci.* **63**, 15.
Mayer, J., and Marshall, N. B. (1956). *Nature* **178**, 1399.
Niijima, A. (1969). *Ann. N. Y. Acad. Sci.* **157**, 690.
Olds, J., and Milner, P. (1954). *J. Comp. Physiol. Psychol.* **47**, 419.
Oomura, Y., Kimura, K., Ooyama, H., Maeno, T., Iki, M., and Kuniyoshi, M. (1964). *Science* **143**, 484.
Russek, M. (1963). *Nature* **197**, 79.
Segura, E. (1962). *Acta Physiol. Latinoam.* **12**, 342.
Smith, G. P., and Epstein, A. N. (1970). *Am. J. Physiol.* (In press).
Smith, M. H. (1966). *J. Comp. Physiol. Psychol.* **61**, 398.
Smith, M. H., and Duffy, M. (1957). *J. Comp. Physiol. Psychol.* **50**, 65.
Sols, A., and Crane, R. K. (1954). *J. Biol. Chem.* **210**, 581.
Strominger, J. L., and Brobeck, J. R. (1953). *Yale J. Biol. Med.* **25**, 383.
Tower, D. B. (1958). *J. Neurochem.* **3**, 185.
Unger, R., and Eisentraut, A. (1964). *Diabetes* **13**, 563.
Waxler, S., and Brecher, G. (1950). *Am. J. Physiol.* **162**, 428.
Wick, A. N., Drury, D. R., Nakada, H. I., and Wolfe, J. B. (1957). *J. Biol. Chem.* **224**, 963.
Wyrwicka, W., and Dobrzecka, C. (1960). *Science* **131**, 805.

The Renin-Angiotensin System in the Control of Drinking

J. T. FITZSIMONS

I. Introduction

The volume of the cells of the body is precisely controlled by mechanisms of intake and excretion. The action of these mechanisms is coordinated through the common variable to which both generally respond, namely, the effective osmotic pressure of the extracellular fluid. A rise in osmotic pressure brought about by loading the animal with certain hypertonic solutions, or by simply depriving the animal of water to drink, causes cellular dehydration (Gilman, 1937) which is detected by hypothalamic osmoreceptors for the release of Antidiuretic Hormone (ADH) (Verney, 1947) and similar receptors for thirst (Andersson, 1953; Andersson and McCann, 1955). The antidiuresis and drinking which ensue lead to restoration of the cells to their normal size, the progress of the restoration being continuously monitored by the two sets of osmoreceptors (Fig. 1).

The extracellular fluid volume is controlled with no less precision, a fact which has become generally known only comparatively recently, and its regulation (like that of the cellular volume) depends

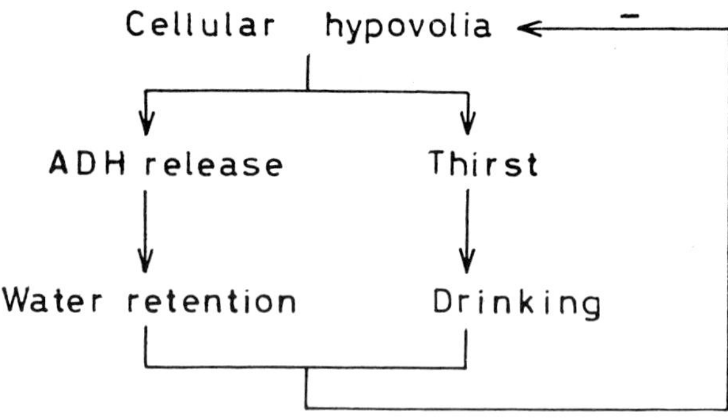

FIG. 1. Mechanisms for the control of cellular volume.

both on drinking and on renal mechanisms. However, very little is known about these mechanisms. On biological grounds it can be argued, of course, that mechanisms for extracellular regulation, and more particularly a thirst mechanism set in action by changes in the extracellular space, must exist. It is inconceivable that a space, which is less than half the volume of the cellular space and which contains the vasculature, through which the huge daily turnover of fluid with the external environment takes place, should be without at least several controlling mechanisms. Yet more than a hundred years after Claude Bernard developed that most seminal of all physiological ideas, the constancy of the internal environment, our understanding of the mechanisms which control the volume of that environment is fragmentary and confused.

II. Control of Extracellular Fluid

The reasons for the confusion about extracellular fluid volume control are several. It is exceedingly difficult to conceive of any single measurement that could give complete information on the volume of a space made up of the following subdivisions: a high-pressure arterial system; a low-pressure system of veins, pulmonary circulation and atria; and an extravascular component, the interstitial fluid. Nor is the task made simpler by the fact that within this space is a mobile phase, the circulating blood, with an elaborate system of controls the purpose of which is to maintain perfusion pressure to the tissues under widely differing conditions. A further complication is that the controlling mechanisms set in action are numerous both for drinking and for excretion. For example, changes in extracellular fluid volume evoke a delayed appetite for Na as well as an immediate thirst (for water).

A number of simplifying assumptions, however, may be made. In the first place, the receptive zone for detecting changes within the space must be located mainly in the vascular compartment, since this is the only part of the space that appears to be endowed with possible receptors. The vasculature also seems to be the likely candidate in view of the absolute necessity of maintaining a sufficient volume of circulating blood for survival of the animal. The interstitial space on the other hand has few if any receptors, but it is in communication with the vasculature through the capillary and lymphatic walls, and the partition of fluid between it and the vasculature is governed by the well-understood physical processes of ultrafiltration and osmosis. Control of the interstitial space could therefore depend on control of the vascular space.

Secondly, it is perfectly clear that the necessary function of the arterial circulation is to distribute blood to the various vascular networks at rates of flow appropriate to the needs of the tissues. This requires mechanisms to modulate cardiac output and vascular tone while at the same time maintaining the blood pressure within reasonable limits. It seems unlikely that the arterial baroreceptors, which are responsible for stabilizing the blood pressure in the face of widely differing hemodynamic situations, should also be the principal source of information on the state of the extracellular fluid volume, except in rather extreme circumstances when the circulation is actually embarrassed by unphysiological changes in extracellular volume. This, however, is not normally the case; extracellular volume is controlled with such precision that it is highly improbable that the normal day-to-day changes in it could affect the output of the heart and therefore alter the arterial baroreceptor discharge beyond its normal range.

Thirdly, the low-pressure side of the circulation contains about 80% of the blood and is 100 to 200 times as distensible as the arterial system (Gauer and Henry, 1963). Therefore moderate changes in extracellular fluid volume mainly affect the low-pressure system and it is here that the receptors for control of extracellular fluid must be sought. Indeed, as long as there is enough blood coming back to the heart there is no reason to suppose that cardiac output or arterial pressure is affected by any other than strictly hemodynamic factors. Moderate extracellular dehydration in no degree alters cardiac output or arterial baroreceptor discharge; it does, however, evoke the volume regulatory responses of thirst and oliguria.

From these arguments we may conclude, as did Gauer and Henry (1963), that the volume of extracellular fluid is measured by stretch receptors in the walls of the distensible low-pressure side of the circulation. It is also probable that only when the changes in volume reach emergency proportions which affect the circulation of blood

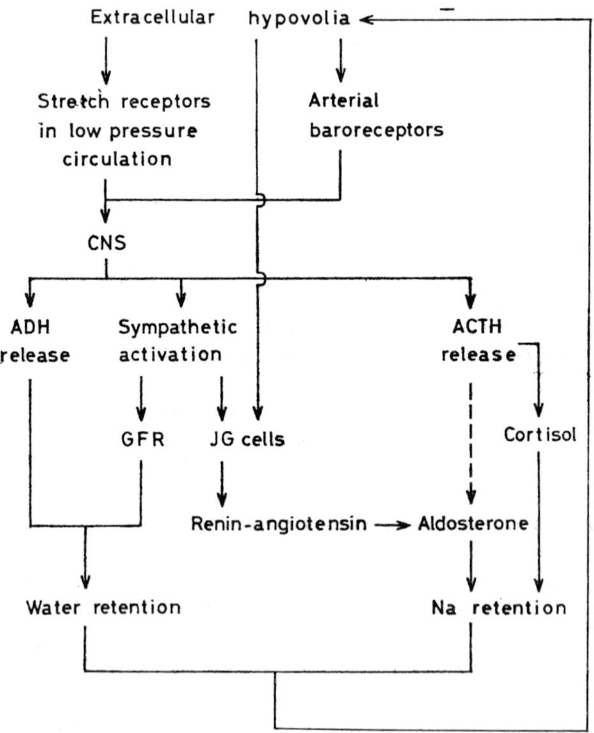

FIG. 2. Some of the renal mechanisms which may be involved in the conservation of water and Na in extracellular dehydration. No attempt has been made in the diagram to indicate the relative importance or physiological significance of the various pathways, nor have all possible mechanisms been included. ADH, antidiuretic hormone; ACTH, adrenocorticotropic hormone; GFR, glomerular filtration rate; JG, juxtaglomerular cells.

do the arterial baroreceptors become involved in the control of volume.

Most experiments on control of extracellular fluid volume relate to the renal aspects of such control; some of the possible mechanisms involving the kidney are summarized in Fig. 2. It must be emphasized that the kidney is primarily an organ concerned with surfeits and that there must be mechanisms of intake to deal with deficits; renal conservation can only slow the rate of dehydration and cannot lead to rehydration. This is not to say that the kidney does not influence intake mechanisms; there is now evidence that it does do this through the intermediary of the renin-angiotensin system.

III. Extracellular Thirst

The evidence for thirst originating in the extracellular compartment is ancient; the clinical association of thirst with hemorrhage, vomiting, or diarrhea has been known for a very long time; such thirst cannot be explained in terms of cellular dehydration, since the fluid lost is essentially isotonic and it is derived exclusively from the extracellular compartment. Wettendorff (1901), though mistaken in his interpretation of the cause of thirst in hemorrhage, was well aware of the association and described a case in which excessive thirst after severe hemorrhage was relieved by abundant injections of 0.8% saline. However, despite its older lineage, the thirst of extracellular dehydration has been the subject of less experimental effort than that of cellular dehydration. Indeed, most of the outstanding advances since Mayer's (1900) and Wettendorff's (1901) pioneering studies on the osmotic pressure of the blood after water privation refer to cellular dehydration. Notable among these studies was Gilman's (1937) demonstration that hypertonic saline, a cellular dehydrating agent, is a more effective stimulus of drinking than is hypertonic urea. Then Verney's (1947) work on the ADH osmoreceptor influenced Andersson (1953) and led to the demonstration that osmotic stimulation or electrical stimulation of the anteromedial hypothalamus in the goat causes drinking. The important studies by Holmes and Gregersen (1950a,b), by Kanter (1953), and by Adolph and his colleagues (1954) were all concerned with cellular dehydration as the stimulus to drinking.

Nevertheless, over the same period of time, a considerable body of evidence indicated that cellular dehydration alone could not account for many of the situations in which drinking occurs. Three types of experiment have established extracellular thirst as an entity distinct from the thirst of cellular dehydration.

A. Sodium Depletion

Removal of the principal extracellular cation together with anion results in a secondary but delayed loss of water from the extracellular compartment, often accompanied by cellular overhydration. The depletion may be brought about in various ways, but at one stage or another there is increased thirst which occurs quite independently of any increased appetite for Na. The means that have been used to produce the depletion are: administration of a salt-free diet in the rat (Swanson *et al.*, 1935; Radford, 1959); the combination of salt-free diet and excessive sweating in man (McCance, 1936); sucrose diuresis and salt-free diet in the dog (Holmes and Cizek, 1951); peritoneal dialysis with isotonic glucose in the dog (Cizek *et al.*, 1951), the rabbit (Huang, 1955) and the rat (Semple, 1952; Falk, 1961); and exteriorization of the salivary ducts in the sheep (Denton and Sabine, 1961).

B. Sequestration of Extracellular Fluid

The intraperitoneal (Fitzsimons, 1961) or subcutaneous (Stricker, 1966) injection of a solution with the same crystalloid composition as extracellular fluid, but with a higher colloid osmotic pressure, results in the simultaneous accumulation of both water and solute by a Starling mechanism at the site of injection, thereby inducing an acute isotonic depletion of the remainder of the extracellular space. The procedure differs from Na depletion in that it produces an immediate extracellular hypovolia instead of an early hypotonicity followed by a secondary and delayed hypovolia, as after Na depletion. Following sequestration there is an immediate thirst for water, and several hours later there is also an increased preference for Na; thirst and increased Na appetite are quite distinct and different responses to the extracellular hypovolia.

C. Hemorrhage and Direct Interference with the Circulation

Thirst has long been regarded as a clinical symptom of hemorrhage, though experimentally there has been disagreement on whether bleeding unaccompanied by any other change causes thirst. Holmes and Montgomery (1951, 1953) reported that human blood donors never complain of thirst. They also found that dogs bled up to 40% of their blood volume did not drink spontaneously, nor was the amount of water the dogs drank after hypertonic saline affected by an accompanying hemorrhage. On the other hand, hemorrhage has been found to cause drinking in the rat (Fitzsimons, 1961; Oatley, 1964) and to cause additional drinking when the hemorrhage is combined with other stimuli of thirst (Oatley, 1964; Fitzsimons and Oatley, 1968; Fitzsimons, 1969b).

Bleeding is an unsatisfactory way of inducing an extracellular fluid deficit because there is a risk of undesirable side effects from the accompanying acute anemia. The volume of circulating fluid is also restored rather rapidly from the interstitial fluid. In order to circumvent the effects of acute anemia and at the same time produce longer lasting hemodynamic changes, the techniques of caval ligation and aortic constriction were devised (Fitzsimons, 1969a) with the expectation that the receptors which underlie extracellular thirst would be activated by the ensuing hemodynamic changes.

IV. Caval Ligation, Constriction of the Aorta or Constriction of the Renal Arteries as a Stimulus to Drinking

Complete ligation of the inferior vena cava in the rat below the hepatic veins and either above or below the renal veins causes a sharp diminution in venous return to the heart, a fall in cardiac output, and a generalized fall in arterial pressure lasting for many hours. The procedure is well tolerated, ascites does not occur, and the growth of the animal and intake of food and water return to normal about one week after the ligation.

From ½ to 1 hr after caval ligation the rat starts to drink water and continues drinking for an hour or two after this (Table I). Animals allowed access to water and 1.8% saline always chose water in the immediate postoperative period, though by about the third postoperative day a considerable preference for saline developed. After ligation

TABLE I

Drinking and Excretion in the 6 Hr Which Followed Ligation of the Abdominal Inferior Vena Cava

Groups	Water Drunk (ml/100 g initial body wt)	Change in Body Wt. (g/100 g initial body wt)	Urine Volume (ml/100 g initial body wt)	Na (μM/100 g initial body wt)	K (μM/100 g initial body wt)
Sham abdominal operation (10)	0.43±0.08	—2.10±0.39	0.86±0.10	47±7.8	187±15.7
Caval ligation (17)	4.20±0.24	+2.77±0.63	0.24±0.08	16±4.7	17±8.5
Caval ligation after adrenalectomy (10)	4.25±0.60	+3.07±0.59	0.08±0.03	6±3.0	9± 6.9
Nephrectomy (15)	1.20±0.45	—0.25±0.20			
Caval ligation after nephrectomy (15)	1.60±0.25	+0.54±0.21			
Ureteric ligation (15)	1.12±0.30	+0.13±0.55			
Caval ligation after ureteric ligation (18)	3.70±0.46	+2.51±0.65			

In Tables I and II, the mean value ± SE of mean with the number of observations in parentheses is given.

above the renal veins there was a marked fall in urine flow and in the rate of electrolyte excretion; impairment of excretion was not so great when the caval obstruction was below the renal veins.

A consequence of drinking coupled with extreme oliguria was that the rat developed a substantial positive fluid balance with a fall, initially, in serum osmotic pressure and Na. The shortness of the interval between ligation and the commencement of drinking and the values for serum osmotic pressure and Na do not support a theory that drinking is a consequence of oliguria. Nevertheless the possibility was excluded by making animals anuric by submitting them to bilateral nephrectomy or to bilateral ureteric ligation (Table I).

Neither procedure alone caused increased drinking; therefore drinking cannot be attributed to failure of excretion. However, it became evident that the kidney has a role to play in drinking after caval ligation, since caval ligation in bilaterally nephrectomized rats was relatively ineffective as a stimulus to drinking, whereas caval ligation combined with bilateral ureteric ligation resulted in drinking.

TABLE II

DRINKING AND EXCRETION IN THE 6 HR WHICH FOLLOWED PARTIAL CONSTRICTION OF THE AORTA ABOVE THE RENAL ARTERIES, CONSTRICTION OF THE RENAL ARTERIES, INFRARENAL AORTIC CONSTRICTION, AORTIC CONSTRICTION AFTER NEPHRECTOMY OR SHAM OPERATION

	Water Drunk (ml/100 g initial body wt)	Change in Body Wt (g/100 g initial body wt)	Urine Volume (ml/100 g initial body wt)	Na (μM/100 g initial body wt)	K (μM/100 h initial body wt)
Aortic constriction above the renal arteries (25)	4.58±0.31	+2.60±0.96	0.53±0.14	13± 4.1	51±12.9
Constriction of the renal arteries (12)	2.18±0.45	+0.91±0.44	0.20±0.05	29± 8.0	20± 6.7
Aortic constriction below the renal arteries (10)	0.64±0.21	—3.19±0.26	2.24±0.31	142±22.7	180±21.2
Aortic constriction after nephrectomy (7)	1.04±0.22	—0.16±0.18			
Sham abdominal operation (10)	0.43±0.08	—2.10±0.39	0.86±0.10	47± 7.8	187±15.7

It now became of interest to determine to what extent the generalized circulatory changes contributed to drinking, since it appeared that alterations confined to the renal circulation might suffice. With this object in mind the abdominal aorta was partially constricted just above the level of the renal arteries; in another series of experiments, both renal arteries were partially constricted. These procedures had much the same effect on drinking and excretion as did caval ligation (Table II). On the other hand, partial constriction of the aorta below the level of the renal arteries or after bilateral nephrectomy had much less effect on drinking.

These experiments show that certain changes in the circulation without any initial alteration in overall fluid and electrolyte balance can cause drinking, an effect best seen in animals with kidneys perfused by their own circulation, though these kidneys need not necessarily be secreting urine.

V. A Renal Thirst Factor

The fact that certain circulatory manipulations are less effective stimuli of drinking after nephrectomy suggests that these procedures cause the kidneys to release a humoral thirst factor into the circulation. To test this hypothesis, simple saline extracts of renal cortex, renal medulla, and liver from the rat were given by intraperitoneal injection to normal and to bilaterally nephrectomized rats. Extracts from the renal cortex caused increased drinking in nephrectomized rats, confirming earlier observations by Nairn and co-workers (1956), and by Asscher and Anson (1963). Contrary to earlier reports the extracts also caused animals with intact kidneys to drink, but the effect was less than in nephrectomized rats.

The renal dipsogen, if not identical with renin, is closely similar to it for the following reasons:

1. Extracts having thirst and pressor actions are found only in the renal cortex.
2. It is impossible to separate dipsogenic and pressor activities of extracts of kidney during the different stages of the fractionation procedure of Haas and his associates (1954) which lead to the production of renin; disappearance of one activity is invariably accompanied by disappearance of the other.
3. Thirst and pressor actions are more marked in nephrectomized rats.
4. Both extractable dipsogenic factor and extractable pressor activity are reduced by treating the rat with saline and desoxycorticosterone acetate for several weeks beforehand.
5. Angiotensin causes rats in water balance to drink water.

VI. Systemic Administration of Angiotensin and Drinking

In view of the fact that most if not all known effects of renin are mediated through angiotensin (Peart, 1965), the action of angiotensin on drinking was tested. The experiments described in this section were performed in collaboration with Simons (Fitzsimons and Simons, 1969).

Angiotensin (val[5]-angiotensin-II-amide, Hypertensin CIBA) dissolved in 0.9% NaCl was infused through a jugular catheter into conscious, unrestrained rats at rates of 0.05 to 3.0 μg kg^{-1} min^{-1} for periods of 30 to 310 min. Control rats were infused with similar volumes of 0.9% NaCl. The animals were allowed to drink freely and continuous records of drinking were made during the experimental period of 6 hr.

Angiotensin caused rats in normal water balance to drink water, the effect being greater after nephrectomy (Fig. 3). It is therefore likely that the dipsogenic effect of renin is mediated through angiotensin. The mean amount of angiotensin needed to initiate drinking was 29.1 $\pm$ 4.6 μg/kg in 20 normal rats, and 15.7 $\pm$ 2.1 μg/kg in 34 nephrectomized rats. The start of drinking and the total amount of water drunk depended more on the amount of angiotensin infused than on the rate at which it was infused.

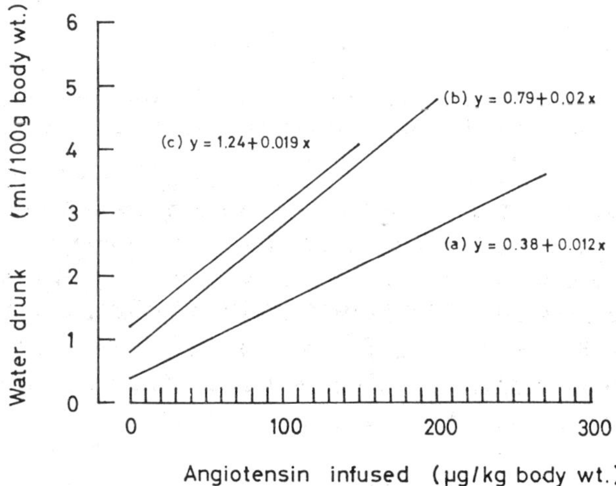

FIG. 3. The regressions of water drunk in 6 hr by (a) normal rats, (b) nephrectomized rats, and (c) nephrectomized-adrenalectomized rats on total amounts of angiotensin infused. Control animals received infusions of 0.9% NaCl. The right-hand limit of the regressions indicates the highest dose of angiotensin for the group in question. In each group, water drunk is positively and significantly correlated with the amount of angiotensin infused: (a) $r = 0.63$, $df = 25$, $p < 0.001$; (b) $r = 0.75$, $df = 42$, $p < 0.001$; and (c) $r = 0.45$, $df = 19$, $p < 0.05$. The regressions (b) and (c) do not significantly differ in slope and are not displaced from each other.

The adrenal cortex does not appear to be involved in the drinking of water caused by angiotensin; infusion of angiotensin into adrenalectomized-nephrectomized rats had the same effect on drinking as infusions into nephrectomized rats. Neither does adrenalectomy interfere with the drinking which follows caval ligation or injection of renin. It is, of course, possible that angiotensin may have an effect on Na appetite through aldosterone (though perhaps not in the rat), since aldosterone in extremely large doses has been shown to increase consumption of isotonic saline (Wolf and Handal, 1966).

There are several ways in which the renin-angiotensin system could cause drinking. First, the setting into action of so powerful a vasoactive system might alter the sensitivity of the vascular stretch receptors to an existing hypovolemia. Second, angiotensin might stimulate drinking by causing a fall in plasma volume (a known stimulus of thirst in the rat (Fitzsimons, 1961) through its action in increasing vascular permeability. It is, however, inherently improbable that a mechanism of this kind would operate within the physiological range of angiotensin secretion, and furthermore, measurements of hematocrit and plasma volume during the infusions lend no support to such a hypothesis (Fitzsimons and Simons, 1969). The third possibility, for which there is recent evidence (Booth, 1968; Epstein *et al.*, 1969) is that renin and/or angiotensin have a direct stimulating action on the hypothalamic drinking centers.

VII. Intracranial Angiotensin and Drinking

The experiments described in this section were carried out in collaboration with Epstein and Simons (Epstein *et al.*, 1969).

In order to investigate a direct action of angiotensin on nervous centers, stainless steel cannulae (OD = 0.57 mm) were chronically implanted in the hypothalamus and adjacent regions of the brain of male rats. Injections of angiotensin were made into conscious, unrestrained animals from a remote microsyringe through about 1 m of polythene tubing connected to a stainless steel injector (OD = 0.31 mm). The injector was placed inside and advanced to the tip of the implanted brain cannula. The animals were in normal water balance and had free access to water and generally to food as well while being tested.

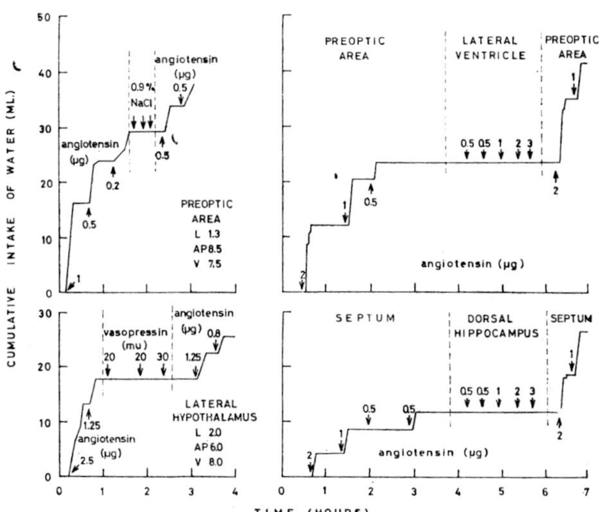

FIG. 4. The cumulative amounts of water drunk by rats given intracranial injections of angiotensin, vasopressin, and 0.9% saline. The stereotaxic coordinates used in implanting the cannulae in the two rats shown on the left are given on the figure; L, mm lateral to the center of the sagittal sinus; AP, mm anterior to the ear bars; V, mm down from the dural surface. Each of the two rats shown on the right had a pair of cannulae the positions of which are indicated on the figure.

Intracranial angiotensin (dissolved in 0.9% NaCl) in doses ranging from 5 ng to 2.5 µg per rat caused animals in normal water balance to drink water (Fig. 4). These doses were between 1/1000 and ½ the mean systemic dose needed to initiate drinking in the nephrectomized rat, and the smaller dose was about 5 pM. Drinking started after a latency of between 10 sec and a few minutes and continued (usually without interruption) for about 10 min. The amounts of water consumed varied from a few milliliters up to about 30 ml and the effects could be obtained on repeated occasions in the same animal with injections separated by intervals of 15 min. The animal was clearly motivated to drink; after the injection it stopped whatever it was doing, proceeded to the water and started drinking. If the water bottle was removed from its usual place in the cage the rat would search for it and lick at any spillage that might have occurred. The rat continued looking for water for at least 10 min after the injection, and if water was restored at the end of this time it started to drink avidly. A starving animal which had just been allowed to start eating, but which had taken only a few mouthfuls of food, stopped eating when injected with angiotensin and then started to drink after the usual latency. On one occasion a sleeping animal was awakened by the injection and started to drink. Intracranial angiotensin was a sufficiently powerful stimulus of drinking to make a rat, which was manually restrained, struggle and overcome its natural fear of man in order to reach the water.

Intracranial renin was also found to stimulate drinking, but the time course of drinking was much slower than after angiotensin. Intracranial vasopressin, adrenaline, 0.9% NaCl, 5% mannitol, thimerosal, and ammonium acetate (the last two substances are mixed with Hypertensin CIBA) had no effect on drinking. Noradrenaline occasionally caused drinking. Intracranial carbachol produced its usual stimulatory effect on drinking; molecule for molecule, angiotensin was at least as potent a dipsogenic substance as carbachol, and at some cannula placements the impression was that angiotensin was a considerably more potent dipsogen. Unlike carbachol, which occasionally produced intense excitement and activity in the rat, angiotensin seemed to be without side effects; its intracranial action appeared to be exclusively dipsogenic, and the drinking which resulted was normal in every respect. Preliminary results indicate that intracranial placements of angiotensin which may cause drinking do not have generalized effects on the circulation. Angiotensin-induced drinking is blocked either by systemic or by intracranial atropine; systemic atropine methyl nitrate was less effective as a blocking agent. The adrenergic blocking agent ethoxybutamoxane, also abolished angiotensin-induced drinking.

Forty-five cannula placements in thirty-five rats have now been studied, but histological verification is available on only about one-third of these placements. Using very large doses of angiotensin (2.0 μg), the area from which positive responses can be elicited is extensive and includes the nucleus accumbens, septum, preoptic area, anterior hypothalamic area, lateral hypothalamus as well as the amygdaloid and ventromedial nuclei. Negative sites were identified in animals with concurrently active positive sites and were negative to doses that exceeded the maximum dose used at the positive site. Negative sites were found in the cerebellum, tegmentum of the midbrain, dorsal hippocampus, posterior hypothalamus, caudate nucleus, lateral ventricle, and frontal cortex. When the amount of angiotensin was reduced to 1/10 or less of these very large doses, the region of maximum sensitivity was found to be confined to the septum and to the medial preoptic and anterior hypothalamic areas; the lateral hypothalamus was less responsive to the smaller doses and other placements yielded negligible drinking.

It may provisionally be concluded that the angiotensin sensitive region is in the medial preoptic area. It is therefore close to Andersson's (1953) osmosensitive region, but is anteromedial to the cholinergic sensitive region in the lateral hypothalamun.

VIII. The Role of the Renin-Angiotensin System in Drinking

Angiotensin is clearly a potent dipsogenic substance, since it causes a rat in normal or positive water balance to drink water. When angiotensin is placed directly into the medial preoptic region a few picomoles are enough to cause quite vigorous drinking. Normally, of course, an increase in endogenous angiotensin does not occur in isolation without some precipitating disturbance within the animal, and these same disturbances, which include hemorrhage, sodium depletion, caval obstruction, aortic constriction, and constriction of the renal arteries, have all been recognized as causes of thirst.

Since all the known extracellular causes of thirst augment secretion of renin, and since angiotensin in doses which are probably physiological causes drinking, it is likely that the renin-angiotensin system intervenes in normal thirst. This is supported by the fact that the effectiveness of an extracellular stimulus of thirst is reduced, though not abolished, by nephrectomy (Fitzsimons, 1969a,b; Blass and Fitzsimons, 1970). It is also noteworthy that exogenous angiotensin causes additional drinking in rats injected with hypertonic saline. Angiotensin therefore does not differ from a number of other extracellular stimuli of drinking which have been shown to add in simple algebraic fashion with cellular stimuli (Fitzsimons and Oatley, 1968). It is highly improbable therefore that endogenous angiotensin does not contribute to the total thirst experienced.

IX. Conclusions

It is now possible to fill in the outlines of a possible mechanism for extracellular thirst, though many of the details remain to be elucidated (Fig. 5). An extracellular deficit causes in the first instance

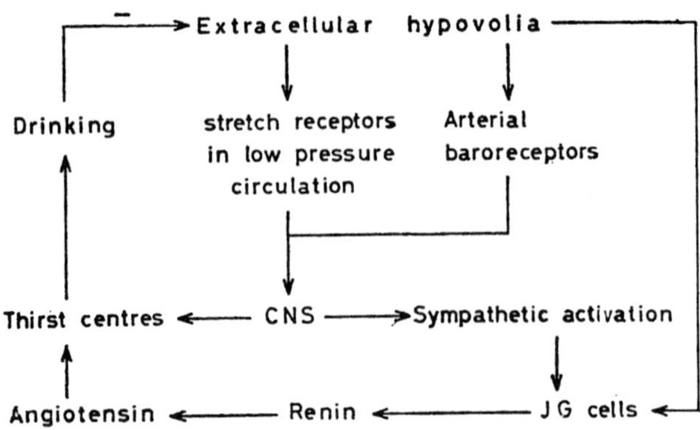

FIG. 5. Possible mechanism for drinking caused by extracellular dehydration. In addition to increased drinking of water, there is also a delayed increase in Na appetite, not indicated on the figure.

a change in sensory information from stretch receptors in the low-pressure side of the circulation. Many of these receptors are to be found in the atria and pulmonary circulation. More severe deficits which interfere with cardiac function also affect arterial baroreceptors, but this does not occur under normal conditions. The altered receptor information, much of it carried in the vagi, directly activates the medial preoptic region which in turn activates the lateral hypothalamus and various limbic structures to cause thirst and drinking. At the medial preoptic level this pathway is distinct from the osmosensitive region which appears to lie more laterally in the hypothalamus (Blass, 1968).

The pathway described operates in the nephrectomized animal. When the kidneys are present, however, the same afferent pathways lead to the reflex release of renin. The vagi and certain arterial baroreceptor nerves have been shown to mediate release of renin following hemorrhage (Hodge *et al.*, 1969), the efferent pathway of the reflex being the sympathetic nerves to the kidney (Vander and Luciano, 1967). A fall in arterial pressure within the kidney itself also causes release of renin which does not depend on the nerves to the kidney (Skinner *et al.*, 1964). It is likely that the threshold of this mechanism is higher than that of the reflex mechanism.

No matter what the mechanism of its release, renin causes increased generation of angiotensin, which then acts directly on the medial preoptic region of the hypothalamus. There it may have a stimulatory action on drinking in its own right, but more likely, it acts synergistically with the nervous information from various parts of the vasculature which caused release of renin in the first place. The renin-angiotensin system therefore enables the animal to meet the threat of extracellular dehydration in at least two ways: first by promoting Na retention by the kidney through the release of aldosterone; secondly, by sensitizing the extracellular thirst mechanism so that amounts of water appropriate to the degree of extracellular dehydration are drunk. The system therefore provides a link between control of output by the kidney and control of input by the the mechanism of thirst.

Acknowledgment

It is a pleasure to acknowledge the collaboration of Miss Barbara J. Simons in the systemic and intracranial angiotensin experiments and that of Dr. A. N. Epstein in the intracranial experiments.

References

Adolph, E. F., Barker, J. P., and Hoy, P. A. (1954). *Am. J. Physiol.* **178**, 538.
Andersson, B. (1953). *Acta Physiol. Scand.* **28**, 188.
Andersson, B., and McCann, S. M. (1955). *Acta Physiol. Scand.* **33**, 333.
Asscher, A. W., and Anson, S. G. (1963). *Nature* **198**, 1097.
Blass, E. M. (1968). *Science* **162**, 1501.
Blass, E. M., and Fitzsimons, J. T. (1970). *J. Comp. Physiol. Psychol.* **70**, 200.
Booth, D. A. (1968). *J. Pharmacol. Exp. Therap.* **160**, 336.
Cizek, L. J., Semple, R. E., Huang, K. C., and Gregerson, M. I. (1951). *Am. J. Physiol.* **164**, 415.
Denton, D. A., and Sabine, J. R. (1961). *J. Physiol.* (London) **157**, 97.
Epstein, A. N., Fitzsimons, J. T., and Simons, B. J. (1969). *J. Physiol.* (London) **200**, 98 P.
Falk, J. L. (1961). *In* "Nebraska Symposium on Motivation" (M. R. Jones, ed.) Univ. of Nebraska Press, Lincoln, pp. 1–33.
Fitzsimons, J. T. (1961). *J. Physiol.* (London) **159**, 297.
Fitzsimons, J. T. (1969a). *J. Physiol.* (London) **201**, 349.
Fitzsimons, J. T. (1969b) *J. Comp. Physiol. Psychol.* **68**, 308.
Fitzsimons, J. T., and Oatley, K. (1968). *J. Comp Physiol. Psychol.* **66**, 450.
Fitzsimons, J. T., and Simons, B. J. (1969). *J. Physiol.* (London). **203**, 45.
Gauer, O. H., and Henry, J. P. (1963). *Physiol. Rev.* **43**, 423.
Gilman, A. (1937). *Am. J. Physiol.* **120**, 323.
Haas, E., Lamfrom, H., and Goldblatt, H. (1954). *Arch. Biochem. Biophys.* **48**, 256.
Hodge, R. L., Lowe, R. D., Ng, K. K. F., and Vane, J. R. (1969). *Nature* **221**, 177.
Holmes, J. H., and Cizek, L. J. (1951). *Am. J. Physiol.* **164**, 407.

Holmes, J. H., and Gregersen, M. I. (1950a). *Am. J. Physiol.* **162**, 326.
Holmes, J. H., and Gregersen, M. I. (1950b). *Am. J. Physiol.* **162**, 338.
Holmes, J. H., and Montgomery, A. V. (1951). *Am. J. Physiol.* **167**, 796.
Holmes, J. H., and Montgomery, A. V. (1953). *Am. J. Med. Sci.* **225**, 281.
Huang, K. C. (1955). *Am. J. Physiol.* **181**, 609.
Kanter, G. S. (1953). *Am. J. Physiol.* **174**, 87.
Mayer, A. (1900). *C.R. Soc. Biol.* **52**, 153.
McCance, R. A. (1936). *Proc. Roy. Soc. Ser. B* **119**, 245.
Nairn, R. C., Masson, G. M. C., and Corcoran, A. C. (1956). *J. Pathol. Bacteriol.* **71**, 155.
Oatley, K. (1964). *Nature* **202**, 1341.
Peart, W. S. (1965). *Pharmacol. Rev.* **17**, 143.
Radford, E. P. (1959). *Am. J. Physiol.* **196**, 1098.
Semple, R. E. (1952). *Am. J. Physiol.* **168**, 55.
Skinner, S. L., McCubbin, J. W., and Page, I. H. (1964). *Circulation Res.* **15**, 64.
Stricker, E. M. (1966). *Am. J. Physiol.* **211**, 232.
Swanson, P. P., Timson, G. H., and Frazier, E. (1935). *J. Biol. Chem.* **109**, 729.
Vander, A. J., and Luciano, J. R. (1967). *Circulation Res. Supplement* II to **20-21**, II 169.
Verney, E. B. (1947). *Proc. Roy. Soc. Ser. B* **135**, 25.
Wettendorff, H. (1901). *Trav. Lab. Physiol. Inst. Solvay* **4**, 353.
Wolf, G., and Handal, P. J. (1966). *Endocrinology* **78**, 1120.

Monoamines of the Hypothalamus as Mediators of Temperature Response

W. FELDBERG

I. Introduction

In an admirable review written this year for the British Medical Bulletin, Cremer and Bligh (1969) summarized our present views about temperature regulation, and in the following paragraph I have made full use of his introductory pages.

We all know that in healthy man the body temperature remains relatively constant throughout life, and it is generally agreed that this constancy of temperature is dependent on the function of the anterior hypothalamus. Information is fed into this region of the brain by afferent nervous pathways from central and peripheral thermosensors. Yet in spite of the many nervous influences which act on the thermoregulatory center and would tend to raise or lower body temperature, it does not change. The center must therefore have a kind of " built-in information " about the level at which temperature is to be set and maintained. This level is called the " set-point " by those working in this field and this concept developed from the work of von Liebermeister (1875). But all the ingenious attempts to explain the working of the set-point by analogy with comparable physical control mechanisms are unable to hide our ignorance about the physiological mechanisms involved. I, too, am unable to explain the working of the set-point in physiological terms; analogies with models taken from physics are never satisfactory in physiology.

The experiments I am going to describe are concerned with the other problem, what Cremer and Bligh (1969) called the " information " fed into the anterior hypothalamus by afferent nervous pathways. The results of our experiments suggest that the nerve fibers which end on the cells of the anterior hypothalamus and thus feed information into this part of the brain are monoaminergic; that is, they release noradrenaline or 5-hydroxytryptamine (5-HT), and these amines by their action on the cells of the anterior hypothalamus either lower or raise body temperature. In my view, a full understanding of this problem, how changes in temperature are brought about, will ultimately lead also to an understanding of the intriguing problem of how changes in temperature are prevented and temperature is kept constant; that is, it will lead to an understanding of the working of the set-point.

To start at the beginning we must go back to 1954 when Vogt as well as Amin and his associates (1954) isolated the three monoamines from extracts prepared from the hypothalamus. The monoamines are natural constituents also of that part of the hypothalamus in which we are especially interested, the anterior hypothalamus.

To locate a substance in a particular structure after it has been found in a tissue extract always seems to me to be a tremendous step. This step was taken by our Swedish colleagues. In beautiful fluorescent microscopic studies they showed that the monoamines are present in neurons, probably as transmitter substances of monoaminergic neurons. In the hypothalamus they appear to be mainly located in nerve fibers and nerve endings but not in nerve cells (Carlsson *et al.*, 1962; Dahlström and Fuxe, 1964; Andén *et al.*, 1965).

The situation resembles that which pertains for acetylcholine in sympathetic ganglia. Here, too, the acetylcholine is located in preganglionic nerve endings and not in the ganglion cells. In the same way as the high acetylcholine content in sympathetic ganglia results from cholinergic neurons which make synaptic connections with the ganglion cells, so the high content of the monoamines in the hypothalamus signifies monoaminergic fibers which synapse in the hypothalamus.

And in the same way as the ganglion cells innervated by cholinergic neurons respond to acetylcholine (i.e., they are cholinoceptive), so the cells of the anterior hypothalamus, innervated by monoaminergic neurons, respond to the monoamines (i.e., they are monoaminoceptive). But here the similarity ends and a number of fundamental differences stand out which have to be kept in mind, particularly when trying to assess drug effects for the role of the monoamines in body temperature.

1. In a sympathetic ganglion all cells are innervated by nerve fibers of one kind which are cholinergic. On the other hand, the cells of the anterior hypothalamus appear to be innervated by two kinds of nerve fibers, the monoaminergic fibers that release 5-HT and the monoaminergic fibers that release noradrenaline. This poses the question whether each cell of the anterior hypothalamus is innervated by both fibers, the one exciting, the other inhibiting the cell, or whether the two types of fibers innervate different cells. In cats, for instance, 5-HT causes shivering when it acts on the anterior hypothalamus, whereas noradrenaline has the opposite effect and abolishes shivering. It is reasonable to assume that in this case the same cell is excited by 5-HT and inhibited by noradrenaline. The situation may, however, vary for different cells, some being innervated by both and others by one or the other of the two types of fibers.

2. Impulses originating from a sympathetic ganglion cell pass uninterrupted to the effector organ. On the other hand, those originating from the anterior hypothalamus traverse several central synapses, and some of the interneurons in the efferent pathway may be monoaminergic as well. Therefore temperature effects produced by drugs which interfere with synthesis, storage, or destruction of the amines need not necessarily act at the synapses in the anterior hypothalamus.

3. The situation is also more complex with regard to the systems activated or inhibited from this region of the brain, compared to the more limited effects elicited by sympathetic nerve stimulation. Apart from the non-shivering thermogenesis, a rise in temperature may result from increased muscle tone, shivering, and skin vasoconstriction, a fall from reduced muscle tone, skin vasodilatation,

TABLE I

AUTHORS WHO FOUND THE ENZYMES FOR BIOSYNTHESIS AND INACTIVATION OF THE MONOAMINES IN THE HYPOTHALAMUS AND ANTERIOR HYPOTHALAMUS

SYNTHESIS

5-Hydroxytryptophan decarboxylase
- Gaddum and Giarman, 1956
- Bogdanski, Weissbach, and Udenfriend, 1957, 1958
- Kuntzman, Shore, Bogdanski, and Brodie, 1961

Dopamine-β-hydroxylase
- Udenfriend and Creveling, 1959

N-Methyltransferase
- McGeer and McGeer, 1964

INACTIVATION

Monoamine oxidase
- Bogdanski, Weissbach, and Udenfriend, 1957
- Weiner, 1960

panting, and sweating. Since different cells of the anterior hypothalamus are probably involved in the various effects which lead to a rise as well as in those which lead to a fall in temperature, the possibility cannot be excluded that, for instance, excitation by 5-HT at one cell results in shivering and at another cell in skin vasodilatation, two effects which would produce opposite changes in temperature. The idea that one amine causes a rise and the other a fall in temperature, though justified in general terms, may therefore in some instances be a too simplified representation.

The experimental findings show that not only the monoamines are present in the hypothalamus, but also the enzymes for their biosynthesis and inactivation. Table I gives the names of the enzymes and of the authors who have found them in the hypothalamus and, in some instances, even in the anterior hypothalamus: the three synthesizing enzymes and the inactivating one, the monoamine oxidase (MAO). These findings provide additional evidence for the physiological function of the monoamines, since if they act as transmitter substances they must not only be present and released, but also synthesized and inactivated in the hypothalamus.

II. Temperature Responses Produced by the Monoamines

To study the temperature responses of the amines when acting on the anterior hypothalamus, they are injected either into the cerebral ventricles, or directly into the anterior hypothalamus by micro-injection when they are effective in smaller doses. Otherwise the effects are the same with both routes of administration. In cats, adrenaline and noradrenaline lower and 5-HT raises body temperature when acting on the anterior hypothalamus (Feldberg and Myers, 1963, 1964a, 1965a). When injected into the cerebral ventricles of unanesthetized cats the temperature response to 5-HT is usually biphasic. Vigorous shivering begins within a minute of the injection and results in a sharp rise in temperature, but later the skin vessels dilate while shivering continues. This leads to arrest of the rise or to a fall in temperature. The fall, which is also associated with tachypnoea though not with panting, varies greatly in different cats. Temperature may even fall below the preinjection level. This happens particularly when larger doses of 5-HT are injected (Kulkarni, 1967; Banerjee *et al.*, 1968a). There is also great variability in the extent of the secondary prolonged rise, not only in different cats but also in the same cat on repeated injections. We have not yet been able to find out what the conditions are that determine the extent of this rise (Feldberg *et al.*, unpublished observations).

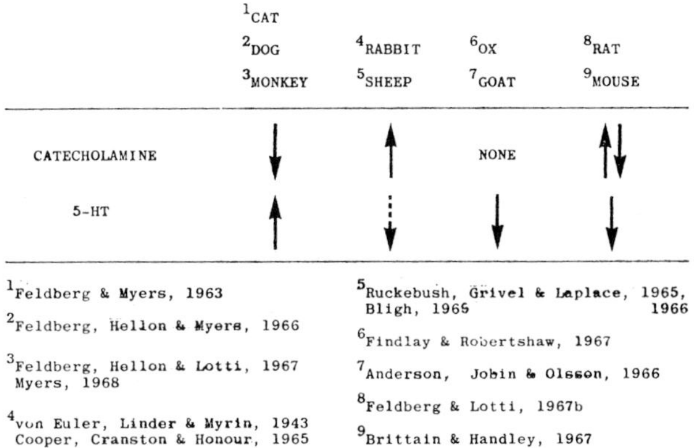

FIG. 1. Effect of catecholamines and of 5-hydroxytryptamine (5-HT), injected into the cerebral ventricles or directly into the anterior hypothalamus, on body temperature, in different species.

How simple the problem would have been if all animals responded to the monoamines in the same way as cats. But they do not. Figure 1 summarizes the temperature effects so far obtained in different species with the monoamines injected into the cerebral ventricles or directly into the anterior hypothalamus.

Dogs and *monkeys* responded like cats. The opposite effects were obtained in *rabbits* and *sheep*, a rise with the catecholamines and a fall with 5-HT, but the fall was usually small and not consistently obtained, as indicated by the dotted arrow in Fig. 1. Thus these species lack an efficient hypothermic monoamine in the hypothalamus. On the other hand, *goats* and *oxen* lack a hyperthermic monoamine in the hypothalamus because the catecholamines had no effect on temperature and 5-HT produced a deep fall in temperature. *Rats* and *mice* responded differently again. Although 5-HT produced a fall in temperature, the catecholamines were not inactive, as in goats and oxen; they, too, lowered body temperature except when injected in small, just threshold, doses which raised temperature.

What is the meaning of these species differences with regard to the physiological role of the monoamines? If these pharmacological effects mimic their central transmitter function in temperature regulation, it would follow that in some species these functions are mediated by the catecholamines and 5-HT, and in others mainly by the catecholamines, and again in others by 5-HT alone; and further, the same monoamine may be used in one species as a central transmitter for raising and in another for lowering temperature. Such great variability of transmitter functions has not been observed elsewhere. Whenever temperature responses differ in different species, this may be the explanation.

III. The Effect of Ambient Temperature on the Responses to the Monoamines

The results so far described were obtained at a neutral ambient temperature, i.e., at room temperature. Bligh, however, pointed out that the environmental temperature may influence the response to the monoamines. In rabbits and sheep, Bligh and Cottle (1969) found that at a low ambient temperature, noradrenaline no longer raised but rather lowered body temperature. To explain this reversal of the response they introduced a diagram of neural connections which allowed for reciprocal inhibition between the pathways for heat production and heat loss. I do not need to go into detail about this diagram because we have not been able to produce a reversal of the noradrenaline response in cats by changing the ambient temperature. The normal fall produced by intraventricular noradrenaline was found to be slightly accentuated at a low ambient temperature (about 10 °C) and greatly attenuated, but not reversed, at a high ambient temperature (about 30 °C) (Feldberg and Hellon, unpublished observations). The attenuation is readily explained by the fact that a high ambient temperature brings about the same changes (dilatation of the skin vessels and reduction in muscle tone) that would otherwise be produced by the noradrenaline when given at neutral or low ambient temperature.

IV. Acetylcholine and Body Temperature

So far we have considered the monoamines only. However, acetylcholine may play a role as well; in fact, Burn and Dutta (1948) suggested that acetylcholine might be involved in the maintenance of body temperature. They had found that atropine and other substances which depress the action of acetylcholine will lower body temperature in mice. This would suggest a temperature-raising effect of acetylcholine. Recently the problem has been taken up by various workers, but the results are not quite uniform. In rats, Lomax and Jenden (1966) obtained a fall in temperature with carbachol given by microinjection into the anterior hypothalamic preoptic area whereas Myers and Yaksh (1968) obtained a rise when they injected acetylcholine or physostigmine, or both together, into the cerebral ventricles. Later, Myers and Sharpe (1968a) obtained the same effect in monkeys, and Bligh and Maskrey (1969) obtained a temperature rise in sheep with intraventricular injections of acetylcholine, physostigmine, and carbachol; the effect was the same at low and high ambient temperature. Myers has explained to me the difference in the results on rats as follows: in the experiments of Lomax and Jenden (1966) the rats had free access to drinking water *ad libitum*. As carbachol injected into the preoptic area or adjacent regions such as the diagonal band of Broca causes the rat to drink large amounts (10 to 40 ml/hr) of water (Fisher and Coury, 1960), the hypothermia may well have resulted from the intake of large volumes of cold water, which is known to lower body temperature.

If this interpretation is correct it would appear that acetylcholine has a hyperthermic action in all animals and does not show the species differences obtained with the monoamines. This would make it unlikely that acetylcholine acts at the same synapse as the monoamines. Rather it would suggest an action on diencephalic cholinergic synapses along the efferent pathway which originates in the anterior preoptic region and activates the heat-producing mechanisms. This view was also suggested by Myers and Yaksh (1969) as the outcome of their results obtained with microinjections of acetylcholine into different parts of the hypothalamus in monkeys. When the injections were made into the anterior preoptic region the hyperthermic effect of acetylcholine was weak, but it was strong when the injections were made into the posterior hypothalamus. Thus acetylcholine appears to act not on the anterior hypothalamus but more caudally at diencephalic synapses in the efferent pathway of the heat-producing mechanisms. We can assume the cells on which the acetylcholine acts to be innervated by cholinergic neurons, which would make them cholinoceptive. The cholinergic neurons either originate in the cell bodies of the anterior preoptic region innervated by the monoaminergic fibers, or they are interneurons.

V. Evidence for the Release of 5-Hydroxytryptamine

In order to establish our theory we must prove that the monoamines are released in association with changes in temperature. So far, only a beginning has been made in this direction, and that only with respect to the release of 5-HT. The reason is that 5-HT can be detected with biological assay methods, for instance, on the rat stomach strip (Vane, 1957) in much smaller amounts than the catecholamines. When the cerebral ventricles of anesthetized cats were perfused with artificial cerebrospinal fluid (CSF), 5-HT was detected in the effluent (Feldberg and Myers, 1966), but the amounts were small, which is not surprising. Surprising is that it could be detected at all because the monoamine oxidase in the hypothalamus will act on the released 5-HT and at once destroy it, similar to the way in which acetylcholine is destroyed by cholinesterase.

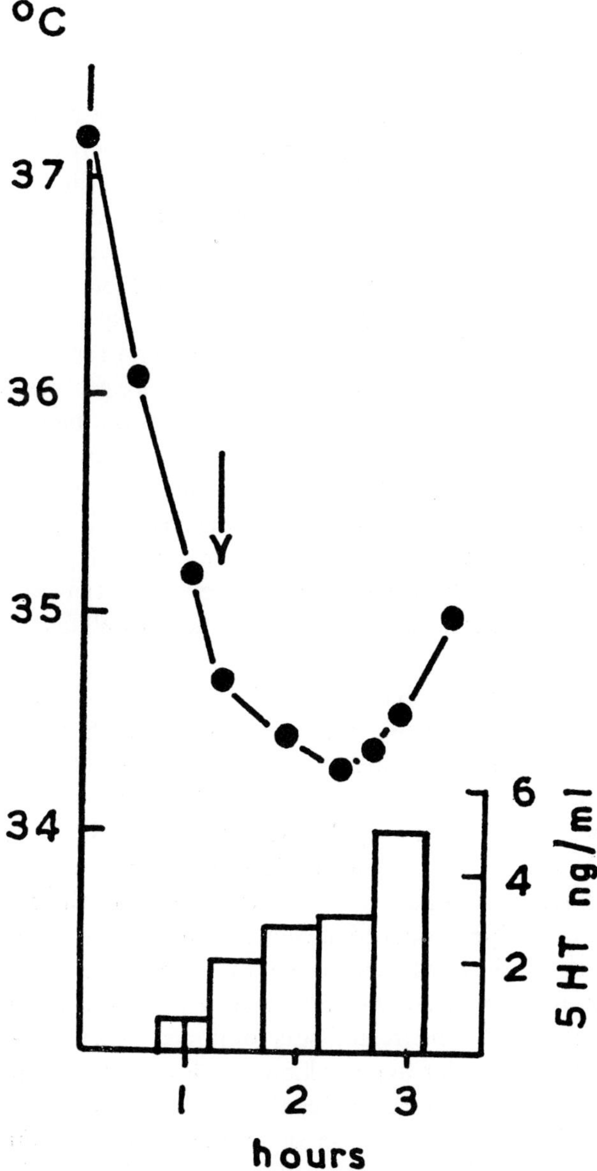

FIG. 2. Record of rectal temperature from a cat anesthetized with intravenous chloralose. Perfusion of third ventricle with artificial CSF at a rate of 0.05 ml/min. Collection of effluent began ½ hr before intraperitoneal injection, indicated by arrow, of 5 mg/kg tranylcypromine. Block diagram, output of 5-HT in μg/ml effluent (From Feldberg and Lotti, 1967a).

To demonstrate the release of acetylcholine from cholinergic nerves its destruction is prevented by inhibiting the cholinesterase with inhibitors like physostigmine. When the same procedure was adopted for 5-HT and the monoamine oxidase was inhibited by the inhibitor tranylcypromine, the 5-HT output in the effluent increased and body temperature rose. Such an experiment is shown in Fig. 2. The temperature record is from a cat anesthetized with chloralose, which produces a fall in temperature. The third ventricle is perfused with artificial CSF and the effluent is assayed for 5-HT on the rat stomach strip. At the time indicated by the arrow, 5 mg/kg tranylcypromine is injected intraperitoneally. As it is absorbed and reaches the brain, the 5-HT released in the hypothalamus is no longer destroyed and its output in the effluent increases, as shown in the block diagrams. Temperature first falls less steeply and then rises as the cat begins to shiver. Without the injection, the 5-HT output would have remained low throughout the experiment and the steep fall in temperature would have continued for some time. In principle, the result of this experiment provides the same kind of evidence that was obtained 30 years ago for the transmitter function of acetylcholine in cholinergic neurons.

Another most elegant piece of evidence was provided by Myers and his co-workers (Myers, 1967a,b; Myers and Sharpe, 1968a,b; Myers *et al.*, 1969) in experiments on unanesthetized monkeys. These experiments can be described as a humoral transmission of temperature effects. What they did was to collect CSF from the third ventricle of one monkey and transfer it to the third ventricle of another monkey. When the donor monkey was made to shiver by being packed in ice its CSF produced shivering and a rise in temperature of the recipient monkey, whereas when the donor was kept in a hot environmental temperature its CSF produced a fall in temperature of the recipient. The same result was obtained when they perfused the region of the anterior hypothalamus with Gaddum's "push-pull" cannula and transferred the effluent from the "pull" cannula to the anterior hypothalamic region of another monkey. Further, the ventricular CSF, as well as the effluent from the "pull" cannula obtained from a cooled monkey, contained 5-HT. However, the amounts of 5-HT detected in these fluids were of the order of nanograms, whereas micrograms are required to produce a rise in temperature on injection into the ventricle or into the anterior hypothalamic region. This thousandfold, or at least hundredfold, difference in magnitude is as yet unexplained; it suggests that 5-HT may not be the sole cause of the rise in temperature produced in the recipient monkeys.

VI. The Hypothermia of Anesthesia

In most animals temperature falls during anesthesia. We all know how important it is to keep the patient warm during anesthesia. What is the explanation of the fall in the light of the new concept? Feldberg and Myers (1964b) suggested that anesthetics act on the anterior hypothalamus and cause there an increased release of all three monoamines, but temperature falls because the action of the released catecholamines predominates.

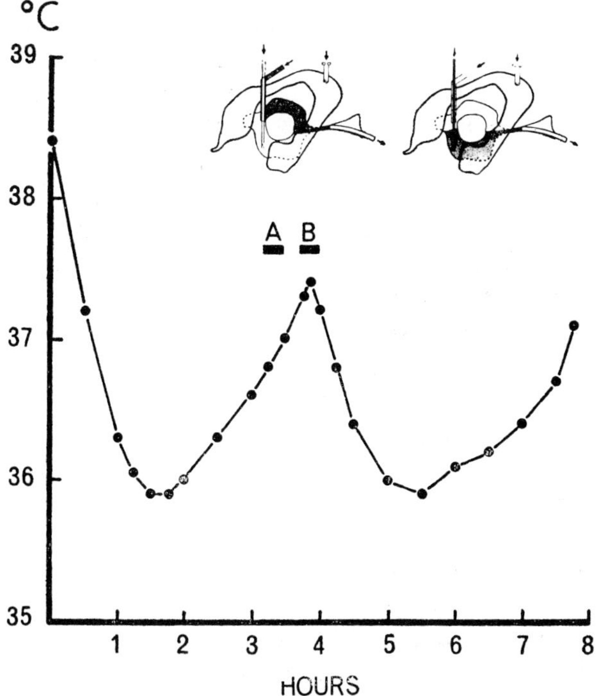

FIG. 3. Record of rectal temperature from a cat anesthetized with intraperitoneal pentobarbitone sodium during perfusion of the cerebral ventricles. Record begins a few minutes after beginning of anesthesia. The horizontal bars indicate perfusion of chloralose 1/2500 for 20 min through the dorsal (A) and ventral (B) part of the third ventricle, with the methods as diagrammatically represented by the inset (From Feldberg and Myers, 1965b).

First we showed that the anesthetics act on the hypothalamus. This has always been assumed but was never proved. Perhaps our best proof was obtained in experiments in which we perfused the anesthetic through the ventral half of the third ventricle, the walls of which contain the hypothalamic nuclei. A double-bore cannula was inserted into the third ventricle in such a way that the opening of the outer tube lay dorsal, that of the inner tube ventral to the massa intermedia which divides the third ventricle. Therefore, fluid delivered throught the outer tube perfused the dorsal, and fluid delivered through the inner tube the ventral half of the third ventricle, and by using one of the tubes for delivering the anesthetic it would pass either only through the dorsal or only through the ventral half of the third ventricle. The effect on temperature of chloralose perfused in this way is shown in Fig. 3.

The cat was anesthetized with pentobarbitone sodium and all parts of the cerebral ventricles were first perfused with artificial CSF. The anesthesia produced a fall in temperature, but as anesthesia lightened, temperature began to rise. Chloralose was then perfused for 15 min through the dorsal half of the third ventricle as indicated by the bar *A*, but temperature continued to rise. But when the chloralose was subsequently perfused through the ventral half, at the bar *B*, also for 15 min, temperature began to fall within a few minutes and had fallen 1.5 °C after 1 hr.

Second we showed that the anterior hypothalamus retained its sensitivity to the amines in anesthesia. This had to be shown because otherwise the theory would not work. Feldberg and Myers (1964b) found that an intraventricular injection of adrenaline prolonged the fall in temperature during anesthesia, whereas an intraventricular injection of 5-HT had the opposite effect. The return of temperature was accelerated.

The *third* piece of evidence, was that the fall did not occur when the anterior hypothalamus had been rendered insensitive to the monoamines by ergotamine. The dotted curve in Fig. 4 shows the typical fall in temperature of an intraperitoneal pentobarbitone sodium anesthesia. The other curve obtained in the same cat shows the effect of the same dose of pentobarbitone sodium three weeks later, but this time it was given in the middle of the experiment, at the arrow *P*, and preceded at the arrow *E* by an intraventricular injection of 100 μg ergotamine. The cat became anesthetized but temperature

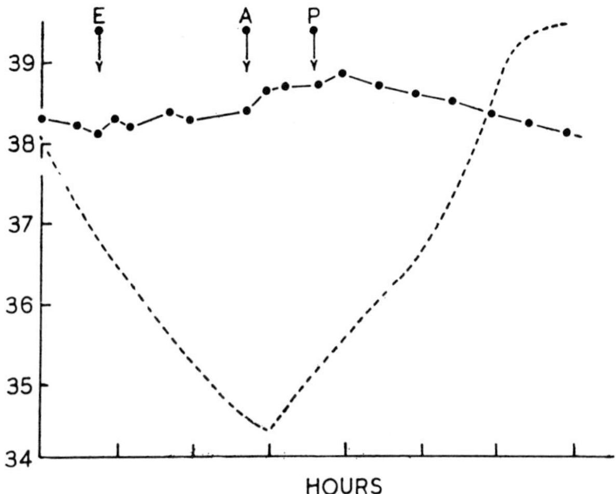

Fig. 4. Records of rectal temperature from the same cat. Dotted curve: intraperitoneal pentobarbitone sodium (33 mg/kg) given 15 min before beginning of record. Curve with solid circles: arrows indicate injections into the cannulated left lateral ventricle of 100 μg ergotamine (E), and of 100 μg adrenaline (A), and intraperitoneal injection of 33 mg/kg pentobarbitone sodium (P) (From Banerjee et al., 1968c).

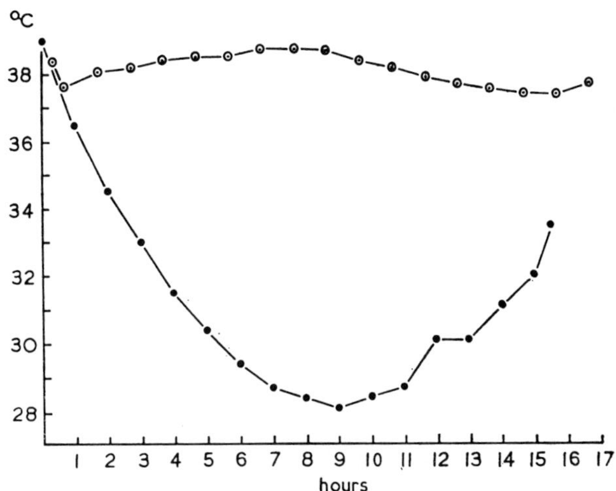

Fig. 5. Records of rectal temperature from the same cat. Both curves begin about 10 min after an intravenous injection of chloralose. Curve with solid circles: chloralose alone. Curve with open circles: 10 mg/kg tranylcypromine injected intraperitoneally 15 min before chloralose (From Feldberg and Lotti, 1967a).

did not fall because the ergotamine had rendered the anterior hypothalamus insensitive to the monoamines. As shown at the arrow *A*, 100 μg adrenaline injected intraventricularly no longer lowered temperature.

The *fourth* piece of evidence was obtained with inhibitors of the monoamine oxidase. In cats, only 5-HT and not noradrenaline appears to be a substrate of the brain monoamine oxidase because inhibitors of the enzyme cause an increase of the 5-HT but not of the noradrenaline content in the brain (Vogt, 1959; Brodie *et al.*, 1959; Funderburk *et al.*, 1962). On the assumption that anesthetics release both noradrenaline and 5-HT but that the temperature-lowering effect of the released noradrenaline normally predominates, one would expect that when the enzyme is inhibited and the released 5-HT is no longer destroyed, its action would be strong enough to counteract the temperature-lowering effect of the released noradrenaline, and therefore temperature would no longer fall in anesthesia. As shown

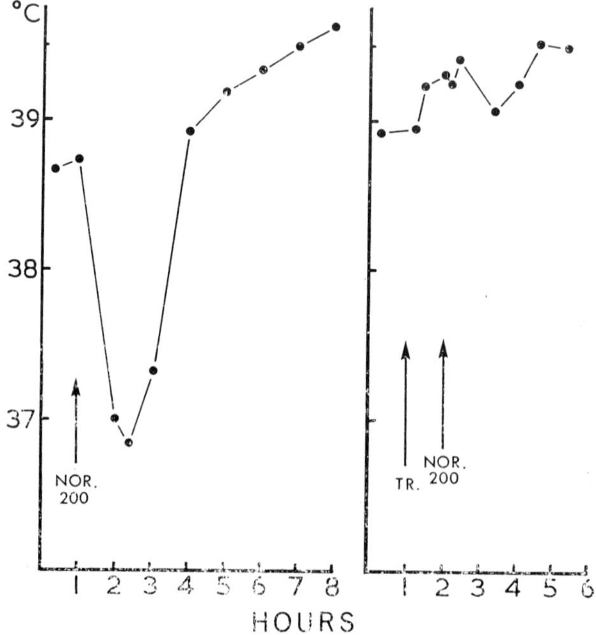

FIG. 6. Records of rectal temperature from a cat. Arrows indicate injection into the cannulated left lateral ventricle of 200 μg noradrenaline (NOR 200) and intraperitoneal injection of 5 mg/kg tranylcypromine (Tr) (From Feldberg and Lang, 1970).

in Fig. 5, this was actually found. The curve with the full circles shows the typical fall of chloralose anesthesia. The fall did not occur in the other curve because 10 mg/kg tranylcypromine had been injected intraperitoneally a few minutes before the chloralose was given. Yet again the cat became fully anesthetized. The same result was obtained with monoamine oxidase inhibitors other than tranylcypromine and when pentobarbitone sodium or the volatile anesthetic halothane was used for anesthetizing the cat (Summers, 1969).

Although the expected result was obtained, the original assumption turned out to be incorrect because intraperitoneal tranylcypromine was found to render the anterior hypothalamus insensitive to noradrenaline, which would fully explain the result. Figure 6 shows the normal hypothermic effect, a fall of about 2 °C, in response to an intraventricular injection of 200 μg noradrenaline, and the small hypothermic effect obtained in the same cat on another day when the noradrenaline injection was preceded by an intraperitoneal injection of 10 mg/kg tranylcypromine.

The insensitivity to noradrenaline may be brought about by the abnormal amounts of undestroyed 5-HT accumulating at the synapses in the anterior hypothalamus after MAO inhibition, but we do not really know how the MAO inhibitors act when preventing the hypothermia of anesthetics. Feldberg and Lang (1970) found that the inhibitors injected intraperitoneally were as effective in preventing the halothane hypothermia as when injected into the cerebral ventricles. This means that even with intraventricular injection they must first be absorbed into the blood stream before they can act. We forget all too easily how well substances are absorbed from the liquor space, nearly half as rapidly as from the subcutaneous tissue (Bhawe, 1958; Draskoci *et al.*, 1960; Feldberg, 1963). After absorption the inhibitors probably act on the anterior hypothalamus and on other parts of the central nervous system as well, and even peripherally. It is thus not proved, though it may well be that the prevention of the hypothermia is due to an action of the inhibitors on the anterior hypothalamus.

Neither can we be sure that the effect actually results from inhibition of monoamine oxidase. Some of the inhibitors used have strong amphetamine-like actions, and it was found (Feldberg and Lang, 1970) that amphetamine in a dose by itself too small (1 mg in a 2½ to 3 kg cat) to affect body temperature would prevent the halothane hypothermia. Again, amphetamine was as effective when given intraperitoneally as when given intraventricularly. However the hypothermia was also prevented by nialamide and pargyline, two inhibitors which have no amphetamine-like actions; thus, in these instances at least, MAO inhibition would appear to be responsible for the effect.

Another observation is worth mentioning in this connection. These inhibitors when given intraperitoneally to cats sometimes produce a condition in which halothane inhalation provokes a lethal fever, either during or immediately after discontinuation of the inhalation (Summers, 1969). This recalls the rare lethal pyrexias described in man during or after an anesthesia. They occur particularly in young people and children and mostly with volatile fluorinated hydrocarbons such as halothane or methoxyfluorane (for references see Summers, 1969). It is tempting to attribute these pyrexias to abnormal release of 5-HT in the anterior hypothalamus, triggered off in man in rare instances by halothane alone and in cats by halothane after MAO inhibition.

As far as the hypothermia of anesthesia is concerned there is one more piece of evidence for the theory that it results from noradrenaline release. Rabbits and sheep lack an efficient hypothermic monoamine in the anterior hypothalamus, and in these animals anesthesia can be produced without an appreciable effect on temperature (Ruckebusch *et al.*, 1965; Feldberg and Lotti, 1967a).

Although all observations described can thus be explained on the assumption that anesthetics release noradrenaline, each of them can also be explained differently, for instance, that anesthetics have a noradrenaline-like action on the anterior hypothalamus. Therefore the theory is anything but proved.

VII. Temperature Responses to Intraventricular Injections of Reserpine

For a number of drugs which affect body temperature, the same problem arises as for the hypothermia of anesthesia, namely, the question of whether the temperature effects result from monoamine release. I shall discuss this problem for one drug, reserpine, which is known to deplete the monoamines. Figure 7 shows that the temper-

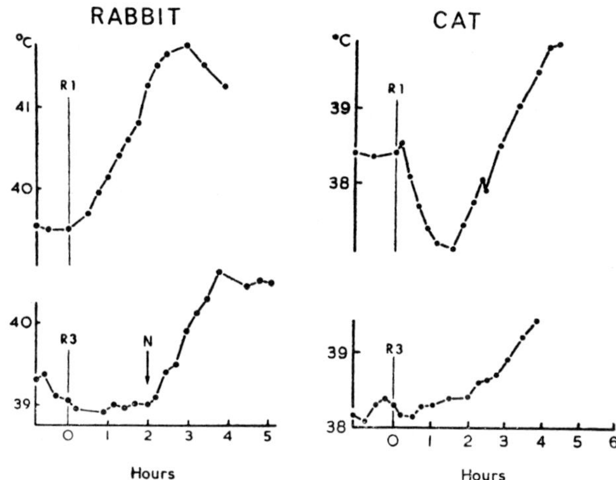

FIG. 7. Records of rectal temperature from a rabbit and a cat. Effect of repeated injections of reserpine (0.5 mg in rabbit and 0.75 mg in cat) at 24-hr intervals into the cannulated left lateral ventricle. R1 indicates first injection; R3, third injection. Arrow (N) indicates intraventricular injection of 200 μg noradrenaline (From Banerjee *et al.*, 1968b).

ature effects produced by reserpine injected into the cerebral ventricles of rabbits and cats are readily accounted for in this way; R 1 shows the temperature rise in a rabbit produced by 0.5 mg reserpine. This effect was first described by Cooper *et al.* (1967) who further showed that such an intraventricular injection depletes the catecholamines in the hypothalamus. The rise no longer occurred with the second or third injection of the same dose of reserpine given at 24-hr intervals. This is shown at R 3 for the third injection.

The rise of the first injection is explained as an effect of released noradrenaline. It no longer occurs with the third injection because noradrenaline is no longer available for release. The adrenergic fibers have been depleted by the preceding injections. This conclusion is supported by the finding that the anterior hypothalamus has retained its sensitivity to noradrenaline; as indicated by the arrow *N* in Fig. 7, 200 μg injected intraventricularly produced its usual hyperthermic response.

The right side of Fig. 7 gives the result obtained on a cat. The first injection of 0.75 mg reserpine, at R 1, produced a biphasic change in temperature, an initial fall followed by a rise. When the same dose of reserpine was injected for a second or third time at 24-hr intervals it no longer produced the initial fall but only a late rise. This is shown for the third injection at R 3. The initial fall of the first injection is explained in the same way as the rise in rabbits, i.e., by release of noradrenaline which lowers temperature in cats. The fall no longer occurs with the third injection because the adrenergic fibers have been depleted of their noradrenaline.

Reserpine also depletes 5-HT, yet the rise did not disappear with repeated injections. This does not necessarily mean, however, that the rise is not due to 5-HT release because intraventricular reserpine has less effect on the 5-HT than on the noradrenaline content of the hypothalamus. Sharman and Vogt (personal communication) determined its monoamine content in cats which had been given two intraventricular injections of 0.5 mg reserpine in the same way as in the experiment of Fig. 7. The noradrenaline was reduced to a mean of 6%, but the 5-HT reduced to between 16 and 43% (mean 27%) only.

VIII. Fever Produced by Pyrogen

Finally a few words about the fever of infectious diseases, the rise in temperature produced by pyrogen. There is good evidence that pyrogen acts on the anterior hypothalamus (Villablanca and Myers, 1965; Adler and Joy, 1965; Cooper *et al.*, 1967), but I think it has a direct action, for if the pyrogen were to act through release of monoamines its action would have to be different in different species according to whether it is noradrenaline or 5-HT that raises temperature when acting on the anterior hypothalamus. The real difficulty, however, arises in species which, like ox and goat, lack a hyperthermic amine in the hypothalamus through which the pyrogen can act. In fact, Cooper and associates (1967) have shown that pyrogen still produces its hyperthermic response in rabbits when the hypothalamus has been depleted of its catecholamines by intraventricular reserpine.

But the story does not end here. Although reserpine does not prevent pyrogen fever when injected intraventricularly, it does so when injected intravenously (Going, 1959; Kroneberg and Kurbjuweit, 1959), but it must be the depletion of the 5-HT which is responsible for this effect, not that of the catecholamines, because their depletion alone, when brought about by alpha-methylparatyrosine, did not abolish pyrogen fever. This was shown by Des Prez, and associates (1966).

But how to explain pyrogen fever in rabbits by release of 5-HT, which in this species lowers temperature when acting on the anterior hypothalamus? The explanation may be that 5-HT has an additional transmitter function at interneurons in the efferent pathway for the temperature-raising mechanisms which originate in the anterior hypothalamus. When these interneurons are activated and release their 5-HT, temperature rises; when they are depleted (and for this to happen with reserpine it has to be injected intravenously), pyrogen no longer raises temperature. Such a transmitter function of 5-HT at interneurons would also account for a number of other hypothermic responses in rabbits (see Feldberg, 1968).

References

Adler, R. D., and Joy, R. J. T. (1965). *Proc. Soc. Exp. Biol. Med.* **119**, 660.

Amin, A. N., Crawford, T. B. B., and Gaddum, J. H. (1954). *J. Physiol.* (London) **126**, 596.

Andén, N. E., Dahlström, A., Fuxe, K., and Larsson, K. (1965). *Life Sci.* (Oxford) **4**, 1275.

Andersson, B., Jobin, M., and Olsson, K. (1966). *Acta Physiol. Scand.* **67**, 50.

Banerjee, U., Burks, T. F., and Feldberg, W. (1968a). *J. Physiol.* (London) **195**, 245.

Banerjee, U., Burks, T. F., Feldberg, W., and Goodrich, C. A. (1968b). *J. Physiol.* (London) **197**, 221.

Banerjee, U., Feldberg, W., and Lotti, V. J. (1968c). *Brit. J. Pharmacol.* **32**, 523.

Bhawe, W. B. (1958). *J. Physiol.* (London) **149**, 169.

Bligh, J., and Cottle, W. H. (1969). *Experientia* **25**, 608.

Bligh, J., and Maskrey, M. (1969) *J. Physiol.* (London) **203**, 55 P.

Bogdanski, D. F., Weissbach, H., and Udenfriend, S. (1957). *J. Neurochem.* **1**, 272.

Bogdanski, D. F., Weissbach, H., and Udenfriend, S. (1958). *J. Pharmacol. Exp. Therap.* **172**, 182.

Brittain, R. T., and Handley, S. L. (1967). *J. Physiol.* (London) **192**, 805.

Brodie, B. B., Spector, S., and Shore, P. A. (1959). *Ann. N. Y. Acad. Sci.* **80**, 609.

Burn, J. H., and Dutta, N. K. (1948). *Nature* **161**, 18.

Carlsson, A., Falck, B., and Hillarp, N. (1962). *Acta Physiol. Scand.* **56**, *Suppl.* 196.

Cooper, K. E., Cranston, W. I., and Honour, A. J. (1965). *J. Physiol.* (London) **181**, 852.

Cooper, K. E., Cranston, W. I., and Honour, A. J. (1967). *J. Physiol.* (London) **191**, 325.

Cremer, J. E., and Bligh, J. (1969). *Brit. Med. Bull.* **25**, 299.

Dahlström, A., and Fuxe, K. (1964). *Acta Physiol. Scand.* **62**, *Suppl.* 232.

Des Prez, R., Helman, R., and Oates, J. A. (1966). *Proc. Soc. Exp. Biol. Med.* **122**, 746.

Draskoci, M., Feldberg, W., Fleischhauer, K., and Haranath, P. S. R. K. (1960). *J. Physiol.* (London) **150**, 50.
Feldberg, W. (1963). " A Pharmacological Approach to the Brain from its Inner and Outer Surface." E. Arnold, London.
Feldberg, W. (1968). *In* " Recent Advances in Pharmacology " (J. M. Robson and R. S. Stacey, eds.). 4th ed. Churchill, London, pp. 349–397.
Feldberg, W., and Lang, W. J. (1970). *Brit. J. Pharmacol.* **38**, 181.
Feldberg, W., and Lotti, V. J. (1967a). *J. Physiol.* (London) **190**, 203.
Feldberg, W., and Lotti, V. J. (1967b). *Brit. J. Pharmacol.* **31**, 152.
Feldberg, W., and Myers, R. D. (1963). *Nature* **200**, 1325.
Feldberg, W., and Myers, R. D. (1964a). *J. Physiol.* (London) **173**, 226.
Feldberg, W., and Myers, R. D. (1964b). *J. Physiol.* (London) **175**, 464.
Feldberg, W., and Myers, R. D. (1965a). *J. Physiol.* (London) **177**, 239.
Feldberg, W., and Myers, R. D. (1965b). *J. Physiol.* (London) **179**, 509.
Feldberg, W., and Myers, R. D. (1966). *J. Physiol.* (London) **184**, 837.
Feldberg, W., Hellon, R. F., and Myers, R. D. (1966). *J. Physiol.* (London) **186**, 413.
Feldberg, W., Hellon, R. F., and Lotti, V. J. (1967). *J. Physiol.* (London) **191**, 501.
Findlay, J. D., and Robertshaw, D. (1967). *J. Physiol.* (London) **189**, 329.
Fisher, A. E., and Coury, J. N. (1960). *Science* **132**, 301.
Funderburk, W. H., Finger, K. F., Drakontides, A. B., and Schneider, J. A. (1962). *Ann. N. Y. Acad. Sci.* **96**, 289.
Gaddum, J. H., and Giarman, N. J. (1956). *Brit. J. Pharmacol.* **11**, 88.
Going, H. (1959). *Arzneimittel Forsch.* **2**, 793.
Kroneberg, G., and Kurbjuweit, H. G. (1959). *Arzneimittel Forsch.* **2**, 536.
Kulkarni, A. S. (1967). *Int. J. Neuropharmacol.* **6**, 325.
Kuntzman, R., Shore, P. A., Bogdanski, D. and Brodie, B. B. (1961). *J. Neurochem.* **6**, 226.
Lomax, P., and Jenden, D. J. (1966). *Int. J. Neuropharmacol.* **5**, 353.
McGeer, P. L., and McGeer, E. G. (1964). *Biochem. Biophys. Res. Commun.* **17**, 502.
Myers, R. D. (1967a). *J. Physiol.* (London) **188**, 50.
Myers, R. D. (1967b). *Physiol. Behav.* **2**, 373.
Myers, R. D. (1968). *In* " Advances in Pharmacology " (S. Garattini and P. A. Shore, eds.). Vol. 6, Part A. Academic Press, New York, pp. 318–321.
Myers, R. D., and Sharpe, L. G. (1968a). *Science* **161**, 572.
Myers, R. D., and Sharpe, L. G. (1968b). *Physiol. Behav.* **3**, 987.
Myers, R. D., and Yaksh, T. L. (1968). *Physiol. Behav.* **3**, 917.
Myers, R. D., and Yaksh, T. L. (1969). *J. Physiol.* (London) **202**, 483.
Myers, R. D., Kawa, A., and Beleslin, D. (1969). *Experientia* **25**, 705.
Ruckebusch, Y., Grivel, M. L., and Laplace, J. P. (1965). *C.R. Soc. Biol.* **159**, 1748.
Ruckebusch, Y., Grivel, M. L., and Laplace, J. P. (1966). *Therapie* **21**, 483.
Summers, R. J. (1969). *Brit. J. Pharmacol.* **37**, 400.
Udenfriend, S., and Creveling, C. R. (1959). *J. Neurochem.* **4**, 350.
Vane, J. R. (1957). *Brit. J. Pharmacol.* **12**, 344.
Villablanca, J., and Myers, R. D. (1965). *Am. J. Physiol.* **208**, 703.
Vogt, M. (1954). *J. Physiol.* (London) **123**, 451.
Vogt, M. (1959). *Pharmacol. Rev.* **11**, 483.
Von Euler, U. S., Lindner, E., and Myrin, S. O. (1943). *Acta Physiol. Scand.* **5**, 85.
Von Liebermeister, C. (1875). Handbuch der Pathologie und Therapie des Fiebers. F.C.W. Vogel, Leipzig.
Weiner, N. (1960). *J. Neurochem.* **6**, 79.

Control of the Cardiovascular System

A. ZANCHETTI

I. Introduction

In reviewing the hypothalamic control of the cardiovascular system there is the risk of making a bothersome list of the cardiovascular effects elicited by electrical stimulation of the hypothalamus. There is no doubt that this technique has been instrumental and perhaps essential in showing the influence of the hypothalamus on circulation and the wide range of this influence. The pioneering contributions of Karplus and Kreidl (1909) at the beginning of the twentieth century, of Ranson and Magoun (1939), and of Hess (1949) can be said to have opened a new field of knowledge.

However, as Ingram aptly stated (1960) in an earlier review of central autonomic mechanisms, " there is danger in thinking of the hypothalamus as a distinct and isolated structural entity. . . . It has direct and indirect nervous connections with many other parts of the brain, including the cerebral cortex, and from the functional viewpoint it probably should not be considered as separate from other diencephalic and lower brain stem mechanisms. . . . The so-called autonomic phenomena with which the hypothalamus seems especially concerned are also related to other parts of the forebrain, as well as the midbrain and hindbrain... The hypothalamus, then, fits into large schemes, some of which are not exclusively autonomic and which involve certain patterns of behavior." The effects of the hypothalamus on cardiovascular function should be viewed as a part of definite behavioral responses in which the hypothalamus plays a major role. The cardiovascular effects of topical hypothalamic stimulation (mostly in animals under anesthesia) should be taken simply as examples (and probably fragmentary ones) of what the hypothalamus is able to do to heart and blood vessels. It should also be kept in mind that these cardiovascular effects are induced in concert with other central structures that are probably involved with the hypothalamus in some rather elaborate regulatory systems. A most difficult though essential task is the demonstration of these complicated mechanisms and functions, which can provide a *raison d'être* for otherwise meaningless cardiovascular signs produced by electrical stimulation of the hypothalamus.

II. Cardiovascular Patterns Elicited by Stimulation of the Hypothalamus

I am not going to review in detail all the circulatory effects that can be elicited by electrical stimulation of various areas of the hypothalamus. I shall rather concentrate on a few patterns of associated cardiovascular signs that are obtained by selective excitation of different areas and for which behavioral correlates have been suggested.

A. Defense Pattern

From the perifornical and the medial forebrain bundle region in the lateral hypothalamus and from parts of the midbrain tegmentum and central gray matter a pattern of cardiovascular changes can be obtained which has been carefully studied by the Uvnäs' and Folkow's groups in Sweden. This pattern consists of arterial pressure increase (Rosén, 1961), sympathetically induced cardiac stimulation with increase in cardiac output (Folkow *et al.*, 1968), heart rate (Rosén, 1961; Folkow *et al.*, 1968), stroke volume (Folkow *et al.*, 1968), contractile force of the heart (Rosén, 1961), and sympathetically mediated vasoconstriction in visceral beds, such as the renal (Feigl *et al.*, 1964) and the intestinal ones (Eliasson *et al.*, 1951; Cobbold *et al.*, 1964), and in the skin (Eliasson *et al.*, 1951). These signs of diffuse sympathetic activation involving also the adrenal medulla (Grant *et al.*, 1958) and intestinal motility (Folkow and Rubinstein, 1965) are associated with selective dilatation of muscle blood vessels (Eliasson *et al.*, 1951; Folkow and Rubinstein, 1965), a response brought about through the so-called cholinergic sympathetic fibers (Uvnäs, 1960). Since Abrahams *et al.* (1960) have shown that stimulation of these hypothalamic and mesencephalic areas in the unanesthetized animal produces coordinated defense reactions, the cardiovascular pattern obtained from stimulation of the same areas under anesthesia has been commonly thought to represent the cardiovascular concomitants of defense behavior, or more broadly, of emotional behavior (Folkow *et al.*, 1965; Abrahams *et al.*, 1964). These views, though current in the physiological and psychological literature, should be approached with some caution. Bolme *et al.* (1967) have recently shown by electrical stimulation of unanesthetized animals that the so-called defense pattern of cardiovascular changes can be evoked at a much lower threshold than behavioral responses and that evoked behavior generally consists in signs of awareness rather than of rage or defense. Recent data on circulation adjustments during natural defense behavior (see below) shed further doubt on the behavioral significance of this hypothalamic cardiovascular pattern.

B. Feeding Pattern

By electrical stimulation of a region in the lateral hypothalamus more dorsally located than the so-called defense area, feeding responses can be induced in the unanesthetized cat. In addition to a vagally-mediated increase in intestinal motility, during anesthesia the same stimuli evoke a moderate increase in arterial pressure, heart rate, and intestinal blood flow, while muscle blood flow is moderately reduced (Folkow and Rubinstein, 1965). This pattern is worth mentioning because of the marked difference in local redistribution of blood flow from the pattern of cardiovascular changes evoked by stimulating the closely localized defense area.

C. Sympatho-Inhibitory Pattern

Folkow's group has shown that electrical stimulation of a strictly localized area in the anterior hypothalamus markedly inhibits the discharge of the sympathetic fibers to the entire cardiovascular system (Folkow *et al.*, 1959; Löfving, 1961; Folkow *et al.*, 1964a). There is a considerable fall in blood pressure associated with dilatation in most vascular areas that have been explored (muscle, skin, intestine, and kidney); both resistance and capacitance vessels are involved. There is also bradycardia, independent of the vagi. This hypothalamic area seems to participate in a more complex sympatho-inhibitory system originating from the rostral parts of the limbic system (anterior cingulate gyrus) and extending to a sympatho-inhibitory area in the medulla (Löfving, 1961). It has been suggested that this system may be responsible for behavioral phenomena or at least for the cardiovascular correlates of behavioral phenomena such as emotional fainting in man and the "playing dead" reactions occurring in some animal species (Löfving, 1961).

III. Mechanisms Regulating the Cardiovascular Functions of the Hypothalamus

The functions of the hypothalamus (and among them those concerning the cardiovascular system) are finely modulated by forebrain structures, particularly those belonging to the limbic system of which the hypothalamus itself is often considered an integral part. There is a vast amount of literature on this subject, which has been recently reviewed (Zanchetti, 1967). I shall deal here with ascending mechanisms of hypothalamic regulation, either those that are reflex in nature or those involving the interrelations between hypothalamus and the lower brain stem. Our group has devoted a long series of investigations to this problem by concentrating on the regulatory mechanisms of the sham rage behavior of the acute thalamic cat. This behavior pattern undoubtedly involves hypothalamic structures and is characterized by striking visceral changes: increase in arterial pressure, piloerection, etc.

The results of our studies are briefly summarized in Fig. 1. First of all, the cardiovascular functions of the hypothalamus (and more broadly, the behavioral patterns to which they belong) are under the reflex control of the cardiovascular receptors which also influence circulatory homeostasis at the medullary level. At both the hypothalamic and medullary levels the sinoaortic baroceptors play an inhibitory

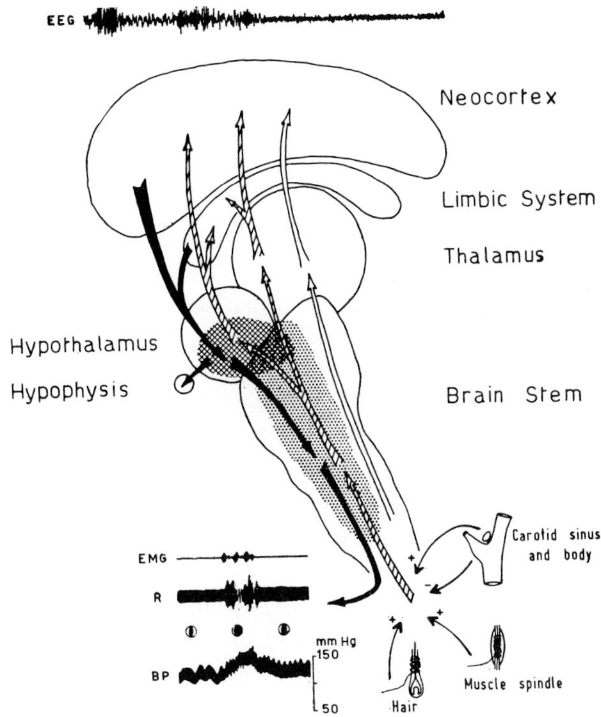

FIG. 1. Interrelationship between brain stem reticular formation and hypothalamic mechanisms for emotional expression and cardiovascular control. Reticular formation indicated by stippled area, hypothalamus by cross-hatched area. Striped arrows indicate ascending influences of various receptors (some outlined in lower right part of the figure) on reticular formation; of reticular formation on hypothalamus; of reticular formation and hypothalamus on limbic system and neocortex. Black arrows indicate descending influences from cortex, limbic system, hypothalamus and brain stem resulting in expression of emotional behavior exemplified in left lower inset (EMG, electromyogram of forelimb; R, respiration; BP, blood pressure). Upper inset (EEG) is electroencephalographic arousal reaction resulting from ascending reticular activation. White arrows indicate specific sensory system. The + and — signs indicate excitatory or inhibitory actions of receptors on reticular formation. Other explanations are given in the text (Modified from Zanchetti, 1967).

role (Bartorelli *et al.*, 1960), and the chemoceptors an excitatory one (Bizzi *et al.*, 1961). However, there are two differences. Both the inhibitory and excitatory influences (which are mostly limited to circulatory and respiratory centers in the medulla) are much more widespread in the hypothalamus, affecting all somatic and visceral signs of sham rage behavior (Bartorelli *et al.*, 1960; Bizzi *et al.*, 1961). In addition, when the baro- and chemo-ceptive reflexes are simultaneously evoked, the inhibitory vasodepressor influence predominates at the medullary level and the excitatory pressor influence is always the overwhelming one at the hypothalamus (Baccelli *et al.*, 1965). Predominance of any kind of excitatory stimuli seems typical of the sham rage preparation; for example, trifling cutaneous stimuli and even stimulation of group I muscle afferents (which are usually without any circulatory and behavioral effect) can trigger violent outbursts of sham rage and pressor fits in the hypothalamic cat (Malliani *et al.*, 1968).

The reflex influences exerted on the hypothalamus and its cardiovascular functions are likely to be mediated through the ascending, activating reticular system. Figure 1 indicates that the interrelationship between hypothalamus and reticular formation of the brain stem is threefold. First of all the mechanisms of sham rage are not limited to the hypothalamus but extend to the midbrain tegmentum and central gray (Carli *et al.*, 1966). Secondly, sham rage and its cardiovascular correlates can occur only when the activating function of the reticular system is not impaired, an observation which implies that the cardiovascular influences of the hypothalamus are reflexly evoked through the reticular system (Malliani *et al.*, 1963). Third, the somatic and visceral manifestations of sham rage result largely from discharges running from the hypothalamus and rostral midbrain tegmentum through the descending reticular system (Carli *et al.*, 1963).

IV. Cardiovascular Changes during Natural Behavior Related to the Hypothalamus

Although the two experimental approaches summarized in Sections II and III are very useful in studying the cardiovascular potentialities of the hypothalamus and their mechanisms of regulation they are both subject to severe limitations. This is particularly true of the interpretation that cardiovascular changes induced by electrical stimulation of the hypothalamus are related to some natural behavior. This interpretation rests on the unverified assumptions that: electrically induced behavior is really comparable to the natural one; anesthesia, which abolishes or severely hampers any natural behavior, may not modify its cardiovascular concomitants; and finally, none of the reported changes results from co-stimulation in the hypothalamus of cells or fibers having no relationship to a given behavior. These difficulties could be overcome only by comparing the cardiovascular patterns induced by hypothalamic stimulation with those occurring during natural behavior more or less directly related to the hypothalamus.

Before taking for granted that one of the most typical circulatory patterns induced by stimulation of the hypothalamus actually represents cardiovascular adjustment during defense behavior, it was necessary to obtain a faithful picture of hemodynamic changes in unanesthetized, unrestrained animals during natural defense or fighting behavior. This information has been recently provided by our group (Adams *et al.*, 1968, 1969; Baccelli *et al.*, 1968). The experimental design consisted of a cage subdivided into two compartments by a movable opaque screen. In one compartment a cat having an electrode chronically implanted in the mesencephalic gray was placed. Electrical stimulation through this electrode invariably elicited attack behavior. This cat was used solely as a stimulus for evoking natural fighting behavior in another cat (the subject of recording) which was placed in the other compartment of the cage. Chronic implantation of electromagnetic probes and the use of suitable integrating amplifiers gave continuous, simultaneous recording of cardiac output, mesenteric blood flow, and blood flow through a hind limb. Blood pressure was also monitored through an indwelling catheter and heart rate was recorded by a cardiotachometer.

Cardiovascular changes were found to be dependent upon the posture and amount of muscle activity of the cat during fighting. In this way two types of fighting were separately studied. In one the cat supported itself on its hind limbs during the base-line period and used them considerably for balance and support during fighting. This was termed supportive fighting. In the other case the animal lay on its side during both base line and fighting, struck with the forelimbs only and did not support itself with the hind limbs or use them during fighting. This type, in which there was little or no muscle activity in the hind limbs, was termed nonsupportive fighting.

During supportive fighting (Fig. 2), there was increase in heart rate and cardiac output, the arterial pressure was only slightly changed, an increase in total peripheral conductance indicated an overall

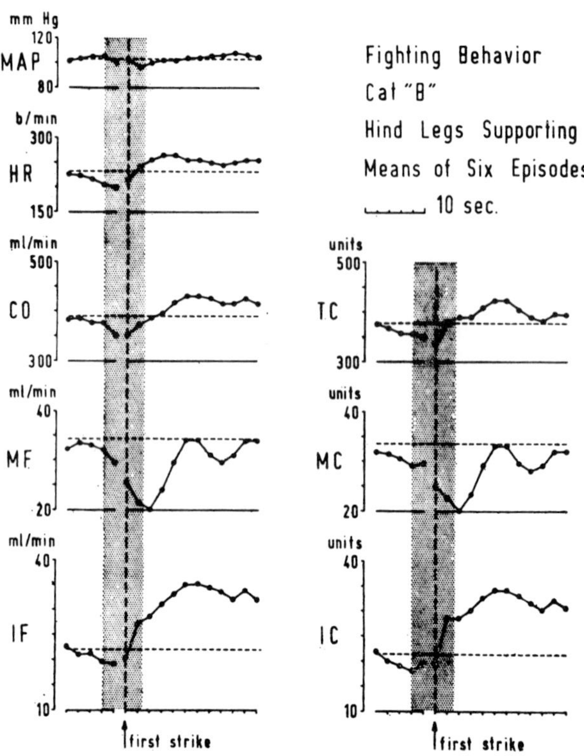

FIG. 2. Cardiovascular changes during supportive fighting behavior in the cat. Each dot on the curves represents the mean of 2-sec measurements performed during six fighting episodes. Shaded area is time during which partition was opened; vertical line indicates beginning of fighting. Horizontal broken line indicates base-line measurements when the cat was quiet prior to the trials. MAP, mean arterial pressure; HR, heart rate; CO, cardiac output; MF, superior mesenteric flow; IF, external iliac flow; TC, total peripheral conductance; MC, superior mesenteric conductance; IC, external iliac conductance (From Baccelli *et al.*, 1968).

trend toward vasodilatation, which was due to a huge vasodilatation in the limbs only partially balanced by a marked vasoconstriction in the mesenteric bed. Iliac vasodilatation was dependent upon muscle activity itself, since if the hind limbs were inactive as in nonsupportive fighing (Fig. 3), there was a vasoconstriction instead. Iliac vasoconstriction occurred during nonsupportive fighting even when the sympathetically controlled portion of the cutaneous circulation (which in the cat is limited to the paw) was mechanically excluded. In these

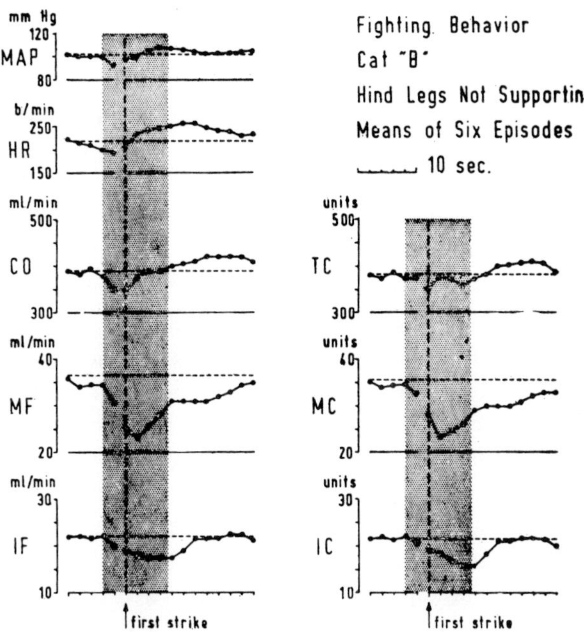

FIG. 3. Cardiovascular changes during nonsupportive fighting behavior in the cat. All explanations as in Fig. 2 (From Baccelli *et al.*, 1968).

conditions iliac vasoconstriction represented purely muscular vasoconstriction. Purely muscular vasoconstriction was also observed in the same cats in trials of expectation or preparation for fighting when the potentially attacking cat was brought close to the subject cat but no attack took place (Adams *et al.*, 1968).

If comparison is made of the cardiovascular changes occurring during natural fighting or defense behavior with those resulting from electrical stimulation of the so-called hypothalamic defense area, it is apparent that several changes are similar (at least in direction) in the two experimental conditions: tachycardia, increased cardiac output, mesenteric vasoconstriction. However, the strong muscle vasodilatation seen during supportive fighting appears to be the result of local metabolic factors related to muscle activity rather than the effect of a central mechanism such as the cholinergic vasodilator system, as might have been expected from experiments of brain stimulation. In fact, if there is a primary nervous effect in addition to the metabolic factors, it would instead appear to be a vasoconstriction such as we have observed in the iliac bed during nonsupportive fighting and preparation for fighting.

Several possibilities can be envisaged to reconcile the results of artificial stimulation of the hypothalamus with the manifestations of natural behavior. One is that sympathetic vasodilator fibers to muscle are also activated during natural fighting, but their action is blocked or largely overwhelmed by concomitant vasoconstriction (Folkow *et al.*, 1964b). This seems rather unlikely because it would make the activity of the sympathetic vasodilators functionally meaningless. A second possibility is that the combination of the various cardiovascular changes evoked by electrical stimulation of the so-called hypothalamic defense area does not really represent a functional pattern and that at least muscle vasodilatation depends on co-stimulation of cells or passing-through fibers having no relationship with defense behavior. Thirdly, it might be suggested that natural defense behavior has little to do with the similar but obviously unnatural defense behavior elicited by electrical stimulation of the hypothalamic and mesencephalic areas from which muscle vasodilatation is also produced. Finally, one should carefully consider that sympathetic muscle vasodilatation might be a visceral correlate of some different behavioral way of reacting in defense, or of some emotional behavior other than fighting and defense, or of some unemotional behavior such as preparation for exercise or arousal from sleep. Behavioral expression is subtler than we often think and we cannot certainly claim that sympathetic muscle vasodilatation is not related to natural behavior simply because it does not seem to be related to natural fighting behavior.

V. Conclusions

This article has mainly tried to emphasize the increasing difficulties and complexities that are met when one leaves the simple approach of listing the various cardiovascular changes elicited by stimulating different portions of the hypothalamus. However, in order to get an integrative and physiologically meaningful view, the investigator should attempt to tie together the cardiovascular and other signs of hypothalamic stimulation in functionally significant patterns. In addition, he should consider these patterns as part of the behavioral function of the hypothalamus and in relation to other brain structures. Finally he should prove whether the cardiovascular patterns artificially induced from manipulation of the hypothalamus are actually found in natural behavior. Unfortunately these approaches are only too recent and difficult. We remain quite far from understanding how the hypothalamus is actually involved in regulating circulation in everyday life conditions. This limitation refers more broadly to neural regulation of circulation. One should not be surprised if after reading this article there are more questions to be asked than have been answered. However, the future development of our knowledge in this field will depend heavily on the pertinence of the questions we are asking ourselves now.

Acknowledgment

The personal research referred to in this article has been sponsored by the Air Force Office of Scientific Research, under Grants AF EOAR 66-47 and 67-41, through the European Office of Scientific Research, United States Air Force, and by Consiglio Nazionale delle Ricerche (Gruppo di Medicina Sperimentale).

References

Abrahams, V. C., Hilton, S. M., and Zbrozyna, A. (1960). *J. Physiol.* (London) **154,** 491.

Abrahams, V. C., Hilton, S. M., and Zbrozyna, A. (1964). *J. Physiol.* (London) **171,** 189.

Adams, D. B., Baccelli, G., Mancia, G., and Zanchetti, A. (1968). *Nature* **220,** 1239.

Adams, D. B., Baccelli, G., Mancia, G., and Zanchetti, A. (1969). *Am. J. Physiol.* **216**, 1226.

Baccelli, G., Guazzi, M., Libretti, A., and Zanchetti, A. (1965). *Am. J. Physiol.* **208,** 708.

Baccelli, G., Mancia, G., Adams, D. B., and Zanchetti, A. (1968). *Experientia* **24,** 1221.

Bartorelli, C., Bizzi, E., Libretti, A., and Zanchetti, A. (1960). *Arch. Ital. Biol.* **98**, 308.

Bizzi, E., Libretti, A., Malliani, A., and Zanchetti, A. (1961). *Am. J. Physiol.* **200**, 923.

Bolme, P., Ngai, S. H., Uvnäs, B., and Wallenberg, L. R. (1967). *Acta Physiol. Scand.* **70**, 334.

Carli, G., Malliani, A., and Zanchetti, A. (1963). *Exp. Neurol.* **7**, 210.

Carli, G., Malliani, A., and Zanchetti, A. (1966). *Boll. Soc. Ital. Biol. Sper.* **42,** 291.

Cobbold, A., Folkow, B., Lundgren, O., and Wallentin, I. (1964). *Acta Physiol. Scand.* **61**, 467.

Eliasson, S., Folkow, B., Lindgren, P., and Uvnäs, B. (1951). *Acta Physiol. Scand.* **23**. 333.

Feigl, E., Johansson, B., and Löfving, B. (1964). *Acta Physiol. Scand.* **62**, 429.
Folkow, B., and Rubinstein, E. H. (1965). *Acta Physiol. Scand.* **65**, 292.
Folkow, B., Johansson, B., and Öberg, B. (1959). *Acta Physiol. Scand.* **47**, 262.
Folkow, B., Langston, J., Öberg, B., and Prerovsky, I. (1964a). *Acta Physiol. Scand.* **61**, 476.
Folkow, B., Öberg, B., and Rubinstein, E. H. (1964b). *Angiologica* (Basel) **1**, 197.
Folkow, B., Heymans, C., and Neil, E. (1965). *In* " Handbook of Physiology " (W. F. Hamilton and P. Dow, eds.). Section 2, Vol. II, pp. 1787–1823. American Physiological Society, Washington, D. C.
Folkow, B., Lisander, B., Tuttle, R. S., and Wang, S. C. (1968). *Acta Physiol. Scand.* **72**, 220.
Grant, R., Lindgren, P., Rosén, A., and Uvnäs, B. (1958). *Acta Physiol. Scand.* **43**, 135.
Hess, W. R. (1949). " Das Zwischenhirn: Syndrome, Lokalisationen, Funktionen." Schwabe, Basel.
Ingram, W. R. (1960). *In* " Handbook of Physiology " (H. W. Magoun, ed.). Section 1, Vol. II, pp. 951–978. American Physiological Society, Washington, D. C.
Karplus, J. P., and Kreidl, A. (1909). *Pfluegers Arch. Ges. Physiol.* **129**, 138.
Löfving, B. (1961). *Acta Physiol. Scand. Suppl.* **53**, 184.
Malliani, A., Bizzi, E., Apelbaum, J., and Zanchetti, A. (1963). *Arch. Ital. Biol.* **101**, 632.
Malliani, A., Carli, G., Mancia, G., and Zanchetti, A. (1968). *J. Neurophysiol.* **31**, 210.
Ranson, S. W., and Magoun, H. W. (1939). *Ergeb. Physiol.* **41**, 56.
Rosén, A. (1961). *Acta Physiol. Scand.* **52**, 291.
Uvnäs, B. (1960). *In* " Handbook of Physiology " (H. W. Magoun, ed.). Section 1, Vol. II, pp. 1131–1162. American Physiological Society, Washington, D. C.
Zanchetti, A. (1967). *In* " The Neurosciences " (G. C. Quarton, T. Melnechuck and F. O. Schmitt, eds.). pp. 602–614. The Rockefeller University Press, New York.

The Hypophysiotropic Area

J. FLAMENT-DURAND and L. DESCLIN

I. Introduction

It has been stated that humoral substances elaborated in the hypothalamus specifically influence the various hypophyseal secretions by way of the portal circulation to the pituitary. This concept resulted from experiments in which pituitary tissue was grafted far from its hypothalamic connections, e.g., under the kidney capsule. Under these conditions the transplant is uniformly composed of slightly eosinophilic cells and loses its functional properties except for the secretion of prolactin (Desclin, 1950; Everett, 1954). When retransplanted under the median eminence, this hypophyseal graft recovers its functional activities, and typical basophils reappear (Harris and Jacobsohn, 1952; Nikitovitch-Winer and Everett, 1958, 1959). These changes can be explained by the contact established between the transplant and the portal system at the level of the median eminence, enabling humoral substances elaborated in the hypothalamus to influence the graft.

This view has been confirmed by experiments in which chemical mediators acting as hormone-releasing factors of the various hypophyseal secretions have been isolated from the hypothalamus. However, the actual sites of elaboration are as yet anatomically uncertain. Many paths, such as investigation of the effects of lesions in different parts of the hypothalamus or of electrical stimulation in local sites of the brain, have been followed in attempting to shed light on this problem (see the article by Mess and associates in this volume).

Another interesting approach is the study of the morphology of hypophyseal grafts stereotaxically implanted, not under the median eminence in contact with the portal vessels, but within the hypothalamus itself (Knigge, 1962; Halász *et al.*, 1962, 1965).

We shall present the results obtained by this method in order to delimit the anatomical sites of elaboration of the mediators acting on the gonadotropic, thyrotropic, and luteotropic functions of the pituitary. We shall also discuss the site of action of estrogens and of reserpine.

II. Morphology of Intrahypothalamic Pituitary Grafts According to their Location in the Hypothalamus

Let us consider the results obtained by studying the morphology of the grafts. When placed outside a definite zone of the hypothalamus, the transplants contain small chromophobes (Fig. 1) like those in pituitaries grafted under the kidney capsule. When located in a territory extending from behind the optic chiasma to the median eminence and including the ventral and median parts of the hypothalamus, the graft contains numerous basophilic cells. This area corresponds to the hypophysiotropic area as defined by Halász and associates (1962, 1965). That part of a graft located in a more restricted zone of this area corresponding to the arcuate nuclei always contains numerous gonadotrophs which transform into castration cells after castration or hypophysectomy (Fig. 2).

III. Attempts to Distinguish between the Areas Influencing the Gonadotrophs and the Thyrotrophs

We have attempted to determine whether the area influencing the gonadotrophs could be distinguished from that which influences the thyrotrophs. As a first step we have determined the cytological response of intrahypothalamic grafts to the administration of pro-

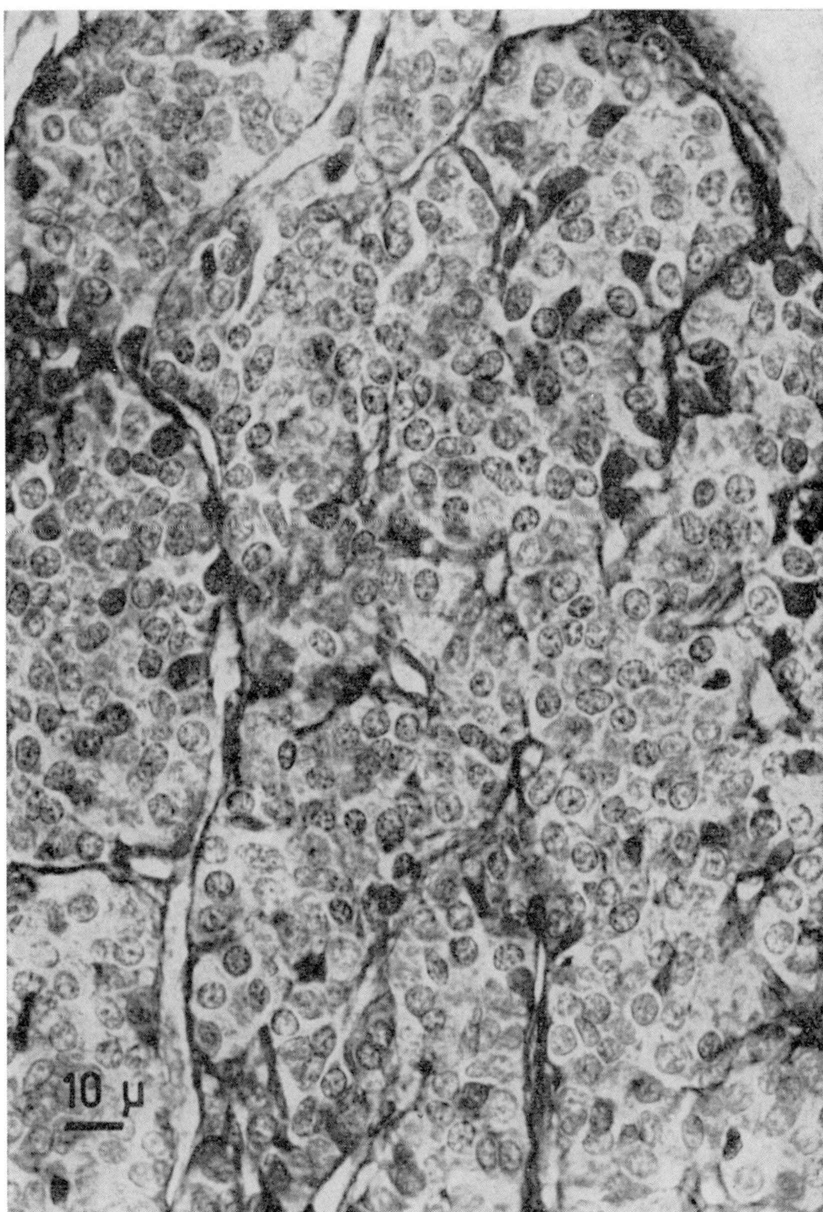

FIG. 1. Part of a pituitary graft stereotaxically implanted outside the hypophysiotropic area in a subsequently hypophysectomized host (stain: Heidenhain's Azan). This part of the graft is composed of small chromophobes.

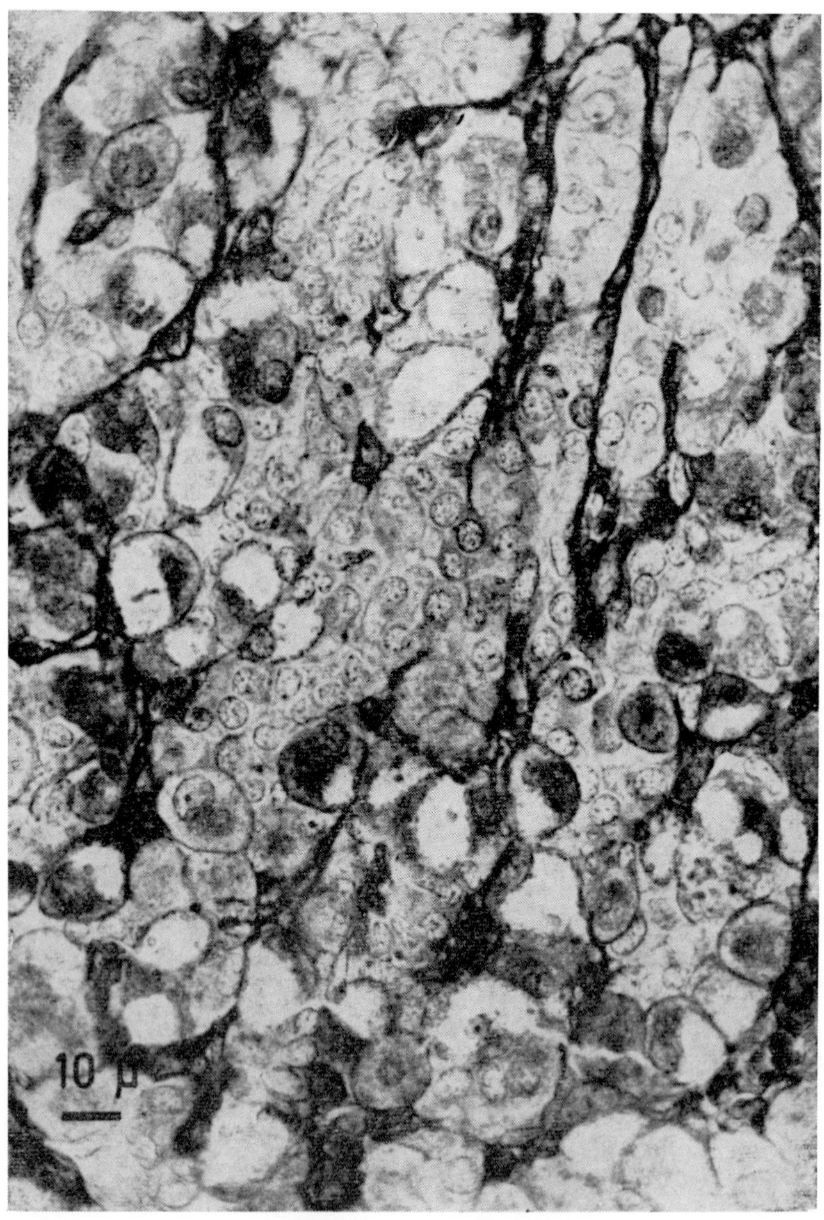

FIG. 2. Portion of a graft located in the tuberal region of a subsequently hypophysectomized host (stain: Heidenhain's Azan). Numerous vacuolated gonadotrophs.

pylthiouracil (PTU) in intact hosts. PTU was given by stomach tube in a daily dose of 20 mg for 15 days. We observed that numerous thyroidectomy cells appeared in the grafts located in a territory including the suprachiasmatic region, the anterior hypothalamic area, and the paraventricular nuclei. Grafts outside this territory or under the kidney capsule were not cytologically modified by administration of PTU (Flament-Durand, 1966). To localize more closely the sites of elaboration of mediators influencing gonadotrophs and thyrotrophs, we have studied grafts in castrated hosts treated after castration with PTU alone or with PTU plus thyroxine. Thyroxine was administered in order to induce regression of thyroidectomy cells and to differentiate them more easily from castration cells. Under these experimental conditions it clearly appeared that the area influencing thyrotrophs is distinct from the gonadotropic area (Flament-Durand and Desclin, 1968). Castration cells appeared in all portions of grafts located in the infundibular recess of the third ventricle in contact with the arcuate nuclei, whereas PTU treatment induced the appearance of thyroidectomy cells in all portions of grafts situated in the suprachiasmatic region, the anterior hypothalamic area, the paraventricular nuclei, and the an-

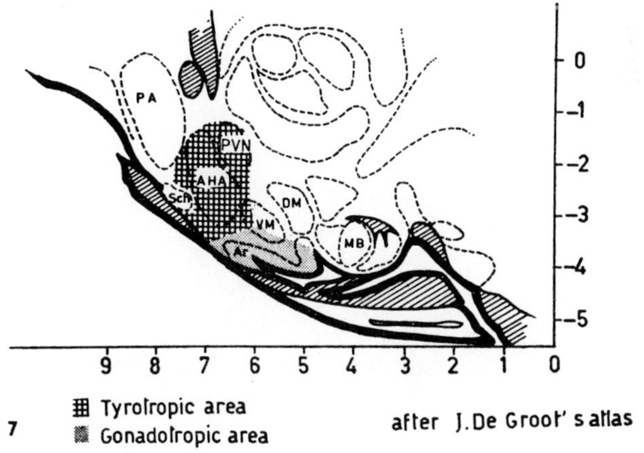

FIG. 3. Diagrammatic representation of the two distinct tropic areas. Abbreviations: AHA, anterior hypothalamic area; Ar, arcuate nucleus; DM, dorsomedial nucleus; MB, mammillary body; PA, preoptic area; PVN, paraventricular nucleus; Sch, suprachiasmatic nucleus; VM, ventromedial nucleus.

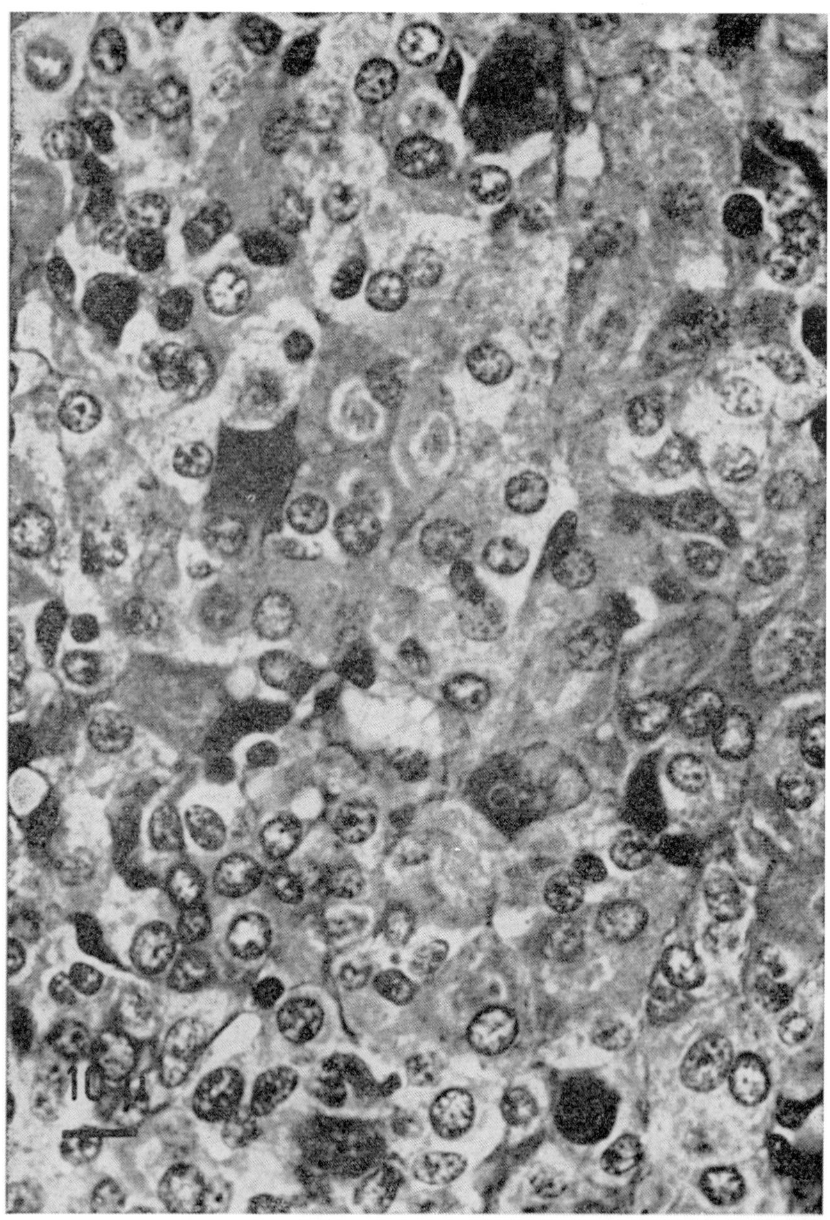

FIG. 4. Pituitary graft located close to the paraventricular nuclei in a spayed host treated with propylthiouracil (stain: Heidenhain's Azan). Numerous vacuolated thyrotrophs.

terior part of the tuberal region (Fig. 3). Morphologically, thyroidectomy cells are different from castration cells: their form remains more angular, their cytoplasm is finely vacuolated, and they stain weakly with aniline blue (Fig. 4). Their thyrotropic nature is well supported by the fact that they disappear completely in hosts treated simultaneously with thyroxine, whereas the appearance of castration cells is not modified by thyroxine; they remain just as numerous and active in grafts located near the arcuate nuclei. Local cytological differences appeared between various parts of the same graft according to their location.

We thus concluded from cytological study of the grafts that the thyrotropic area extends more anteriorly and more dorsally than the gonadotropic area. These areas in the hypothalamus are relatively diffuse, but it appears that the region of the paraventricular nuclei acts particularly on thyrotrophs whereas the tuberal region close to the arcuate nuclei acts more directly on gonadotrophs. The importance of the region of the paraventricular nuclei in the control of the thyrotropic function of the pituitary is supported by data found in the literature (Greer, 1951; Greer and Erwin, 1954, 1956; Moll *et al.*, 1961; Averill *et al.*, 1961).

A. Gonadotropic Function of the Grafts

Let us now consider the activity of grafts in hypophysectomized hosts.

In spite of morphological signs of high secretory activity, gonadotropic function of the grafts appeared to be very poor. Among 93 grafted females, only 8 showed return of estrus cycles after hypophysectomy. Placed with males, only 6 of them mated (as demonstrated by the presence of sperm in vaginal smears). However, no female became pregnant or pseudopregnant. The ovaries of these 8 animals were larger than those of the acyclic hosts or of hypophysectomized controls. They contained many follicles but no corpora lutea. The grafts in this group of 8 cyclic rats were located in the gonadotropic area. Interestingly some of the grafts in the acyclic rats were located in the same area and were full of active gonadotrophs. In spite of this appearance of high secretory activity they were unable to restore the function of the ovaries, which were atrophic and comparable to those of hypophysectomized controls.

In male rats, 4 of 23 grafted and hypophysectomized hosts had normal testes. Three out of the four had a pituitary graft located in the gonadotropic area; the graft, in spite of the normal appearance of the gonads, still contained numerous castration cells. These results seem to indicate difficulty in release of gonadotropins, as judged by the weakness of their action on peripheral receptors; in addition, alteration of the feedback mechanisms was also observed.

B. Thyrotropic function of the Grafts

When we examined thyrotropic function, we observed that grafts located in the thyrotropic area were able to restore thyroid weight and morphology regardless of contact with the portal system. Administration of PTU is definitely goitrogenic, and this action is mediated by all grafts located in the thyrotropic area regardless of contact with the portal vessels. This result is in contradiction with the conclusions of Halász and his associates (1965).

C. Luteotropic Function of the Grafts

Let us now consider the luteotropic function of intrahypothalamic grafts. Whereas pituitary transplants under the kidney capsule induce pseudopregnancy (Quilligan and Rothchild, 1960), thus confirming the release of prolactin by pituitary cells separated from their hypothalamic contact, the intrahypothalamic grafts did not have this effect. Non hypophysectomized hosts remained regularly cyclic or exhibited periods of prolonged estrus.

IV. Action of Estrogens on Pituitary Grafts

Our next study was directed toward the action of estrogens on intrahypothalamic grafts. The mechanism by which estrogens promote the development of stimulated prolactin cells has not yet been established. The process might be through the mediation of the central nervous system, but direct action on hypophyseal cells is also possible (Nicoll and Meites, 1962). In order to localize the possible site of action of estrogens on the central nervous system, we injected estradiol and studied the sensitivity of grafts located in various parts of the brain. Un derthese conditions, all grafts contained stimulated prolac-

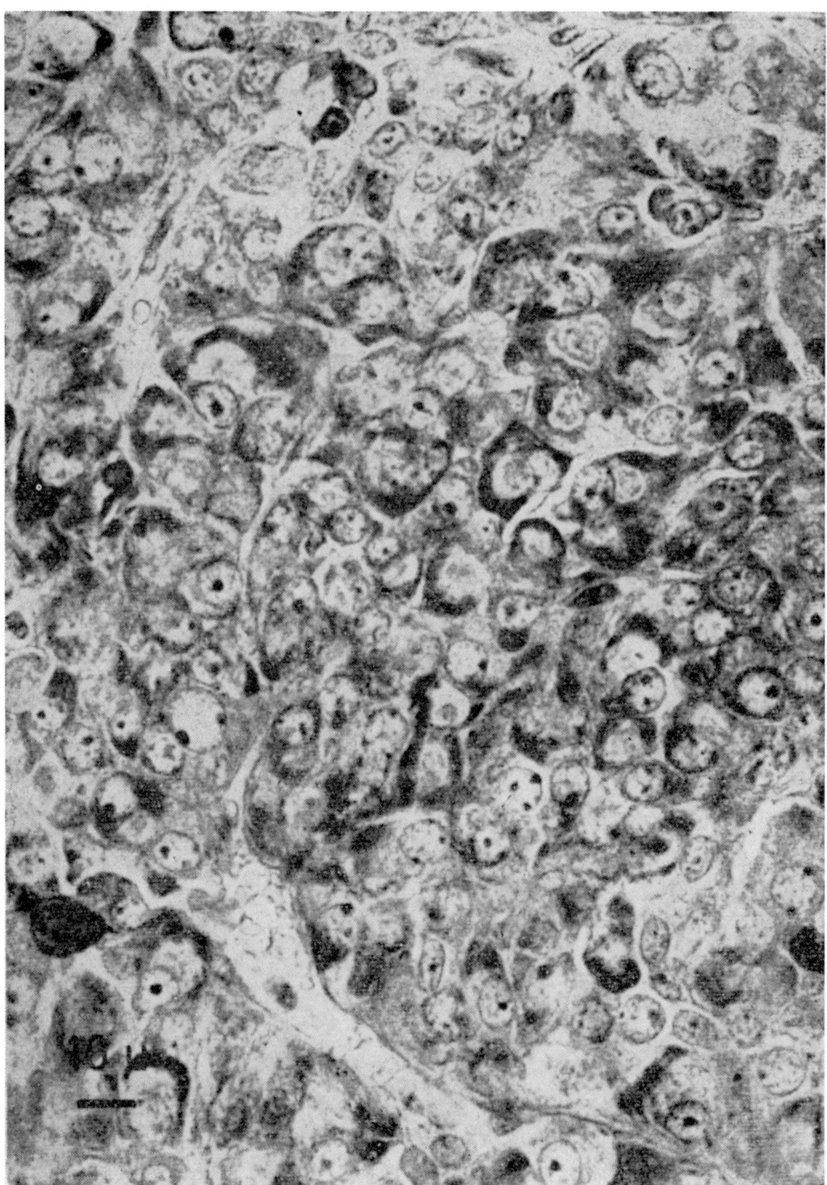

FIG. 5. Intracerebral graft in a host treated daily with 15 μg of estradiol for 15 days (stain: Dominici). Numerous stimulated prolactin cells.

tin cells, regardless of their location within or outside the hypothalamus (Fig. 5) or under the kidney capsule. These results indicate favorably that estrogens act directly on pituitary cells, but they do not exclude the possibility of an action on the Prolactin Inhibiting Factor (PIF). Although highly stimulated, these intrahypothalamic grafts were unable to maintain active corpora lutea after hypophysectomy (Desclin and Flament-Durand, 1966). This inability to exert a luteotropic function was demonstrated in a study of hosts bearing a pituitary graft in the hypothalamus or under the kidney capsule. Hosts were treated daily with 15 μg estradiol benzoate for 15 days. They were then hypophysectomized and sacrificed five days later. The grafts under the kidney capsule were able to maintain active corpora lutea after hypophysectomy and the vagina was mucified. Grafts located in the central nervous system were unable to exert this luteotropic action; the corpora lutea degenerated and the vagina keratinized. However, the cytological appearance of the grafts was the same regardless of their location, suggesting that the pituitary cells were actively synthetising prolactin. Thus contact with the hypothalamus seems to have prevented the release of prolactin from the grafts but not its synthesis and the poor effect of the intracerebral graft on receptors indicates difficulty in the relesae of the hormone.

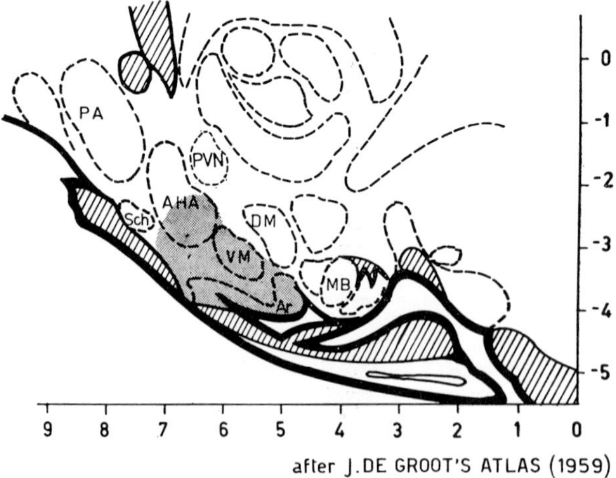

FIG. 6. Diagrammatic representation of the area where reserpine influences the cytology of the pituitary graft. Abbreviations: PA, preoptic area; PVN, paraventricular nucleus; Sch, suprachiasmatic nucleus; VM, ventromedial nucleus; DM, dorsomedial nucleus; Ar, arcuate nucleus; MB, mammillary body.

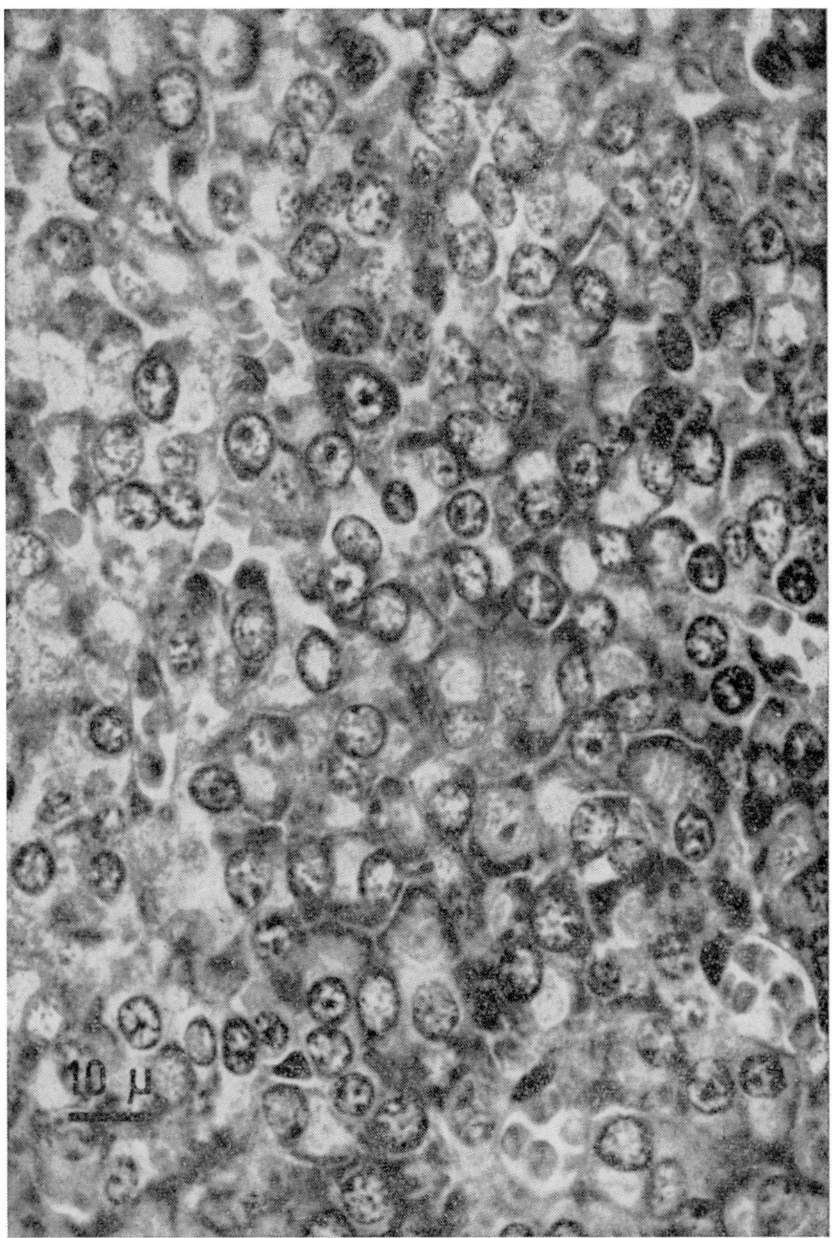

FIG. 7. Tuberal portion of a pituitary graft in a host treated with reserpine (stain: Dominici).

V. Action of Reserpine on Pituitary Grafts

Next we studied the influence of reserpine on pituitary grafts into various parts of the brain, on pituitary *in situ*, and on grafts under the kidney capsule. Reserpine was given subcutaneously at a daily dose of 50 μg per 100 g of body weight for nine consecutive days before autopsy. Reserpine, like estrogens, is claimed to depress PIF content of the hypothalamus (Ratner and Meites, 1964; Ratner *et al.*, 1965). However, the results obtained with reserpine are in sharp contrast to those observed with estrogens: reserpine modifies the cytology of only those grafts located in a definite zone of the hypothalamus (Fig. 6) corresponding to the hypophysiotropic area described by Halász *et al.* (1962, 1965). In this area, reserpine induced the appearance of highly stimulated prolactin cells resembling those seen during lactation in the pituitary *in situ*. These cells show a large Golgi apparatus surrounded by a basophilic rim that is well demonstrated by Dominici staining (Fig. 7); with Herlant's tetrachromic method, erythrosinophilic granules can be demonstrated. Grafts located outside this territory are not modified by reserpine. Under the influence of reserpine, the level of stimulation of the prolactin cells observed in grafts located in that definite hypothalamic zone is higher than that observed in grafts under the kidney capsule, i.e., those cells separated from the influence of PIF. This difference cannot be explained solely on the basis of suppression of PIF. Results obtained with reserpine seem to imply intervention of a Prolactin Releasing Factor (PRF). The existence of such a PRF has already been postulated (Mishkinsky *et al.*, 1968). A preliminary report of our results with reserpine has been published (Desclin and Flament-Durand, 1969).

Acknowledgment

These studies have been supported by a grant from the Fond National de la Recherche scientifique.

References

Averill, R. L. W., Purves, H. D., and Sirett, N. E. (1961). *Endocrinology* **69**, 735.
De Groot, J. (1959). *J. Comp. Neurol.* **113**, 389.
Desclin, L. (1950). *Ann. Endocrinol.* (Paris) **11**, 656.
Desclin, L., and Flament-Durand, J. (1966). *Z. Zellforsch.* **69**, 274.
Desclin, L., and Flament-Durand, J. (1969). *J. Endocrinol.* **43**, LIX.
Everett, J. W. (1954). *Endocrinology* **54**, 685.

Flament-Durand, J. (1966). *C. R. Acad. Sci.* **262**, 297.
Flament-Durand, J., and Desclin, L. (1968). *J. Endocrinol.* **41**, 531.
Greer, M. A. (1951). *Proc. Soc. Exp. Biol. Med.* **77**, 603.
Greer, M. A., and Erwin, H. L. (1954). *J. Clin. Invest.* **33**, 938.
Greer, M. A., and Erwin, H. L. (1956). *Endocrinology* **58**, 665.
Halász, B., Pupp, L., and Uhlarik, S. (1962). *J. Endocrinol.* **25**, 147.
Halász, B., Pupp, L., Uhlarik, S., and Tima, L. (1965). *Endocrinology* **77**, 343.
Harris, G. W., and Jacobsohn, D. (1952). *Proc. Roy. Soc.* **139**, 263.
Knigge, K. M. (1962). *Am. J. Physiol.* **202**, 387.
Mishkinsky, J., Khazen, K., and Sulman, F. G. (1968). *Endocrinology* **82**, 611.
Moll, J., De Wied, D., and Kranendonk, G. H. (1961). *Acta Endocrinol.* **38**,330.
Nicoll, C. S., and Meites, J. (1962). *Endocrinology* **70**, 272.
Nikitovitch-Winer, M., and Everett, J. W. (1958). *Endocrinology* **63**, 916.
Nikitovitch-Winer, M., and Everett, J. W. (1959). *Endocrinology* **65**, 357.
Quilligan, E. J., and Rothchild, I. (1960). *Endocrinology* **67**, 48.
Ratner, A., and Meites, J. (1964). *Endocrinology* **75**, 377.
Ratner, A., Talwalker, P. K., and Meites, J. (1965). *Endocrinology* **77**, 315.

Site of Production of Releasing and Inhibiting Factors

B. MESS, M. ZANISI [1], and L. TIMA [1]

I. Introduction

In the history of endocrinology, about two to three decades ago there was a period characterized by the detection of the different adenohypophyseal hormones. Nearly every year one or another hormone of the anterior pituitary was described. Many of them, e.g., the alleged " diabetogenic hormone, " the " medullotropic hormone, " or the " parathyrotropic hormone " proved to be secondary effects of other tropic hormones or a consequence of some interglandular interplay. The past 15 years of neuroendocrinology might be characterized by the detection of the hypothalamic hypophysiotropic principles. In 1955 Saffran and associates described the first releasing factor, the Corticotropin Releasing Factor (CRF), and in the subsequent 12 years three different fractions of CRF and six more releasing factors were described. In the same period three inhibiting factors were also found in hypothalamic extracts;

[1] Ford Foundation Fellow.

TABLE I

HYPOTHALAMIC RELEASING AND INHIBITING FACTORS

Hypothalamic Factors	Abbreviations	References
Corticotropin Releasing Factor	CRF	Saffran *et al.* (1955)
Fractions of CRF	α_1-CRF	Guillemin *et al.* (1960)
	α_2-CRF	Schally *et al.* (1962)
	β-CRF	Schally *et al.* (1962)
Thyrotropin Releasing Factor	TSHRF	Shibusawa *et al.* (1959)
Luteinizing Hormone Releasing Factor	LHRF	McCann *et al.* (1960)
Growth Hormone Releasing Factor	GHRF	Franz *et al.* (1962)
Follicle-Stimulating Hormone Releasing Factor	FSHRF	Igarashi and McCann (1964)
Melanocyte-Stimulating Hormone Releasing Factor	MSHRF	Taleisnik and Orias (1965)
Prolactin Releasing Factor (pigeon)	PRF	Kragt and Meites (1965)
Prolactin Inhibiting Factor (mammals)	PIF	Pasteels (1962)
Melanocyte-Stimulating Hormone Inhibiting Factor	MSHIF	Kastin (1965)
Growth Hormone Inhibiting Factor	GHIF	Krulich *et al.* (1967)

these inhibiting factors act antagonistically to the releasing factors. Table I summarizes the various findings. Taking into consideration the possible dichotomy of the different " releasing " and " synthesizing " factors as proposed by some authors, the group of hypothalamic hypophysiotropic principles may grow to a considerably high number.

II. Site of Neurons Producing Releasing and Inhibiting Factors

Authors generally agree that all these principles are present in highest concentration in the stalk median eminence region. However, the fact that these principles are localized in high concentration in the stalk median eminence does not prove that they are synthesized in this hypothalamic region. It seems to be more probable that these hypothalamic factors, or at least part of them, may be synthesized in regions apart from the stalk median eminence and are stored and released only by this circumscribed small part of the hypothalamus (Mess and Martini, 1968).

A. Role of the Hypophysiotropic Area in the Production of Releasing and Inhibiting Factors

The observation that a half-moon-shaped area of the hypothalamus, called the hypophysiotropic area, is able without any contact with higher brain areas to maintain nearly normal anterior pituitary function gives strong support for the assumption that releasing and inhibiting factors are synthesized in this area. This area, as described by Halász and his co-workers (1962) in our laboratory, includes the suprachiasmatic, paraventricular, periventricular, anterior hypothalamic, arcuate, and premammillary nuclei. Also the medial halves of the ventromedial nuclei are part of the hypophysiotropic area. Since the nature of this hypophysiotropic area was also thoroughly investigated by Flament-Durand and Desclin, who give an excellent survey of this problem in their article in this volume, only the most important data will be presented here to lend support of the proposed idea that releasing and inhibiting factors are synthesized inside the hypophysiotropic area.

Transplantation of adenohypophyseal tissue in a series of intrahypothalamic, intracerebral but extrahypothalamic, or extracerebral

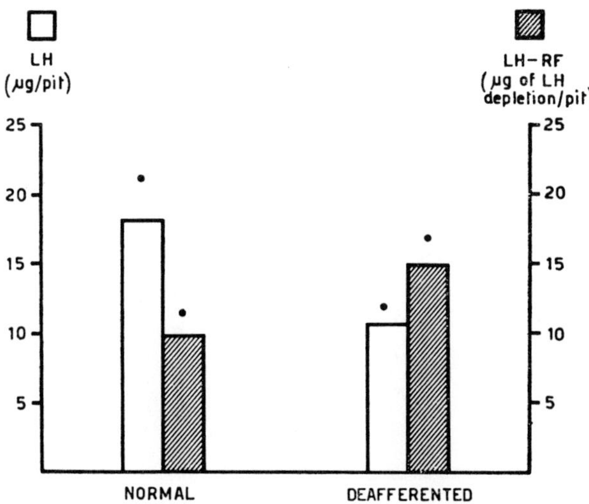

Fig. 1. Effect of hypothalamic deafferentation on pituitary LH and hypothalamic LHRF content in male rats (8 days after operation).

sites of hypophysectomized hosts unequivocally proved that the normal histological structure of the grafted pituitary was preserved only in transplants localized inside the aforementioned hypophysiotropic area. These results show that the trophism of the differentiated (basophilic and acidophilic) adenohypophyseal cells is assured only in transplants placed in this part of the brain (Halász *et al.*, 1962). The body growth of the hypophysectomized pituitary-grafted rats was nearly normal only if the graft was localized in the hypophysiotropic area (Halász *et al.*, 1963). The histological structure and function of the target glands were nearly normal in animals bearing pituitary homograft inside the hypophysiotropic area (Halász *et al.*, 1965).

All these results suggest that neural elements of the hypophysiotropic area are able to maintain the trophism and hormone secretion of the adenohypophysis. The relative autonomy of these nervous structures, i.e., the question of whether this area in itself is capable of exerting all types of stimulating influences on the anterior pituitary tissue, was further investigated by the same group of authors. They

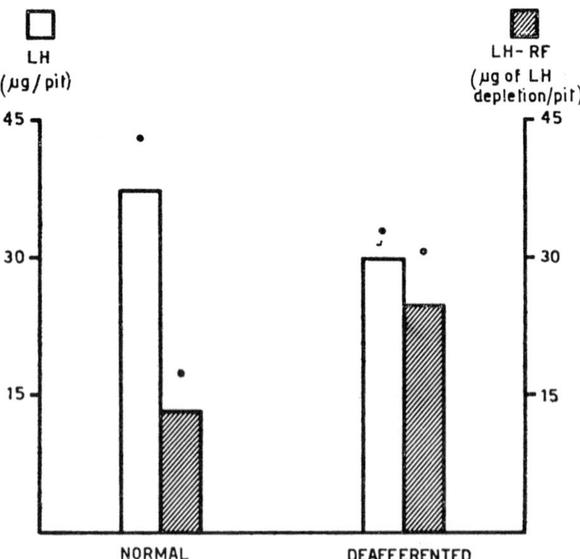

FIG. 2. Effect of hypothalamic deafferentation on pituitary LH and hypothalamic LHRF content in male rats (15 days after operation).

surgically isolated the hypophysiotropic area by means of a bayonette-shaped knife applied into the Horsley-Clarke stereotaxic apparatus. The isolated, or "deafferented," hypophysiotropic area was found to be able to maintain normal basal function of the tropic hormone-target organ systems, except that of the ovary; ovulation and other rhythmic endocrine phenomena were seriously altered in the "deafferented" animals (Halász and Pupp, 1965). Authors conclude that the hypophysiotropic area acts directly on the pituitary by its tropic and/or releasing substances carried there by the portal circulation, but that higher neuron structures exert a regulatory influence on it.

The assumption of the Halász group that the hypophysiotropic area is responsible for the production of tropic or releasing substances was directly proved by the experiments of Tima *et al.* (1969). They applied the technique of isolation, or "deafferentation," of the hypophysiotropic area as described by Halász and Pupp (1965). After one or two weeks the surviving rats were killed, and the isolated hypophysiotropic area was quickly removed and extracted with 0.1 *N* hydrochloric acid. The neutralized acidic extract of the hypophysiotropic area was injected into the carotid artery of normal male recipients. Twenty minutes following injection, the pituitaries of the recipients were homogenized in physiologic saline; the Luteinizing Hormone (LH) content was assayed by the Parlow (1961) procedure; the Follicle-Stimulating Hormone (FSH) content of these pituitaries was assayed by the Steelman and Pohley (1953) method. These procedures are generally known as the "pituitary depletion methods" for the evaluation of releasing factors and were originally described by Dávid *et al.* (1965). The same techniques were used in the experiments aimed at investigating the finer localization of the nuclei that manufacture the different releasing factors inside the hypophysiotropic area (see below). Luteinizing Hormone Releasing Factor (LHRF) concentration of the deafferented hypophysiotropic area was increased 8 days after operation, while LH content of the pituitary showed a slight decrease (Fig. 1). The increase in hypothalamic LHRF concentration was even higher 15 days following deafferentation; two weeks following deafferentation, pituitary LH stores were reduced to the same extent as those found 8 days after deafferentation (Fig. 2). These results might be interpreted by assuming that LHRF is synthesized within the isolated hypophysiotropic area. The trigger mechanisms for the release of LHRF located outside the hypophysiotropic area being completely excluded, all LHRF produced is stored in the hypothalamic island (Tima *et al.*, 1969).

The experiments of Tima *et al.* (1969) also provided convincing evidence in favor of the assumption that the Follicle-Stimulating Hormone Releasing Factor (FSHRF) is synthesized by the paraventricular area (see below). FSHRF content of the deafferented hypothalamus was significantly depressed 8 days after operation. No change, however, was observed in the pituitary stores of FSH (Fig. 3). An even bigger reduction of hypothalamic FSHRF concentration was found 15 days following deafferentation. The decrease in pituitary FSH content was again insignificant (Fig. 4).

The probable cause of the decreased hypothalamic FSHRF concentration was evident after the histological investigation of the brains. The operation excluded the greater part of the paraventricular area and almost the whole paraventricular nucleus from the hypothalamic island. Therefore none or a considerably reduced amount of FSHRF-producing elements (see below) were present in the deafferented hypothalamus. These findings give support to the assumption

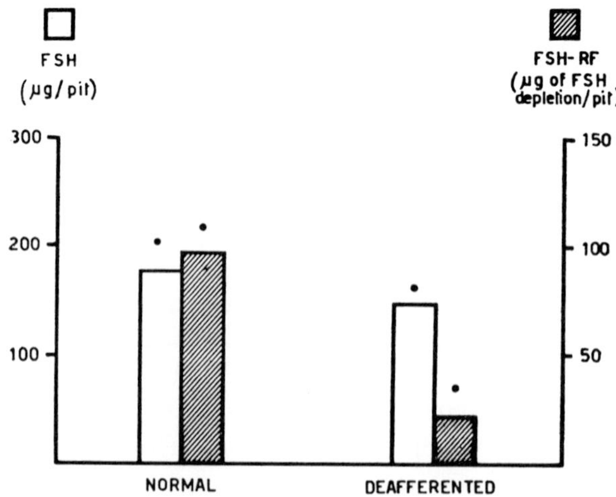

FIG. 3. Effect of hypothalamic deafferentation on pituitary FSH and hypothalamic FSHRF content in male rats (8 days after operation).

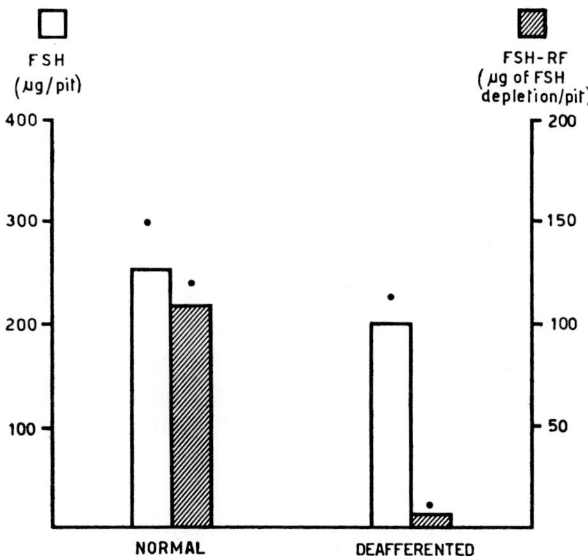

FIG. 4. Effect of hypothalamic deafferentation on pituitary FSH and hypothalamic FSHRF content in male rats (15 days after operation).

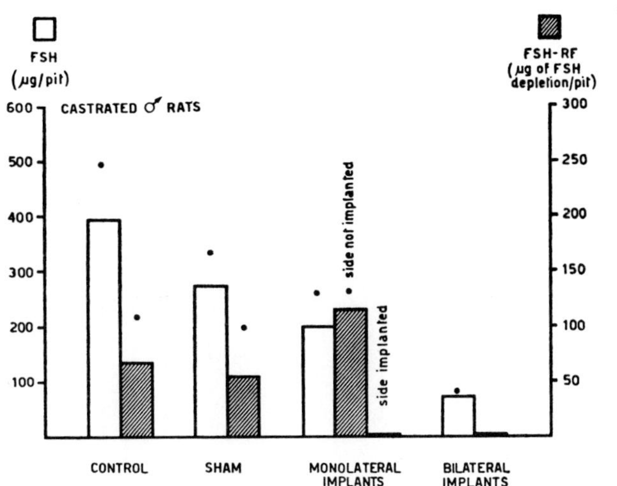

FIG. 5. Effect of implants of cycloheximide in the paraventricular area on pituitary FSH and hypothalamic FSHRF content in castrated male rats.

that FSHRF is really synthesized by the neurons of the paraventricular area. The observation that FSH is still present in nearly normal concentration in the pituitary in spite of the complete lack of hypothalamic FSHRF suggests the possible existence of an FSH-synthesizing factor independent of FSHRF.

The fact that the paraventricular nuclei might produce FSHRF was also proved by Zanisi and Martini (1969). They implanted cycloheximide, an inhibitor of protein synthesis, into the paraventricular nuclei and evaluated the changes in FSHRF content of extracts prepared from the median eminence of implanted animals; FSH stores of the anterior pituitary were also measured. Both were significantly depressed following bilateral implantation of the drug into the paraventricular region. Unilateral implantation caused only a decrease in FSHRF stores in the ipsilateral half of the median eminence; no change was observed in the FSH content of the pituitary (Fig. 5). It was concluded from these experiments that local inhibition of protein synthesis in the paraventricular area inhibits the synthesis of FSHRF.

Since no change could be observed in either hypothalamic LHRF concentration or the pituitary LH stores following bilateral im-

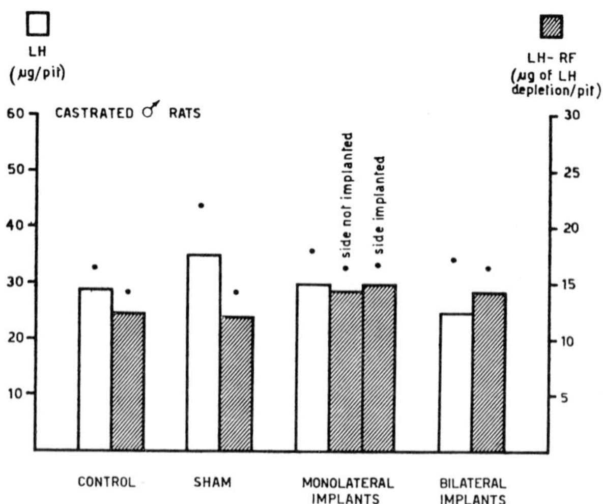

FIG. 6. Effect of implants of cycloheximide in the paraventricular area on pituitary LH and hypothalamic LHRF content in castrated male rats.

plantation of cycloheximide (Fig. 6), this clearly indicates that blockade of protein synthesis in the paraventricular nuclei inhibits exclusively the production of FSHRF, but not that of LHRF. It further proves that FSHRF synthesis occurs only in the paraventricular area.

B. Finer Localization of Areas Producing Different Releasing and Inhibiting Factors inside the Hypophysiotropic Area

1. Experiments Performed by Extracting Releasing and Inhibiting Factors from Different Regions of the Hypothalamus

The next and most interesting question was that of ascertaining whether the production of the different hypothalamic principles is localized in different nuclei within the hypophysiotropic area. The fact that localized electrolytic lesions which destroy different hypothalamic nuclei within the hypophysiotropic area (but not having altered the stalk median eminence region) may separately inhibit the secretion of one or the other tropic hormone presents evidence for the assumption that the production of the different releasing and inhibiting factors is bound to different hypothalamic areas. Experiments of this character were presented by many authors with respect to practically all tropic hormones (see for reference Szentágothai *et al.*, 1968).

Direct evidences were obtained by two different methods of approach. The first generally used technique for this purpose was the section of the hypothalamus into small portions and measurement of the releasing or inhibiting activity of these portions. The greatest disadvantage of this approach lies in the difficulty and variability of macroscopic dissection of fresh hypothalamic tissue. The boundaries of the different areas can hardly be standardized by this way of dissection. Data obtained by the use of this technique are summarized in Table II. It can be seen that most of the authors agree that releasing factors are produced only inside the hypothalamus. Thyrotropin Releasing Factor (TSHRF) was found by Averill and Kennedy (1967) also in the cerebrospinal fluid, but this might easily be interpreted by the close anatomical relation between the optic and infundibular recess of the third ventricle and the hypophysiotropic area of the hypothalamus. Endröczi and Hilliard (1965) des-

TABLE II

LOCALIZATION OF PRODUCTION OF RELEASING AND INHIBITING FACTORS

Releasing Factors	Hypothalamic Areas: Pituit. Stalk	SME	POA	SON	SCH	PV	Arc	LHA	MB	Whole Hypothal. except SME	Mes	CSF	Cerebral Cortex	References
CRF	0	*	0	*	0	–	0	–	–	0	0	0	–	Brizzee and Eik-Nes (1961) (dog)
	0	*[3]	0	0	0	0	0	0	0	*[3]	0	0	0	Witorsch and Brodish[1] (1968) (rat)
	*	–	0	0	0	0	0	0	0	0	0	0	0	Porter (1968) (rat)
TSHRF	0	*	0	0	0	0	0	0	0	*	0	0	–	Guillemin *et al.* (1965) (sheep)
	0	**	0	0	0	0	0	0	0	*	0	*	–	Averill and Kennedy (1967) (rat, sheep)
FSHRF	*	**	–	–	–	–	**	–	–	0	0	0	–	Watanabe and Mc-Cann (1968) (rat)
LHRF	0	**	0	*	*	*	*	*	*	0	0	0	–	McCann (1962) (rat)
	*	**	*	*	*	0	*	0	*	0	*	0	–	Endröczi and Hilliard (1965) (dog, rabbit)
	0	*	*	0	*	0	0	0	0	0	0	0	0	Moszkowska and Kordon (1965) (rat)
	0	**	–	*	*	0	–	–	–	0	0	0	–	Yamashita (1966) (dog)
	0	0	0	0	0	0	0	*	0	0	0	0	0	Barry *et al.* (1966) (guinea pig)
MSHRF	0	*	0	–	0	**	0	0	0	0	0	0	–	Taleisnik *et al.* (1966) (rat)
	0	*	0	–	0	0	0	0	0	0	0	0	–	Taleisnik and Tomatis (1967) (rat)
MSHIF	0	–	0	**	0	0	0	0	0	0	0	0	0	Taleisnik and Tomatis (1967) (rat)
GHRF	0	**	0	0	0	**[2]	0	0	0	0	0	0	0	Schally *et al.* (1968) (cat)

[1] Experiments made by hypothalamic lesion or pituitary stalk transection.
[2] Experiments made by paraventricular lesions.
[3] Not clearly indicated which region of the hypothalamus was damaged.

SYMBOLS:
- ** More releasing factor (in relation to *).
- * Less releasing factor (in relation to **).
- – No releasing factor present.
- 0 Not investigated.

ABBREVIATIONS: SME, stalk median eminence; POA, preoptic area; SON, supraoptic nucleus; SCH, suprachiasmatic nucleus; PV, paraventricular nucleus; Arc, arcuate nucleus; LHA, lateral hypothalamic area; MB, mammillary bodies; Mes, mesencephalon; CSF, cerebrospinal fluid.

cribed LHRF content of the mesencephalon and of the amygdala complex in the dog and rabbit. However, these results still remain questionable because, in the same experiments, LHRF activity was found not only in the aforementioned extrahypothalamic structures, but also in the anterior as well as in the posterior part of the hypothalamus; in addition, the criterion used by these authors to measure LHRF (the increase of ovarian progesterone secretion of the pregnant or pseudopregnant rabbit) is not the most specific test for assaying LHRF. Strangely enough, Barry *et al.* (1966) found LHRF exclusively in the lateral hypothalamic area in the guinea pig. This might be due to species difference or to some other unknown factors; at any rate their observations are unique.

With respect to the exact intrahypothalamic localization of production of the other releasing factors, the data found in the literature are very confusing. CRF is synthesized according to the results of Porter (1968) exclusively in the distal portion of the pituitary stalk, while Witorsch and Brodish (1968) found CRF activity to be diffusely present in the whole hypothalamus as well as in the stalk median eminence. In contrast to this, Brizzee and Eik-Nes (1961) localize CRF production in the stalk median eminence and in the supraoptic region. Authors agree that TSHRF, besides being present in the stalk median eminence region is also present in the entire hypothalamus (Guillemin *et al.*, 1965; Averill and Kennedy, 1967). FSHRF is synthesized, according to Watanabe and McCann (1968), in the stalk median eminence and in its adjacent area, including the arcuate nuclei. This is in contradiction to the data of Tima *et al.* (1969) and Zanisi and Martini (1969), obtained by other techniques, who consider the paraventricular region to be the place of FSHRF synthesis. The most contradictory observations could be found with respect to LHRF production. It is worthy of mention, however, that most authors consider the site of LHRF synthesis to be in the anterobasal region of the hypothalamus (in the supraoptic and suprachiasmatic nuclei). The production of the releasing factor that controls Melanocyte-Stimulating Hormone (MSHRF) secretion might be localized into the paraventricular area (Taleisnik *et al.*, 1966), while supraoptic nuclei would produce an antagonistic factor, the so-called Melanocyte-Stimulating Hormone Inhibiting Factor (MSHIF) (Taleisnik and Tomatis, 1967). Schally *et al.* (1968) described retardation of body growth and disappearance of Growth

Hormone Releasing Factor (GHRF) in stalk median eminence extracts following bilateral lesions of the paraventricular nuclei in the kitten. This gives fairly reliable evidence that the paraventricular region might be responsible for the production of GHRF. Besides this single but direct observation, many indirect and conflicting data are available concerning the localization of the neurons producing GHRF. Using the technique of putting lesions into the hypothalamus, the posterior part of the median eminence (Krulich *et al.*, 1965) and the ventromedial hypothalamic nuclei (Han and Liu, 1966; Frohman and Bernardis, 1968) were found to be responsible for the secretion of GHRF. Joseph and Knigge (1968), however, deny any role of the ventromedial nuclei in the regulation of Growth Hormone (GH) secretion and consequently in the production of GHRF.

No direct or exact data are available concerning the sites of production of other releasing and inhibiting factors (Growth Hormone Inhibiting Factor, Prolactin Inhibiting Factor, Prolactin Releasing Factor).

2. *Experiments Performed by Lesioning Different Hypothalamic Regions*

Another technique for investigating the exact location of neurons that synthesize the different hypophysiotropic principles is the use of animals bearing localized, circumscribed lesions in different areas of the hypothalamus. This technique has been used by Mess and co-workers (1967). Animals were killed 5 to 7 days after placement of different types of hypothalamic lesions; the content of different releasing factors in the hypothalamic extract of the lesioned animals was assayed by the method of Dávid *et al.* (1965) as described in detail before. For bioassay of thyrotropin a modified McKenzie (1958) test (Yamazaki *et al.*, 1963) was used. In order to evaluate CRF content, the stalk median eminence extracts were injected intravenously into female rats that had been pretreated 4 hr previously with 25 μg/100 g body weight of dexamethasone. The recipient animals were killed 15 min following injection of the extracts and their plasma corticosterone levels were evaluated (Fraschini *et al.*, 1966).

The use of animals bearing electrolytic hypothalamic lesions has two main advantages. The hypothalamic damage is circumscribed and can be localized exactly so that the hypothalamic areas producing the different releasing factors can be delineated. The other advantage

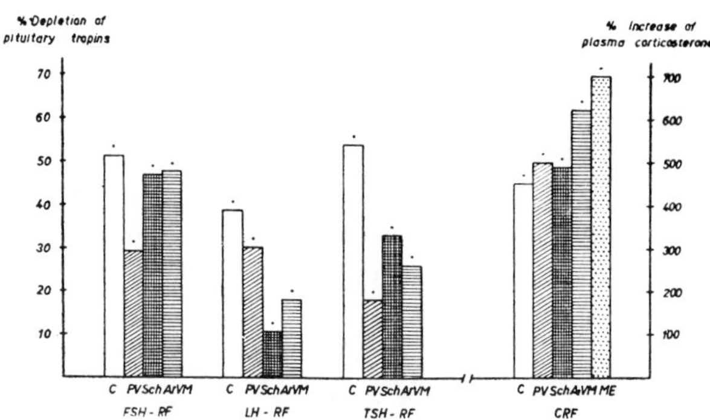

FIG. 7. Effect of localized lesions on the hypothalamic concentration of different releasing factors in male rats. C, controls; PV, lesion in the paraventricular area; Sch, lesion in the suprachiasmatic region; ArVM, lesion in the arcuate-ventromedial area; ME, lesion in the median eminence.

of this procedure is that all releasing factors are investigated simultaneously in the hypothalamus of the same groups of control and lesioned animals.

Four types of lesions were used in this study. These were localized (1) in the suprachiasmatic area (which includes the suprachiasmatic nuclei and most of the anterior hypothalamic nucleus); (2) in the paraventricular area (region including the magnocellular paraventricular nucleus); (3) in the region called the arcuate-ventromedial area (including the posterior part of the ventromedial hypothalamic nucleus and the arcuate nucleus anterior to the mammillary body); and (4) in the median eminence (lesions destroyed the anterior and central part of the median eminence with concomitant interruption of the pituitary stalk).

The results of these experiments are briefly summarized in Fig. 7. FSHRF content of the stalk median eminence extract was significantly depressed exclusively in rats bearing lesions in the paraventricular region. Destruction of the suprachiasmatic as well as of the arcuate-ventromedial area caused no change in the FSHRF production. Hypothalamic LHRF concentrations were equally diminished following lesions of the suprachiasmatic and of the arcuate-ventromedial area. An insignificant reduction was observed in the group of animals bearing a lesion in the paraventricular region. This is the very probable consequence of the extension of the lesions somewhat more ventrally, i.e., toward the suprachiasmatic region. TSHRF was nearly equally and considerably decreased in the stalk median eminence following all types of lesions. No change, or even a moderate increase, of CRF concentration was observed in the different groups of animals bearing hypothalamic lesions. This increase in CRF concentration was particularly evident in the group with median eminence lesions, and was somewhat less pronounced in rats bearing lesions in the arcuate-ventromedial area.

The interpretation of this latter finding might be based on the impossibility for animals with damage in the portal circulation to deliver hypothalamic hypophysiotropic principles to the anterior pituitary. The other question as to why CRF production could not be inhibited by any type of hypothalamic lesion might be interpreted in two different ways. Either CRF is synthesized in a zone that has not been lesioned or the structures that manufacture this mediator have a very large distribution; therefore, those regions that have not been lesioned can very rapidly compensate the loss of the lesioned areas. The second possibility seems to be more probable, and some

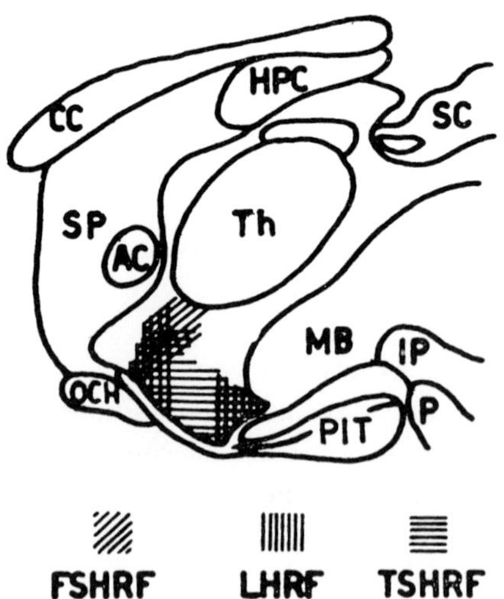

FIG. 8. Localization of the hypothalamic areas manufacturing FSHRF, LHRF and TSHRF. AC, anterior commissure; CC, corpus callosum; HPC, hippocampal commissure; IP, interpeduncular nucleus; MB, mammillary bodies; OCH, optic chiasm; P, pons; PIT, pituitary; SC, superior colliculus; SP, septum pellucidum; Th, thalamus.

data in the literature (D'Angelo *et al.*, 1964; Brodish, 1968) lend support to this assumption.

Our results have demonstrated, on the other hand, that TSHRF production is distributed over a broad and diffuse region of the hypothalamus. This is in direct parallelism with the findings of Guillemin *et al.* (1965) and Averill and Kennedy (1967), whose experiments showed that an extract prepared from the whole hypothalamus after elimination of the median eminence had nearly the same TSHRF activity as the stalk median eminence itself. Physiological data obtained by electrical stimulation of different hypothalamic areas (D'Angelo and Snyder, 1963) also showed no circumscribed localization of the structures regulating TSH secretion.

The areas producing FSHRF and LHRF were those most exactly defined; this can be seen on the schematic representation of the hypothalamic regions producing the different releasing factors (Fig. 8). FSHRF production was found exclusively in the paraventricular area, while LHRF is synthesized in two separated areas (in the suprachiasmatic as well as in the arcuate-ventromedial area). The secretion of TSHRF is bound to a more extended area of the hypothalamus, including nearly the whole hypophysiotropic area. The case might be similar to CRF production, as discussed before.

C. Effect of Hypothalamic Lesions outside the Hypophysiotropic Area on the production of the Thyrotropin Releasing Factor

All the experiments quoted so far provide good evidence that hypothalamic hypophysiotropic principles are synthesized in the hypophysiotropic area; apparently structures localized within this area are involved in the production of the different releasing and inhibiting factors. These results, however, do not exclude the possibility that these factors might also be synthesized outside the hypophysiotropic area. This assumption seems to be most relevant with respect to the production of TSHRF, since this releasing factor has the most extended representation in the hypophysiotropic area; i.e., it is diffusely produced in this entire area. For investigation of this question, bilateral lesions were put into two hypothalamic regions outside the hypophysiotropic area and the TSHRF concentration of the stalk median eminence extract was assayed by means of the intracarotid technique, described above. Dorsomedial nuclei, which lie

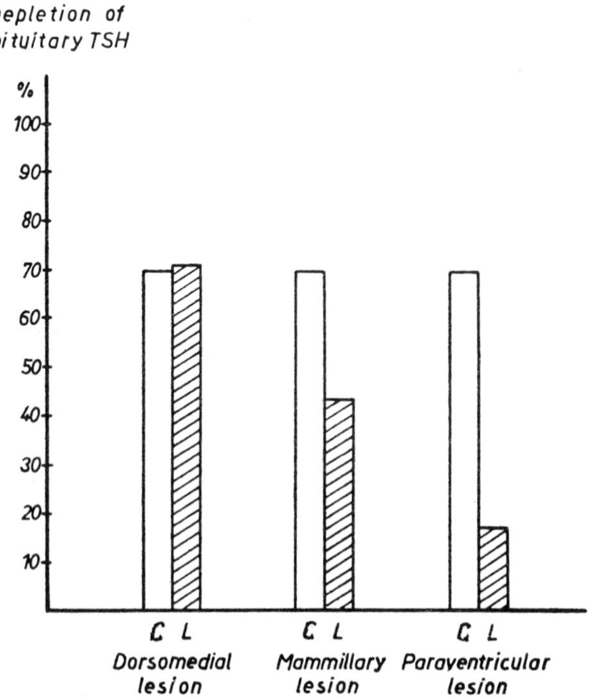

FIG. 9. Effect of lesions performed outside the hypophysiotropic area on hypothalamic TSHRF concentration in male rats. C, controls; L, lesioned.

dorsal to the hypophysiotropic area and mammillary nuclei, which are at the caudal edge of this area, were chosen for this purpose.

TSHRF stores of the stalk median eminence remained unchanged 7 days following lesion of the dorsomedial nuclei. A moderate decrease of hypothalamic TSHRF content was observed in animals bearing mammillary lesions. The slight decrease of TSHRF concentration found following lesions of the mammillary nuclei is very probably due to the fact that lesions of the mammillary bodies very often extended into the premammillary region, which is already a part of the hypophysiotropic area (Fig. 9). These latter results give clear-cut evidence for our assumptions that releasing factors (TSHRF in this particular case) are synthesized by the hypophysiotropic area of the hypothalamus and that hypothalamic structures outside this area are not directly involved in the production of hypothalamic mediators (Mess, 1969).

III. Possible Role of Extrahypothalamic Structures in the Regulation of the Production and/or of the Release of Releasing and Inhibiting Factors

A bulk of convincing experimental evidence proves that hypothalamic nuclei outside the hypophysiotropic area, or even a series of extrahypothalamic structures, exert important influences upon the different tropic hormone-target organ systems. These data do not contradict the proposed localization of the synthesis of the different releasing and inhibiting factors within the hypophysiotropic area. Extrahypothalamic structures, belonging mainly to the limbic system (e.g., amygdala complex, habenular region, and hippocampus) or to the lower brain stem (e.g., reticular formation of the mesencephalon) might influence the rate of synthesis or release of these different hypothalamic mediators, either by direct or indirect neuronal connections with the hypophysiotropic area or by some indirect interglandular interplay. The releasing and the inhibiting factors must then be considered the final link between the hypothalamus and the adenohypophyseal cells. The hypophysiotropic area represents in this way the final common pathway between the central nervous system and the hypophyseal-target organ unit.

References

Averill, R. L. W., and Kennedy, T. H. (1967). *Endocrinology* **81**, 113.

Barry, J., Biserte, G., Lefranc, G., Leonardelli, J., and Moschetto, Y. (1966). *C. R. Acad. Sci.* (Paris) **263**, 536.

Brizzee, K. R., and Eik-Nes, K. B. (1961). *Endocrinology* **68**, 166.

Brodish, A. (1968). *Excerpta Medica Intern. Congr. Ser.* **157**, 79.

D'Angelo, S. A., and Snyder, J. (1963). *Endocrinology* **73**, 75.

D'Angelo, S. A., Snyder, J., and Grodin, J. M. (1964). *Endocrinology* **75**, 417.

Dávid, M. A., Fraschini, F., and Martini, L. (1965). *C. R. Acad. Sci.* (Paris) **261**, 2249.

Endröczi, E., and Hilliard, J. (1965). *Endocrinology* **77**, 667.

Franz, J., Haselbach, C. H., and Libert, O. (1962). *Acta Endocrinol.* **41**, 336.

Fraschini F., Motta, M., and Martini, L. (1966). *In* "Methods in Drug Evaluation" (P. Mantegazza and F. Piccinini, eds.), pp. 424–457. North-Holland Publishing Company, Amsterdam.

Frohman, L. A., and Bernardis, L. L. (1968). *Endocrinology* **82**, 1125.

Guillemin, R., Schally, A. V., Andersen, R., Lipscomb, H. S., and Long, J. M. (1960). *C. R. Acad. Sci.* (Paris) **250**, 4462.

Guillemin, R., Sakiz, E., and Ward, D. N. (1965). *Proc. Soc. Exp. Biol. Med.* **118**, 1132.

Halász, B., and Pupp, L. (1965). *Endocrinology* **77**, 553.

Halász, B., Pupp, L., and Uhlarik, S. (1962). *J. Endocrinol.* **25**, 147.

Halász, B., Pupp, L., Uhlarik, S., and Tima, L. (1963). *Acta Physiol. Acad. Sci. Hung.* **23**, 287.

Halász, B., Pupp, L., Uhlarik, S., and Tima, L. (1965). *Endocrinology* **77**, 343.

Han, P. W., and Liu, A. C. (1966). *Am. J. Physiol.* **211**, 229.

Igarashi, M., and McCann, S. M. (1964). *Endocrinology* **74**, 446.

Joseph, S. A., and Knigge, K. M. (1968). *Neuroendocrinology* **3**, 309.

Kastin, A. J. (1965). *Program 47th Meeting Endocrine Soc.*, p. 98.

Kragt, C. L., and Meites, J. (1965). *Endocrinology* **76**, 1169.

Krulich, L., Dhariwal, A. P. S., and McCann, S. M. (1965). *Program 47th Meeting Endocrine Soc.*, p. 21.

Krulich, L., Lackey, R. W., and Dhariwal, A. P. S. (1967). *Fed. Proc.* **26**, 316.

McCann, S. M. (1962). *Am. J. Physiol.* **202**, 395.

McCann, S. M., Taleisnik, S., and Friedman, H. M. (1960). *Proc. Soc. Exp. Biol. Med.* **104**, 432.

McKenzie, J. M. (1958). *Endocrinology* **63**, 372.

Mess, B. (1969). *In* " Progress in Endocrinology " (C. Gual, ed.), pp. 564–570. Excerpta Medica, Amsterdam.

Mess, B., and Martini, L. (1968). *In* " Recent Advances in Endocrinology " (V. H. T. James, ed.). 8th ed., pp. 1–49. J. and A. Churchill, London.

Mess, B., Fraschini, F., Motta, M., and Martini, L. (1967). *In* " Hormonal Steroids " (L. Martini, F. Fraschini, and M. Motta, eds.), pp. 1004–1013. Excerpta Medica, Amsterdam.

Moszkowska, A., and Kordon, C. (1965). *Gen. Comp. Endocrinol.* **5**, 596.

Parlow, A. F. (1961). *In* " Human Pituitary Gonadotropins " (A. Albert, ed.), pp. 300–326. C. C. Thomas, Springfield, Ill.

Pasteels, J. L. (1962). *C. R. Acad. Sci.* (Paris) **254**, 2664.

Porter, J. C. (1968). *Fed. Proc.* **27**, 217.

Saffran, M., Schally, A. V., and Benfey, B. G. (1955). *Endocrinology* **57**, 439.

Schally, A. V., Lipscomb, H. S., Long, J. M., Dear, W. E., and Guillemin, R. (1962). *Endocrinology* **70**, 478.

Schally, A. V., Sawano, S., Müller, E. E., Arimura, A., Bowers, C. Y., Redding, T. W., and Steelman, S. L. (1968). *In* " Growth Hormone " (A. Pecile and E. E. Müller, eds.), pp. 185–203. Excerpta Medica, Amsterdam.

Shibusawa, K., Nishi, K., and Abe, C. (1959). *Endocrinol. Japon.* **6**, 31.

Steelman, S. L., and Pohley, F. M. (1953). *Endocrinology* **53**, 604.

Szentágothai, J., Flerkó, B., Mess, B., and Halász, B. (1968). " Hypothalamic Control of the Anterior Pituitary." Akadémiai Kiadó, Budapest.

Taleisnik, S., and Orias, R. (1965). *Am. J. Physiol.* **208**, 293.

Taleisnik, S., and Tomatis, M. E. (1967). *Endocrinology* **81**, 819.

Taleisnik, S., Orias, R., and De Olmos, J. (1966). *Proc. Soc. Exp. Biol. Med.* **122**, 325.

Tima, L., Motta, M., and Martini, L. (1969). *Program 51st Meeting Endocrine Soc.* p. 194.

Watanabe, L., and McCann, S. M. (1968). *Endocrinology* **82**, 664.

Witorsch, R., and Brodish, A. (1968). *Fed. Proc.* **27**, 217.

Yamashita, K. (1966). *J. Endocrinol.* **35**, 401.

Yamazaki, E., Sakiz, E., and Guillemin, R. (1963). *Experientia* **19**, 480.

Zanisi, M., and Martini, L. (1969). *Program 51st Meeting Endocrine Soc.*, p. 202.

Chemistry and Physiological Aspects of Hypothalamic Releasing and Inhibiting Factors

S. M. McCANN

I. Introduction

Several articles in this volume have documented fully the imposing array of evidence which indicates that the hypothalamus regulates the secretion of the adenohypophyseal hormones. It is also clear that the anterior pituitary is a gland under neural control but lacking a secretomotor innervation. Instead there is a specialized vascular link between the hypothalamus and the pituitary, the hypophyseal portal system of veins. These take origin in the median eminence and stalk and carry capillary blood from these regions to the sinusoids of the anterior lobe. Hinsey (1937), Green and Harris (1949), and others suggested that hypothalamic control over the pituitary might be mediated by means of specific neurohormones which are liberated into the primary capillary plexus of the hypophyseal portal vessels and pass down the pituitary stalk to reach the sinusoids of the anterior lobe, there to trigger release of particular pituitary hormones. This prophetic suggestion has born fruit, and it is now evident as a result of many investigations over the past 10 to 15 years that a whole new family of neurohormones called hypophysiotropic hormones or hypothalamic releasing and hypothalamic inhibiting factors governs the secretion of each hormone from the adenohypophysis.

There appears to be at least one neurohormone for each pituitary hormone and in some instances both releasing and inhibiting factors appear to exist, thereby providing a dual control over the pituitary hormone (McCann and Porter, 1969). The nomenclature used specifies first the pituitary hormone affected and follows this by the term releasing factor. Corticotropin Releasing Factor (CRF) was the first such factor to be described. Shortly thereafter a Luteinizing Hormone Releasing Factor (LHRF) a Follicle-Stimulating Hormone Releasing Factor (FSHRF), a Thyrotropin Releasing Factor (TSHRF), and a Growth Hormone Releasing Factor (GHRF) were described. In contrast to these pituitary hormones which are under the influence of a stimulatory factor from the hypothalamus, prolactin is under inhibitory hypothalamic control, which is mediated by a Prolactin Inhibiting Factor (PIF). More recently, evidence has accrued to suggest the presence of dual factors influencing Melanocyte-Stimulating Hormone (MSH) discharge, an MSH-Releasing Factor (MSHRF), and an MSH-Inhibiting Factor (MSHIF). There is evidence also for an inhibitor of growth hormone release, Growth Hormone Inhibiting Factor (GHIF). We will discuss briefly the evidence for the existence, physiological significance, and chemical nature of these new hypothalamic hormones.

II. The Assay of Releasing and Inhibiting Factors

One of the major problems which has confronted investigators in this field has been the development of suitable bioassays for these new hypothalamic hormones. The general types of assays are summarized in Table I. In the case of *in vivo* assays, of course it would be preferable to monitor the secretion rate of the pituitary hormone in question. This is not possible at this time, so that less direct indices of pituitary hormone release have had to be employed, such as measurement of blood levels of pituitary hormones or output of target gland hormones. Normal animals can be used to assay releasing factors, provided the output of the hormone to be studied is not

TABLE I

METHODS OF ASSAY OF RELEASING FACTORS

A. *In Vivo:* Administer releasing factor under study and measure

1. increased blood level of pituitary hormone, or change in target organ secondary to increased pituitary secretion
2. change in pituitary hormone content in
    - *a.* normal animals
    - *b.* animals with pharmacological blockade
        - (1) target hormone-treated
        - (2) central nervous system depressant-treated (i.e., barbiturate, morphine)
    - *c.* animals with lesions in the median eminence
    - *d.* animals with a cannula in anterior pituitary gland for intrapituitary infusion

B. *In Vitro:* Add releasing factor and measure

1. hormone released from pituitary tissue
2. hormone content stored in pituitary in
    - *a.* short-term incubation
    - *b.* long-term tissue or organ culture

---

responsive to the manipulation of the animal and to the mere injection of tissue extracts. This situation may prevail in the case of Follicle-Stimulating Hormone (FSH) and Luteinizing Hormone (LH), but most certainly does not hold for most other pituitary hormones, including Adrenocorticotropic Hormone (ACTH), Growth Hormone (GH), prolactin, and possibly MSH (see McCann *et al.*, 1968 for references). For this reason, various drugs have been employed to block the nonspecific responses.

A variety of central nervous system depressant drugs are useful in the assay of CRF (McCann and Porter, 1969) and have recently also been employed in assay of GHRF (McCann and Porter, 1969). The disadvantages here are the question of a possible action of the drug on the pituitary directly and the likelihood that the blockade of nonspecific release will be incomplete. Similarly, inhibition of tropic hormone release by target gland hormones has been used to suppress nonspecific release of ACTH (McCann and Porter, 1969). This treatment has the disadvantage that it may suppress the responsiveness of the pituitary to CRF (Russell *et al.*, 1970). This certainly is the case for thyroxine and TSHRF (Vale *et al.*, 1968), but apparently large doses of gonadal steroids do not suppress the responses to gonadotropin releasing factors and may even enhance them (Ramirez and McCann, 1963).

Theoretically, an animal with a lesion in the median eminence which eliminates neural control over the gland should constitute an ideal assay animal for the characterization of releasing factors. One of the problems with this technique is the question of the completeness of the lesion. Also, because of the necessary interruption of some portal vessels temporarily, the blood supply to the gland may be impaired for a while. It has recently been shown (Porter *et al.*, 1967) that most of these animals have a normal pituitary blood flow two days after lesions. However, with time, these animals may lose their sensitivity to releasing factors and/or pituitary hormones. This is true for CRF (McCann, 1957). Nonetheless this type of assay when properly employed has given very valuable information, and if care is taken to ensure completeness of the lesion, it is probably the most rigorous *in vivo* assay for releasing factors.

Still another type of *in vivo* assay has been the infusion of extract directly into the anterior pituitary. The problem is that part of the infusate may enter the interstices of the gland and could lead to nonspecific effects. Recent work makes it apparent that nonspecific results can occur if the rate of infusion is so fast that extract can flow retrograde along the needle track into the hypothalamus or if concentrations of extract are so high that cellular injury is produced (Dhariwal *et al.*, 1969).

It is now clear from results obtained in many laboratories that the releasing factors can be assayed *in vitro* in either short-term or long-term incubation experiments. Dose-response relationships can be established and under certain conditions there appears to be little interference by other tissue extracts. One advantage is that effects *in vitro* must be exerted directly on the gland. It is clear that high doses of a variety of agents can cause release of CRF or FSH in such a system, but the doses required are so great that they are unlikely to be encountered in hypothalamic extracts (McCann and Porter, 1969).

The precision encountered with both *in vivo* and *in vitro* methods of assay of releasing factors leaves much to be desired. This lack of precision may be attributed not only to the variability of pituitary response to a given releasing factor, but also to the variability in the assay of the released pituitary hormone. It is apparent from the foregoing discussion that there are advantages and disadvantages to each assay so far devised. It is probably advisable not to rely solely on a single type of assay before drawing a firm conclusion (McCann and Porter, 1969).

III. The Physiological Significance of Releasing and Inhibiting Factors

Hypothalamic extracts from a variety of mammalian species, including man, have been found to possess hormone releasing and inhibiting activities, and these hypothalamic extracts are active in rats, rabbits, monkeys and even in a few instances in man. One recent example of the effect of hypothalamic extracts on release of

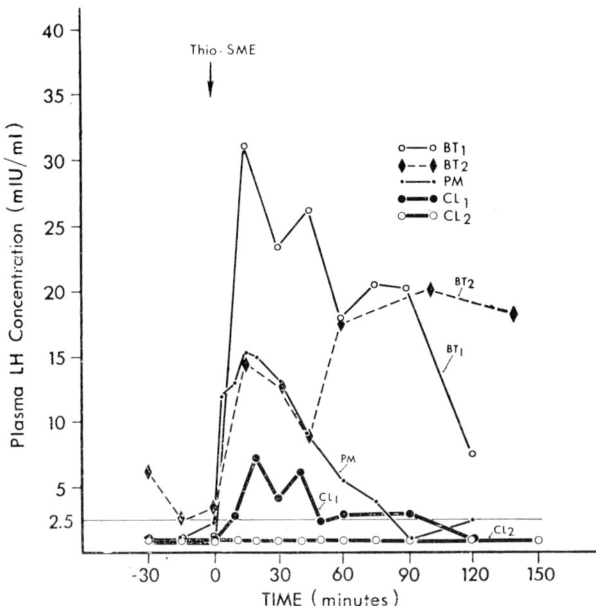

FIG. 1. Effect on LH release in human infants of intravenous injection of ovine hypothalamic extract treated with thioglycollate (Thio-SME) to inactivate neurohypophyseal polypeptide hormones. All patients responded with an increase in plasma LH. There was no response in one patient to a lower dose of extract (CL2), and none of the patients responded to an injection of equivalent amounts of a cerebral cortex extract (From Root *et al.*, 1969).

pituitary hormones in man is provided by the observation that these extracts, within a few minutes of their intravenous injection (Fig. 1), will increase the radioimmunoassayable LH in plasma of children with chromosomal abnormalities (Root *et al.*, 1969). Extracts derived from the cerebral cortex were ineffective.

The mere demonstration of these activities in hypothalami does not, of course, constitute proof of their physiological significance. Several lines of investigation indicate that these factors are probably of prime significance in conveying information from the hypothalamus to the pituitary gland. One type of experiment has been to impose a physiological condition in the animal which would alter the release of the pituitary hormone effected by a particular hypothalamic factor. Then the effect of this intervention on the hypothalamic content of the hypothalamic hormone is assessed. This type of experiment has been carried out extensively and the results indicate that, in general, acute or chronic alterations in the release of pituitary hormone can result in alterations in the content of stored hypothalamic factor. Some examples of this type of experiment are shown for LH and LHRF in Table II. This type of experiment, although implicating the hypothalamic factor in the regulation of pituitary hormone secretion, leaves a good bit to be desired because it is im-

TABLE II

CONDITIONS IN WHICH AN ALTERATION IN HYPOTHALAMIC LHRF HAS BEEN REPORTED IN RATS

Sex	Condition	Effect on LH Release	Effect on Hypothalamic LHRF
Male	Castration	+	+
	Testosterone, injection	—	—
	Testosterone, hypothalamic implant	—	—
Female	Spaying	+	—
	Estrogen, injection	—	—
	Estrogen, hypothalamic implant	—	—
	Enovid	—	—
	Proestrus	+	—
	Puberty	+	—
	Suckling	—	—
	Light	?	+ (sheep)

+ = increase
— = decrease

possible to be sure whether the change in hypothalamic content of the factor is due to altered synthesis or release of the factor, or a combination of the two (McCann and Porter, 1969).

If one could demonstrate the releasing factors in peripheral blood and show that the blood level of the factors varied under situations which alter pituitary hormone release, then strong evidence for the physiological significance of these factors would be at hand. To date it has not been possible to demonstrate releasing factors in peripheral blood of animals with intact pituitaries; however, after hypophysectomy a number of the factors appear in peripheral blood in sufficient quantities to be measurable. These activities have been reported to disappear after lesions in the median eminence, which should eliminate the source of the releasing factor. CRF activity was first reported in blood of hypophysectomized rats, and this was followed by reports of LHRF, GHRF and FSHRF activity in this blood (McCann and Porter, 1969). Most reports indicate that the level is not measurable immediately after hypophysectomy, but that it does rise after a period of time to detectable albeit rather low levels. There is some evidence that these levels of circulating releasing factors can be altered in the hypophysectomized animals by imposition of stimuli which alter pituitary hormone secretion in intact animals (Table III). For example, insulin-induced hypoglycemia, which is a stimulus for growth hormone release, caused the appearance of increased levels of GHRF in the blood of hypophysectomized rats

TABLE III

RELEASING FACTORS OBSERVED IN BLOOD OF HYPOPHYSECTOMIZED RATS

Factor	Stimuli which alter its level	Response to stimulus
CRF	Corticoids [1]	—
LHRF	LH, gonadal steroids [2]	—
FSHRF	Light [3]	+
	Testosterone [4]	—
	Reserpine [4]	—
GHRF	Hypoglycemia,[5] cold [6]	+

[1] Brodish (1962).
[2] Preliminary experiments of Nallar and Antunes-Rodriguez.
[3] Negro-Vilar *et al.* (1968a).
[4] Negro-Vilar *et al.* (1968b).
[5] Krulich and McCann (1966).
[6] Müller *et al.* (1967).

(Krulich and McCann, 1966). Stimuli which alter FSH release have been shown to alter the level of circulating FSHRF in blood of chronically hypophysectomized animals (Negro-Vilar *et al.*, 1968a, b).

Considerable gonadotropin secretion occurs when multiple pituitary grafts are placed into hypophysectomized hosts, and the degree of maintenance is proportional to the number of such grafts (Gittes and Kastin, 1966). Recent studies indicate that the residual function of the grafted gland is at least in part caused by circulating gonadotropin releasing factors, since median eminence lesions in the hypophysectomized-grafted animals led to regression of testicular and accessory organ weight (Beddow and McCann, 1969). Earlier it had been observed in a small series of animals that implants of testosterone in the basal tuberal region of hypophysectomized-grafted rats inhibited gonadotropin secretion, presumably because the androgen blocked the release from the hypothalamus of gonadotropin releasing factors (Smith and Davidson, 1967).

Why does LHRF (and a number of other releasing factors) appear in the blood of chronically hypophysectomized rats? The answer is not available, but in this situation there is reduction in output of target gland hormones. Thus negative feedback of gonadal steroids would be reduced, which should stimulate LHRF release and synthesis. Furthermore, if a negative feedback of LH on its own secretion also exists (McCann *et al.*, 1968), then this would also be eliminated by hypophysectomy, which could augment LHRF discharge.

Conclusive proof for the physiological significance of releasing factors would be available if their content in hypophyseal portal vessel blood could be shown to be higher than that in peripheral blood and to vary under conditions which vary the output of the target pituitary hormone. Porter was the first to show that CRF activity could be obtained in blood dripping from the cut stalk of the dog and that this activity was higher in stalk blood than in peripheral blood (Porter and Rumsfeld, 1956). More recently there have been reports of LHRF (Fink, 1967) and TSHRF (Averill *et al*,. 1966) activity in blood collected from the cut stalk of the rat.

Porter has recently developed a new technique for infusion of substances into individual portal vessels which have been microcannulated. With this technique he has been able to evoke acute increases in LH discharge from the pituitary after the infusion of LHRF (Porter and Mical, 1969).

IV. Interaction between Releasing Factors and Target Gland Hormones

If one accepts the significance of these hypothalamic releasing factors, it is important to examine the interplay between these new hormones and the target gland hormones which feed back to modify pituitary secretion (see McCann and Porter, 1969, and McCann *et al.*, 1968 for references). Although all the evidence is not at hand, it appears now that most target gland hormones feed back both at the hypothalamic and at the pituitary level to modify release of pituitary hormones. The relative importance of these two sites for feedback seems to vary, depending on the pituitary hormone in question. In the case of thyroxine the principal feedback appears to be at the pituitary level, and it is possible to block the response to TSHRF by infusion of thyroxine. In the case of the gonadal steroids the principal feedback actions appear to lie at the hypothalamic level, where both positive and negative feedback are operative. Even in this situation, some action of gonadal steroids directly at the pituitary level is demonstrable; however, it is difficult to block the response of the pituitary gland to LHRF or FSHRF by the administration of gonadal steroids, either *in vivo* or *in vitro*. In the case of adrenal steroids, perhaps a middle ground exists, since there is evidence for both hypothalamic and pituitary sites of feedback, but recent work clearly indicates that the response of the pituitary gland to CRF can be suppressed by adrenal steroids (Russell *et al.*, 1969).

Considerable attention has been focused recently on the possibility that pituitary hormones themselves feed back, in an autofeedback, to alter the secretion of the pituitary hormone in question (McCann and Porter, 1969; Martini *et al.*, 1968). These autofeedbacks may occur at least in part at the hypothalamic level, since implants of the pituitary hormones in the hypothalamus will suppress pituitary hormone secretion and also alter the content of the stored releasing factor in the hypothalamus. As in the case of target gland hormones, it is also possible that the autofeedback action may occur at least in part at the pituitary level.

V. Characteristics of Action of Releasing Factors

It has been considered axiomatic that at least one of the primary actions of the releasing factors is to stimulate release of a particular pituitary hormone from its cellular storage site. This view is supported by the very rapid release of hormones which follows injection of releasing factors into the circulation. For example, ACTH release occurs within a minute or two after application of stressful stimuli or injection of CRF. Thyrotropin (TSH) and GH release have also been shown to occur within a few minutes after intravenous injection of TSHRF and GHRF, whereas LH and FSH release occur within 5 or 10 min or less after administration of LHRF and FSHRF, respectively.

Since depletion of pituitary content of GH and FSH has been reported to occur shortly after giving the appropriate releasing factor, it would appear that in some instances release exceeds new synthesis of hormone, leading to a depletion of stores. Conversely, suckling-induced depletion of pituitary prolactin is blocked by PIF. These are strong arguments for a primary action of these agents on the release of hormone.

There is now little doubt that these neurohormones also influence synthesis of pituitary hormones, either by a primary action or secondary to release of stored hormones.

VI. Lack of Interaction between Luteinizing Hormone Releasing Factor and Other Releasing Factors

One additional important characteristic of the action of the releasing factors appears to be their specificity of action. For example, in the case of the LH-secreting cell, its release of hormone *in vitro* is uninfluenced by the addition of GHRF, GHIF, CRF, or FSHRF (unless contaminated with LHRF). These other releasing factors also fail to alter the release of LH, which is stimulated by the LHRF (Crighton *et al.*, 1969). If this specificity holds for each pituitary cell, it means that release of a given pituitary tropin is controlled primarily by the rate of discharge of its releasing factor into hypophyseal portal blood and is uninfluenced by discharge of other factors.

VII. The Synaptic Transmitter Which Releases Gonadotropin Releasing Factors

Influences from the rest of the central nervous system impinge on the secretory neurons which secrete LHRF. A variety of drugs which might be expected to interfere with transmission across adrenergic or cholinergic synapses are capable of blocking ovulation (Everett, 1964). Alterations in hypothalamic catecholamine, monoamine oxidase, and cholinesterase concentrations with the states of the estrous cycle or following altered titers of gonadal steroids have recently been observed (McCann and Porter, 1969; McCann *et al.*, 1968). Furthermore, drugs which deplete brain stores of catecholamines block Pregnant Mare Serum (PMS) -induced ovulation in the rat (Coppola *et al.*, 1966). Therefore it is possible that cholinergic and adrenergic synapses are involved at certain points in the transmission of information from the central nervous system to the pituitary. At one time it was thought that acetylcholine or epinephrine might be the transmitter agent released into the portal vessels to trigger gonadotropin release; however, current evidence renders this view untenable. The presence of serotonin- and noradrenaline-containing neurons in the hypothalamus has recently been demonstrated by fluorescence microscopy (Hillarp *et al.*, 1966). In particular, dopamine-containing neurons originate in the vicinity of the arcuate nucleus and end near the primary plexus of the hypophyseal portal vessels in the median eminence (Fuxe and Hökfelt, 1967). The content of dopamine in these neurons is altered in situations associated with altered gonadotropin secretion (Fuxe *et al.*, 1967). Thus the possibility was raised that dopamine might be LHRF or FSHRF. Recent evidence indicates that this is not the case, since dopamine failed to augment FSH (Kamberi and McCann, 1969) or LH release (Schneider and McCann, 1969) by pituitaries incubated *in vitro*. At high doses the amount of FSH and LH recovered from the medium was reduced, but this was caused by destruction of the hormone.

Although dopamine is not a gonadotropin releasing factor, it appears likely that it may play a role as a synaptic transmitter to stimulate release of LHRF from the secretory neurons. This possibility has been revealed by incubating median eminence tissue together with pituitaries *in vitro*. The addition of dopamine increased the release of LH from the combined tissue, whereas serotonin, epinephrine, or norepinephrine in equivalent doses were ineffective (Schneider and McCann, 1969) (Fig. 2). Control studies revealed that dopamine failed to potentiate the action of LHRF in releasing LH from pituitaries incubated alone, so the conclusion drawn was that dopamine evoked a release of LHRF from the hypothalamic fragments. Dopamine acts on α-adrenergic receptors in evoking LHRF release,

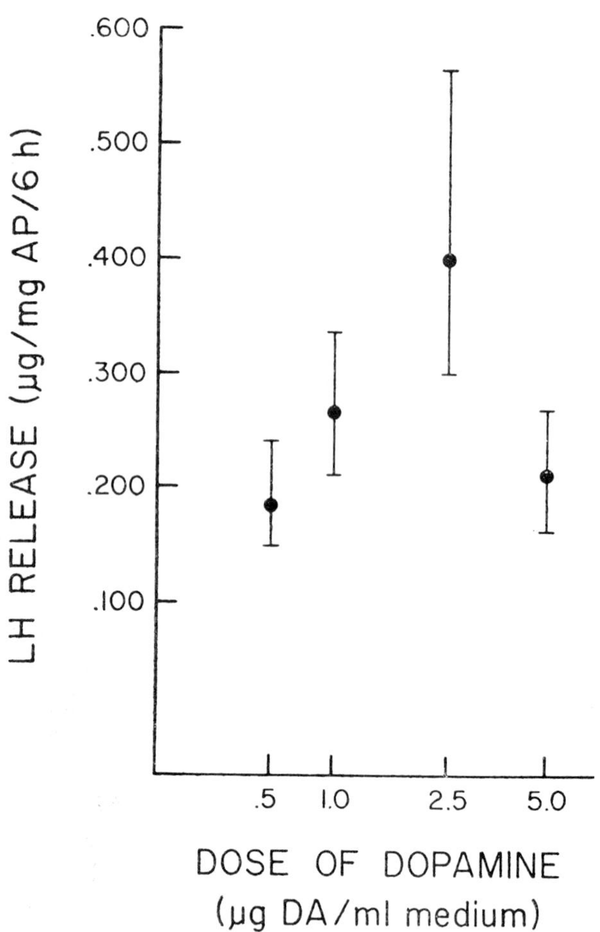

FIG. 2. Dose response relationship between dose of dopamine and LH released into the medium of anterior pituitaries incubated with stalk median eminence. The dose of dopamine is on a logarithmic scale. Vertical bars give the 95% confidence limits.

since the response to the drug was blocked by α- and not by β-receptor-blocking drugs. Recent experiments in which dopamine was infused directly into the third ventricle of rats indicate that it has an LH-releasing action *in vivo* as well. More recently we have shown that dopamine can enhance the release of FSHRF from median eminence incubated *in vitro*, and therefore dopamine may be the synaptic transmitter for release of FSHRF as well as LHRF (Kamberi, *et al.*, 1969).

VIII. Chemistry of Releasing Factors

In spite of much effort in a number of laboratories over the past 10 to 15 years the chemical nature of the releasing factors remains elusive. We will attempt to summarize very briefly the progress to date (for references see McCann *et al.*, 1968; McCann and Porter, 1969; Harris *et al.*, 1966; Schally *et al.*, 1968; McCann, 1970).

Considerable progress has been made in purification of releasing factors with the utilization of hypothalamic tissue as the starting material. Since most factors appear to be concentrated in the stalk median eminence region, the use of stalk median eminence rather than whole hypothalamic extracts gives an advantage in subsequent attempts at purification.

Most workers have defatted the frozen or lyophilized tissue prior to extraction, and nearly all workers employ extraction with dilute acid as the initial step. Few data can be found to indicate whether or not these methods are effective in extracting all of the activity.

Several chromatographic procedures have been used to accomplish further purification of the releasing factors and to separate them from each other. The most widely used initial step is gel filtration on Sephadex G-25, a technique introduced to separate substances according to their molecular size. There is a characteristic pattern

TABLE IV

SOME CHEMICAL CHARACTERISTICS OF RELEASING AND INHIBITING FACTORS

Factor	Heat Stability	Effect of Enzymes: Pepsin	Effect of Enzymes: Trypsin	Effect of Thioglycollate	Order of Elution from Sephadex	Retention on CMC *
CRF	+	+	+	—	3	+
FSHRF	+	+	+	?	4	+
LHRF	+	+	+	—	7	+
TSHRF	+	±	±	?	5	+
GHRF	+	+	+	±	1	+
GHIF	?	+	+	?	2	+
MSHRF	?	+	+	?	4	+
MSHIF	?	+	+	?	9	?
PIF	+	?	?	?	6	+
Vasopressin	+	—	+	+	8	+

* CMC = carboxymethyl cellulose

of elution of the factors from Sephadex columns (Table IV). The gonadotropin releasing factors are the last releasing factors eluted from Sephadex. FSHRF is first eluted and this is followed by LHRF. It has been difficult to separate PIF from LHRF by gel filtration; however, PIF tends to emerge just prior to LHRF. Vasopressin, which is also found in stalk median eminence extracts, emerges just after LHRF. Gel filtration provides a good separation of the early emerging CRF and GHRF from the later emerging gonadotropin releasing factors (Table IV).

Further purification has been obtained by ion-exchange chromatography on carboxymethyl cellulose, but unless the fractions are desalted either by extraction with glacial acetic acid and by a preliminary pass through the column or by the use of phenol or by chromatography on Dowex and Amberlite Cg4B columns, some of the releasing factor activity will appear at the void volume ofthe column. Following desalting, all three of the factors are well retai ned on carboxymethyl cellulose. Chromatography on carboxymethyl cellulose has been used successfully to purify the factors and to separate them from residual contaminating activities.

A variety of other procedures have been employed in several laboratories for further purification and most of the factors have now been obtained in a highly purified state.

Methods of isolation were adapted from those in use to purify peptides and most of the factors have been shown to be inactivated by proteolytic enzymes. Consequently it was reasonable to postulate that the releasing factors were polypeptide hormones. On the basis of their migration rates on Sephadex, molecular weights were estimated to lie between 1000 and 2500. It is now realized that factors other than molecular size affect mobility on Sephadex, and therefore these estimates of molecular weight should be viewed with caution. Amino acid compositions of the purified factors were published and they were all thought to be basic peptides.

Although CRF was originally thought to be related to vasopressin, it now appears unlikely that any of these hypothalamic hormones has a structure similar to that of vasopressin or oxytocin. Hypothalamic CRF and LHRF are not inactivated by treatment with thioglycollate under conditions which inactivate both vasopressin and oxytocin by reduction of the disulfide bond (Ramirez and McCann, 1964).

What then can we say about the chemical nature of these factors? First, they can be purified and separated from each other. It appears quite likely that each is a distinct chemical entity. This is certainly an important first step. All the factors appear to be heat stable, at least to 100 °C for 10 min in crude or partially purified extracts. Most of the factors are inactivated by proteolytic digestion. They appear to be relatively small molecules. Whether they are simple peptides as originally supposed or more complex compounds containine peptidic linkages remains to be determined.

IX. Conclusions

As a result of the work of many investigators during the past 15 years, it is now clear that the final common pathway between hypothalamus and anterior pituitary gland is bridged by a family of neurohormones. These agents are secreted into the hypophyseal portal vessels and act directly and specifically on the adenohypophysis to increase or decrease the release of each pituitary tropic hormone. Admittedly, fragmentary evidence suggests that variations in the rate of secretion of the factors is responsible in large part for variations in secretion rate of anterior pituitary hormones in response to stimuli. The major site of feedback action of target gland hormones and the pituitary hormones themselves appears to be at the hypothalamic level to alter the rate of release of releasing factors; however, it is clear that this feedback of target gland hormones is also exerted at the pituitary level.

If an animal is hypophysectomized for a period of time, releasing factors appear in the peripheral circulation which may be responsible at least in part for the residual function of the pituitary grafted to a site distant from the median eminence. Recent evidence indicates that dopamine may be the synaptic transmitter involved in the release of FSHRF and LHRF from neurosecretory neurons. Although the precise chemical nature of releasing factors remains elusive, they have been prepared in highly purified form and separated chemically from each other; thus it is clear that they represent distinct chemical entities. All the factors appear to be small molecules and most of them require the presence of peptide bonds for their activity.

Acknowledgment

This study was supported by the U.S. Public Health Service, Research Grant AM 10073-04, and by a grant from the Ford Foundation.

References

Averill, R. L. W., Salaman, D. F., and Worthington, W. C. Jr., (1966). *Nature* **211**, 144.
Beddow, D. A., and McCann, S. M. (1969). *Endocrinology* **84**, 595.
Brodish, A. (1962). *Endocrinology* **71**, 298.
Coppola, J. A., Leonardi, R., and Lippman, W. (1966). *Endocrinology* **78**, 225.
Crighton, D. B., Schneider, H. P. G., and McCann, S. M. (1969). *J. Endocrinol.* **44**, 405.
Dhariwal, A. P. S., Russell, S. M., McCann, S. M., and Yates, F. E. (1969). *Endocrinology* **84**, 544.
Everett, J. W. (1964). *Physiol. Rev.* **44**, 373.
Fink, G. (1967). *Nature* **215**, 159.
Fuxe, K., and Hökfelt, T. (1967). *In* " Neurosecretion " (F. Stutinsky, ed.), pp. 165–177. Springer Verlag, Berlin.
Fuxe, K., Hökfelt, T., and Nilsson, O. (1967). *Life Sci.* **6**, 2057.
Gittes, R. F., and Kastin, A. J. (1966). *Endocrinology* **78**, 1023.
Green, J. D., and Harris, G. W. (1949). *J. Physiol.* (London) **108**, 359.
Harris, G. W., Reed, M., and Fawcett, C. P. (1966). *Brit. Med. Bull.* **22**, 266.
Hillarp, N. A., Fuxe, K., and Dahlström, A. (1966). *Pharmacol. Rev.* **18**, 727.
Hinsey, J. C. (1937). *Cold Spring Harbor Symp. Quant. Biol.* **V**, 269.
Kamberi, I., and McCann, S. M. (1969). *J. Reprod. Fert.* **18**, 153.
Kamberi, I., Schneider, H. P. G., and McCann, S. M. (1969). *Progr. 51st Meeting Endocrine Soc.*, p. 157.
Krulich, L., and McCann, S. M. (1966). *Proc. Soc. Exp. Biol. Med.* **122**, 668.
Martini, L., Fraschini, F., and Motta, M. (1968). *Recent Progr. Hormone Res.* **24**, 439.
McCann, S. M. (1957). *Endocrinology* **69**, 664.
McCann, S. M. (1970). *In* " Chemistry and Assay of the Hypothalamic Hypophysiotropic Hormones " (J. Meites, ed.), pp. 90-102. Williams and Wilkins Co., Baltimore.
McCann, S. M., and Porter, J. C. (1969). *Physiol. Rev.* **49**, 240.
McCann, S. M., Dhariwal, A. P. S., and Porter, J. C. (1968). *Ann. Rev. Physiol.* **30**, 589.
Müller, E. E., Arimura, H., Saito, T., and Schally, A. V. (1967). *Proc. Soc. Exp. Biol. Med.* **125**, 874.
Negro-Vilar, A., Dickerman, E., and Meites, J. (1968a). *Endocrinology* **83**, 939.
Negro-Vilar, A., Dickerman, E., and Meites, J. (1968b). *Endocrinology* **83**, 1349.
Porter, J. C., and Mical, R. S. (1969). *Fed. Proc.* **28**, 317.
Porter, J. C., and Rumsfeld, H. W., Jr., (1956). *Endocrinology* **58**, 359.
Porter, J. C., Dhariwal, A. P. S., and McCann, S. M. (1967). *Endocrinology* **80**, 679.
Ramirez, V. D., and McCann, S. M. (1963). *Endocrinology* **73**, 193.
Ramirez, V. D., and McCann, S. M. (1964). *Am. J. Physiol.* **207**, 441.
Root, A. W., Smith, G. P., Dhariwal, A. P. S., and McCann, S. M. (1969). *Nature* **221**, 570.
Russell, S. M., Dhariwal, A. P. S., McCann, S. M., and Yates, F. E. (1969). *Endocrinology* **85**, 512.
Schally, A. V., Arimura, A., Bowers, C. Y., Kastin, A. J., Sawano, S., and Redding, T. W. (1968). *Recent Progr. Hormone Res.* **24**, 497.
Schneider, H. P. G., and McCann, S. M. (1969). *Endocrinology* **85**, 121
Smith, E. R., and Davidson, J. M. (1967). *Endocrinology* **80**, 725.
Vale, W., Burgus, R., and Guillemin, R. (1968). *Neuroendocrinology* **3**, 34.

Concerning the Mechanisms of Action of Hypothalamic Releasing Factors on the Adenohypophysis

M. JUTISZ, M. P. DE LA LLOSA, A. BÉRAULT, and B. KERDELHUÉ

I. Introduction

This article deals with some information now available about the cellular mechanism of action of hypothalamic hormones on the adenohypophysis.

It is known that regulation of the secretion of adenohypophyseal hormones is a complex phenomenon in which other factors besides hypothalamic hormones may play an important role. For these and other reasons, and in order to simplify and limit the material covered, we shall consider almost exclusively the results obtained by *in vitro* studies. Considering the kind of problem we deal with, this procedure seems to be much more appropriate than the *in vivo* methods, and indeed the use of viable pituitary tissue removed from the influence of the hypothalamus seems preferable to operation on the whole animal. The danger in drawing conclusions solely from an *in vitro* study is that the results thus obtained may not be applicable to a situation in which the gland remains *in situ.* While admitting this limitation, we feel that *in vitro* studies can provide some interesting and useful data for solving our problems.

Three kinds of methods are generally used with *in vitro* studies: (1) short-term incubations, from a few minutes to about 2 hr; (2) long-term incubations, from 2 hr to about 24 hr; (3) tissue culture, from a few days to several weeks.

II. Factors Affecting the in vitro Release of Anterior Pituitary Hormones

Large variations in the amount of anterior pituitary hormones released *in vitro* by pituitaries of different types of rats are usually observed. These variations result from the operation of factors known to alter the secretion of anterior pituitary hormones *in vivo.* Thus, age, sex, estrous cycle, environmental and other stimuli, stresses of various kinds, removal of target organs, pretreatment with substances producing either stimulatory or inhibitory effects on hormone secretion, etc., may influence not only the hormonal content of a pituitary gland but also the capacity of the isolated gland to respond to a stimulus.

It is not the purpose of this article to discuss this aspect of the control of the endocrine functions of the pituitary. These problems have been treated in other articles of this volume and reviewed recently, in different books and articles (see Harris and Donovan, 1966; Meites and Nicoll, 1966; Gray and Bacharach, 1967; Guillemin, 1967; McCann *et al.*, 1968; Martini *et al.*, 1968; Schally *et al.*, 1968a).

A. Dose-Response Relationships between Hypothalamic Hormones and the Release of Anterior Pituitary Hormones

One of the characteristics of the action of most hypothalamic hormones is a log dose-response relationship between the amounts of these substances added to a medium and the release of adenohypophyseal hormones. This has been observed for the following hormones: Follicle-Stimulating Hormone Releasing Factor (FSHRF) by Mittler and Meites (1966) and Jutisz and de la Llosa (1967a); Luteinizing Hormone Releasing Factor (LHRF) by Piacsek and Meites (1966) and Jutisz *et al.* (1967); Thyrotropin Releasing Factor (TSHRF) by Guillemin *et al.* (1963) and Sinha and Meites (1965); Growth Hormone Releasing Factor (GHRF) by Dickerman *et al.* (1969); and Prolactin Inhibiting Factor (PIF) by Kragt and Meites

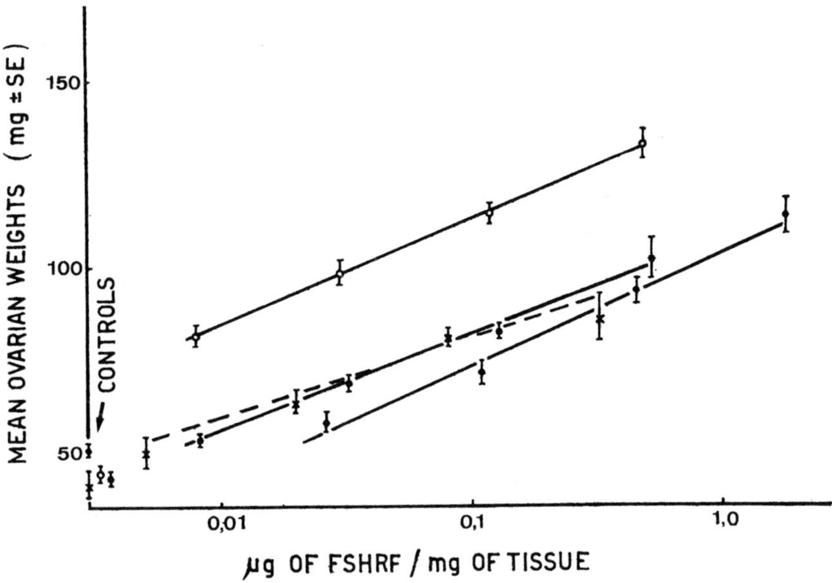

Fig. 1. Release of FSH *in vitro* (in terms of ovarian weights in the augmentation assay) from rat pituitaries as a function of graded doses of a highly purified FSHRF. Pituitary halves from four different types of rats were incubated for 2-hr periods: ● — ●, normal males; ∗ — ∗, castrated males; o — o, castrated males treated with testosterone propionate; × - - - ×, ovariectomized females treated with estradiol benzoate and progesterone (From Jutisz and de la Llosa, 1968a).

(1967). Only Corticotropin Releasing Factor (CRF) does not seem to follow this rule.

Figure 1 shows the release of Follicle-Stimulating Hormone (FSH) in terms of ovarian weights in the augmentation assay (Steelman and Pohley, 1953) as a function of graded doses of FSHRF (Jutisz and de la Llosa, 1968a). Pituitary halves from four different types of rats were used: normal males, castrated males, castrated males treated with testosterone propionate, and ovariectomized females treated with estradiol benzoate and progesterone. The slopes of all curves are very similar and only the sensitivity of the pituitaries in the various groups differs.

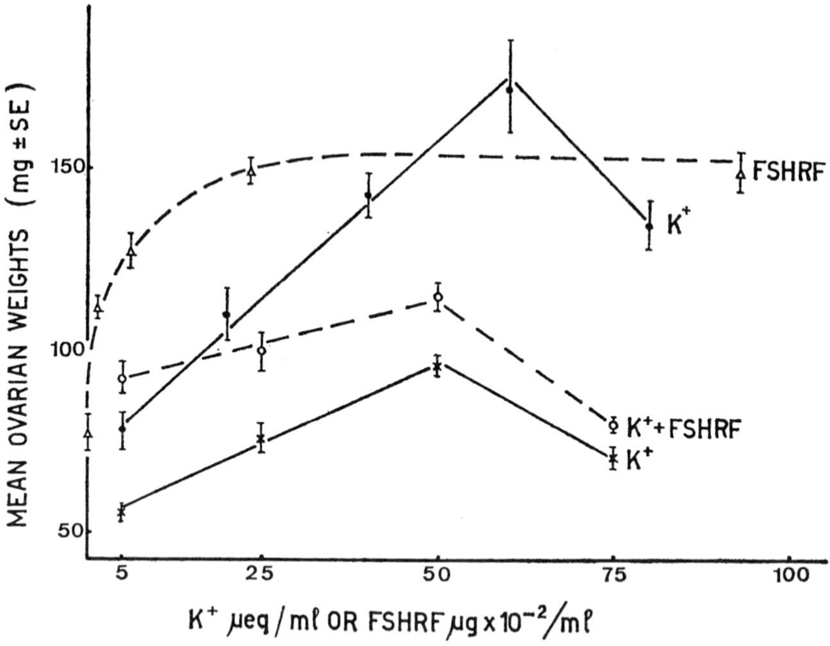

FIG. 2. Release of FSH *in vitro* (in terms of ovarian weights in the augmentation assay) from rat pituitaries as a function of the concentration of potassium in the incubation medium (normal concentration: 5.9 μeq of K^+/ml). Pituitary halves of ovariectomized female rats treated with estradiol benzoate and progesterone were incubated for 2-hr periods. In experiment FR 167 (△ - - - △) pituitary halves were incubated in a normal Krebs-Ringer medium containing graded doses of FSHRF. In experiment FR 168 (o - - - o) a constant dose of FSHRF (93 ng/mg of tissue) was added to the incubation media in which the concentration of K^+ was gradually increased (From Jutisz and de la Llosa, 1968a).

B. Effect of an Increased Concentration of K^+ in the Incubation Medium on the Release of Some Anterior Pituitary Hormones

Douglas and Rubin (1961) reported that a high concentration of potassium in the perfusion medium stimulates the release of catecholamines from the adrenal medulla. Similarly, an excess of K^+ in the incubation medium greatly accelerates the rate of release of vasopressin from rat neurohypophysis (Douglas, 1963; Douglas and Poisner, 1964). By increasing the concentration of K^+ in the external medium and removing an equivalent amount of Na^+, the nonspecific release from the rat pituitary gland of the following hormones was promoted: Luteinizing Hormone (LH) by Samli and Geschwind (1967a, 1968), Wakabayashi *et al.* (1968), and Bérault (1969); FSH by Jutisz *et al.* (1968) and Jutisz and de la Llosa (1968a); and Thyrotropin (TSH) by Vale and Guillemin (1967).

Figure 2 shows the release of FSH (in terms of ovarian weights in the augmentation assay) by rat pituitary halves as a function of the concentration of potassium in the incubation medium. Maximal release was obtained with a concentration of about 60 meq of K^+ (approximately $10 \times K^+$ concentration of a normal Krebs-Ringer medium). A higher $[K^+]$ (75 meq) had a much smaller effect on the release of FSH. When high $[K^+]$ media and FSHRF were both present, the two effects were additive.

The release of neural hormones by a high $[K^+]$ was attributed by Douglas (1963) to the depolarization of specific cell membranes. The same hypothesis has been accepted for LH release by Samli and Geschwind (1968).

C. Requirement of Ca^{2+} for the Release of Some Adenohypophyseal Hormones

The action of many hormones *in vitro* on their target organs requires the presence of Ca^{2+} in the external medium. Thus, in the absence of Ca^{2+} the effects of the following hormones *in vitro* are inhibited: the action of Adrenocorticotropic Hormone (ACTH) on the adrenals, reported by Birmingham *et al.* (1953) and Peron and Koritz (1958); acetylcholine and noradrenaline on the submaxillary gland, by Douglas and Poisner (1963); epinephrine on the parotid gland, by Rasmussen and Tenenhouse (1968); LH on the corpus luteum, by Hermier and Jutisz (1969). Samli and Geschwind (1967a, 1968) reported for the first time that Ca^{2+} was necessary for LH release from rat pituitary gland stimulated by hypothalamic extract or by a high [K^+] medium. Wakabayashi *et al.* (1968) claimed that the effect of high [K^+] but not LHRF was blocked in a Ca^{2+}-free medium. Ca^{2+} has been found to be necessary for the release of TSH by TSHRF or high [K^+] (Vale *et al.*, 1967; Vale and Guillemin, 1967).

With de la Llosa we studied the effects of Ca^{2+} elimination on the release of FSH. We found first (Jutisz and de la Llosa, 1968b) that elimination of Ca^{2+} from the media used for preincubation

TABLE I

Requirement of Ca^{2+} for Release of FSH *in vitro* from Rat Pituitary Glands (results of FSH assay [1])

No.	Conditions of Incubations [2]	Mean Ovarian Weights, mg ± SE	
		Experiment FR 205 [3]	Experiment FR 209 [4]
1	Control, normal KRB (5.9 mM K^+)	39.7 ± 4.2	45.7 ± 1.9
2	Normal KRB + FSHRF (100 ng/mg)	71.9 ± 3.9	72.7 ± 5.7
3	KRB, 59.0 mM K^+	85.1 ± 3.6	79.8 ± 6.2
4	EDTA-treated, Ca^{2+}-free KRB	42.6 ± 3.4	46.9 ± 2.3
5	Medium No. 4 + FSHRF (100 ng/mg)	42.4 ± 4.2	44.1 ± 3.9
6	Medium No. 4, 59.0 mM K^+	45.0 ± 3.2	40.7 ± 2.2
7	EDTA-treated, 2.72 mM Ca^{2+} KRB	41.2 ± 2.9	41.1 ± 3.0
8	Medium No. 7 + FSHRF (110 ng/mg)	79.3 ± 5.7	82.8 ± 8.8
9	Medium No. 7, 59.0 mM K^+		76.8 ± 5.6
	Assay control (HCG 20 IU) [5]	32.0 ± 1.3	38.8 ± 4.9
	Standard I (NIH-FSH-S3, 50 μg)	89.1 ± 4.1	74.6 ± 4.6
	Standard II (NIH-FSH-S3, 100 μg)	141.0 ± 4.9	132.6 ± 7.6

[1] In terms of ovarian weights in the augmentation assay (Steelman and Pohley, 1953).

[2] In experiments No. 4–9, preincubations were carried out for 45 min in Ca^{2+}-free Krebs-Ringer bicarbonate (KRB) media containing 0.72 mM of EDTA. In the same experiments, all incubation media contained 0.18 mM of EDTA. Furthermore, media 4–6 were Ca^{2+}-free and media 7–9 contained 2.72 mM of Ca^{2+}.

[3] Twenty pituitary halves of normal male rats per flask were incubated for 2-hr periods. Each assay rat (5 per group) was injected with the medium corresponding to an equivalent of 7.1 mg of tissue.

[4] Fifteen (per flask) pituitary halves of ovariectomized female rats treated with estradiol benzoate and progesterone were incubated for 2-hr periods. Each assay rat (5 per group) was injected with the incubation medium corresponding to an equivalent of 8.1 mg of tissue.

[5] HCG, Human Chorionic Gonadotropin.

and incubation produced no inhibition of FSH release from pituitary tissue incubated with FSHRF, so that under these conditions the membrane may still retain Ca^{2+} (Samli and Geschwind, 1968). Inhibition of FSH release occurred only in a medium containing ethylenediaminotetraacetic acid (EDTA) (Jutisz and de la Llosa, 1970). Table I shows the results obtained when incubation was performed in Ca^{2+}-free medium containing EDTA. The stimulatory effects of FSHRF and high $[K^+]$ on FSH release were inhibited. In order to ensure that EDTA did not exert some detrimental action in addition to complexing Ca^{2+}, enough Ca^{2+} was added after preincubation with EDTA to bring the concentration of Ca^{2+} to that normally present in Krebs-Ringer medium. As shown in Table I, the release of FSH was restored. Thus Ca^{2+} was found to be necessary for the releasing action of at least three hypothalamic hormones, but its exact role is not yet established.

D. Effect of Inhibitors of Protein and RNA Synthesis on the Release of Anterior Pituitary Hormones

Actinomycin D, an inhibitor of RNA synthesis, does not affect the releasing activity of the following hypothalamic hormones: LHRF, reported by Jutisz *et al.* (1966), Samli and Geschwind (1967b), and Crighton *et al.* (1968); FSHRF, by Jutisz and de la Llosa (1967b) and Schally *et al.* (1967); TSHRF by Schally and Redding (1967) and Vale *et al.* (1968). On the other hand, TSH release from the pituitary gland is under dual control by TSHRF and by thyroxine (T_4) and triiodothyronine (T_3) as reported by Guillemin *et al.* (1963) and Sinha and Meites (1966). Thyroid hormones antagonize the action of TSHRF and so inhibit TSH release. Actinomycin D prevents the inhibitory effect of T_3 and T_4 on TSH release (Schally and Redding, 1967; Vale *et al.*, 1968). Only the releasing action of GHRF is inhibited by actinomycin (Schally *et al.*, 1968b). Watanabe *et al.* (1968) recently reported that actinomycin and puromycin completely inhibited the release of FSH in the presence of hypothalamic extract. This discrepancy cannot be explained at the present time.

Inhibitors of protein biosynthesis, such as puromycin or cycloheximide, fail to inhibit the action of any of the hypothalamic hormones studied, as reported for LHRF by Jutisz *et al.* (1966), Samli and Geschwind (1967b), and Crighton *et al.* (1968); for FSHRF by Jutisz

TABLE II

EFFECT OF PUROMYCIN ON THE RELEASE OF LH BY LHRF [1]

No.	Treatment	μg LH Released/mg of Tissue [2]	Relative Potencies [3]
1	LHRF (1.3 μg/mg tissue)	1.85 (1.11–3.08)	
2	Puromycin (90 μg/ml)		0.46 (0.29–0.71)
	LHRF (1.2 μg/mg tissue)	0.96 (0.57–1.59)	
3	LHRF (3.5 μg/mg tissue)	1.13 (0.63–2.01)	
4	Puromycin (90 μg/ml)		0.45 (0.27–0.73)
	LHRF (2.9 μg/mg tissue)	0.44 (0.24–0.81)	
5	LHRF (0.4 μg/mg tissue)	0.55 (0.33–0.89)	
6	Puromycin (90 μg/ml)		0.64 (0.37–1.11)
	LHRF (0.4 μg/mg tissue)	0.36 (0.21–0.61)	
7	LHRF (0.4 μg/mg tissue)	0.59 (0.33–1.03)	
8	Puromycin (90 μg/ml)		0.74 (0.40–1.40)
	LHRF (0.4 μg/mg tissue)	0.42 (0.24–0.73)	

[1] In experiments Nos. 1 to 4, 10 (Nos. 1 and 2) and 6 (Nos. 3 and 4) pituitary halves (per flask) of ovariectomized female rats treated with estradiol benzoate and progesterone were incubated for 2 hr. In Experiments Nos. 5 to 8, 15 pituitary halves (per flask) of normal male rats were incubated for 6 hr.

[2] In terms of NIH-LH-S3 with 95 % confidence limits. LH was assayed by the method of Parlow (1961).

[3] With 95 % confidence limits.

and de la Llosa (1967b, 1968a); for TSHRF by Vale *et al.* (1968); and for CRF by Estep *et al.* (1966). No data are available with respect to GHRF and PIF. As in the case of actinomycin D, cycloheximide prevents or even reverses the inhibition by T_4 of the release of TSH normally produced by TSHRF (Vale *et al.*, 1968; see also *in vivo* studies by Bowers *et al.*, 1968a,b).

It was reported from our laboratory that puromycin partially interfered with the releasing activity of LHRF (Jutisz *et al.*, 1966) and that puromycin and actinomycin D interfered with the releasing activity of FSHRF (Jutisz and de la Llosa, 1967a,b). Table II shows some recent results obtained with puromycin and LHRF. Pituitary halves of normal male rats and of ovariectomized female rats treated with estradiol benzoate and progesterone were used and incubations were performed lasting 2 and 6 hr, respectively. Assays of LH in the incubation media were done by the ovarian ascorbic acid depletion (OAAD) assay (Parlow, 1961) using the four-point method. In spite of the low precision of this test (due to the strain of rats used) an apparent inhibition of 30–50% of the stimulatory action of LHRF was observed (in the last two experiments the inhibition is statistically not significant).

The following experiment shows that reduction of the amount of FSH released into a medium containing an antibiotic should not be necessarily regarded as due to an inhibition of the activity of FSHRF. Figure 3 shows the release of FSH (in terms of ovarian weights in the augmentation assay) as a function of time when two sets of rat pituitary halves were incubated either with FSHRF alone or with FSHRF and cycloheximide (Jutisz and de la Llosa, 1968a). The release of FSH in the presence of the antibiotic was nearly maximal after 30–45 min; release then proceeded very slowly. In the absence of cycloheximide, there was a progressive release of FSH. As an explanation of these results, one can suggest the existence of two compartments in the appropriate pituitary cell: the first (e.g.,

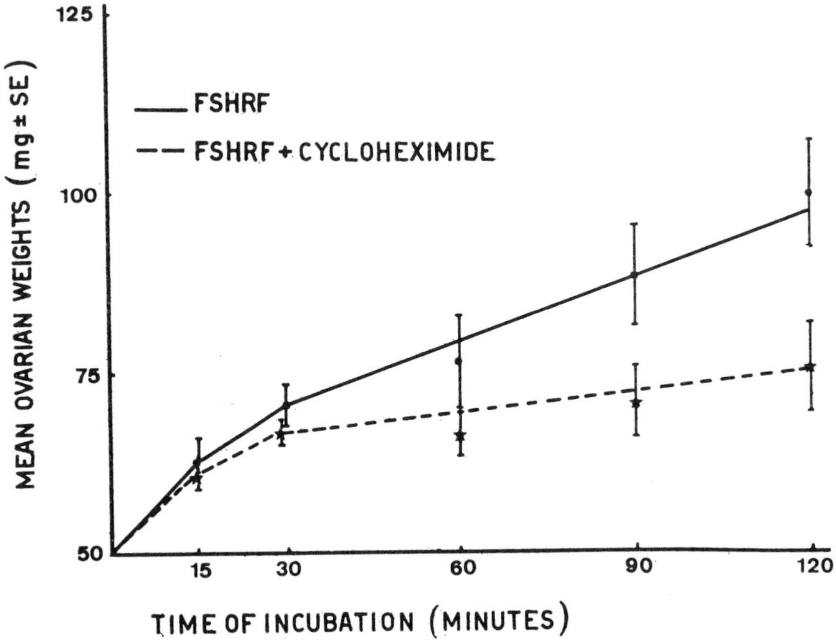

FIG. 3. Release of FSH *in vitro* (in terms of ovarian weights in the augmentation assay) as a function of time. Anterior pituitary halves of ovariectomized female rats treated with estradiol benzoate and progesterone were incubated for 2 hr, either with a highly purified FSHRF (160 ng/mg tissue) or with cycloheximide (5 μg/ml) and FSHRF (160 ng/mg tissue) (From Jutisz and de la Llosa, 1968a).

FSH granules) containing FSH directly available for the release and the second containing FSH which must move to the preceding compartment before release. If the passage from the second to the first compartment is dependent on the synthesis of FSH, then the results in Fig. 3 are readily understandable, for in the presence of cycloheximide the synthesis of FSH is inhibited and the only available FSH for release comes from the first compartment. In the incubation without cycloheximide, the release of FSH induced the *de novo* synthesis of this hormone, as will be shown later. Increase in the amount of FSH in the second compartment allowed transfer of hormone to, and consequently release from, the first compartment.

In summary, inhibitors of proteins and RNA synthesis do not affect the release of pituitary hormones other than growth hormone. It is postulated that the stimulatory effects of hypothalamic factors on the release of pituitary hormones do not require synthesis of an intermediate protein or RNA. On the other hand, the inhibition of the effects of TSHRF by thyroid hormones may be mediated through one or several substances, the synthesis of which can be inhibited by antibiotics.

E. Does Cyclic 3',5'-Adenosine Monophosphate Mediate the Action of Hypothalamic Hormones?

Since the discovery of cyclic 3',5'-adenosine monophosphate (cAMP) by Rall and co-workers (1957), this nucleotide " has been established as an intracellular second messenger mediating many of the actions of a variety of different hormones " (Sutherland *et al.*, 1965; Robison *et al.*, 1968). The following experiments were carried out on the basis of the possibility of cAMP being an intermediate of the action of FSHRF (Jutisz and de la Llosa, 1969).

Figure 4 shows the release of FSH (in terms of ovarian weights in the augmentation assay) as a function of graded doses of cAMP as compared with graded doses of FSHRF. The two curves have almost the same slope. Other derivatives of adenosine, such as ATP and 5'-AMP, do not release FSH under the same conditions.

It is known from the investigations of the Sutherland's group (Sutherland *et al.*, 1965; Robison *et al.*, 1968) that the level of cAMP depends upon the activities of at least two enzymes: adenyl cyclase, which catalyses the formation of cAMP from ATP, and phospho-

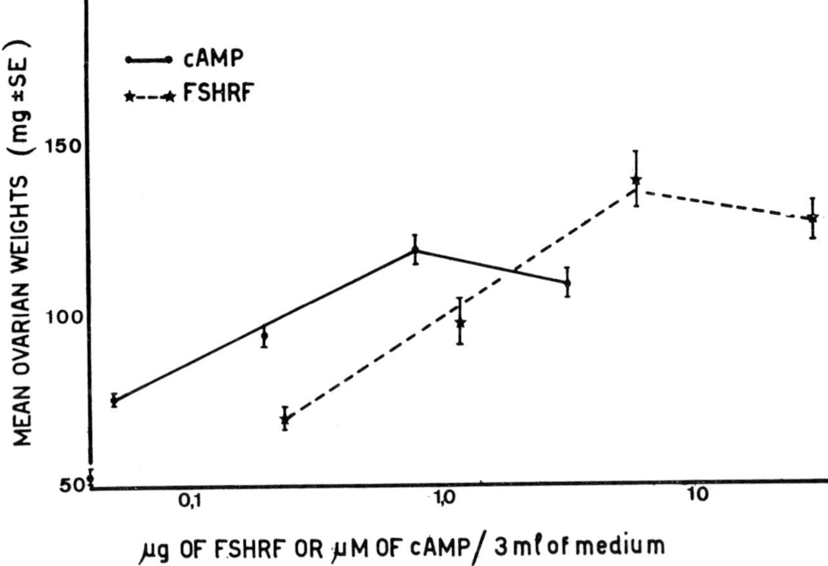

FIG. 4. Release of FSH *in vitro* (in terms of ovarian weights in the augmentation assay) from rat pituitaries as a function of graded doses of cAMP and graded doses of a highly purified FSHRF. Pituitary halves of ovariectomized female rats treated with estradiol benzoate and progesterone were incubated for 2-hr periods (From Jutisz and de la Llosa, 1969).

diesterase, which inactivates cAMP. Phosphodiesterase can be inhibited by the methylxantines such as caffeine and theophylline. Consequently this inhibition leads to the accumulation of cAMP and to the enhancement of a physiological effect if cAMP is an intermediate in the action of a hormone.

Table III shows some results obtained on the release of FSH when rat pituitary halves were incubated with FSHRF, cAMP, and theophylline. Two doses of FSHRF (experiment 224) and two of cAMP were used. Although the first dose of FSHRF was only slightly (but significantly) potentiated by theophylline, the second dose of FSHRF was strongly potentiated. In the second experiment (229) both FSHRF and cAMP were potentiated by theophylline.

In preliminary experiments we checked the effect of cAMP and of theophylline on the release of LH from rat pituitaries. Our results suggest that cAMP may also mediate the action of LHRF (Jutisz and associates, unpublished observations).

TABLE III

POTENTIATION BY THEOPHYLLINE OF THE EFFECT OF FSHRF AND CYCLIC AMP ON THE RELEASE OF FSH *in vitro* FROM RAT PITUITARIES (RESULTS OF FSH ASSAY [1]).[2]

No.	Treatment [3]	Dose/3 ml of Incubation Medium	Mean Ovarian Weights, mg±SE [4] Experiment FR 224	Experiment FR 229
1	Incubation control	– – –	57.8± 2.5	55.0±3.7
2	FSHRF	1.2 μg	96.5± 4.3 (**)	80.8±6.9 (**)
3	FSHRF	6.5 μg	139.9± 6.2 (**)	
4	Cyclic AMP	0.2 μM	92.9± 3.3 (**)	66.0±2.7 (ns)
5	Cyclic AMP	0.8 μM	93.5± 4.4 (**)	
6	Theophylline	7.5 μM	62.2± 2.4 (ns)	52.5±4.2 (ns)
7	FSHRF + Theophylline	1.2 μg +7.5 μM	111.1± 3.2 (**)	125.2±6.2 (**)
8	FSHRF + Theophylline	6.5 μg +7.5 μM	190.9± 8.4 (**)	
9	Cyclic AMP + Theoph.	0.2 μM+7.5 μM	82.5±12.0 (*)	78.5±3.0 (**)
10	Cyclic AMP + Theoph.	0.8 μM+7.5 μM	96.2± 3.7 (**)	
Assay control (HCG 20 IU)			43.7± 2.8	39.4±2.9
Standard I (NIH-FSH-S3, 50 μg)			87.9± 7.7	89.8±2.2
Standard II (NIH-FSH-S3, 100 μg)			172.7± 6.6	177.1±6.1

[1] In terms of ovarian weights in the augmentation assay (Steelman and Pohley, 1953).
[2] From Jutisz and de la Llosa (1969).
[3] Sixteen pituitary halves (per flask) of ovariectomized female rats treated with estradiol benzoate and progesterone were incubated for 2-hr periods. Each assay rat (5 per group) was injected with the incubation medium corresponding to an equivalent of 12 mg (exp. FR 224) and 15 mg (exp. FR 229) of tissue.
[4] Significance of the responses was tested *vs.* incubation controls: (ns), not significant; (*), significant ($p = 0.05$); (**), highly significant ($p = 0.01$).

Cehovic *et al.* (1968) reported in a short note that cAMP and cyclic iso-AMP were able to release TSH from rat pituitaries incubated *in vitro*. A recent publication of Wilber *et al.* (1969) reports similar results.

The findings outlined above suggest that at least one of the biological effects of three hypothalamic factors (FSHRF, LHRF, and TSHRF) may consist of the activation of an enzyme (adenyl cyclase) located on the adenohypophyseal cell wall. However, for more definite conclusions, it will be necessary to assay cAMP specifically in order to ascertain that its amount in pituitary tissue increases as a function of graded doses of hypothalamic hormones and that this increase in cAMP parallels the physiological effect of hypothalamic factors.

III. Releasing Factors and the Synthesis of Adenohypophyseal Hormones

A. Do Releasing Factors Stimulate the Synthesis or the Activation of Adenohypophyseal Hormones?

We shall consider in this section the possible action of releasing factors (FSHRF, LHRF, TSHRF, GHRF, CRF), with exclusion of PIF, on the synthesis of anterior pituitary hormones. It has been postulated by several laboratories that the hypothalamus influences the synthesis of these hormones as well as their release. Deuben and Meites (1964, 1965) reported an increase in Growth Hormone (GH) content in tissue cultures of rat pituitaries after the addition of rat hypothalamic extracts. These findings have been confirmed in short-term incubations by other laboratories (Schally *et al.*, 1965, 1968b; Symchowicz *et al.*, 1966). Similar results concerning the enhancement of the synthesis of other pituitary hormones by hypothalamic extracts have been obtained; for example Solomon and McKenzie (1966) reported that crude hypothalamic extracts increase incorpora-

TABLE IV

Total Amount of LH after a 2-hr Incubation of Rat Pituitaries with LHRF[1]

No.	Treatment	Total LH, μg/mg of Tissue[2]	Relative Potencies[3]
1	Control	4.37 (2.85– 6.72)	
2	LHRF (0.7 μg/mg tissue)	5.99 (3.75– 9.55)	1.52 (0.80–2.87)
3	Control	6.35 (3.69–10.93)	
4	LHRF (0.5 μg/mg tissue)	8.18 (5.14–13.01)	1.42 (0.80–2.54)
5	Control	3.78 (2.42– 5.91)	
6	LHRF (0.8 μg/mg tissue)	5.32 (3.14– 9.01)	1.41 (0.91–2.20)
7	Control	2.26 (1.29– 3.97)	
8	LHRF (0.7 μg/mg tissue)	3.83 (2.32– 6.35)	1.63 (1.01–2.65)

[1] Ten pituitary halves (per flask) of ovariectomized female rats treated with estradiol benzoate and progesterone were incubated for 2 hr.

[2] In terms of NIH-LH-S3 with 95% confidence limits. LH was assayed by the method of Parlow (1961).

[3] With 95% confidence limits.

tion of ^{14}C amino acids into TSH. The concept that TSHRF can stimulate synthesis of TSH has been supported by several authors (Sinha and Meites, 1966; Bowers *et al.*, 1967).

Many results obtained *in vitro* in our laboratory also suggested that releasing factors such as FSHRF (reported by Jutisz and de la Llosa, 1967a and Jutisz *et al.*, 1968) and LHRF (reported by Jutisz *et al.*, 1967) induce synthesis (or activation) of FSH and LH. In a recent note, Corbin and Daniels (1968) envisaged the possibility of the presence of a hypothalamic FSH-Synthesizing Factor, different from FSHRF.

Table IV records some recent results showing an increase in the total amount of LH after incubation with LHRF (Bérault, 1969; Jutisz and associates, unpublished observations). Pituitary halves of ovariectomized female rats treated with estradiol benzoate and progesterone were incubated for 2-hr periods without or with LHRF. At the end of the incubation the tissues were extracted in their supernatants in order to obtain the total amount of LH, and assays were performed using the OAAD test (Parlow, 1961). As can be seen from Table IV, the amount of LH in LHRF-treated incubations was always greater than in controls, but the low precision of the assay method did not always permit the establishment of statistically significant differences.

In agreement with our results, Midgley *et al.* (1968) showed recently, using a radioimmunoassay method, that incubation of rat pituitaries with a hypothalamic extract increased the total amount of LH. However, these findings conflict with a claim that in a 4-hr incubation of rat pituitaries, hypothalamic extracts had no effect on the incorporation of either ^{14}C-leucine or ^{14}C-glucosamine into LH (Samli and Geschwind, 1967a,b). It is possible that the method of isolation of LH, by means of an LH-antiserum, used by Samli and Geschwind (1967b) was not specific enough for this purpose. Another possibility is that LHRF, besides being concerned with the release of LH, might participate not in the synthesis of LH but in the activation of a precursor of LH. Although it has been shown that both LHRF and FSHRF failed to increase the amount of LH and FSH when incubated with pituitary homogenates (Jutisz *et al.*, 1968), it might be that the process of activation requires the integrity of the cellular structure.

B. Effect of an Increase in $[K^+]$ in the Incubation Medium on the *de novo* Synthesis of Follicle-Stimulating Hormone and Luteinizing Hormone

It seemed interesting to determine whether the possible effect of releasing factors on the synthesis of adenohypophyseal hormones was primary or whether it arose secondarily as a result of release. The fact that media with high $[K^+]$ can stimulate the release of some pituitary hormones was used in this study. Table V (Jutisz and de la Llosa 1968a) shows the total amount of FSH obtained after a 2-hr incubation of rat pituitary halves, either with FSHRF in a normal Krebs-Ringer medium or in a medium enhanced with K^+ to a concentration of 51.2 mM, or finally in the high $[K^+]$ medium containing FSHRF. Relative potencies indicate that the total amount of FSH was significantly increased in FSHRF-treated pituitaries as

TABLE V

Total Amount of FSH after a 2-Hr Incubation of Rat Pituitaries in Different Conditions [1]

No.	Treatment	Total FSH, μg/mg of Tissue [2]: Experiment FR 153 [3]	Experiment FR 157 [4]
1	Control, normal KRB (5.9 mEq K+)	30.6 (26.2–35.8)	29.0 (26.2–32.2)
2	Normal KRB+FSHRF (130 and 110 ng/mg)	38.5 (32.3–45.4)	33.4 (30.2–37.0)
3	KRB, 51.2 mEq K+	38.6 (33.0–45.1)	33.2 (29.8–37.0)
4	Medium No. 3+FSHRF (130 and 110 ng/mg)	48.3 (39.4–59.2)	38.0 (33.7–42.9)
Relative potencies with 95% confidence limits			
	Groups 2/1	1.29 (1.10–1.52)	1.16 (1.04–1.28)
	Groups 3/1	1.31 (1.12–1.53)	1.15 (1.02–1.28)
	Groups 4/1	1.63 (1.34–1.99)	1.32 (1.16–1.49)

[1] From Jutisz and de la Llosa (1968a).

[2] In terms of NIH-FSH-S3 with 95% confidence limits. FSH was assayed by the Steelman-Pohley (1953) method.

[3] Pituitary halves (128) of ovariectomized female rats treated with estradiol benzoate and progesterone were randomized and divided into four equal groups. After 30 min of preincubation, incubation proceeded for 4 hr, but the medium was replaced after 2 hr of incubation.

[4] Pituitary halves (60) of the same type of rat as in experiment FR 153 were randomized and divided into four equal groups. After 30 min of preincubation, incubation proceeded for 2 hr.

TABLE VI

TOTAL AMOUNT OF LH AFTER A 2-HR INCUBATION OF RAT PITUITARIES IN A NORMAL KREBS-RINGER MEDIUM AND A HIGH $[K^+]$ MEDIUM [1]

No.	Treatment	Total LH, μg/mg of Tissue [2]	Relative Potencies [3]
1	Control	2.26 (1.28– 3.97)	
2	KRB 59, mEq K^+	2.42 (1.12– 5.20)	1.27 (0.67–2.41)
3	Control	6.65 (3.69–11.99)	
4	KRB 59, mEq K^+	6.35 (4.09– 9.87)	0.98 (0.50–1.98)
5	Control	5.55 (3.05– 9.05)	
6	KRB 59, mEq K^+	5.63 (3.54– 8.95)	1.02 (0.67–1.54)

[1] Ten pituitary halves (per flask) of ovariectomized female rats treated with estradiol benzoate and progesterone were incubated.

[2] In terms of NIH-LH-S3 with 95% confidence limits. LH was assayed by the method of Parlow (1961).

[3] With 95% confidence limits.

well as in pituitaries incubated in high $[K^+]$ media. When FSHRF and high $[K^+]$ were both present, the effects were additive.

Table VI (Bérault, 1969; Jutisz and associates, unpublished observations) shows some preliminary results of similar experiments in which LH was assayed after a 2-hr incubation of rat pituitary halves (ovariectomized females treated with estradiol benzoate and progesterone) in a high $[K^+]$ medium. It seems evident from these results that incubation in a high $[K^+]$ medium, in contrast to the preceding results, did not increase the total amount of LH. If these preliminary results are confirmed, this could mean that a difference exists in the mechanism by which the two gonadotropins, FSH and LH, are secreted. In the first case, only the release of FSH is stimulated by FSHRF; as FSH is released the synthesis of further FSH is initiated. For the time being, the mechanism by which LH is secreted cannot be explained unambiguously. At least two major possibilities must still be envisaged: (1) a dual action of LHRF on the release and synthesis of LH; (2) an action of LHRF on the release of LH and possibly on the activation of an LH precursor. In the latter case the activation of LH precursor, but not the release of active LH, may provide the stimulus that controls the synthesis of LH.

IV. Conclusions

It seems possible now to draw some conclusions and to advance some hypotheses as to the mechanism of action of hypothalamic factors, except PIF for which only few data are available. The results discussed in the present article and other results obtained *in vivo* indicate that hypothalamic releasing factors exert their primary effects on the release of hormones from the pituitary gland. With the exception of GHRF the effects on release are not abolished by inhibitors of proteins and RNA synthesis. These results suggest that the release of pituitary hormones does not involve *de novo* synthesis of some protein or nucleic acid.

The release of at least three pituitary hormones (FSH, LH, TSH) is characterized, as is the release of many other hormones (Rasmussen and Tenenhouse, 1968), by the following features:

1. Cyclic AMP is the probable intermediate between a releasing factor and release.
2. A high K^+ concentration in the external medium can function as a nonspecific stimulus for release.
3. The effects of a releasing factor and high $[K^+]$ are dependent upon the presence of Ca^{2+} in the external medium.

Several laboratories have reported that high external $[K^+]$ leads to a significant increase in the level of cAMP in different tissues and to an enhancement of physiological response (Lundholm *et al.*, 1967; Sattin and Rall, 1967; Rasmussen and Tenenhouse, 1968). However, according to Zor *et al.* (1969), an increase in $[K^+]$ will not enhance the concentration of cAMP in the pituitary tissue.

According to many authors (see Rasmussen and Tenenhouse, 1968), Ca^{2+} in association with cAMP may play an active intracellular role in activating some enzymes. Nevertheless this is probably not the only mechanism by which Ca^{2+} contributes to the regulation of the release of pituitary hormones (Samli and Geschwind, 1968).

There is no evidence from the reported data that the synthesis of pituitary hormones is under the direct control of the releasing factors. On the contrary, it has been shown that synthesis (or activation) of FSH can be enhanced by a nonspecific effect such as high K^+ concentration. The release of pituitary hormones seems to induce (perhaps through a " short " feedback mechanism) further synthesis.

This hypothesis does not exclude the existence of an intermediary step between synthesis and release, involving storage of a hormone in an inactive state. The results obtained with LH suggest that a releasing factor may also play a role in the activation of a hormonal precursor.

Acknowledgments

It is a pleasure to acknowledge the help of Dr. B. T. Donovan in the preparation of the English text. We express our thanks to Mr. G. Vassent for the statistical analysis of our results and to Mrs. G. Ribot for her valuable technical assistance.

References

Bérault, A. (1969). Thesis, University of Paris.

Birmingham, M. K., Elliott, F. H., and Valere, P. H. L. (1953). *Endocrinology* **53**, 687.

Bowers, C. Y., Schally, A. V., Reynolds, G. A., and Hawley, W. D. (1967). *Endocrinology* **81**, 741.

Bowers, C. Y., Lee, K. L., and Schally, A. V. (1968a). *Endocrinology* **82**, 75.

Bowers, C. Y., Lee, K. L., and Schally, A. V. (1968b). *Endocrinology* **82**, 303.

Cehovic, G., Marcus, I., Vengadabady, S., and Posternak, T. (1968). *C. R. Soc. Physiol.* (Geneva) **3**, 135.

Corbin, A., and Daniels, E. L. (1968). *Experientia* **24**, 1260.

Crighton, D. B., Watanabe, S., Dhariwal, A. P. S., and McCann, S. M. (1968). *Proc. Soc. Exp. Biol. Med.* **128**, 537.

Deuben, R. R., and Meites, J. (1964). *Endocrinology* **74**, 408.

Deuben, R. R., and Meites, J. (1965). *Proc. Soc. Exp. Biol. Med.* **118**, 409.

Dickerman, E., Negro-Vilar, A., and Meites, J. (1969). *Neuroendocrinology* **4**, 75.

Douglas, W. W. (1963). *Nature* **197**, 81.

Douglas, W. W., and Poisner, A. M. (1963). *J. Physiol.* (London) **165**, 528.

Douglas, W. W., and Poisner, A. M. (1964). *J. Physiol.* (London) **172**, 1.

Douglas, W. W., and Rubin, R. P. (1961). *J. Physiol.* (London) **159**, 40.

Estep, H., Blaylock, K., Mullinax, F., Brown, R., and Butts, E. (1966). *Program 48th Meeting Endocrine Soc.*, p. 28.

Gray, C. H., and Bacharach, A. L., eds. (1967). "Hormones in Blood." Vol. I. Academic Press, New York.

Guillemin, R., (1967). *Ann. Rev. Physiol.* **29**, 313.

Guillemin, R., Yamazaki, E., Gard, D. A., Jutisz, M., and Sakiz, E. (1963). *Endocrinology* **73**, 564.

Harris, G. W., and Donovan, B. T., eds. (1966). "The Pituitary Gland." Vols. I–III. Butterworths, London.

Hermier, C., and Jutisz, M. (1969). *Biochim. Biophys. Acta* **192**, 96.

Jutisz, M., and de la Llosa, M. P. (1967a). *Endocrinology* **81**, 1193.

Jutisz, M., and de la Llosa, M. P. (1967b). *C. R. Acad. Sci.* (Paris) Ser. D, **264**, 118.

Jutisz, M., and de la Llosa, M. P. (1968a). *Bull. Soc. Chim. Biol.* **50**, 2521.

Jutisz, M., and de la Llosa, M. P. (1968b). *Excerpta Med. Intern. Congr. Ser.* **157**, 137.

Jutisz, M., and de la Llosa, M. P. (1969). *C R. Acad. Sci.* (Paris) Ser. D, **268**, 1636.

Jutisz, M., and de la Llosa, M. P. (1970). *Endocrinology* **86**, 761.

Jutisz, M., Bérault, A., Novella, M. A., and Chapeville, F. (1966). *C. R. Acad. Sci.* (Paris) Ser. D, **263**, 664.

Jutisz, M., Bérault, A., Novella, M. A., and Ribot, G. (1967). *Acta Endocrinol.* (Copenhagen) **55**, 481.
Jutisz, M., Bérault, A., and de la Llosa, M. P. (1968). *In* "Pharmacology of Hormonal Polypeptides and Proteins" (N. Back, L. Martini, and R. Paoletti, eds.), pp. 138–147. Plenum Press, New York.
Kragt, C. L., and Meites, J. (1967). *Endocrinology* **80**, 1170.
Lundholm, L., Rall, T. W., and Vamos, N. (1967). *Acta Physiol. Scand.* **70**, 127.
Martini, L., Fraschini, F., and Motta, M. (1968). *Recent Progr. Hormone Res.* **24**, 439.
McCann, S. M., Dhariwal, A. P. S., and Porter, J. C. (1968). *Ann. Rev. Physiol.* **30**, 589.
Meites, J., and Nicoll, C. S. (1966). *Ann. Rev. Physiol.* **28**, 57.
Midgley, A. R., Jr., Gay, V. L., Caligaris, L. C. S., Rebar, R. W., Monroe, S. E., and Niswender, G. D. (1968). *In* "Gonadotropins 1968" (E. Rosemberg, ed.), pp. 307–312. Geron-X, Los Altos, Calif.
Mittler, J. C., and Meites, J. (1966). *Endocrinology* **78**, 500.
Parlow, A. F. (1961). *In* "Human Pituitary Gonadotropins" (A. Albert, ed.), pp. 300–326. Charles C. Thomas, Springfield, Ill.
Peron, F. G., and Koritz, S. B. (1958). *J. Biol. Chem.* **233**, 256.
Piacsek, B. E., and Meites, J. (1966). *Endocrinology* **79**, 432.
Rall, T. W., Sutherland, E. W., and Berthet, J. (1957). *J. Biol. Chem.* **224**, 463.
Rasmussen, H., and Tenenhouse, A. (1968). *Proc. Nat. Acad. Sci. U. S.* **59**, 1364.
Robison, G. A., Butcher, R. W., and Sutherland, E. W. (1968). *Ann. Rev. Biochem.* **37**, 149.
Samli, M. H., and Geschwind, I. I. (1967a). *Program 49th Meeting Endocrine Soc.*, p. 58.
Samli, M. H., and Geschwind, I. I. (1967b). *Endocrinology* **81**, 835.
Samli, M. H., and Geschwind, I. I. (1968). *Endocrinology* **82**, 225.
Sattin, A., and Rall, T. W. (1967). *Fed. Proc.* **26**, 707.
Schally, A. V., and Redding, T. W. (1967). *Proc. Soc. Exp. Biol. Med.* **126**, 320.
Schally, A. V., Steelman, S. L., and Bowers, C. Y. (1965). *Proc. Soc. Exp. Biol. Med.* **119**, 208.
Schally, A. V., Saito, T., Arimura, A., Sawano, S., Bowers, C. Y., White, W. F., and Cohen, A. I. (1967). *Endocrinology* **81**, 882.
Schally, A. V., Arimura, A., Bowers, C. Y., Kastin, A. J., Sawano, S., and Redding, T. W. (1968a). *Recent Progr. Hormone Res.* **24**, 497.
Schally, A. V., Müller, E. E., and Sawano, S. (1968b). *Endocrinology* **82**, 271.
Sinha, D., and Meites, J. (1965). *Neuroendocrinology* **1**, 4.
Sinha, D., and Meites, J. (1966). *Endocrinology* **78**, 1002.
Solomon, S. H., and McKenzie, J. M. (1966). *Endocrinology* **78**, 699.
Steelman, S. L., and Pohley, F. M. (1953). *Endocrinology* **53**, 604.
Sutherland, E. W., Øye, I., and Butcher, R. W. (1965). *Recent Progr. Hormone Res.* **21**, 623.
Symchowicz, S., Peckham, W. D., Oneri, R., Korduba, C. A., and Perlman, P. L. (1966). *J. Endocrinol.* **35**, 379.
Vale, W., and Guillemin, R. (1967). *Experientia* **23**, 855.
Vale, W., Burgus, R., and Guillemin, R. (1967). *Experientia* **23**, 853.
Vale, W., Burgus, R., and Guillemin, R. (1968). *Neuroendocrinology* **3**, 34.
Wakabayashi, K., Watanabe, S., Crighton, D. B., Ashworth, R., and Dhariwal, A. P. S. (1968). *Excerpta Med. Intern. Congr. Ser.* **157**, 137.
Watanabe, S., Dhariwal, A. P. S., and McCann, S. M. (1968). *Endocrinology* **82**, 674.
Wilber, J. F., Peake, G. T., and Utiger, R. D. (1969). *Endocrinology* **84**, 758.
Zor, U., Kaneko, T., Schneider, H. P. G., McCann, S. M., and Field, J. B. (1969). *J. Clin. Invest.* **48**, 93a.

Control of Adrenocorticotropin and Melanocyte-Stimulating Hormone Secretion

W. F. GANONG

I. Introduction

In this article, I propose to do three things:

1. Briefly summarize the generally accepted view of hypothalamic control of pituitary function with particular reference to Adrenocorticotropic Hormone (ACTH) and Melanocyte-Stimulating Hormone (MSH).

2. Comment on several areas of current active research and controversy relative to the control of ACTH secretion, research which has raised some questions about the correctness of widely held views of the ACTH regulatory mechanism.

3. Discuss in some detail recent research in my laboratory that suggests the existence in the dog of a hypothalamic adrenergic system that appears to inhibit the secretion of ACTH.

Control of the secretion of the ten hormones found in the pituitary gland has been the subject of reviews in a number of comprehensive treatises on neuroendocrinology (Nalbandov, 1963; Martini and Ganong, 1966, 1967; Ganong and Martini, 1969; McCann and Porter, 1969).

II. Hypothalamic Control of Pituitary Secretion

The mode of control of the two posterior pituitary hormones, oxytocin and vasopressin, is neural. They are formed in supraoptic and paraventricular neurons and secreted from the endings of these neurons in response to action potentials propagated along the axons of these neurons (Sawyer and Mills, 1966). They represent hormones secreted by the brain into the general circulation. The six generally accepted anterior lobe hormones, ACTH, Growth Hormone (GH), Thyrotropin (TSH), Follicle-Stimulating Hormone (FSH), Luteinizing Hormone (LH), and prolactin, all appear to be at least in part under hypothalamic control via a neurovascular mechanism. The

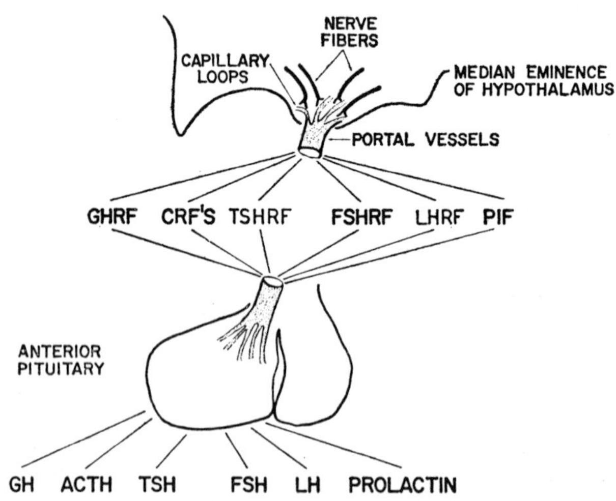

FIG. 1. Diagrammatic summary of releasing and inhibitory factors concerned with the regulation of anterior pituitary secretion. GHRF, Growth Hormone Releasing Factor; CRF, Corticotropin Releasing Factor; TSHRF, Thyrotropin Releasing Factor; FSHRF, Follicle-Stimulating Hormone Releasing Factor; LHRF, Luteinizing Hormone Releasing Factor; PIF, Prolactin Inhibiting Factor (From Ganong and Kragt, 1969).

neurovascular hypothesis holds, of course, that pituitary-influencing factors or hormones liberated from nerve endings in the median eminence enter the hypophyseal portal vessels and pass directly to the anterior pituitary, where they regulate its secretion. The principal factors are summarized in Fig. 1. The mode of control of most of the hormones appears to be excitatory, via releasing factors, but the control of prolactin secretion is inhibitory via Prolactin Inhibitory Factor (PIF). There may be a prolactin stimulating factor in some species, and a growth hormone inhibitory factor has been described which presumably operates in conjunction with Growth Hormone Releasing Factor (GHRF) (McCann and Porter, 1969).

It is worth pointing out that secretion of hormones, or at least of chemical agents that act on other cells, is hardly unique to neurons in the hypothalamus. Most and probably all neurons in the nervous system liberate substances such as acetylcholine and norepinephrine at their endings which act on other cells to excite or inhibit them. In some instances (e.g., the sympathetic nervous system) these mediators not only act locally but also enter the general circulation (Ganong, 1969). It seems to me that the important point is that neurons in general liberate chemical agents in response to electrical activity reaching their axon terminals, and it is merely the distance these subtances travel (and whether or not they enter the circulation) that differs from one location to another.

III. Melanocyte-Stimulating Hormone Secretion

The other two hormones found in the mammalian pituitary are α- and β-MSH. The mode of control of the secretion of these polypeptides is still undetermined. There are in addition singular differences between the situation in amphibia, reptiles, and fish on the one hand, and birds and mammals on the other. These differences make it wise to consider the two groups separately.

In amphibia and related classes of animals, certain facts are established:

1. MSH expands melanophores and, along with direct neural control of melanophores in some species, plays a role in color changes that help the animals to adapt to their environment (Landgrebe and Mitchell, 1966).

2. The chemistry of MSH has not been studied in as great detail as it has in mammals, and data are not available on the exact structure.

3. The hypothalamic control of MSH secretion is generally inhibitory in nature. Lesions of the hypothalamus or transplantation of the pituitary to a distant site result in skin darkening.

4. The cells of the intermediate lobe appear to be innervated by two sets of nerve fibers, one containing small catecholamine-like granules and another containing large, "peptidergic" granules. There may be innervation by fibers containing small vesicles as well. On the basis of a number of experiments it has been postulated that the secretion of the intermediate lobe is under the control of these inhibitory nerve fibers rather than agents reaching the gland in the blood stream (Etkin, 1967).

In mammals, the situation is very different:

1. There are no melanophores and of course no normal changes in skin coloration comparable to those in amphibia and fish. Prolonged administration of large doses of MSH can produce skin darkening by an action on melanocytes, and the hyperpigmentation seen in Addison's disease and certain tumors appears to be due to increased circulating MSH (Abe *et al.*, 1967). Numerous more physiologic functions have been suggested, including an action on potentials in the spinal cord (Krivoy *et al.*, 1963), an action on lipid metabolism (Raben *et al.*, 1961), a natriuretic action (Orias and Johnson, 1969), an action on fluid balance in the eye (Dyster-Aas and Krakau, 1964) and a number of actions on behavior (for references see De Wied, 1969). However, it seems fair to state that as yet no physiologic function for MSH has been established. In addition, the intermediate lobe is rudimentary or absent in a number of mammalian species including adult humans (Wingstrand, 1966), although in some species there are MSHs in the anterior lobe.

2. In mammals the chemical nature of the intermediate lobe hormones is known. There are two MSHs: α-MSH, a tridecapeptide that has a relatively constant structure from species to species,

and β-MSH, a larger peptide that shows considerable variation in structure from one species to another. β-MSH circulates, but α-MSH has not been detected in the blood stream. Circulating MSH levels parallel those of ACTH to the degree that they are both increased in Addison's disease and inhibited by glucocorticoids (Abe *et al.*, 1967).

3. There is considerable evidence that in mammals, as in amphibia, the tonic hypothalamic control of MSH secretion is inhibitory (Etkin, 1967).

4. The mode of control of MSH secretion in mammals remains undetermined. Although control by direct innervation of secretory cells has by no means been ruled out, two factors claimed to influence MSH secretion have been extracted from hypothalamic tissue. One of these, an MSH-Releasing Factor (MSHRF) has been reported to decrease pituitary MSH content and increase circulating MSH (for references see McCann and Porter, 1969). The other, an MSH-Inhibiting Factor (MSHIF) has been reported to increase pituitary MSH and decrease circulating MSH (Kastin *et al.*, 1969). It is interesting that circulating MSH and pituitary MSH content are not always reciprocally related; for instance, darkness diminishes and pinealectomy increases pituitary MSH content, but these procedures do not appear to alter circulating MSH (Kastin *et al.*, 1969). It seems clear that only future research, preferably carried out with the new immunoassay methods for MSH, will settle the exact mode of the control of MSH secretion in mammals.

IV. Adrenocorticotropic Hormone Secretion: Established Components

A diagrammatic summary of the established components involved in the regulation of ACTH secretion is presented in Fig. 2. According to this view, one would expect the rate of ACTH secretion at any given time to be determined primarily by two factors, the magnitude of the hypothalamic stimulation of Corticotropin Releasing Factor (CRF) secretion, and the magnitude of the negative feedback effect of circulating cortisol. The ACTH-cortisol interaction constitutes a

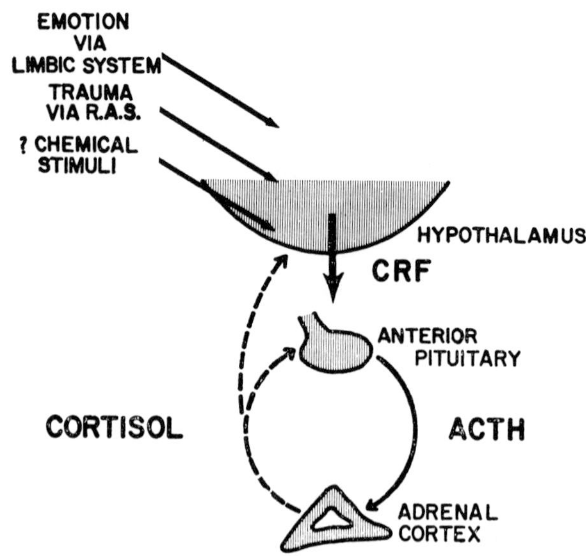

FIG. 2. Diagrammatic summary of established pathways involved in the control of ACTH secretion (From Ganong, 1969).

closed-loop feedback mechanism that operates to stabilize the rate of ACTH secretion at rest. The illustrated neural components are converging excitatory inputs which increase CRF secretion. They represent an open-loop component which provides a mechanism by which the feedback control can be overridden to bring about increased ACTH secretion in emergencies. These emergencies are the myriad, diverse stresses known to have the ability to increase ACTH secretion. For completeness, a neural imput should probably be added to account for the diurnal rhythm in ACTH secretion. This rhythm appears to be driven by a biological clock in the temporal lobes of the brain (for references see Ganong, 1963).

The afferent fibers responsible for the increase in ACTH secretion produced by a physical stimulus such as a leg fracture probably travel in the lateral spinothalamic tract. The increase is prevented not only by cord section (for references see Mangili *et al.*, 1966), but also by hemisection of the cord on the side opposite to the injury (Gibbs, 1969). From the spinal cord, the pathways mediating the response have been traced into the midbrain (Gibbs, 1969). It is known that collaterals from the spinothalamic tracts enter the reticular formation, and stimulation of the reticular formation (Mason, 1958), as well as the dorsal longitudinal bundle of Schütz and the mammillary peduncle (Ganong *et al.*, 1965) increases ACTH secretion. Therefore it seems reasonable to conclude that the impulses in the ascending pathways enter the reticular core of the brain and traverse it to the ventral hypothalamus, where they initiate CRF release.

There is general agreement that stimulation of the amygdala increases ACTH secretion (see Mangili *et al.*, 1966, for references). The amygdaloid region is known to be concerned with the genesis and expression of emotions. Some years ago, Smulekoff and I (Ganong, 1961) found that amygdaloidectomy in dogs abolished the eosinopenic response to the emotional stimulus of immobilization without altering the response to surgical stress. These experiments should be repeated with more modern methods of evaluating ACTH secretion, but taken along with the results of the stimulation experiments they support the view that afferents from the limbic system mediate the marked increases in ACTH secretion produced by emotional stimuli.

Other stimuli may act directly on the hypothalamus. There is good evidence that in rats, for instance, ether acts directly on the median eminence region to increase ACTH secretion (Greer and Rockie, 1968).

V. Unsettled Problems

The overview of the control of ACTH secretion presented above still has a number of gaps in it. Current research activities in many laboratories are directed at filling these gaps, with particular reference to the final isolation and chemical analysis of CRF, the actual site and mechanism of CRF release, the site at which cortisol and ACTH feed back to inhibit ACTH secretion, and the functional role of the neural systems that appear to inhibit ACTH secretion.

A. The Chemistry of Corticotropin Releasing Factor

It is nagging fact that although its existence has been postulated for 20 years, the exact chemistry of CRF is still unknown. Four factors with CRF activity have been characterized in extracts of posterior lobe tissue: vasopressin, β-CRF, which appears to be related to vasopressin, and α_1-CRF and α_2-CRF, compounds related to α-MSH (for references see McCann and Porter, 1969). However, the relationship of these factors to the CRF found in the hypothalamus is unsettled, and it is the hypothalamic CRF that is believed to be of major physiologic importance. Current evidence suggests that the hypothalamic CRF is a peptide that has some similarity to vasopressin but differs from it significantly (McCann and Porter, 1969). The nature of CRF is discussed in more detail in the article by McCann in this volume. The results of further research in the field are awaited with great interest.

B. The Site of Corticotropin Releasing Factor Release

There has been uncertainty about whether in certain species the median eminence is the actual site of the release of CRF. Brodish (1963) reported that lesions anywhere along the base of the hypothalamus from the optic chiasm to the infundibular recess produced partial inhibition of ACTH secretion in the rat, and on the basis of this observation he concluded that in this species the hypothalamic system controlling ACTH was diffuse rather than discrete. Kendall and Allen (1968) found that lesions of the median eminence failed to lower circulating corticosterone levels in rats with pituitary transplants in the kidneys, whereas removal of the whole hypothalamus did (Kendall, personal communication). Redgate (1969) has also made a point of his failure to obtain increases in circulating ACTH following stimulation of the median eminence region in experiments in cats in which stimulation in other parts of the brain stem was effective. In view of these uncertainties it is pertinent to emphasize the localization to the median eminence of the site that appears to be concerned with CRF secretion in the dog. In the data summarized in Fig. 3, small lesions in the midportion of the median eminence inhibited stress-induced ACTH secretion, while comparable lesions anterior and posterior to this region had no inhibitory effect (Ganong, 1959). The more anterior lesions selectively inhibited TSH secretion, while the more posterior ones caused gonadal atrophy without inhibiting TSH or ACTH secretion. Furthermore, in the dog at least, stimulation of the median eminence region does increase ACTH secretion (Ganong, 1959; Goldfien and Ganong, 1962; Ganong *et al.*, 1965).

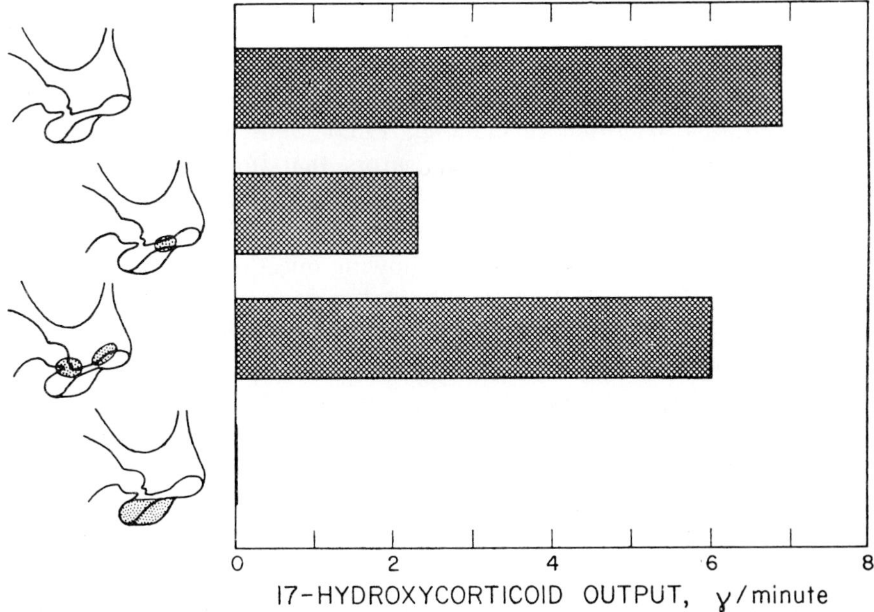

FIG. 3. Output of 17-hydroxycorticoids (μg/min) from the right adrenal in surgically stressed dogs. Samples collected 15 min after start of trauma under ether anesthesia. Values are means in 15 normal dogs, 6 dogs with anterior median eminence lesions, 6 dogs with other lesions, and 4 hypophysectomized dogs. The areas destroyed or ablated are indicated by the shaded areas superimposed on the diagrams of sagittal sections of the hypothalamus on the left side of the figure (From Ganong, 1959).

C. The Site of Cortisol Feedback

A major unsettled problem in pituitary-adrenal physiology is the site at which cortisol and related corticoids feedback to inhibit ACTH secretion. It is worth pointing out in this regard that the site responding to a low circulating corticoid level may not be the same as that responding to a high level. The increase in ACTH secretion that is produced by unilateral adrenalectomy is generally believed to be due to a decrease in the circulating corticoid level, and this increase is prevented by lesions of the median eminence. However, the atrophy of the adrenals produced by corticoid injections in animals with median eminence lesions is identical to that produced in normal controls (see Ganong, 1963, for references). Almost all reported attempts to localize the site of corticoid feedback have involved study of the mechanisms responsive to high corticoid levels.

Many investigators initially assumed that the pituitary was the site of action of the corticoid feedback. Evidence supporting this conclusion included data suggesting inhibitory effects of corticoids on pituitaries *in vitro* (see Ganong, 1963). Subsequently, however, it was shown by numerous investigators that implants of cortisol and other corticoids in the median eminence inhibited ACTH secretion (see Mangili *et al.*, 1966, for references). In addition, inhibition of ACTH secretion was reported following injection or implantation of corticoids in the midbrain, the septal region, and even the amygdaloid nuclei (Dallman and Yates, 1968; Davidson and Feldman, 1967; Mangili *et al.*, 1966; Kawakami *et al.*, 1968). On the basis of these reports, the pendulum swung toward the view that the nervous system was the site of corticoid feedback, and this view is still widely held. However, other interpretations of the implant data are possible. Implants in the median eminence region are always subject to the criticism that the steroid is absorbed into the portal vessels and transported to the pituitary. It is true that corticoid implants in the pituitary do not inhibit ACTH secretion, but this could be due to the fact that the steroid reaches only a small portion of the pituitary cells. The ACTH-inhibiting implants in the septal region, midbrain, and amygdaloid nuclei are some distance from the median eminence. However, Kendall and his associates (1969) have pointed out that all these implants are near the ventricular system, and he and his associates have presented evidence that intraventricular injections of corticoids inhibit ACTH secretion. He has advanced the hypothesis that the implanted steroids enter the cerebrospinal fluid and diffuse to the region of the median eminence. It has been reported that ependymal cells in this region look as if they transported substances from the cerebrospinal fluid to the portal vessels (see Kobayashi and Matsui, 1969).

In the meantime, attention has returned to the pituitary as a site of corticoid feedback. Pollock and La Bella (1966) have reported that ACTH release from incubated pituitaries is inhibited by cortisol, and Fleischer and Vale (1968) have found that the stimulatory effect of vasopressin on ACTH secretion *in vitro* is inhibited by dexamethasone. Kendall and Allen (1968) made the interesting observation that in rats with pituitaries transplanted to the kidneys, implants of corticoids in the median eminence did not depress ACTH secretion. In these transplanted animals, small amounts of systemically injected steroids did produce inhibition. Finally, Russell *et al.* (1969) have found that in rats given dexamethasone systemically, injection of CRF directly into the pituitary fails to increase ACTH secretion.

These observations have clearly started the pendulum swinging back to the pituitary as the major site of corticoid feedback. Of course they do not rule out an additional action on the nervous system, and other actions of corticoids on the brain are well established (Beyer and Sawyer, 1969). However, it seems fair to state that in view of the possibility of diffusion from implants to the median eminence via the cerebrospinal fluid and transport down the portal vessels to the pituitary, the burden of proof now rests on those who argue that corticoids act on the brain to produce inhibition of ACTH secretion.

It also should be mentioned here that ACTH appears to inhibit its own secretion via a " short loop " feedback action on CRF secretion from the hypothalamus. The subject of " short loop " feedback has recently been reviewed in detail (Motta *et al.*, 1969).

D. Neural Systems that Inhibit the Secretion of Adrenocorticotropic Hormone

There have been scattered reports for a number of years that suggest the presence of neural systems capable of inhibiting ACTH secretion. Perhaps the most consistent finding has been the observation that stimulation of the hippocampus and portions of the pons and midbrain inhibit ACTH secretion (for references see Ganong, 1963; Mangili *et al.*, 1966). However, it is the work on hypothalamic isolation and deafferentation that has focused attention on the physiologic role of inhibitory systems.

If the inputs that converge on the ventral hypothalamus all increase CRF secretion, as shown in Fig. 2, interruption of all neural input to this region would be expected to reduce ACTH secretion to a low level. This does not occur. Edgahl (1960) was the first to show that ACTH output was actually elevated in animals in which the hypothalamus had been isolated by removal of all brain tissue around it. We subsequently confirmed and extended this finding. A summary of our experiments is shown in Fig. 4. Elevated levels of corticoid secretion have also been reported in rats in which the hypothalamus has been isolated by cutting all its neural connections with a stereotaxic knife (Halász, 1969). These results suggest the presence of a tonic inhibitory input to the hypothalamus which holds ACTH secretion in check.

Not only do the hypothalamic island preparations have an elevated ACTH output, but ACTH output is also still appreciable in animals from which the hypothalamus has been removed (isolated pituitary preparations). This is true even when at an earlier date the pituitary stalk has been cut and a plate inserted between the pituitary and the

	17 HYDROXYCORTICOID OUTPUT ($\mu g/min$)	
	TRAUMA	*AFTER ACTH*
(1) CRANIOTOMY (4 DOGS)	10.8 ± 2.0	13.1 ± 3.0
(2) HYPOTHALAMIC ISLAND (5 DOGS)	6.8 ± 1.5	9.2 ± 1.3
(3) ISOLATED PITUITARY, HIND BRAIN IN (4 DOGS)	5.4 ± 0.6*	8.4 ± 1.2
(4) HYPOPHYSECTOMIZED (2 DOGS)	0.2*	6.6
(5) ISOLATED PITUITARY, HIND BRAIN OUT (6 DOGS)	3.6 ± 0.8*	8.2 ± 1.4
(6) STALK SECTION (5 DOGS)	5.9 ± 1.6	10.6 ± 1.1

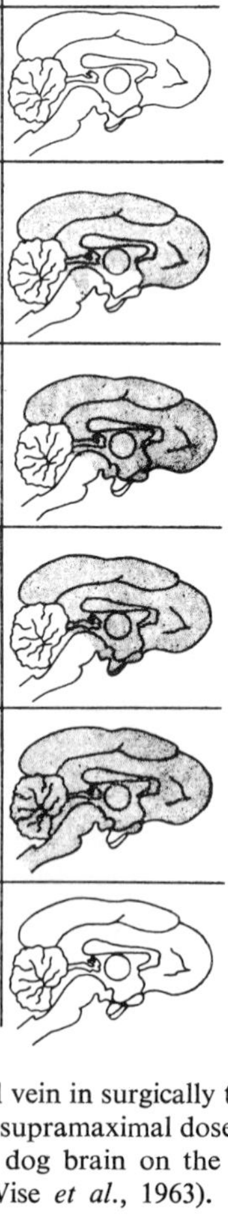

* Value significantly different from group (1), $p < 0.01$

Fig. 4. Output of 17-hydroxycorticoid in the right adrenal vein in surgically traumatized dogs 4 hr after various brain operations, and 4 min after a supramaximal dose of ACTH. Shaded areas in the diagrams of sagittal sections of the dog brain on the right show the extent of the brain removal in each group (From Wise *et al.*, 1963).

hypothalamus so that all neural elements below the plate degenerate (Wise *et al.*, 1964). Moderately elevated corticoid levels have also been reported in monkeys (Kendall and Roth, 1969) and rats (Dunn and Critchlow, 1969) with isolated pituitaries. This suggests that the hypothalamus may exert a moderate tonic inhibitory effect on the pituitary at rest. However, the data are open to other interpretations, including the possibility that in these severely traumatized animals, a CRF-like circulating " wound hormone " of some sort is acting directly on the pituitary.

It should be emphasized that in rats with hypothalamic deafferentation, ACTH secretion is elevated in the resting state. There is interesting evidence that these animals respond to stressful stimuli with an additional increase in ACTH secretion. This includes, remarkably enough, a response to the presumably purely psychic stress of restraint (Halász, 1969). The explanation of these stress responses is currently being investigated.

E. Morphology and Chemical Anatomy of Pathways Converging on the Hypothalamus

As noted above, the major excitatory pathways to the hypothalamus from the spinal cord and the limbic system have been mapped in a preliminary fashion. However, the detailed anatomy of these fiber systems and their ultimate connections remain unknown. Furthermore, the evidence summarized above indicates that an inhibitory input reaches the hypothalamus. Where does it come from and how does it mediate inhibition? What are the chemical mediators at the synaptic junctions in the excitatory and inhibitory pathways? These are questions that are just beginning to be answered.

It is pertinent in this regard to mention the prominent system of dopaminergic neurons that end in concentrated fashion around the capillary loops of the primary plexus of the portal hypophyseal vessels. The morphological features of this and the noradrenergic system of neurons in the hypothalamus, which are described in detail in the article by Fuxe and Hökfelt in this volume, make them prime candidates for a role in neuroendocrine control.

To investigate the possibility that dopamine acts directly on the pituitary, Van Loon and Kragt in my laboratory incubated this catecholamine with rat pituitary glands. They found (Van Loon and Kragt, 1970) that 20 μg of dopamine per pituitary had no effect on the release of ACTH into the medium, but 200 μg per pituitary reduced the amount of ACTH in the medium at the end of the 6-hr incubation period. However, when rat ACTH was mixed with dopamine and incubated for 6 hr, the activity of the ACTH was reduced; in other words, this dose of dopamine appears to chemically inactivate ACTH. Similar results were obtained with FSH. When pituitaries were preincubated with dopamine and then exposed to hypothalamic extract in fresh medium, ACTH and FSH release were the same as in controls. Thus we were unable to obtain any evidence for a direct effect of dopamine on pituitary secretion.

VI. Evidence for a Hypothalamic Adrenergic System that Inhibits Adrenocorticotropic Hormone Secretion

My associates and I have become involved in the question of the inhibition of ACTH secretion and the role in this inhibition of hypothalamic catecholamines in the course of studies on the mechanism by which α-ethyltryptamine inhibits stress-induced ACTH secretion in the dog. This interesting drug has an indole structure. It is a monoamine oxidase inhibitor and antidepressant which also has a prompt pressor effect and presumably liberates catecholamines. Its ACTH-inhibiting activity, which was first discovered by Tullner and Hertz (1964) is prompt in onset. We routinely restress the dogs by performing a laparotomy 25 min after injection of the drug and measure

TABLE I

EFFECT OF INTRAVENOUS ADMINISTRATION OF α-ETHYLTRYPTAMINE, α-METHYLTRYPTAMINE, AND SALINE ON ACTH SECRETION IN PENTOBARBITAL-ANESTHETIZED DOGS

			Adrenal Venous 17-Hydroxycorticoid Output (μg/min)		
	Dose	No. of dogs	Before Drug	After Drug	After ACTH *
α-Ethyltryptamine	20 mg/kg	6	9.0±2.3	2.7±1.5	9.5±2.0
α-Methyltryptamine	10 mg/kg	4	9.2±0.6	2.5±1.2	9.9±0.8
Saline	2 ml/kg	6	9.3±1.2	9.8±1.1	10.2±1.2

* 100–1000 mU i.v.

17-hydroxycorticoid output in adrenal venous blood 30 min after injection. As shown in Table I, ACTH promptly overcomes the blockade of adrenal secretion, and the sensitivity of the adrenal to large and small doses of ACTH is normal in dogs that have received the drug (Ganong and Boryczka, unpublished observations). The methyl congener of the drug also inhibits ACTH secretion. Saline injections have no effect and the response to the laparotomy is as great as the response to the initial stress. Alpha-ethyltryptamine also inhibits the increase in ACTH secretion produced by insulin hypoglycemia (Ganong, 1965), but most of our investigations have been limited to studies of surgical stress.

We initially attempted to localize the site at which α-ethyltryptamine acts to inhibit ACTH secretion by determining the adrenal response to stimulation of various parts of the nervous system before and after administration of the drug. We found that the increase in ACTH secretion produced by stimulation of the femoral nerve, medulla, and various portions of the midbrain was blocked by the drug, but the response to hypothalamic stimulation was not (Ganong *et al.*, 1965). In other, previously unpublished experiments, Hardcastle and I found that the increase in ACTH secretion produced by stimulation of the amygdaloid nuclei and related limbic areas was also overcome by α-ethyltryptamine (Table II). These data suggest that α-ethyltryptamine acts on the hypothalamus, possibly by an action on the CRF-secreting neurons.

In other studies, Lorenzen and I investigated the effects of drugs

TABLE II

EFFECT OF α-ETHYLTRYPTAMINE (10 mg/kg, i.v.) ON THE ADRENOCORTICAL RESPONSE TO ELECTRICAL STIMULATION OF THE LIMBIC SYSTEM IN SIX PENTOBARBITAL-ANESTHETIZED DOGS [1]

	Adrenal Venous 17-Hydroxycorticoid Output (μg/min)
Stimulated before drug	8.3±1.1
Stimulated after drug	2.1±1.2
ACTH, 100 mU, i.v.	10.5±0.6

[1] In 4 dogs the electrode was in the amygdala; in one it was in the orbital frontal cortex and in one it was in the suprachiasmatic area. Stimulus 60 cps, 1-msec pulses, 3.4–11 V. Adrenal venous blood collection started after 5 min of stimulation and 5 min after ACTH injection.

with one or another of the properties of α-ethyltryptamine to determine which of its properties correlated with its ACTH-inhibiting activity. We found (Lorenzen and Ganong, 1967) that indole structure did not appear to correlate with the inhibition, since 5-hydroxytryptamine, 5-hydroxytryptophan, tryptamine, and tryptophan did not produce the same prompt inhibition of ACTH secretion. Neither did inhibition of monoamine oxidase or antidepressant activity, since the monoamine oxidase inhibitors iproniazid and pargyline and the tricyclic antidepressants imipramine and amitryptyline were inactive. We found that the monoamine oxidase inhibitor tranylcypromine inhibited 17-hydroxycorticoid output when administered in large doses, but this was found to be due to a direct effect of the compound on the 17α-hydroxylase activity of the adrenal; circulating blood ACTH levels were unaffected by the drug (Johnson *et al.*, 1967).

On the other hand, we found that the sympathomimetic agents amphetamine and methamphetamine inhibited ACTH secretion. This inhibition correlated with the magnitude of the pressor response produced by these drugs, and a similar correlation was observed with α-ethyltryptamine; in individual animals in which the pressor response was low, the inhibition was generally reduced or absent. Clopane, a cyclic sympathomimetic compound, and 2-aminoheptane, an aliphatic sympathomimetic compound, also inhibited ACTH secretion, but only when given in doses large enough to produce a marked pressor response. Finally, we found (Ganong *et al.*, 1967) that α-ethyltryptamine failed to inhibit ACTH secretion when its pressor effect was prevented by bleeding the animal after its administration.

These observations led us to hypothesize that any rise in blood pressure would inhibit ACTH secretion if it were of sufficient magnitude. The inhibition was presumed to be mediated via the carotid and aortic baroreceptors. There was considerable precedent for this view. A decline in blood pressure increases and a rise in blood pressure inhibits vasopressin secretion, and these responses are mediated via the arterial baroreceptors and receptors in the atria and great veins (Share, 1969). In addition, a drop in blood pressure in the carotid sinuses is known to increase ACTH secretion (Biglieri and Ganong, 1961; Redgate, 1968). This effect must be due to reduction of inhibition because the baroreceptors are stretch receptors and a decline in blood pressure decreases the rate at which they discharge.

However, the results of two new series of experiments have forced us to reappraise the hypothesis that it is the pressor activity of α-ethyltryptamine that is responsible for its inhibitory effect on ACTH secretion. In the first series of experiments we tested the ACTH-inhibiting effect of catecholamines and their precursors. We found (Van Loon *et al.*, 1969) that L-dopa in a dose of 50 mg/kg produced inhibition of ACTH secretion (Table III). This dose produces a marked pressor response and may act by liberating catecholamines; the minimum effective dose is increased to 100 mg/kg by treatment with the tyrosine hydroxylase inhibitor α-methyl-*p*-tyrosine, and is reduced to 10 mg/kg by treatment with the monoamine oxidase inhibitor pargyline. However, intravenous dopamine and norepinephrine failed to inhibit ACTH secretion (Table III) even though both agents produced marked pressor responses. These latter two compounds do not enter the brain, whereas L-dopa does (Wurtman, 1966). The effect of L-dopa was also tested in the presence of the α-adrenergic blocking agent phenoxybenzamine. Despite complete abolition of the pressor response (Fig. 5) the inhibition of ACTH secretion was still present. This result is difficult to explain except in terms of a central action of the L-dopa that is independent of any effect on blood pressure.

TABLE III

EFFECT OF INTRAVENOUS ADMINISTRATION OF L-DOPA, DOPAMINE AND NOREPINEPHRINE ON ACTH SECRETION IN PENTOBARBITAL-ANESTHETIZED DOGS

			Adrenal Venous 17-Hydroxycorticoid Output (μg/min)		
Drug	Dose	No. of Dogs	Stress before Drug	Stress after Drug	ACTH [3]
L-Dopa	50 mg/kg	12	9.2±0.9	2.6±0.7	9.0±1.0
L-Dopa	10 mg/kg	5	8.9±1.5	8.6±1.7	9.0±1.6
L-Dopa	10 mg/kg after pargyline [1]	6	9.2±1.8	2.7±0.9	9.6±1.3
L-Dopa	50 mg/kg after α-methyl-*p*-tyrosine [2]	4	8.2±1.4	7.7±1.2	9.0±2.9
L-Dopa	100 mg/kg after α-methyl-*p*-tyrosine [2]	5	8.2±0.5	2.3±1.1	7.3±0.8
Dopamine	140 μg/kg	4	12.2±3.1	11.7±1.1	11.3±2.1
Norepinephrine	33 μg/min	4	9.0±2.0	9.6±3.3	10.8±3.0

[1] 25 mg/kg, 43 and 20 hr previously.
[2] 100 mg/kg, 20 hr previously.
[3] 1 IU i.v.

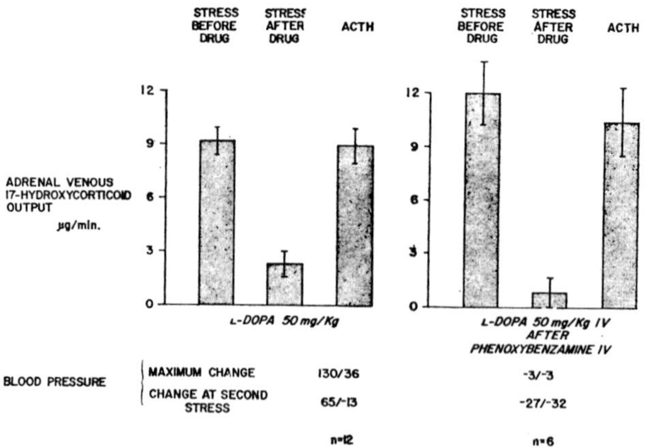

FIG. 5. Lack of effect of phenoxybenzamine on the inhibition of ACTH secretion produced by L-dopa. N = number of dogs in each case (From Van Loon *et al.*, 1969).

Additional experiments which point toward a central action for catecholamines in inhibiting ACTH secretion involve injection of drugs into the third ventricle immediately over the median eminence. The results of these experiments to date are summarized in Table IV. L-Dopa in the ventricle produced striking inhibition in three of eight animals. Control injection of acid saline at a pH comparable to that of the dopa solution produced no inhibition. Dopamine in the ventricle had a clear-cut inhibitory effect and control acid saline injections at the pH of the dopamine solution had none. Inhibition was also produced by α-ethyltryptamine in the ventricle in a dose that was a fraction of the systemically effective dose. Furthermore, this intraventricular dose had no effect on blood pressure. The catecholamine-releasing agent, tyramine, was also found to produce inhibition of ACTH secretion. Thus a catecholamine precursor, a catecholamine, and two drugs that are capable of releasing catecholamines inhibit ACTH secretion when injected into the third ventricle. Additional experiments are under way. The doses of the agents used to date are large and we do not as yet have any data indicating whether it is dopamine or norepinephrine that mediates the inhibition. However, these results plus the fact that there are prominent adrenergic systems in the ventral hypothalamus make it reasonable to advance the working hypothesis that in the dog a hypothalamic adrenergic system inhibits ACTH secretion.

TABLE IV

EFFECT OF INTRAVENTRICULAR ADMINISTRATION OF DRUGS ON ACTH SECRETION IN PENTOBARBITAL-ANESTHETIZED DOGS

Drug	Total Dose [1]	No. of Dogs	Adrenal Venous 17-Hydroxycorticoid Output (μg/min)		
			Stress before Drug	Stress after Drug	ACTH [2]
L-Dopa	20 mg	8	9.5±1.0	5.3±1.6	8.3±1.0
Acid saline control (pH 1.0)	0.5 ml	4	10.8±1.4	6.9±0.9	7.3±1.1
Dopamine	5 mg	9	12.2±1.1	3.7±1.1	10.9±1.2
Acid saline control (pH 4.2)	0.5 ml	10	9.1±1.2	6.5±0.6	7.5±1.1
α-Ethyltryptamine	8 mg	8	12.8±1.5	4.1±1.2	12.7±2.1
Tyramine	20 mg	8	10.5±0.8	1.6±1.1	9.7±1.5

From Van Loon *et al.*, 1969.

[1] Administered over ½ hr into the third ventricle.

[2] 100–1000 mU i.v.

It still could be, of course, that a rise in blood pressure acts via the arterial baroreceptors to inhibit ACTH secretion, possibly by way of the hypothalamic adrenergic system. The results with intravenous norepinephrine and dopamine infusions make this conclusion a bit tenuous, but the inhibition, if it exists, is probably not great and is balanced off against the excitatory input from other sources. The effects of carotid sinus nerve stimulation on ACTH secretion and similar experiments need to be performed to settle this point.

VII. Conclusions

In this article, I have presented a summary of current views of the regulation of pituitary secretion with particular reference to MSH and ACTH. The position of MSH is unique in that in amphibia, fish and reptiles, in which it has an established function, the chemistry of MSH and certain aspects of its secretion have not been studied in detail. In mammals, MSH-releasing and MSH-inhibiting factors of hypothalamic origin have been prepared and refined assay techniques for the study of pure MSH peptides are available, but the hormone has no established function.

In the case of ACTH, much is known about the regulatory mechanism, but there are unsettled problems. There is great current interest in the site at which corticoids feed back to inhibit ACTH secretion, with considerable new data pointing toward the pituitary as a major site of this action. There must also be neural systems that inhibit ACTH secretion. Evidence is presented that in the dog, one of these systems is an adrenergic system in the hypothalamus.

Acknowledgment

This article includes results of experiments carried out in collaboration with numerous collaborators, including G. Van Loon, A. King, C. L. Kragt, L. C. Lorenzen, B. L. Wise, L. Hilger, R. Cohen, and A. T. Boryczka. The research was supported by USPHS Grants AM 06704 and AM 05613.

References

Abe, K., Nicholson, W. E., Liddle, G. W., Island, D. P., and Orth, D. N. (1967). *J. Clin. Invest.* **46**, 1609.

Beyer, C., and Sawyer, C. H. (1969). *In* " Frontiers in Neuroendocrinology 1969 " (W. F. Ganong and L. Martini, eds.). Oxford Univ. Press, New York, pp. 255–287.

Biglieri, E. G., and Ganong, W. F. (1961). *Proc. Soc. Exp. Biol. Med.* **106**, 806.

Brodish, A. (1963). *Endocrinology* **73**, 727.

Dallman, M. F., and Yates, F. E. (1968). *Mem. Soc. Endocrinol.* **17**, 39.

Davidson, J. M., and Feldman, S. (1967). *Acta Endocrinol.* (Copenhagen) **55**, 240.

De Wied, D. (1969). *In* " Frontiers in Neuroendocrinology 1969 " (W. F. Ganong and L. Martini, eds.). Oxford Univ. Press, New York, pp. 97–140.

Dunn J., and Critchlow, V. (1969). *Life Sci.* **8**, 9.

Dyster-Aas, K., and Krakau, C. T. (1964). *Endocrinology* **74**, 255.

Egdahl, R. A. (1960). *Acta Endocrinol.* (Copenhagen) **35**, *Suppl.* 151, 49.

Etkin, W. (1967). *In* " Neuroendocrinology " (L. Martini and W. F. Ganong, eds.). Vol. II. Academic Press, New York, pp. 261–282.

Fleischer, N., and Vale, W. (1968). *Endocrinology* **83**, 1232.

Ganong, W. F. (1959). *In* " Comparative Endocrinology " (A. Gorbman, ed.). Wiley, & Sons, New York, pp. 187–201.

Ganong, W. F. (1961). *In* " Physiology of Emotions " (A. Simon, ed.). Charles C. Thomas, Springfield, Ill., pp. 64–71.

Ganong, W. F. (1963). *In* " Advances in Neuroendocrinology " (A. Nalbandov, ed.). Univ. Illinois Press, Urbana, Ill., pp. 92–129.

Ganong, W. F. (1965). *In* " Proc. II Intern. Congr. Endocrinology " (S. Taylor, ed.). *Excerpta Med.*, Intern. Congr. Ser. **83**, 624–628.

Ganong, W. F. (1969). " Review of Medical Physiology." 4th ed. Lange Medical Publications, Los Altos, Calif.

Ganong, W. F., and Kragt, C. L. (1969). *In* " Reproduction in Domestic Animals " (H. H. Cole and P. T. Cupps, eds.). 2nd ed. Academic Press, New York, pp. 155–185.

Ganong, W. F., and Martini, L., eds. (1969). " Frontiers in Neuroendocrinology 1969 " Oxford Univ. Press, New York.

Ganong, W. F., Wise, B. L., Shackelford, R., Boryczka, A. T., and Zipf, B. (1965). *Endocrinology* **76**, 526.

Ganong, W. F., Boryczka, A. T., Lorenzen, L. C., and Egge, A. S. (1967). *Proc. Soc. Exp. Biol. Med.* **125**, 558.

Gibbs, F. (1969). *Am. J. Physiol.* **217**, 78.

Goldfien, A., and Ganong, W. F. (1962). *Am. J. Physiol.* **202**, 205.

Greer, M. A., and Rockie, C. (1968). *Endocrinology* **83**, 1247.

Halász, B. (1969). *In* " Frontiers in Neuroendocrinology 1969 " Oxford Univ. Press, New York, pp. 307–342.

Johnson, P. C., Lorenzen, L. C., Biglieri, E. G., and Ganong, W. F. (1967). *Endocrinology* **80**, 510.

Kastin, A. J., Schally, A. V., Viosca, S., and Miller, M. C., III (1969). *Endocrinology* **84**, 20.

Kawakami, M., Seto, K., and Yoshida, K. (1968). *Neuroendocrinology* **3**, 349.

Kendall, J. W., and Allen C. (1968). *Endocrinology* **82**, 397.

Kendall, J. W., and Roth, J. G. (1969). *Endocrinology* **84**, 686.

Kendall, J. W., Grim, Y., and Shimshak, G. (1969). *Endocrinology* **85**, 200.

Kobayashi, H., and Matsui, T. (1969). *In* " Frontiers in Neuroendocrinology 1969 " (W. F. Ganong and L. Martini, eds.). Oxford Univ. Press, New York, pp. 3–46.

Krivoy, W. F., Lane, M., and Kroeger, D. C. (1963). *Ann. N. Y. Acad. Sci.* **104**, 312.

Landgrebe, F. W., and Mitchell, G. M. (1966). *In* " The Pituitary Gland " (G. W. Harris and B. T. Donovan, eds.). Vol. III. Butterworths, London, pp. 41–58.

Lorenzen, L. C., and Ganong, W. F. (1967). *Endocrinology* **80**, 889.

Mangili, G., Motta, M., and Martini, L. (1966). *In* " Neuroendocrinology " (L. Martini and W. F. Ganong, eds.). Vol. I. Academic Press, New York, pp. 297–370.

Martini, L., and Ganong, W. F., eds. (1966-1967) " Neuroendocrinology." Academic Press, New York.

Mason, J. W. (1958). *Endocrinology* **63**, 403.

McCann, S. M., and Porter, J. C. (1969). *Ann. Rev. Physiol.* **49**, 240.

Motta, M., Fraschini, F., and Martini, L. (1969). *In* " Frontiers in Neuroendocrinology 1969 " (W. F. Ganong and L. Martini, eds.). Oxford Univ. Press, New York, pp. 211–253.

Nalbandov, A. V., ed. (1963). " Advances in Neuroendocrinology." Univ. Illinois Press, Urbana, Ill.

Orías, R., and Johnson, R. L. Jr., (1969). *Fed. Proc.* **28**, 573.

Pollock, J. J., and La Bella, F. S. (1966). *Can. J. Physiol. Pharmacol.* **44**, 549.

Raben, M. S., Landolt, R., Smith, F. A., Hofman, K., and Yajima, H. (1961). *Nature*, **189**, 681.

Redgate, E. S. (1968). *Endocrinology* **82**, 704.

Redgate, E. S. (1969). *Fed. Proc.* **28**, 438.

Russell, S. M., Dhariwal, A. P. S., McCann, S. M., and Yates, F. E. (1969). *Endocrinology* **85**, 512.

Sawyer, W. H., and Mills, E. (1966). *In* " Neuroendocrinology " (L. Martini and W. F. Ganong, eds.). Vol. I. Academic Press, New York, pp. 187–216.

Share, L. (1969). *In* " Frontiers in Neuroendocrinology 1969 " (W. G. Ganong and L. Martini, eds.). Oxford Univ. Press, New York, pp. 183–210.

Tullner, W. W., and Hertz, R. (1964). *Proc. Soc. Exp. Biol. Med.* **116**, 837.

Van Loon, G. R., and Kragt, C. L. (1970). *Proc. Soc. Exp. Biol. Med.* (In press).

Van Loon, G. R., Hilger, L., Cohen, R., and Ganong, W. F. (1969). *Fed. Proc.* **28**, 438.

Wingstrand, K. G. (1966). *In* " The Pituitary Gland " (G. W. Harris and B. T. Donovan, eds.). Vol. III. Butterworths, London, pp. 1–27.

Wise, B. L., Van Brunt, E. E., and Ganong, W. F. (1963). *Proc. Soc. Exp. Biol. Med.* **112**, 792.

Wise, B. L., Van Brunt, E. E., and Ganong, W. F. (1964). *Proc. Soc. Exp. Biol. Med.* **116**, 306.

Wurtman, R. J. (1966). " Catecholamines." Little, Brown and Co., Boston.

Hypothalamic Control of Thyrotropin

J. M. McKenzie, P. R. Adiga, and S. H. Solomon

I. Evidence for Existence of Hypothalamic Control

In recent years hypothalamic control of pituitary Thyrotropin (TSH) increasingly has been the subject of review (e.g., Brown-Grant, 1960; D'Angelo, 1963; Greer, 1957; Harris, 1948, 1955; Reichlin, 1966; Szentágothai *et al.*, 1962). Initially the question of concern was whether or not there was such control.

A. Pituitary Stalk Section Experiments

Uotila (1939, 1940) was among the first to obtain evidence indicating that the hypothalamus exerted an influence on the thyroid-pituitary axis. He showed that transection of the pituitary stalk in the rat did not affect the histological appearance of the thyroid; however, placing the animal in a cold environment failed to produce the histological features of thyroid gland activation that normally were to be expected. His conclusion (Uotila, 1940) was that " . . . the thyrotropic function of the anterior hypophysis has a basic secretory rhythm which is for the most part controlled humorally by variations in the organism's thyroxine level without the mediation of the hypothalamo-hypophyseal pathways. Hypothalamic impulses through the pituitary stalk . . . can under certain circumstances modify this basic secretory rhythm." This verdict was not readily accepted because of difficulty experienced in reproducing such experiments where stalk-section was involved. Eventually, however, the importance of preventing restitution of the hypothalamus-pituitary portal vessels was recognized (Harris, 1955) and consequences (to the thyroid-pituitary axis) of complete interruption of the hypothalamic portal system were established:

1. Thyroid gland function is reduced, but not to the level found with hypophysectomy.
2. There is no impairment of " suppressive " effects of the administration of thyroid hormone; e.g., injection of thyroxine still leads to marked reduction of thyroid gland function, secondary to suppression of secretion of thyrotropin.
3. Certain stimuli (reduction of ambient temperature is perhaps most commonly used) are no longer effective in enhancing thyroid function.

That these consequences are not due simply to a nonspecific reduction in pituitary function, secondary to ischemia, has been refuted by several indirect lines of evidence (Reichlin, 1966); in any case the concept of hypothalamic control of pituitary-thyrotropin function was further strengthened by experiments with pituitary grafts. For instance, a graft functioning inadequately at a remote site recovered virtually total function when regrafted under the median eminence (Nikitovitch-Winer and Everett, 1958).

B. Hypothalamic Lesion Experiments

Another approach to identifying the importance of the hypothalamus for a normal thyroid and pituitary interaction was the use of techniques whereby fairly discrete electrolytic lesions were produced in the hypothalamus and other suprahypophyseal areas. Greer (1957) in particular exploited this procedure and concluded that an appropriate lesion of the hypothalamus reduced the ability of the pituitary to enhance secretion of thyrotropin in response to a lowered concentration of circulating thyroid hormone. The required corollary of these data was provided by several laboratories; i.e., it was shown that electrical stimulation of specific areas in the diencephalon led to enhancement of thyroid function (D'Angelo and Snyder, 1963; Harris and Woods, 1958; Shizume *et al.*, 1962) with associated decrease in pituitary thyrotropin content and increase in the concentration of circulating thyrotropin (D'Angelo and Snyder, 1963).

II. Mechanism of Hypothalamic Control

By these different means a hypothalamic influence on the thyroid-pituitary axis was recognized and accepted, and reviews became more concerned with the mechanism by which the control was exerted. While a neurohumoral mediator was long postulated (Harris, 1955), other considered possibilities were that the hypothalamus might control blood flow to the hypophysis (Harris, 1948; Worthington, 1960) or, specifically, the proportion of thyroid hormone passed down to the portal vascular system (Brown-Grant, 1960; Purves, 1964). While a subsidiary role of these two potential mechanisms has not been disproved, the existence of a chemical transmitter substance, probably the primary mediator of hypothalamic influence on pituitary thyrotropin, is now established.

A. Thyrotropin Releasing Factor

In a valuable review published in 1966, Reichlin referred to the substance currently known as Thyrotropin Releasing Factor or TSHRF, and stated that " the hypophysiotropic-hormonal specificity of this material, its chemical structure, proof that it circulates in portal vein blood, and its mechanism of action on the hypophysis are problems as yet unsolved."

Of these four aspects only the last (mechanism of action) will be explored in the present review; indeed, of the other three only " chemical structure " is a subject on which much further substantive, though not necessarily conclusive, data have been reported. In brief, the works of Guillemin (Guillemin *et al.*, 1965; Guillemin, 1967) and Schally (Schally *et al.*, 1968) and their respective colleagues have indicated that what was thought to be a simple small peptide is apparently a more complex, though small, molecule containing equimolar quantities of only three amino acids, viz., histidine, proline, and glutamic acid; the TSHRF activity of the preparation is resistant to proteolytic enzymes (Schally *et al.*, 1966; Guillemin, 1967). Most recently (Burgus *et al.*, 1969a) acetylated derivatives, not finitely identified, of the synthetic tripetide Glu-His-Pro were found to have TSHRF-like activity. The significance of these most interesting findings must await further developments, since the acetylated preparations reputed to be biologically active were mixtures of products and were potent *in vitro* and *in vivo* only in doses 100 to 1.000 times the minimum required with the purest TSHRF preparations extracted from hypothalami[1] (Guillemin *et al.*, 1965; Schally *et al.*, 1966).

III. Mode of Action of Thyrotropin Releasing Factor

Some current studies on the mechanism of action of TSHRF may be considered under a few arbitrary headings:

1. Is inhibition of TSHRF by thyroid hormone exerted at the level of the hypothalamus, the adenohypophysis, or both? What is the mechanism of the inhibition?
2. Does TSHRF directly stimulate synthesis of thyrotropin as well as release of the hormone?
3. What is the mediatory role of cyclic 3′5′-AMP in the action of TSHRF?

Of the first question, only a portion is readily answered; there is no doubt that thyroxine and triiodothyronine can act in their negative feedback role at the level of the hypophysis, to influence the secretion of thyrotropin. This is established with a host of studies (well reviewed by Reichlin, 1966) involving intrapituitary microinjections, isolated pituitaries, hypothalamic lesions, and combinations of pituitary extrasellar grafting and hypothalamic lesions. The last type of experiment (Reichlin, 1963) would seem effectively to rule out an influence of thyroid hormone on TSHRF which might reach an isolated pituitary by way of the general circulation (Flor-

[1] Recently the structure of ovine (Burgus *et al.*, 1969b) and porcine (Bøler *et al.*, 1969) TSHRF was reported to be PCA (or pyro) -Glu-His-Pro (NH_2); this material has been synthesized and biological actions and potency found to be identical to the purified, extracted, natural products.

sheim *et al.*, 1963; Kendall *et al.*, 1967). A similar body of data (see Reichlin, 1966) bears on the matter of hypothalamic involvement in the thyroid-pituitary feedback system, but conclusions are rather more equivocal. One factor leading to uncertainty in interpretation of some experiments is the possibility of thyroid hormone, when inserted by microinjection into the hypothalamus, tracking down into the pituitary; thus an apparent direct effect on the hypothalamus might be in fact on the pituitary. However, Knigge (1964) reported that injection of thyroxine in the preoptic area of the cat (i.e., sufficiently remote from the pituitary stalk to obviate that possibility) markedly suppressed thyroid function.

The conclusion from these experiments dealing with hypothalamus and pituitary seems to be that in the control of secretion of thyrotropin, both areas, at least in some species, are sensitive to the concentration of available thyroxine. Interaction between the two controlling mechanisms and a conclusion as to which is important in a given disturbance of thyroid-pituitary balance remain uncertain.

A. Inhibition of Thyrotropin Releasing Factor by thyroid Hormone

One report, using an entirely different technique, attempts to deal with the question of the site of feedback action of thyroxine; this is the work of Sinha and Meites (1965), who assessed by bioassays the concentration of TSHRF in the hypothalamus and thyrotropin in the pituitary in rats subjected to thyroidectomy or injected with thyroxine. Results associated with thyroidectomy indicated that there were increases in the concentration of both TSHRF and thyrotropin, but injection of thyroxine led to no change in TSHRF concentration despite a fourfold decrease in pituitary concentration of thyrotropin. Conclusions were that the increase in secretion of thyrotropin after thyroidectomy might be due to enhanced synthesis and secretion of TSHRF, whereas thyroxine would inhibit by acting directly in the pituitary. Speculations arising from static measurements of concentrations can, of necessity, be accepted only as such; however, the last suggestion has gained support from other studies made in the same laboratory and elsewhere (Sinha and Meites, 1966; Guillemin *et al.*, 1965; Bowers *et al.*, 1968, a, b; Schally and Redding, 1967).

A uniform finding has been that preincubation of pituitary tissue with thyroxine leads to inhibition of the effect of a subsequently added preparation of TSHRF. Unfortunately the dose of thyroxine required has been of the order of 10^{-6} *M* thyroxine, which is approximately 10^5-fold greater than estimates of free thyroxine concentration in circulating blood (Robbins and Rall, 1967); it seems unlikely that pituitary tissue *in vivo* would concentrate thyroxine from the circulation to this extent (Reichlin, 1966). While there is no knowledge of concentrations of thyroxine achieved *in vitro* or *in vivo* at whatever the active site or sites may be these considerations indicate a need for caution in accepting that the *in vitro* findings reflect a physiological *in vivo* counterpart.

Similar general uniformity occurs with reports that antibiotics inhibiting protein or DNA-dependent RNA synthesis prevent the inhibition of the TSHRF effect, either *in vitro* or *in vivo*, by thyroid hormone (Bakke and Lawrence, 1968; Bowers *et al.*, 1968 a, b; Schally and Redding, 1967; Vale *et al.*, 1968). The obvious conclusion is that the effect of thyroxine or triiodothyronine in inhibiting TSHRF is mediated by a specific protein. Again, however, the obvious conclusion must remain, at best, a reasonable hypothesis until further substantiated, since the antibiotics used (actinomycin D, puromycin and cycloheximide) are all known to have actions other than inhibition of protein or RNA synthesis.

TABLE I

^{14}C-AMINO ACIDS INCORPORATION INTO PITUITARY PROTEIN: INFLUENCE OF PUROMYCIN AND OF HYPOTHALAMIC EXTRACTS (HE)

Preparation [1]	Protein / Pituitary (mg)	Protein (dpm/mg)	Pituitary (dpm/mg)
Control	0.24	31,577	7,623
HE	0.14	62,149	8,483
Puromycin	0.22	720	120
HE + Puromycin	0.12	727	86

[1] Puromycin was added 15 min before addition of HE and mixed ^{14}C-amino acids; incubation were continued for an additional 60 min before extraction of protein by the method of Siekewitz (1952) and measurement by the Folin reaction (Lowry *et al.*, 1951).

B. Thyrotropin Releasing Factor and Synthesis of Thyrotropin

The second question raised earlier pertained to the possibility of TSHRF affecting synthesis as well as relase of the hormone. First, it seems clear that thyrotropin released *in vitro* by the action of TSHRF on the pituitary is that previously synthesized and stored

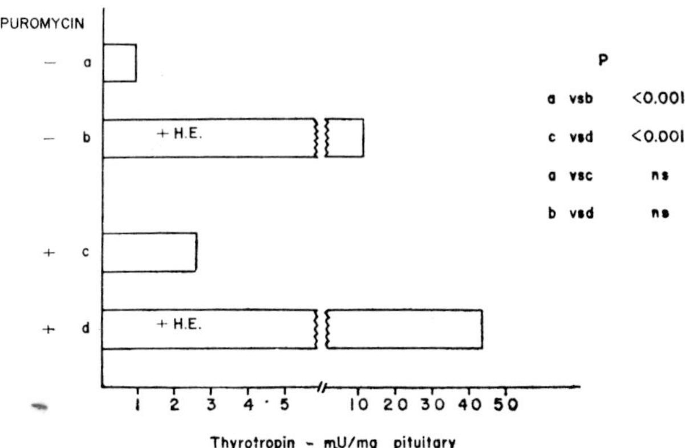

FIG. 1. Thyrotropin release *in vitro*: lack of effect of puromycin. Methods of incubation of rat pituitary halves, bioassay of thyrotropin and statistical analyses were as described previously (Solomon and McKenzie, 1966). Medium was removed and replaced hourly; data are with medium ± puromycin during the third hour of incubation. Data in Table I were obtained in the same experiment (HE, hypothalamic extract).

in the gland. Data permitting that conclusion are shown in Table I (where puromycin is shown to inhibit incorporation of ^{14}C-labeled amino acids into pituitary protein) and Fig. 1; in the figure are depicted stimulation of release of thyrotropin by addition of an extract of hypothalamus to pituitary halves *in vitro*; no effect of puromycin on this stimulation is shown. Similar data have been reported from other laboratories (Wilber and Utiger, 1968; Vale *et al.*, 1968) and indeed it may be a general biological principle that secretion is not dependent on protein synthesis (Bauduin *et al.*, 1967). Consequently it is clear that inhibition of protein synthesis does not interfere with stimulation of thyrotropin release.

On the other hand, depletion of the pituitary content of thyrotropin *in vivo* will be associated with fresh synthesis of the hormone, to maintain a supply of thyrotropin. However, whether synthesis is a direct effect of TSHRF or an eventual consequence of release of thyrotropin is not clearly established. Sinha and Meites (1966) measured concentrations of thyrotropin in the medium and gland, with pituitaries incubated in medium that contained hypothalamic

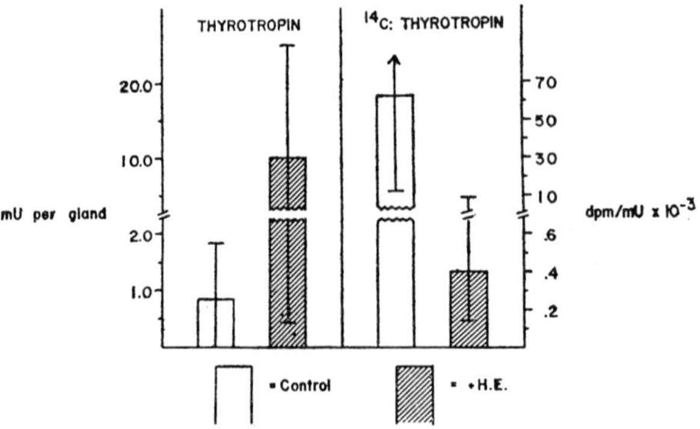

FIG. 2. Thyrotropin and associated ^{14}C in medium of rat pituitary incubations: influence of hypothalamic extract (HE). Pituitary halves were incubated, with or without HE, as described in the legend to Fig. 1; mixed ^{14}C-amino acids were present. A thyrotropin fraction of protein was obtained by successive filtration of medium on Sephadex G-25, alcohol-salt percolation (Bates *et al.*, 1959) and refiltration after addition of nonradioactive amino acids.

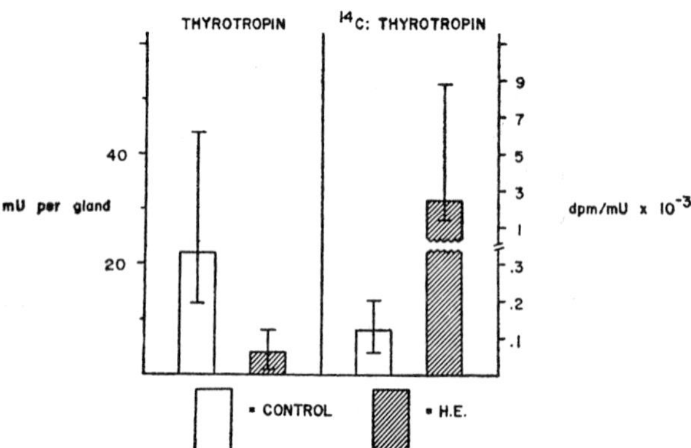

FIG. 3. Thyrotropin and associated ^{14}C in rat pituitary halves after inbucation with ^{14}C-amino acids; influence of hypothalamic extract (HE). Pituitary halves were homogenized in 1% albumin solution and the extracts processed as described in the legend to Fig. 2.

extract for the last two of four to six days of culture; they found the glands and medium contained 50 to 97% more thyrotropin than there was in control fresh glands. The duration of these experiments (two days with hypothalamic extract) was such that a direct effect of TSHRF on thyrotropin synthesis could hardly be considered proven; rather the findings indicate that the experimental conditions must have been adequate to mimic *in vivo* circumstances.

We have attempted to approach the problem by assessing the acute influence of a hypothalamic extract on the incorporation of ^{14}C-labeled amino acids into protein in thyrotropin fractions of gland and medium with an *in vitro* pituitary incubation; Figs. 2 and 3 illustrate some of the results. ^{14}C-labeled amino acids were present for 1 hr of incubation, as was hypothalamic extract in appropriate flasks. When the medium was filtered on Sephadex G-25 and the protein so obtained subjected to percolation as a means of concentrating and partly purifying thyrotropin (Bates *et al.*, 1959), the proportion of thyrotropin to ^{14}C was much less in the stimulated system, although the quantity of thyrotropin was much greater (Fig. 2). These data are in keeping with the concept that thyrotropin released into the medium was that stored in the gland and was not freshly synthesized.

On the other hand, the glands that had been exposed to hypothalamic extract contained less thyrotropin than did control glands, but the ratio of biological activity to radioactivity was much greater than in the control preparations. It is tempting to conclude that the hypothalamic extract stimulated synthesis of thyrotropin which was largely, if not entirely, retained in the pituitary during the 1 hr of incubation. However, a greater ratio of ^{14}C to thyrotropin throughout the processes of partial purification of thyrotropin can have at least three interpretations:

1. There was stimulation of incorporation of amino acids into a protein which " travelled " with thyrotropin through ethanol-salt percolation.
2. There was a basal incorporation of ^{14}C-amino acids, unstimulated by the extract, into either such a protein or into thyrotropin itself, with sufficient fall in total thyrotropin through release of the hormone that the ratio ^{14}C/thyrotropin increased to the degree shown.
3. There was truly stimulation of synthesis of thyrotropin.

The fall in total pituitary thyrotropin brought about by the hypothalamic extract was not sufficient to make the second possibility acceptable as the sole explanation, but no doubt it was a factor. Since relatively crude extract of hypothalamus was used, other pituitary hormones might have been affected, in terms of both release and synthesis, and at least gonadotropins would be expected in the thyrotropin fraction of the percolation system (Bates *et al.*, 1959). We have no check on the degree to which this occurred, but it might be logical to expect that if gonadotropin synthesis was stimulated, so also was thyrotropin synthesis. While stimulation by a hypothalamic extract of large quantities of enzyme or structural or other proteins seems unlikely, we have no knowledge of to what degree the thyrotropin fraction was contaminated by such components. No doubt if pure TSHRF were used in the experiments we report here, the resulting data would be less ambiguous, but even then the synthesis of proteins other than the end product presumably must occur if thyrotropin synthesis is enhanced.

C. Thyrotropin Releasing Factor and Cyclic AMP

The possible role of cyclic AMP in mediating effects of TSHRF is perhaps the most recently recognized problem in this area. Wilber *et al.* (1969) described the release of thyrotropin from rat pituitaries *in vitro* under the influence of dibutyryl cyclic AMP or theophylline (which would be expected to build up the intracellular concentration of cyclic AMP). From the same laboratory Steiner and his colleagues (1969) reported a rapid increase in the concentration of cyclic AMP in the rat pituitary upon addition to the medium of theophylline

TABLE II

Effects of Stalk Median Eminence Extracts (SME) on Isolated Porcine Pituitary Cells in Vitro

Preparation	Thyrotropin Release (Assay response, %)	Adenyl Cyclase [1] (cAMP produced, cpm)
Control	Diln, 1 : 10 149; 172	208; 340
+SME	Diln, 1 : 40 283; 269	3321; 4471

[1] Adenyl cyclase was assessed by measuring ^{32}P-labeled cyclic AMP (cAMP) produced from α-^{32}P-labeled ATP. Cells and SME prepared by Dr. C. Kudo.

or an extract of stalk median eminence (SME) and there was an associated release of thyrotropin (and growth hormone). Activity of adenyl cyclase, the enzyme that produces cyclic AMP from ATP, was enhanced by SME, but activity of cyclic AMP phosphodiesterase (the enzyme hydrolysing cyclic AMP) was not affected. In Table II are shown similar data in an experiment using isolated cells of porcine pituitary and a porcine SME; there was clear-cut stimulation of thyrotropin release and of adenyl cyclase activity. Similar data regarding adenyl cyclase and cyclic AMP were reported by Zor and co-workers (1969), who studied release of Luteinizing Hormone (LH) from rat pituitaries, as stimulated by ovine hypothalamic extract. Both Steiner (1969) and Zor (1969) with their colleagues noted that enhancement of the concentration of cyclic AMP in the pituitary by SME was not dependent upon the presence of Ca^{++} in the medium; an interesting observation, since release of thyrotropin by hypothalamic extract requires Ca^{++} (Vale and Guillemin, 1967; Steiner *et al.*, 1969). Perhaps of related significance are the observations that an increase in concentration of K^{+} will stimulate *in vitro* release of thyrotropin (Vale and Guillemin, 1967), but will not enhance the concentration of cyclic AMP (Zor *et al.*, 1969). It appears there-

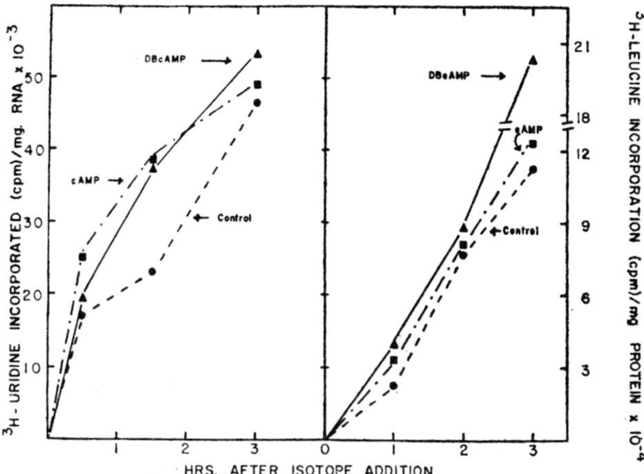

FIG. 4. Influence of cyclic AMP (cAMP) and dibutyryl cyclic AMP (DBcAMP) on ^{3}H-uridine and ^{3}H-leucine incorporation into RNA and protein, respectively, in bovine anterior pituitary slices. Experimental conditions are the same as in Fig. 5 except that preincubation before ^{3}H-uridine addition was omitted.

fore that by whatever mechanism Ca^{++} and K^{+} exert their influence on the release of thyrotropin *in vitro*, it is subsequent to an action of cyclic AMP. That the effect of TSHRF in releasing thyrotropin is mediated by cyclic AMP appears probable from the data reviewed above, but proof of this must await, among other things, repetition of the experiments with pure TSHRF.

In our laboratory we are engaged in studies of effects of TSHRF and cyclic AMP on aspects of pituitary metabolism in addition to the release of thyrotropin. As shown in Fig. 4, cyclic AMP (10^{-3} M) and dibutyryl cyclic AMP (10^{-3} M) enhanced incorporation of radioactive precursor into RNA and protein in bovine pituitary slices. Of other nucleotides similarly tested at 10^{-3} M, AMP, ATP, and cyclic 2′3′-AMP had no significant effect, but cyclic 2′3′-guanosine monophosphate (GMP) also stimulated, particularly protein incorporation of ^{3}H-leucine; ADP was inhibitory.

A purified TSHRF preparation stimulated precursor incorporation into bovine pituitary RNA and protein, and this effect of TSHRF was markedly dose-dependent, being stimulatory at 1.25 μg/ml but

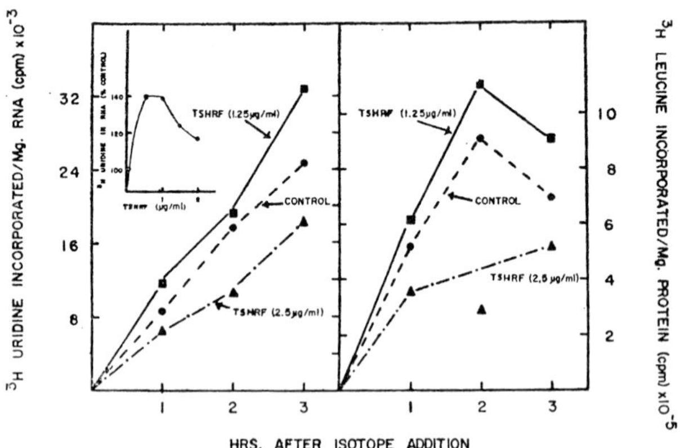

FIG. 5. Influence of TSHRF on ^{3}H-uridine and ^{3}H-leucine incorporation into RNA and protein, respectively, in bovine anterior pituitary slices. Slices incubated in the presence of TSHRF in Krebs-Ringer bicarbonate buffer (pH 7.4) containing 0.4% glucose at 37 °C for 1 hr before ^{3}H-uridine addition and 2 hr before ^{3}H-leucine addition, and incubation continued. Approximately equal amount of slices withdrawn at 1, 2, and 3 hr after isotope addition and specific activity of total RNA and protein determined. RNA was extracted and estimated by the method of Fleck and Munro (1962), and protein by the method described elsewhere (Adiga *et al.*, 1966). The inset shows data similarly obtained, but from another experiment, which illustrates dose-dependency.

inhibitory at 2.5 μg/ml (Fig. 5). This observation may have to be kept in mind in comparing data from different laboratories, particularly if other end points of TSHRF effects (especially release of thyrotropin!) are similarly inhibited with higher-than-stimulatory doses. In fact such a biphasic response of an *in vitro* pituitary preparation to synthetic oxytocin and lysine-8-vasopressin, both reported as stimulating release of thyrotropin, was described earlier (La Bella, 1964); however, there evidently has been no success in confirming these findings (Guillemin, 1967). In exploring the mechanism of these effects, DNA-dependent RNA polymerase activity (Weiss, 1960) in isolated nuclei of bovine pituitaries was assayed. This was found to have an Mg^{++}-activated and a $(NH_4)_2SO_4$—dependent, Mn^{++}-activated component, similar to the enzyme of liver nuclei described by Widnell and Tata (1964); both these enzyme activities, particularly the former, were stimulated by incubation of the pituitary slices with 10^{-3} *M* dibutyryl cyclic AMP for 2 to 3 hr before isolation

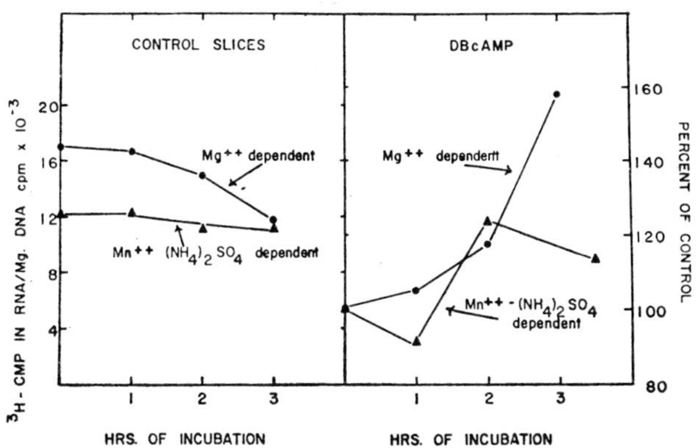

FIG. 6. Effect of preincubation of bovine anterior pituitary slices in the presence of dibutyryl cyclic AMP (DBcAMP) on RNA polymerase activities in isolated nuclear preparations. Slices were incubated in Krebs-Ringer bicarbonate buffer (pH 7.4) containing 0.4% glucose and 1 mM dibutyril cyclic AMP for different time intervals. Slices were chilled and washed with 0.25 *M* sucrose (2°–4°C), homogenized in 0.32 *M* sucrose containing 0.003 *M* Mg^{++}; nuclear fraction isolated by differential centrifugation and purified by sedimentation through 2.2 *M* sucrose–0.003 *M* Mg^{++}. Nuclear preparation suspended in 0.25 *M* sucrose–0.001 *M* Mg^{++} used for RNA polymerase activity measurements performed as outlined by Hamilton *et al.* (1968). Nuclear DNA content was measured in cold 0.2 *N* $HClO_4$ -washed nuclei according the procedure of Burton (1956).

of the nuclei (Fig. 6). Similar concentrations of ATP and cyclic 2′3′-GMP had no significant influence on either enzyme activity; 10^{-3} *M* cyclic 3′5′-AMP enhanced Mg^{++}-stimulated activity to 111%.

Understanding of the significance of these influences on RNA polymerase should be advanced by further studies presently under way in this laboratory, but it will also require a more general comprehension of the biological role of the enzyme. The RNA produced by Mg^{++}-stimulated RNA polymerase is the ribosomal type, whereas the product of the Mn^{++}-stimulated $(NH_4)_2SO_4$-dependent enzyme is more DNA-like. Particularly the stimulation of RNA was reported as an early effect of thyroid hormone, growth hormone, and testosterone on rat liver (Widnell and Tata, 1966), and of estrogen on rat uterus (Hamilton *et al.*, 1968). There has been no other report of modification of activity of RNA polymerase of pituitary nuclei, but if releasing factors indeed enhance the pituitary concentrations of cyclic AMP then the effects of dibutyryl cyclic AMP on pituitary RNA polymerase reported here may indicate that hypothalamic factors employ a mechanism similar to that of the other hormones in influencing their target tissues.

Recently it was suggested that hormones that have their action mediated by cyclic AMP might influence genetic mechanisms in their target tissues by means of phosphorylation of histones (Langan, 1968), since the cyclic nucleotide at 10^{-7} *M* caused a fourfold to sixfold increase in the rate of histone phosphorylation catalyzed by a liver enzyme preparation. Histones can inhibit RNA synthesis in a cell-free system (Huang and Bonner, 1962) and phosphorylation of histones may derepress activity of the genetic operon (Kleinsmith *et al.*, 1966).

Thus activation of RNA polymerase by dibutyryl cyclic AMP (illustrated in Fig. 6) might, by some such mechanism, be due to unmasking of DNA template, although other modes of enhancement of the enzyme activity more directly are conceivable. Stimulation of release of thyrotropin by TSHRF, as indicated earlier, does not require neosynthesis of RNA or protein. Consequently the speculations in the preceding paragraph are irrelevant to this action of the hypothalamic hormone. However, if the indications that TSHRF directly stimulates synthesis of thyrotropin are correct, it is in this area that relevance of the speculations should develop.

ACKNOWLEDGMENTS

We are grateful to Dr. A. P. S. Dhariwal and S. M. McCann, Dallas, Texas, for the extracts of bovine hypothalamus used in experiments, the results of which are shown in Table I and Figs. 1, 2, and 3. We also thank Dr. R. Guillemin for a purified TSHRF preparation (ovine) used in the experiments indicated in Fig. 5.

The work reported in this article and carried out in the senior author's laboratory was supported by the Medical Research Council of Canada (MT 884), U.S. Public Health Service (AM 04121), and the Foundations' Fund for Research in Psychiatry (65-318).

REFERENCES

Adiga, P. R., Rao, P. M., Hussa, R. O., and Winnick, T. (1966). *Biochemistry* **5**, 3850.

Bakke, J. L., and Lawrence, N. (1968). *European J. Pharmacol.* **2**, 308.

Bates, R. W., Garrison, M. M., and Howard, T. B. (1959). *Endocrinology* **65**, 7.

Bauduin, H., Reuse, J., and Dumont, J. E. (1967). *Life Sci.* **6**, 1723.

Bøler, J., Enzmann, F., Folkers, K., Bowers, C. Y., and Schally, A. V. (1969). *Biochem. Biophys. Res. Commun.* **37**, 705.

Bowers, C. Y., Lee, K. L., and Schally, A. V. (1968a). *Endocrinology* **82**, 75.

Bowers, C. Y., Lee, K. L., and Schally, A. V. (1968b). *Endocrinology* **82**, 303.

Brown-Grant, K. (1960). *Brit. Med. Bull.* **16**, 165.

Burgus, R., Dunn, T. F., Ward, D. N., Vale, W., Amoss, M., and Guillemin, R. (1969a). *C. R. Acad. Sci.* (Paris) **268**, 2116.

Burgus R., Dunn, T. F., Desiderio, D., and Guillemin, R. (1969b), *C. R. Acad. Sci.* (Paris) **269**, 1870.

Burton, K. (1956). *Biochem. J.* **62**, 316.

D'Angelo, S. A. (1963). *In* "Advances in Neuroendocrinology" (A. V. Nalbandov, ed.). Univ. of Illinois Press, Urbana, Ill., pp. 158–205.

D'Angelo, S. A., and Snyder, J. (1963). *Endocrinology* **73**, 75.

Fleck, A., and Munro, H. N. (1962). *Biochim. Biophys. Acta.* **55**, 571.

Florsheim, W. H., Austin, N. S., and Velcoff, S. M. (1963). *Endocrinology* **72**, 817.

Greer, M. A. (1957). *Recent Progr. Hormone Res.* **13**, 67.

Guillemin, R. (1967). *Ann. Rev. Physiol.* **29**, 313.

Guillemin, R., Sakiz, E., and Ward, D. N. (1965). *Proc. Soc. Exp. Biol. Med.* **118**, 1132.

Hamilton, T. H., Widnell, C. C., and Tata, J. R. (1968). *J. Biol. Chem.* **243**, 408.

Harris, G. W. (1948). *Physiol. Rev.* **28**, 139.

Harris, G. W. (1955). "Neural Control of the Pituitary Gland". Arnold, London.

Harris, G. W., and Woods, J. W. (1958). *J. Physiol.* (London) **143**, 246.

Huang, R. C., and Bonner, J. (1962). *Proc. Nat. Acad. Sci. U.S.* **48**, 1216.

Kendall, J. W., Shimoda, S. I., and Greer, M. A. (1967). *Neuroendocrinology* **2**, 76.

Kleinsmith, L. J., Allfrey, V. G., and Mirsky, A. E. (1966). *Proc. Nat. Acad. Sci. U.S.* **55**, 1182.

Knigge, K. M. (1964). *In* "Major Problems in Neuroendocrinology" (E. Bajusz and G. Jasmin, eds.). S. Karger, Basel, pp. 261–285.

LaBella, F. S. (1964). *Can. J. Physiol. Pharmacol.* **42**, 75.

Langan, T. A. (1968). *Science* **162**, 579.

Lowry, O. H., Rosebrough, N. J., Farr, A. L., and Randall, R. J. (1951). *J. Biol. Chem.* **193**, 265.

Nikovitch-Winer, M., and Everett, J. W. (1958). *Endocrinology* **63**, 916.

Purves, H. D. (1964). *In* "The Thyroid Gland" (R. Pitt-Rivers and W. R. Trotter, eds.). Vol. II. Butterworths, London, pp. 1–38.

Reichlin, S. (1963). *In* " Thyrotropin " (S. C. Werner, ed.). Charles C. Thomas, Springfield, Ill., pp. 56–67.
Reichlin, S. (1966). *In* " Neuroendocrinology " (L. Martini and W. F. Ganong, eds.). Vol. I. Academic Press, New York, pp. 445–536.
Robbins, J., and Rall, J. E. (1967). *In* " Hormones in Blood " (C. H. Gray and A. L. Bacharach, eds.). Vol. I. Academic Press, New York, pp. 383–490.
Schally, A. V., and Redding, T. W. (1967). *Proc. Soc. Exp. Biol. Med.* **126**, 320.
Schally, A. V., Bowers, C. Y., Redding, T. W., and Barrett, J. F. (1966). *Biochem. Biophys. Res. Commun.* **25**, 165.
Schally, A. V., Arimura, A., Bowers, C. Y., Kastin, A. J., Sawano, S., and Redding, T. W. (1968). *Recent Progr. Hormone Res.* **24**, 497.
Shizume, K., Matsuda, K., Irie, M., Iino, S., Ishii, J., Nagataki, S., Matsuzaki, F., and Okinaka, S. (1962). *Endocrinology* **70**, 298.
Siekewitz, P. (1952). *J. Biol. Chem.* **195**, 549.
Sinha, D., and Meites, J. (1965). *Neuroendocrinology* **1**, 4.
Sinha, D. K., and Meites, J. (1966). *Endocrinology* **78**, 1002.
Solomon, S. H., and McKenzie, J. M. (1966). *Endocrinology* **78**, 699.
Steiner, A. L., Peake, G. T., Utiger, R., and Kipnis, D. (1969). *J. Clin. Invest.* **48**, 80a abstract).
Szentágothai, J., Flerkó, B., Mess, B., and Halász, B. (1962). "Hypothalamic Control of the Anterior Pituitary." Akadémiai Kiadó, Budapest.
Uotila, U. U. (1939). *Endocrinology* **25**, 605.
Uotila, U. U. (1940). *Endocrinology* **26**, 129.
Vale, W., and Guillemin, R. (1967). *Experientia* **23**, 855.
Vale, W., Burgus, R., and Guillemin, R. (1968). *Neuroendocrinology* **3**, 34.
Weiss, S. B. (1960). *Proc. Nat. Acad. Sci. U.S.* **46**, 1020.
Widnell, C. C., and Tata, J. R. (1964). *Biochim. Biophys. Acta* **87**, 531.
Widnell, C. C., and Tata, J. R. (1966). *Biochem. J.* **98**, 621.
Wilber, J. F., and Utiger, R. D. (1968). *Proc. Soc. Exp. Biol. Med.* **127**, 488.
Wilber, J. F., Peake, G. T., and Utiger, R. D. (1969). *Endocrinology* **84**, 758.
Worthington, W. C. Jr., (1960). *Endocrinology* **66**, 19.
Zor, U., Kaneko, T., Schneider, H. P. G., McCann, S. M., and Field, J. B. (1969). *J. Clin. Invest.* **48**, 93a.

Control of Follicle-Stimulating Hormone and Luteinizing Hormone Secretion

B. FLERKÓ

I. Tonic Mechanism of Gonadotropin Secretion

A. HYPOPHYSIOTROPIC AREA

In the anatomy department of the Pécs University Medical School, Halász and his associates (1962, 1965) have been dealing with the problem of which part of the brain contains nerve cells producing the different inhibiting and releasing (hypophysiotropic) factors. It has been found by Halász *et al.* (1962, 1965) and later by Flament-Durand (1965) that only anterior pituitary grafts situated in the medial basal part of the hypothalamus retained their normal cytological characteristics. The finding (Halász *et al.*, 1962, 1965; Knigge, 1962) that such pituitary grafts not only release amounts of Follicle-Stimulating Hormone (FSH) and Luteinizing Hormone (LH) sufficient to maintain the normal function of the gonads, but also contain well-granulated gonadotropic basophilic cells, supports the view that hypophysiotropic factors influence both synthesis and release of FSH and LH. This is also indicated by the observations of Kobayashi *et al.* (1963a). The experiment of Evans and Nikitovitch-Winer (1969) also gives credence to this theory. The latter workers were the first to show in *in vivo* experiments that, in addition to a releasing effect, the hypophysiotropic factors must also have a sustained basal effect on the development of cytologic characteristics of individual anterior pituitary cells and on their ability to synthesize and secrete their particular tropic hormone.

The half-moon-shaped area (Fig. 1), which maintained the normal cytological characteristics and tropic hormone function of anterior pituitary grafts transplanted into it, was termed the " hypophysiotropic area " (HTA) by Halász *et al.* (1962). In order to explore the functional capacity of the HTA, Halász and Pupp (1965) developed a technique for interrupting neural connections of the HTA without altering its contact with the pituitary gland.

With respect to the Gonadotropic Hormones (GTH), the findings following the neural isolation of HTA were as follows: in male rats GTH secretion was fairly well maintained after the interruption of all neural connections of the HTA (Halász and Pupp, 1965; Halász *et al.*, 1967). Testicular weight and histology were nearly normal; only a slight decrease in seminal vesicle weight was observed. Voloschin *et al.* (1968) have published similar findings recently. There was no gonadal atrophy in female rats, but FSH and LH secretion appeared to be seriously altered (Halász and Pupp, 1965; Halász and Gorski, 1967). Such animals were not able to ovulate; their ovaries were polyfollicular without corpora lutea. The vaginal smears indicated permanent estrus (for an explanation of the permanent estrus syndrome, see Flerkó, 1968).

Thus, interruption of the neural connections of the HTA interferes with the pituitary gonadotropic function in the female but not in the male. This difference might be related to the fact that FSH and LH secretion is tonic in the male and cyclic in the female (for details see Flerkó, 1966). It seems likely that the HTA is responsible for tonic gonadotropin secretion, and that cyclic release depends on neural afferents to the area.

B. Tubero-infundibular Tract

The HTA corresponds to the distribution of the nerve cells from which the fine-fibered tubero-infundibular tract to the mediane eminence arises. Szentágothai (1964) succeded in Golgi-staining of this tract a few years ago. The fibers of the tubero-infundibular tract originate mainly from the small cells of the arcuate nucleus. They enter the pituitary stalk and terminate on or in the immediate neighborhood of the capillary loops connected with the portal vessels (Fig. 1).

The functional significance of the nerve endings of this tract on the capillary loops draining into the portal vessels is easily under-

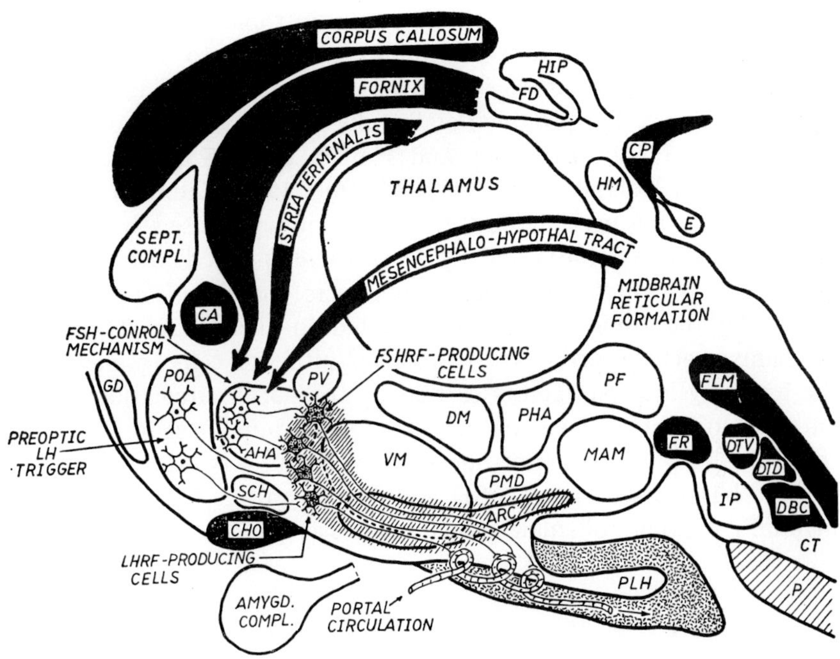

FIG. 1. Schematic representation of the hypothalamic and limbic control mechanisms for FSH and LH secretion. Arrows show the direction of the blood flow in the hypothalamo-hypophyseal portal system, the special capillary loops of which penetrate the median eminence. The FSHRF- and LHRF- producing neurons of the " tonic mechanism " are indicated by the neural units with dotted bodies. They are situated in the hypophysiotropic area (hatched) and their nerve endings terminate on the capillary loops of the portal system. Perikarya of the neurons belonging to the hypothalamic " cyclic mechanism " (preoptic LH-trigger and FSH-control mechanism) are represented by clear bodies located in the preoptic and anterior hypothalamic area, respectively. Abbreviations used: AHA, anterior hypothalamic area; Amygd. compl., amygdaloid complex; ARC, arcuate nucleus; CA, anterior commissure; CHO, optic chiasma; CP, posterior commissure; CT, nucleus centralis tegmenti; DBC, decussatio brachiorum conjunctivorum; DM, dorsomedial nucleus; DTD, decussatio tegmenti dorsalis; DTV, decussatio tegmenti ventralis; E, pineal gland (epiphysis); FD, fascia dentata; FLM, fasciculus longitudinalis medialis; FR, fasciculus retroflexus; GD, gyrus diagonalis; HIP, hippocampus; HM, medial habenular nucleus; IP, interpeduncular nucleus; MAM, mammillary nucleus; P, pons; PF, nucleus parafascicularis thalami; PHA, posterior hypothalamic area; PLH, posterior lobe of the hypophysis; PMD, dorsal premammillary nucleus; POA, preoptic area; PV, paraventricular nucleus; SCH, suprachiasmatic nucleus; Sept. compl., septal complex; VM, ventromedial nucleus.

stood. The endings liberate the hypophysiotropic factors produced by the neurons of the tubero-infundibular tract into the portal circulation. Electron microscopy of the surface zone of the median eminence and of the proximal stalk fully supports this view (Kobayashi *et al.*, 1963b; Bradbury and Harris, 1964; Szentágothai and Halász, 1964; Röhlich *et al.*, 1965; Monroe, 1967; Zambrano, 1968).

The above findings support the suggestion proposed by Barraclough and Gorski (1961) and by Flerkó (1962) that two levels exist in the neural control of gonadotropin secretion. One level is represented by the tonic mechanism situated in the HTA. The tonic mechanism acts directly on the anterior pituitary cells by means of the hypophysiotropic factors carried by the portal circulation to the anterior lobe. This level of neural control cannot secure by itself the cyclic output of gonadotropic hormones, but it is able to maintain a tonic basal discharge of FSH and LH in an amount sufficient to maintain ovarian follicular growth and estrogen secretion.

II. Cyclic Mechanism of Gonadotropin Secretion

The second or higher level in the neural control of gonadotropin secretion would include all brain structures outside the HTA which can modulate (enhance or inhibit) the activity of the HTA neurons that produce Follicle-Stimulating Hormone and Luteinizing Hormone Releasing Factors (FSHRF and LHRF). Brain structures of this type appear to be concentrated in the preoptic, anterior, and lateral area of the hypothalamus as well as in the limbic system and epithalamoepiphyseal complex of the brain.

A. Hypothalamic Parts of the Cyclic Mechanism

1. Regio Preoptica

The center for the neural mechanism that triggers the burst of LH secretion responsible for ovulation (Fig. 1) is probably located in the preoptic area (Bunn and Everett, 1957; Critchlow, 1958). Stimulation involving electrolysis with electrodes of ferrous alloys, combined with hypothalamic deafferentation, has been used by Everett and Radford (1961), Everett *et al.* (1962), and Tejasen and Everett (1967) in mapping the localization of LH-releasing neural

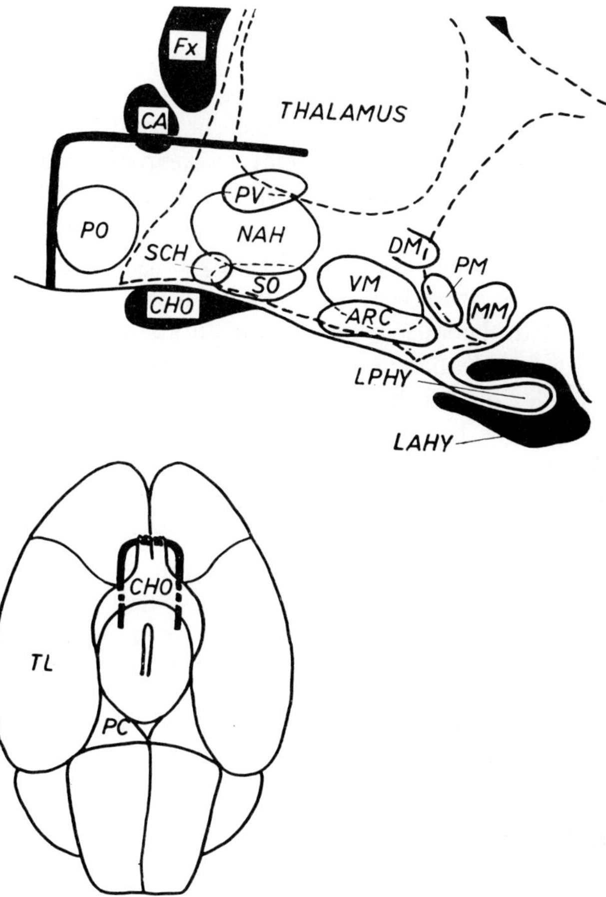

FIG. 2. Schematic sagittal drawing of the brain, illustrating deafferentation of the preoptic area and the same cut as seen from the base of the brain. Heavy lines indicate the cut. Fx, fornix; LAHY, anterior lobe of hypophysis; LPHY, posterior lobe of hypophysis; MM, medial mammillary nucleus; NAH, anterior hypothalamic area; PC, cerebral peduncle; PM, premammillary nucleus; PO, preoptic area; SO, supraoptic nucleus; TL, temporal lobe. For other abbreviations see Fig. 1.

structures approaching the HTA from the rostral side of the brain. On the basis of these investigations it was postulated that a diffuse system of LH-releasing neurons originates throught the septal complex, converges as it enters the medial preoptic area and anterior hypothalamus, and rapidly assumes a restricted basal location as it reaches the HTA.

In recent studies, Halász (1969) investigated the question of where the critical afferents for ovulation originate. The anterior, lateral, and superior connections of the preoptic area were interrupted bilaterally in adult female rats (Fig. 2) and the occurrence of ovulation was tested. It was extremely difficult to keep these animals alive. If the surgical procedure was performed in one step, all the animals died within a few days. Therefore deafferentation was carried out in two steps, 7–10 days apart, performing only half the cut at one time. The mortality rate was still more than 90%, but the surviving animals ovulated. Tubal ova were found in six of the rats with histologically verified deafferentation, and the ovaries of all ten animals contained fresh corpora lutea. However, the number of ova seen in the oviducts was less than in the controls, and the animals with preoptic isolation exhibited irregular vaginal cycles. These findings indicate that the neurogenic stimulus that causes the release of LHRF in amounts necessary for ovulation comes from the preoptic area itself. The findings, however, that ovulation was subnormal and that the rats had irregular cycles indicates that neural structures outside the preoptic area are also involved in the control of entirely normal, cyclic, ovulatory output of gonadotropic hormones.

2. Anterior Hypothalamic Area

In addition to the control of the ovulatory gonadotropin surge, control of FSH secretion by the HTA requires that afferent impulses reach it from the anterior hypothalamic area. Many years ago it was found that electrolytic lesion of the anterior hypothalamic area prevented compensatory ovarian hypertrophy following the removal of one ovary (D'Angelo and Kravatz, 1960; Flerkó and Bárdos, 1961). Similarly, Halász and Gorski (1967) reported that no compensatory ovarian hypertrophy occurred in rats bearing a frontal cut that separated the anterior hypothalamic area from the HTA. This indicated that the neural structures responsible for the occurrence of enhanced FSH output after hemicastration are located outside the HTA. Such an assumption is consistent with the view that estrogen-sensitive neurons occur in the anterior hypothalamic area and are involved in the negative feedback action of estrogen on FSH release (Flerkó and Szentágothai, 1957).

Another aspect of this negative neurohormonal feedback (i.e., that rats with anterior hypothalamic lesion showed less inhibition of FSH secretion following estrogen treatment than did non lesioned animals) was reported more than ten years ago (Flerkó 1957a,b; Flerkó and Bárdos, 1960). The assumption that there are estrogen-sensitive neurons in the anterior hypothalamic area is supported by the finding that individual neurons in the anterior hypothalamus accumulate estrogen (Michael, 1962; Attramadal, 1964), and that the anterior hypothalamus shows a pattern of uptake and retention of estradiol-^{3}H which is similar to the pattern found in the uterus, vagina, and anterior pituitary (Kato and Villee, 1967; Flerkó *et al.*, 1969).

On the basis of these and similar findings (Donovan and van der Werff ten Bosch, 1959a,b; Bogdanove and Schoen, 1959; Hohlweg and Daume, 1959; Krejci and Critchlow, 1959; Littlejohn and De Groot, 1963; Fendler and Endröczi, 1966), an FSH-control mechanism (Fig. 1) was assumed in the anterior hypothalamic area through which estrogen exerts its influence on FSH secretion.

3. *Premammillary Region*

Kordon (1967) also localized FSH-inhibiting structures in the anterior hypothalamus. Simultaneously, however, he postulated a nerve mechanism in the premammillary region which would continuously stimulate FSHRF-producing neurons. This assumption was based on the following experimental findings: (1) in the experiments of Corbin and Schottelius (1960), premammillary lesions in infantile rats retarded puberty; (2) in Kordon's (1967) experiments, premammillary lesions were followed in adult rats by reduction in the number of FSH-producing pituitary cells, by inhibition of follicular growth, and atrophy of the ovarian interstitial gland; (3) on the other hand, premammillary stimulation induced anovulatory cycles and persistent vaginal cornification quite similar to that observed after anterior hypothalamic lesions; (4) furthermore, in rats with persistent estrus after anterior hypothalamic lesions, formation of corpora lutea could be induced by bilateral electrolytic lesions placed in the premammillary region. Unilateral premammillary lesions or bilateral injuries to other hypothalamic regions have never been followed by formation of corpora lutea in Kordon's (1967) experiments. On the basis of the presence of corpora lutea of various ages he supposed that ovulation would occur repeatedly after bilateral premammillary lesions in rats with persistent estrus.

In Kordon's (1967) opinion, premammillary stimulatory and anterior hypothalamic inhibitory impulses conveyed to the FSHRF-producing neurons would maintain the normal cyclic output of FSHRF and FSH. If this train of thought were valid, anterior hypothalamic lesions would result in an enhanced FSH-estrogen output because of the absence of inhibitory impulses (when unimpeded stimulatory afferentations from the premammillary region reach FSHRF-producing neurons). Enhanced estrogen levels would inhibit the output of LHRF (and LH) in amounts sufficient to cause ovulation. This would result in a permanent anovulatory state in these rats. When one added bilateral premammillary lesions (eliminating impulses stimulating FSHRF output and consequent augmentation of FSH-estrogen action) the unimpeded activity of the LHRF-producing neurons would again permit the release of LH in amounts sufficient for ovulation.

Attractive as the Kordon's hypothesis may be, results of recent experimental findings (Illei-Donhoffer *et al.*, 1970) do not support it because: (1) not only bilateral but also unilateral premammillary lesions have been followed by formation of corpora lutea from ruptured follicles. The specificity of premammillary lesions is further diminished by the fact that formation of true corpora lutea was also induced by extensive hypothalamic lesions, which destroyed all hypothalamic structures indispensable to the normal ovulatory release of gonadotropic hormones; (2) formation of corpora lutea from ovulated follicles was found only one or two days after the second hypothalamic lesion. All corpora lutea present in the ovaries beyond this period of time were either persistent or regressive or, when newly forming, not true but atretic. This and the fact that normal, or at least approximately regular, vaginal cycling did not occur in any of the rats after the second hypothalamic lesion speak strongly against the assumption that premammillary lesions would lead to repeated ovulation, i.e., to restitution of more or less normal cycles.

4. Lateral Hypothalamic Area

Observations from various laboratories (Porter *et al.*, 1957; Barry *et al.*, 1961; 1962; Barraclough and Cross, 1963; Cross, 1964a) have suggested that certain nerve cells in the lateral hypothalamic area are sensitive to sex steroids and contribute to the cyclic mechanism. Cross (1964b) assumed that the cells of the arcuate nuclei, a part of the HTA, would be maintained in a state of normal FSH- and

LH-releasing activity by tonic impulses from the lateral hypothalamic area. This assumption was not supported by the findings of Halász and Gorski (1967), whose rats bearing HTA isolated from the lateral hypothalamic area ovulated normally whereas the separation of the preoptic and anterior hypothalamic area from the HTA abolished ovulation and compensatory ovarian hypertrophy.

B. Extrahypothalamic Parts of the Cyclic Mechanism

1. Limbic System

a. Amygdaloid nuclear complex. Koikegami and associates (1954) were the first to report that stimulation of medial amygdalar nuclei induces ovulation. Olfactory activity has been implicated in the pharmacological induction of ovulation in the rabbit (Sawyer, 1955). Ovulation was elicited by electrical excitation of the medial amygdalar nuclei in the cat (Shealy and Peele, 1957) and in rats with constant estrus induced by continuous illumination (Bunn and Everett, 1957). Electrical stimulation of the medial amygdala at parameters that induced ovulation was found by Hayward *et al.* (1964) to produce an immediate increase in ovarian progestin output. Control stimulation of closely adjacent regions of the temporal lobes produced neither a rise in progestin output nor ovulation.

There seems to be a negative progesterone feedback on amygdalar neurons which is able to raise the activity of the LHRF-producing neurons. In the experiment of Kawakami *et al.* (1966), LH output evoked by amygdalar activation induced progesterone secretion, which in turn lowered the excitability of the amygdala. These data suggest that the amygdalar complex contains neurons stimulating LHRF and LH production. However, these extrahypothalamic afferents seem not to be indispensable to the simple ovulatory release of LH. This is indicated not only by the findings of Halász (1969) but also by those of Sawyer (1959). Sawyer showed that lesions in the amygdala and in the septum pellucidum did not block copulation-induced ovulation, although reduced gonadotropin production after the bilateral ablation of the amygdala in the male has been reported by Yamada and Greer (1960).

On the other hand, the amygdalar complex seems to contain neurons that permanently inhibit FSHRF and FSH output, at least in infantile rats. Krejci and Critchlow (1959) observed marked uterine stimulation in immature rats with amygdalar lesions. Later, Elvers and Critchlow (1960) reported that effective amygdalar lesions associated with increased uterine weight shared a common area of destruction involving parts of the medial nuclei and an area between them containing the convergent fibers of the stria terminalis. The same result was found also by Ganong (1961) after lesions of the amygdala.

b. Midbrain reticular formation. Experiments performed to investigate the mechanism of the ovulation-blocking effect of morphine, chlorpromazine, and reserpine pointed to the possibility of an involvement of the midbrain reticular formation in the LH-releasing mechanism (Barraclough and Sawyer, 1955, 1957; Sawyer *et al.*, 1955).

In the experiment of Sas *et al.* (1965), lesions in the periaqueductal gray matter of the mesencephalon of the male rat resulted in a " castration-like " pituitary and an increase of the weight of the testicles (which is probably due to an increased secretion of FSH). In the investigation of Appeltauer *et al.* (1966), animals bearing lesions in the periaqueductal gray matter had a potentiation of the inhibiting effects of estrogen on gonadotropin secretion. This resulted in a marked decrease in the weights of the testicles and of the ventral prostate, with more marked regressive changes in the Leydig cells than in the seminiferous epithelium. Since the administration of reserpine to rats bearing lesions in the periaqueductal gray matter has been found to bring about an increase of the weight of the testicles, it may be inferred that the caudal pole of the limbic system may regulate either activating or inhibiting influences that control gonadotropin secretion.

c. Septum complex. Ovulation by stimulation of the septum pellucidum has been elicited in constant estrous rats, induced by continuous illumination (Bunn and Everett, 1957). Everett (1964) later found that in the septal complex (including the medial and lateral septal nuclei, the nucleus accumbens, and the caudal portion of the medial paraolfactory area), it was necessary to use somewhat larger currents (i.e., forming larger stimulative focus than in the preoptic area of the atropine-blocked proestrous rat), in order to induce ovulation. On the basis of these results it was postulated that a diffuse system of LH-releasing neurons originates throughout the septal complex and converges as it enters the medial preoptic area.

d. Hippocampal formation. According to Kawakami *et al.* (1966), stimulation of the hippocampus also facilitates the release of LH, yielding an increased secretion of progesterone. In turn, the elevated progesterone level raises the excitability of the hippocampus so that enhancement by afferent impulses of hippocampal activity can be more easily induced. This suggests the existence of a positive progesterone feedback on the hippocampal neuronal level, in turn stimulating the LHRF-producing cells in the HTA.

The limbic system receives afferent impulses primarily through olfactory and somesthetic pathways. It is therefore easy to understand that olfactory and various tactile stimuli play a considerable part in the regulation of gonadotropin secretion (for details see Flerkó, 1966).

2. *Epithalamo-Epiphyseal Complex*

Another extrahypothalamic part of the brain that participates in the control of gonadotropin secretion is the epithalamo-epiphyseal complex: lesions situated in or in the neighborhood of the epithalamic region resulted in reduced fertility and in decreased pituitary prolactin content (De Groot, 1962).

Recent experimental findings of Fraschini *et al.* (1968a,b) have shown that subjecting adult male rats to pinealectomy caused testicular hypertrophy, enhanced the weight of the ventral prostate and of the seminal vesicles, and increased pituitary LH stores. These data indicated that pinealectomy stimulated synthesis and release of LH and suggested that the pineal gland would exert an inhibitory influence on LH secretion.

In order to determine whether such an effect of the pineal gland was taking place through a direct action on the pituitary gland or via the mediation of nervous structures, fragments of pineal tissue or crystals of melatonin and of 5-hydroxytryptophol were implanted stereotaxically in the median eminence, in the midbrain reticular formation, and in the pituitary gland of castrated male rats. Placement of pineal fragments or of indole compounds in these two areas was followed by a significant reduction of pituitary LH stores, but melatonin was unable to reduce pituitary LH when directly implanted in the pituitary gland.

It is suggested by these results that these indole compounds may play a role in the control of LH secretion, possibly by acting on specific receptors localized in the median eminence and in the midbrain reticular formation. Since light exerts a marked influence on pineal function, it is reasonable to assume that the influence of light might be transmitted, at least partly, to the tonic mechanism of gonadotropin control by indole compounds that are normally synthetized in the pineal gland (see also the articles by Wurtman and by Fraschini and Martini in this volume).

References

Appeltauer, L. C., Reissenweber, N. J., Domínguez, R., Griñó, E., Sas, J., and Benedetti, W. L. (1966). *Acta Neuroveget.* (Wien) **29**, 75.

Attramadal, A. (1964). *Excerpta Med. Intern. Congr. Ser.* **83**, 612.

Barraclough, C. A., and Cross, B. A. (1963). *J. Endocrinol.* **26**, 339.

Barraclough, C. A., and Gorski, R. A. (1961). *Endocrinology* **68**, 68.

Barraclough, C. A., and Sawyer, C. H. (1955). *Endocrinology* **57**, 329.

Barraclough, C. A., and Sawyer, C. H. (1957). *Endocrinology* **61**, 341.

Barry, J., Torre, J. F., and Slimane-Taleb, S. (1961). *C. R. Soc. Biol.* **155**, 2144.

Barry, J., Torre, J. F., and Leonardelli, J. (1962). *C. R. Soc. Biol.* **156**. 613.

Bogdanove, E. M., and Schoen, H. C. (1959). *Proc. Soc. Exp. Biol. Med.* **100**, 664.

Bradbury, S., and Harris, G. W. (1964). *In* " Brain-Thyroid Relationships " (M. P. Cameron and M. O'Connor, eds.), pp. 3–16. J. and A. Churchill, London.

Bunn, J. B., and Everett, J. W. (1957). *Proc. Soc. Exp. Biol. Med.* **96**, 369.

Corbin, A., and Schottelius, B. A. (1960). *Proc. Soc. Exp. Biol. Med.* **103**, 208.

Critchlow, B. V. (1958). *Am. J. Physiol.* **195**, 171.

Cross, B. A. (1964a). *Symp. Soc. Exptl. Biol.* **18**, 157.

Cross, B. A. (1964b). *Excerpta Med. Intern. Congr. Ser.* **83**, 513.

D'Angelo, S. A., and Kravatz, A. S. (1960). *Proc. Soc. Exp. Biol. Med.* **104**, 130.

De Groot, J. (1962)." Proc. 22nd Intern. Congr. Physiol. Sci." Vol. I, Part II, pp. 623–624. Excerpta Medica, Amsterdam.

Donovan, B. T., and van der Werff ten Bosch, J. J. (1959a). *J. Physiol.* **147**, 78.

Donovan, B. T., and van der Werff ten Bosch, J. J. (1959b). *J. Physiol.* **147**, 93.

Elvers, M., and Critchlow, B. V. (1960). *Am. J. Physiol.* **198**, 381.

Evans, J. S., and Nikitovitch-Winer, M. B. (1969). *Neuroendocrinology* **4**, 83.

Everett, J. W. (1964). *In* " Major Problems in Neuroendocrinology " (E. Bajusz and G. Jasmin, eds.), pp. 346–366. S. Karger, Basel.

Everett, J. W., and Radford, H. M. (1961). *Proc. Soc. Exp. Biol. Med.* **108**, 604.

Everett, J. W., Radford, H. M., and Holsinger, J. (1962). *Excerpta Med. Intern. Congr. Ser.* **51**, 24.

Fendler, K., and Endröczi, E. (1966). *Neuroendocrinology* **1**, 129.

Flament-Durand, J. (1965). *Endocrinology* **77**, 446.

Flerkó, B. (1957a). *Endokrinologie* **34**, 202.

Flerkó, B. (1957b). *Arch. Anat. Microscop. Morphol. Exp.* **46**, 159.

Flerkó, B. (1962). *In* " Hypothalamic Control of the Anterior Pituitary " (J. Szentágothai, B. Flerkó, B. Mess, and B. Halász, eds.), pp. 192–265. Akadémiai Kiadó, Budapest.

Flerkó, B. (1966). *In* "Neuroendocrinology" (L. Martini and W. F. Ganong, eds.). Vol. I., pp. 613–668. Academic Press, New York.
Flerkó, B. (1968). *In* "Endocrinology and Human Behaviour" (R. P. Michael, ed.), pp. 119–138. Oxford University Press, Oxford.
Flerkó, B., and Bárdos, V. (1960). *Acta Endocrinol.* **35**, 375.
Flerkó, B., and Bárdos, V. (1961). *Acta Endocrinol.* **36**, 180.
Flerkó, B., and Szentágothai, J. (1957). *Acta Endocrinol.* **26**, 121.
Flerkó, B., Mess, B., and Illei-Donhoffer, A. (1969). *Neuroendocrinology* **4**, 164.
Fraschini, F., Mess, B., and Martini L. (1968a). *Endocrinology* **82**, 919.
Fraschini, F., Mess, B., Piva, F., and Martini, L. (1968b). *Science* **159**, 1104.
Ganong, W. F. (1961). *In* "Control of Ovulation" (C. A. Villee, ed.), p. 183. Pergamon Press, Oxford.
Halász, B., (1969). *In* "Frontiers in Neuroendocrinology 1969" (W. F. Ganong and L. Martini, eds.), pp. 307–342. Oxford University Press, New York.
Halász, B., and Gorski, R. A. (1967). *Endocrinology* **80**. 608.
Halász, B., and Pupp, L. (1965). *Endocrinology* **77**, 553.
Halász, B., Pupp, L., and Uhlarik, S. (1962). *J. Endocrinol.* **25**, 147.
Halász, B., Pupp, L., Uhlarik, S., and Tima, L. (1965). *Endocrinology* **77**, 343.
Halász, B., Florsheim, W. H., Corcorran, N. L., and Gorski, R. A. (1967). *Endocrinology* **80**, 1075.
Hayward, J. N., Hilliard, J., and Sawyer, C. H. (1964). *Endocrinology* **74**, 108.
Hohlweg, W., and Daume, E. (1959). *Endokrinologie* **38**, 46.
Illei-Donhoffer, A., Tima, L., and Flerkó, B. (1970). *Acta Biol. Acad. Sci. Hung.* (In press).
Kato, J., and Villee, C. A. (1967). *Endocrinology* **80**, 567.
Kawakami, M., Seto, K., and Yoshida, K. (1966). *Japan. J. Physiol.* **16**, 254.
Knigge, K. M. (1962). *Am. J. Physiol.* **202**, 387.
Kobayashi, T., Kobayashi, T., Kigawa, T., Mizuno, M., and Amenomori, Y. (1963a). *Endocrinol. Japon.* **10**, 16.
Kobayashi, T., Kobayashi, T., Yamamoto, K., and Inatomi, M. (1963b). *Endocrinol. Japon.* **10**, 69.
Koikegami, H., Yamada, T., and Usui, K. (1954). *Folia Psychiat. Neurol. Japon.* **8**, 7.
Kordon, C. (1967). *Arch. Anat. Microscop. Morphol. Exp. Suppl.* 3-4, **56**, 458.
Krejci, M. E., and Critchlow, B. V. (1959). *Anat. Record* **33**, 300.
Littlejohn, M., and De Groot, J. (1963). *Fed. Proc.* **22**, 571.
Michael, R. P. (1962). *Science* **136**, 322.
Monroe, B. C. (1967). *Z. Zellforsch.* **76**, 405.
Porter, R. W., Cavanaugh, E. B., Critchlow, B. V., and Sawyer, C. H. (1957). *Am. J. Physiol.* **189**, 145.
Röhlich, P., Vigh, B., Teichmann, I., and Aros, B. (1965). *Acta Biol. Acad. Sci. Hung.* **15**, 431.
Sas, J., Griño, E., Benedetti, W. L., Appeltauer, L. C., and Domínguez, R. (1965). *Acta Morphol. Acad. Sci. Hung.* **13**, 377.
Sawyer, C. H. (1955). *Am. J. Physiol.* **180**, 37.
Sawyer, C. H. (1959). *J. Exp. Zool.* **142**, 227.
Sawyer, C. H., Critchlow, B. V., and Barraclough, C. A. (1955). *Endocrinology* **57**, 345.
Shealy, C. N., and Peele, T. L. (1957). *J. Neurophysiol.* **20**, 125.
Szentágothai, J. (1964). *Progr. Brain Res.* **5**, 135.
Szentágothai, J., and Halász, B. (1964). *Nova Acta Leopoldina* **28**, 227.
Tejasen, T., and Everett, J. W. (1967). *Endocrinology* **81**, 1387.
Voloschin, L., Joseph, S. A., and Knigge, K. M. (1968). *Neuroendocrinology* **3**, 387.
Yamada, T., and Greer, M. A. (1960). *Endocrinology* **66**, 565.
Zambrano, D. (1968). *Neuroendocrinology* **3**, 141.

The Control of Gonadotropin Secretion in the Human

P. FRANCHIMONT [1] and J. J. LEGROS [2]

I. Introduction

All people working on gonadotropins agree that mechanisms of secretion differ with respect to sex and, for women, with respect to the state of gonadal function. Therefore this article will be divided into three parts: mechanism of secretion in postmenopausal women, in women before the menopause, and in men.

Gonadotropins were assayed by radioimmunoassay (Franchimont, 1966a,b,c, 1968a,b).

---

[1] Associate, Faculty of Medicine, University of Liège, Belgium.
[2] Research Fellow, F.N.R.S.

II. Postmenopausal Women

A. Basal Mean Values

The basal mean values of serum gonadotropins are 88.0 ± 32.2 (S.D.) mIU-IRP$_2$/ml for Follicle-Stimulating Hormone (FSH) and 47.6 ± 20.3 mIU-IRP$_2$/ml for Luteinizing Hormone (LH). [3]

With the exception of a single case, all values of serum FSH levels studied have been higher than those observed during the menstrual cycle. It must be stressed that these high values are found many years after the beginning of menopause, as many as 25 years after. In these women serum LH values are often equal to or higher than the highest mean level observed at mid-cycle peak.

B. Some Mechanisms of Regulation of Gonadotropin Secretion in Postmenopausal Women

1. *Action of Estrogens*

Estradiol benzoate (5 mg) was injected intramuscularly into five postmenopausal women. This estrogen induced significant and fast drop in LH amounts and a delayed FSH decrease, which is less noticeable (Fig. 1). These results in menopausal women are in agreement with the experimental findings of Bogdanove (1964), Taleisnik and McCann (1961), Parlow (1964) and Gans and Van Rees (1962). These authors demonstrated in castrated rats that estrogens block the synthesis as well as the release of LH from the pituitary. Estrogens can also decrease FSH release by the pituitary, but require much higher doses than those effective for the inhibition of LH release. Odell and Swerdloff (1968) have confirmed the drop in FSH and LH during oral estrogen treatment in postmenopausal women. Peterson *et al.* (1968) found similar results in men. However, it is interesting to remember that estrogens inhibit only FSH in eugonadal young women and may stimulate LH release, as demonstrated by Swerdloff and Odell (1969).

[3] Second International Reference Preparation (IRP$_2$) was prepared from postmenopausal women's urine and assigned unitage on the basis of bioassay. The preparation was kindly provided by Dr. Bangham, Division of Biological Standards, Medical Research Council, Mill Hill, London.

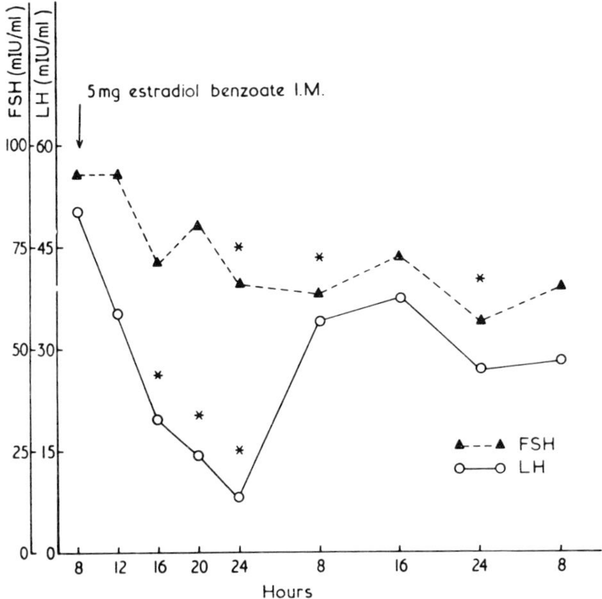

FIG. 1. Effect of intramuscular (I.M.) injection of estradiol benzoate on serum FSH and LH levels. Each point represents the mean value obtained from five individual values. The asterisk indicates a statistical difference with respect to starting mean level.

The reason for this striking difference in eugonadal women on one hand and postmenopausal women and normal men on the other is not apparent. It may be that the dose and the kind of estrogen used in each study influence the pituitary response. The effect of small amounts of estradiol benzoate on gonadotropin levels as tested by radioimmunoassays have not yet been reported.

2. *Action of Natural Progesterone*

Progesterone (25 mg) injected intramuscularly did not bring about any modification in FSH and LH during the first 24 hr after its administration to three postmenopausal women. There is a slight increase in FSH in each patient afterwards but only one had a rise in LH (Fig. 2). Similarly in castrated rats progesterone alone does not modify the serum FSH and LH levels or change the gonadotropin content of the pituitary. However, when associated with small

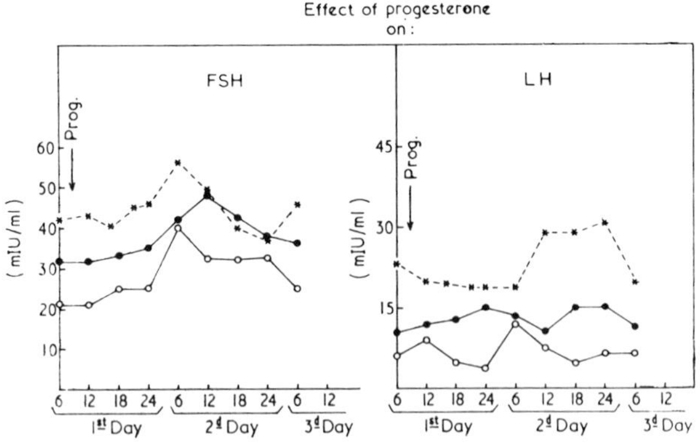

FIG. 2. Effect of progesterone on serum FSH and LH levels in three postmenopausal women.

amounts of estrogens, progesterone induces a rapid fall of FSH and LH levels (McCann and Ramirez, 1964).

3. *Action of Progestagens*

Intramuscular norethisterone enanthate,[4] (200 mg) was tested in five postmenopausal women. The serum FSH and LH levels are illustrated in Fig. 3. FSH and LH were measured three times a day for three days before the injection. After the injection LH quickly fell and remained practically undetectable for one month. After two months, LH was still low (about 4 mIU-IRP_2/ml). FSH levels also decreased but more slowly and rarely down to an undetectable level. Twenty days after injection, the FSH level increased and the starting values were obtained again two months later.

Other women injected with norethisterone enanthate showed the same behavior of FSH and LH serum levels. Evidently the progestagen alone partially inhibits the secretion of FSH and completely inhibits the secretion of LH for a variable time, always longer than one month. This observation may partially explain the antiovulatory effect of this substance when injected intramuscularly in eugonadal women.

[4] Preparation kindly provided by Dr. Ufer, Schering, Berlin.

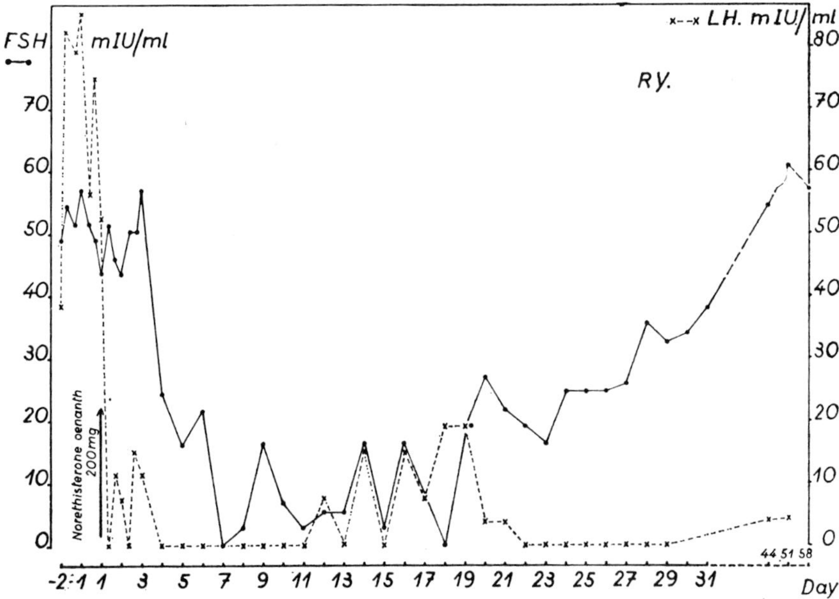

FIG. 3. Effect of intramuscular injection of norethisterone enanthate on serum FSH and LH levels in one postmenopausal woman. The arrow indicates the time of injection (RY., 62 years old) (From Franchimont *et al.*, 1970).

III. Eugonadal Women

A. FOLLICLE-STIMULATING HORMONE AND LUTEINIZING HORMONE LEVELS THROUGHOUT THE MENSTRUAL CYCLE

The variations of serum concentrations of FSH and LH have been followed daily throughout the menstrual cycle together with the urinary output of FSH, LH, estrogens, and pregnandiol. Sixteen young women without any history of menstrual abnormality have thus been investigated. Figure 4 represents one of these cycles with the basal temperature, the serum FSH, the urinary FSH output, the serum LH, the urinary LH output, the urinary output of estrogens (method of Brown *et al.*, 1958) and the urinary output of pregnandiol (method of Henri and Thevenet, 1959).

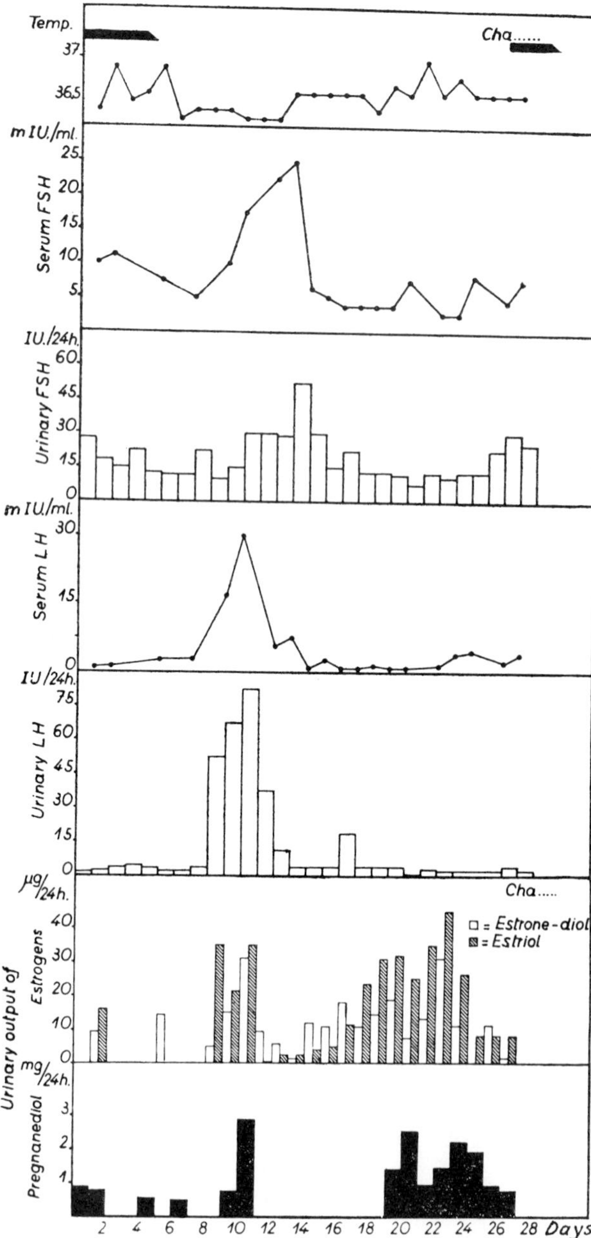

FIG. 4. Serum FSH and LH levels and urinary FSH, LH estrogens and pregnandiol output during the menstrual cycle of a normal woman (Cha. JO., 24 years old).

The following points have been observed:

1. In all cases there are FSH and LH peaks in both serum and urine at mid-cycle. The FSH urinary mid-cycle peak was not found by Fukushima *et al.* (1964) or by Vorys *et al.* (1965), contradicting reports of MacArthur *et al.* (1958, 1964), Rosenberg and Keller (1965), Loraine and Bell (1968), who used bioassays, and of those workers who used radioimmunoassays (Franchimont, 1966a,b; Midgley, 1967; Midgley and Jaffe, 1966, 1968; Faiman and Ryan, 1967a,b; Rosselin and Dolais, 1967; Saxena *et al.*, 1968; Cargille *et al.*, 1969). In many cases, the LH peak seems to come before the FSH increase. The LH peak occurs before or at the moment of the lowest basal body temperature.
2. The FSH level is higher during the follicular phase than during the luteal phase. These findings are in agreement with those of Midgley and Jaffe (1966, 1968), Rosselin and Dolais (1967). We have not found any significant difference in the levels of LH between the follicular and the luteal periods. These levels are very low in any case.
3. The highest amount of LH (in both serum and urine) was found at the time of the highest preovulatory urinary estrogen output. Burger *et al.* (1968) did not find exactly the same results. In their experiments the initial increase in urinary estrogens preceded that of serum LH, the day of maximum LH concentration occurring either on the same day as the mid-cycle peak of estrogen value or 1 to 2 days later. Catt (1970) has shown that the plasma estradiol peak precedes the plasma LH peak by at least 1 to 3 days but that the LH peaks are usually preceded by a fall in plasma estradiol from the previously high level.
4. In most cases during the luteal phase we have observed no modification of LH and FSH serum and urinary concentrations. However, in 4 of 16 cases we found a second LH peak around the 19th to 22nd day. This was much lower than the mid-cycle peak. There was no accompanying changes in concentration of FSH.
5. The ratio of FSH to LH is modifiying throughout the cycle. It is highest at the begining of the cycle and lowest at the mid-cycle LH peak. It rises in the luteal phase (Table I).

TABLE I

RATIO FSH[1] *vs.* LH[1] DURING THE MENSTRUAL CYCLE

							Peak [2] of LH							
Days	—12	—10	—8	—6	—4	—2	0	+2	+4	+6	+8	+10	+12	+14
FSH/LH	12.2	9.5	2.3	1.8	1.2	0.9	0.5	0.7	1.1	1.8	3.4	2.1	1.5	2.0

[1] mIU HMG-IRPI/ml.

[2] During the day 0, the serum LH level observed at mid-cycle is maximum.

On the basis of these results we can speculate on the role of the gonadotropins in the events of the human menstrual cycle.

1. The high concentration of FSH during the follicular phase might be related to the development of ovarian follicle. The high ratio FSH/LH would indicate that the FSH action is of primary importance.
2. The constant events appearing at mid-cycle (i.e., peaks of FSH and LH one day before the rise in the basal temperature) suggest that these peaks are etiologically related to ovulation. The FSH peak could influence the two last states of the meiosis which take place during ovulation.
3. The preovulatory estrogens are excreted in the urine at the same time at which the serum and urinary LH peaks appear. The relationship between these two phenomena must be studied.
4. No constant variation of gonadotropins are observed during the luteal phase. It should be emphasized that the functional activity of the corpus luteum is independent of gonadotropin secretion. The physiological mechanism of the inconstant second peak is not yet clear.

B. ACTION OF CONTRACEPTIVES ON SECRETION OF GONADOTROPINS

Radioimmunological measurement of both FSH and LH has demonstrated that the nonsequential contraceptive containing a combination of estrogen and progestagen suppressed the early FSH elevation and the mid-cycle LH and FSH peaks (Ross *et al.*, 1966;

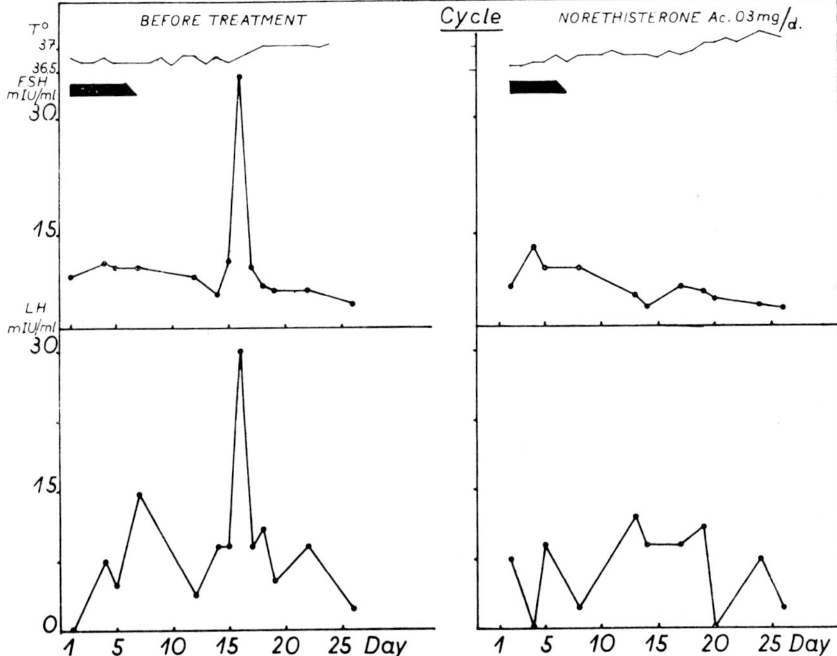

FIG. 5. FSH and LH levels in serum of one normal woman during two successive cycles. During the first cycle no contraceptive treatment was given, and mid-cycle FSH and LH peaks were found. During the second cycle this woman received 0.3 mg norethisterone acetate daily, and no FSH and LH mid-cycle peaks were observed.

Cargille *et al.*, 1968). The sequential contraceptives seem to block only FSH and allow or perhaps stimulate the release of LH (Swerdloff and Odell, 1969).

Progestagen alone (norethisterone acetate 0.3 mg per day) suppresses the mid-cycle FSH and LH peak according to the three cases studied before and during treatment (Fig. 5). This observation must be related to the suppressing action of the intramuscular injection of norethisterone enanthate, a long-acting progestagen, on FSH and LH levels (see Section II, B, 3).

Our findings are in agreement with the recent data obtained in rats by Schally and co-workers (1968). For them, progestagens significantly depress plasma LH, as do estrogens alone. But the progestagen-estrogen combination given daily suppressed plasma LH levels in most cases more efficiently that did estrogen alone.

IV. Men

A. Basal Mean Values

Radioimmunoassays for FSH and LH were used to determine the concentration of these gonadotropins in sera of 40 healthy men aged from 18 to 35. The average value for FSH is 4.14 $\pm$ 2.67 mIU-IRP_2/ml. Urinary output of FSH in 25 of these subjects was 2.21 IU-IRP_2 per 24 hr. No evidence of diurnal or other rhythmic variations in serum concentrations was found in daily samples, taken six times a day (4, 8 and 12 A.M. and 4, 8, and 9 P.M.).

LH was also estimated in the same subjects and the average serum level was 13.13 $\pm$ 9.11 mIU-IRP_2/ml. The urinary output of LH was 12.08 IU-IRP_2/24 hr.

Serum LH levels showed only very slight variations. In three of four subjects studied we observed a small decrease of the concentration between midday and 4 P.M. (Franchimont, 1968b). These results must be related with the similar variation of plasma testosterone level described by Burger *et al.* (1968) and Saxena *et al.* (1968).

In 24 students, aged 18 to 22, we have attempted to find a relation between serum FSH and LH levels and the plasma concentration of testosterone (testosterone was evaluated by the method of Palem *et al.*, 1969). We have not found any definite relationship (Fig. 6).

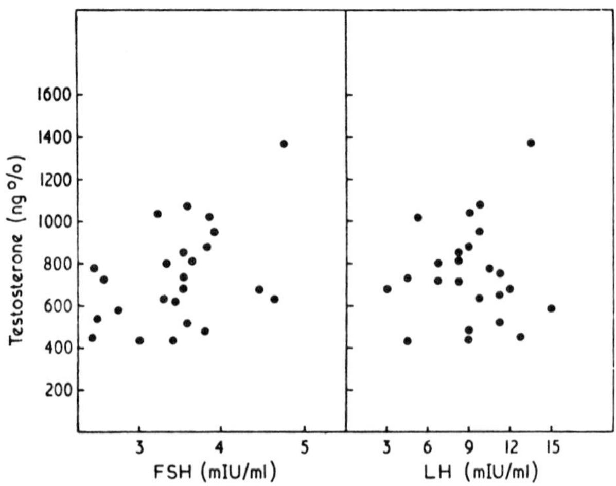

Fig. 6. There is no relationship between serum FSH and LH levels and the plasma concentration of testosterone in normal males.

One has to keep in mind that such variations in FSH and LH are rather small and that such a relationship is hence rather difficult to establish.

B. Mechanisms of Regulation

1. *Action of Testosterone*

Testosterone propionate (50 mg) injected intramuscularly in the male induces but few variations in FSH, whereas there is a marked and durable decrease in LH (Fig. 7). These facts observed in man were very recently confirmed by Peterson *et al.* (1968).

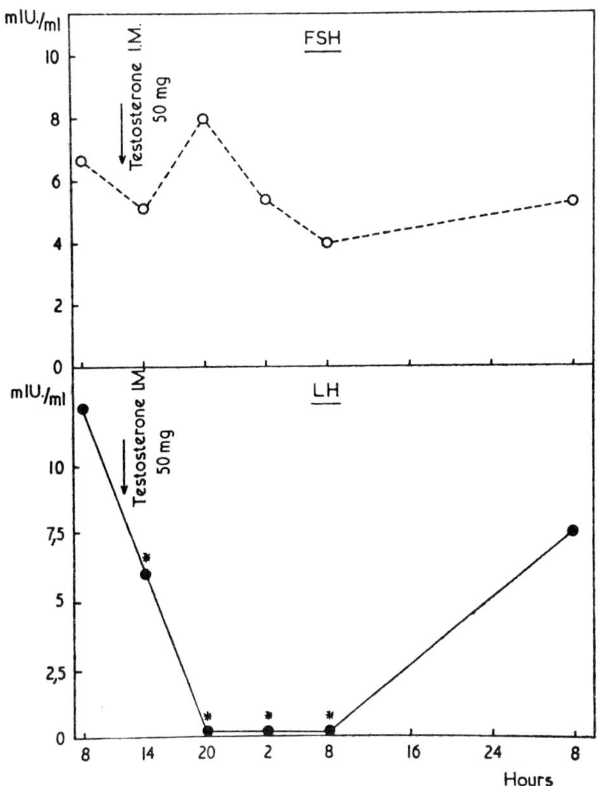

Fig. 7. Effect of intramuscular (I.M.) injection of testosterone propionate on serum FSH and LH levels. Each point represents the mean value obtained from five individual values.

Our own experiments in man bring forth facts similar to those observed by McCann and Ramirez (1964) and Bogdanove (1964) in the rat. McCann and Ramirez (1964) have clearly demonstrated that testosterone has an inhibitory feedback action on LH in the castrated male rat.

Bogdanove (1964) has shown that in the normal rat and castrated male rat, testosterone decreases both the synthesis and the release of pituitary LH in blood. On the other hand, FSH increases in the pituitary, but its blood concentration is not significantly altered.

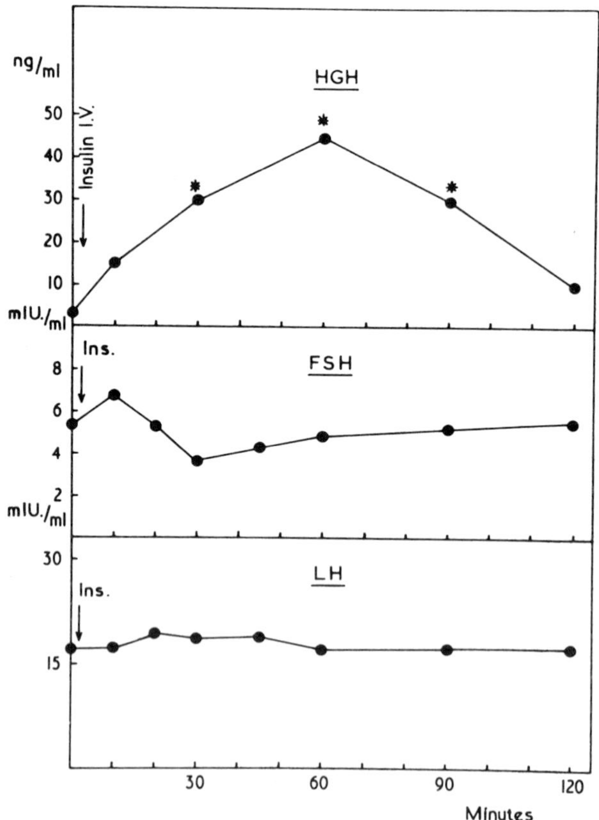

FIG. 8. Insulin hypoglycemia in men causes an elevation of HGH but has no effect on plasma FSH and LH. The asterisk indicates a statistical difference from the starting level.

2. *Action of Insulin Hypoglycemia*

Insulin hypoglycemia induces an increase of Adrenocorticotropic Hormone (ACTH) and Human Growth Hormone (HGH) secretion. A few authors think that hypoglycemia acts as a nonspecific stress, stimulating all hypothalamic structures (Mahfouz, 1958; Hearn *et al.*, 1961). With regard to gonadotropin secretion, this mechanism cannot be demonstrated. Indeed, insulin injection induces an increase of plasma growth hormone level but does not modify the levels of FSH and LH (Fig. 8). Therefore the insulin hypoglycemia test is of no value in the exploration of gonadotropin secretion.

3. *Action of Dexamethasone*

We know that dexamethasone inhibits secretion of ACTH and in some conditions also inhibits the secretion of HGH (Frantz and Rabkin, 1964; Franchimont, 1966b). To observe the action of dexamethasone on gonadotropin levels, 4 mg of dexamethasone were injected in 15 men. FSH and LH were assayed in serum every 10 min for 2 hr, but the 4 mg of dexamethasone induced no statistical modification of FSH and LH levels.

4. *Action of Posterior Pituitary Hormones*

Since 1944 (Nelson, 1944) vasopressin has been known to be a corticotropin releasing substance and has been suspected of being the physiologic hypothalamic corticotropin releasing factor. Other authors have shown a stimulation of growth hormone with pitressin and synthetic lysine-vasopressin (Meyer and Knobil, 1966; Gagliardino *et al.*, 1967; Eddy, 1968). With Legros (Legros and Franchimont, 1968; Franchimont and Legros, 1969), we have examined the effects of posterior pituitary extract, of a synthetic lysine-vasopressin, and of synthetic oxytocin on gonadotropin levels. Three groups of seven young normal men were injected intravenously with posterior pituitary extract (Choay, 2 units), with synthetic lysine-vasopressin (2 units), and with synthetic oxytocin (2 units). The low doses of these drugs have produced no or very slight side effects.

HGH, LH, and FSH were assayed in serum after each of these treatments (Fig. 9). The posterior pituitary extract induced an increase of HGH (about 600% of the starting mean value), FSH, and LH serum levels (about 200% of the starting mean values).

To the contrary, 2 units of synthetic lysine-vasopressin did not modify the HGH, FSH, and LH levels. Other authors using higher doses of vasopressin injected intramuscularly (Brostoff *et al.*, 1968; Czarny *et al.*, 1968; Yalow *et al.*, 1969) have found no constant HGH response. Synthetic oxytocin does not modify serum LH and HGH but induces an increase of plasmatic FSH 10 min after the intravenous injection.

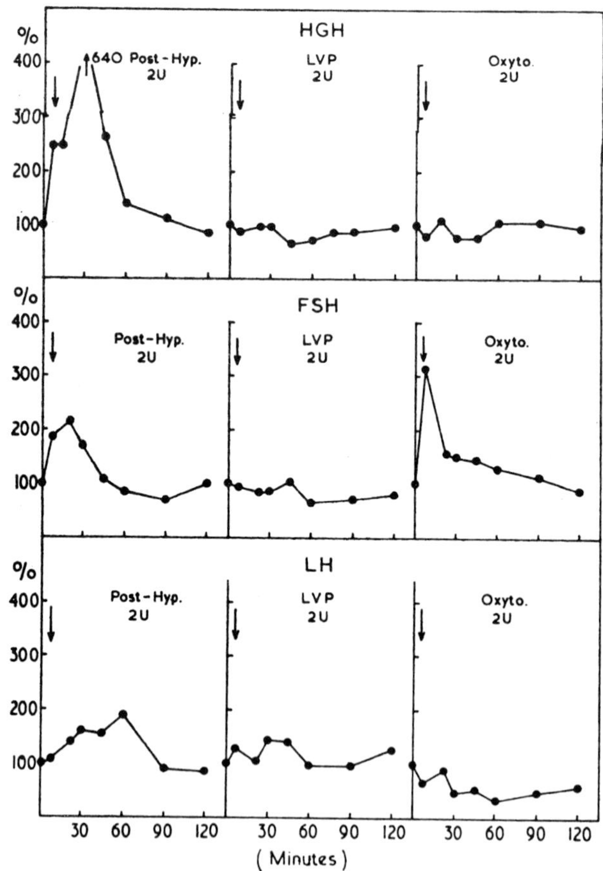

FIG. 9. Effect of posterior pituitary extract (Post-Hyp.), synthetic lysine-vasopressin (LVP) and synthetic oxytocin (oxyto.) on serum HGH, FSH, and LH levels. Each curve represents the mean response obtained from seven individual responses. Results are expressed in percentage of the basal value.

To explain the stimulatory effect of posterior pituitary extract, the lack of effects of synthetic lysine-vasopressin on the levels of the three hormones, and the lack of effect of synthetic oxytocin on HGH and LH release, three hypotheses are proposed: (1) posterior pituitary extracts contain releasing factors or other substances that can induce pituitary tropin release; (2) lysine-vasopressin is not active because it is not the physiological human hormone (which is arginine-vasopressin); the action of posterior pituitary extract could be due to arginine-vasopressin present in the extract; (3) there would be a synergic action of arginine-vasopressin and oxytocin capable of producing pituitary stimulation while each single hormone would not have any effect.

C. Pathological Conditions: Indirect Proof of Feedback Mechanisms

1. *Klinefelter Syndrome*

In our studies, very high serum FSH levels were always found in 14 cases of Klinefelter disease, proved by biopsy and karyotype (48 mIU HMG-IRP$_2$/ml). On the other hand, LH ranged between normal and increased values (Fig. 10).

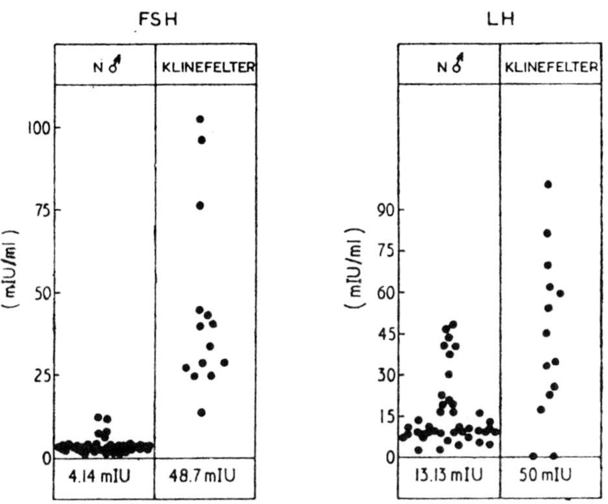

Fig. 10. FSH and LH levels in serum of normal male subjects and in serum of 14 patients with Klinefelter disease.

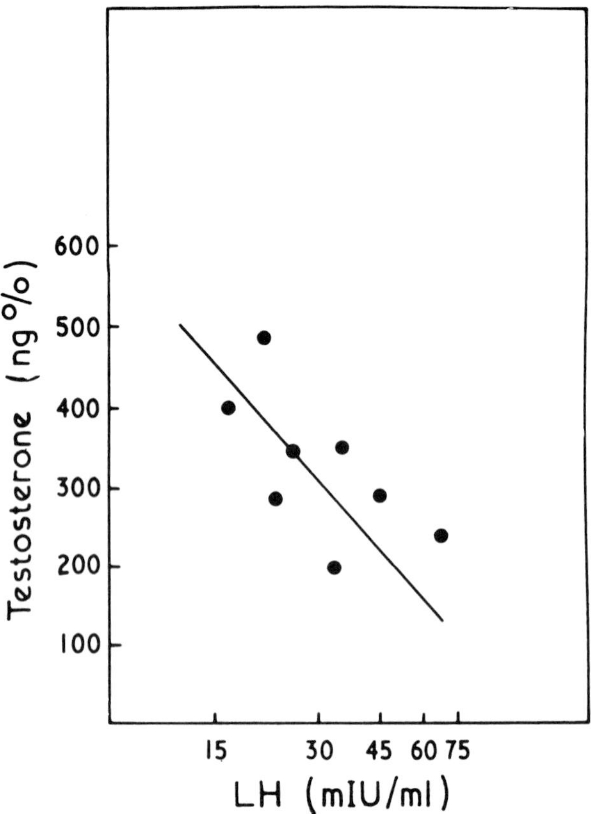

FIG. 11. There is a relationship between serum LH levels and testosterone concentration in the Klinefelter syndrome.

We believe that this high FSH output can be explained by the constant histological lesion in this disease, i.e., hyalinized seminal tubes with absence of the germinative cells. The pituitary gland reacts to this poor development of the germinative cells by increasing the output of FSH. For LH, the variations in the degree of insufficiency of the Leydig cells might account for the variation of LH. Thus there is a linear relationship between serum LH levels and the plasma concentrations in testosterone, determined at the same time (Fig. 11). This experiment seems to suggest that levels of LH depend on the endocrine insufficiency of the testes in Klinefelter syndrome; it also seems to prove that the increase of serum levels of testicular androgens induces a negative feedback effect on LH.

2. *Azoospermia and Oligospermia*

Eleven cases of azoospermia and 12 cases of oligospermia due to abnormal spermatogenesis proved by biospy were studied.[5] In all these cases, the diagnosis of Klinefelter's syndrome, Reifenstein's syndrome, hypogonadotropic eununchoidism, Sertoli-cell-only syndrome, cryptorchidism, and mumps orchitis were excluded. The etiologic mechanisms are not known.

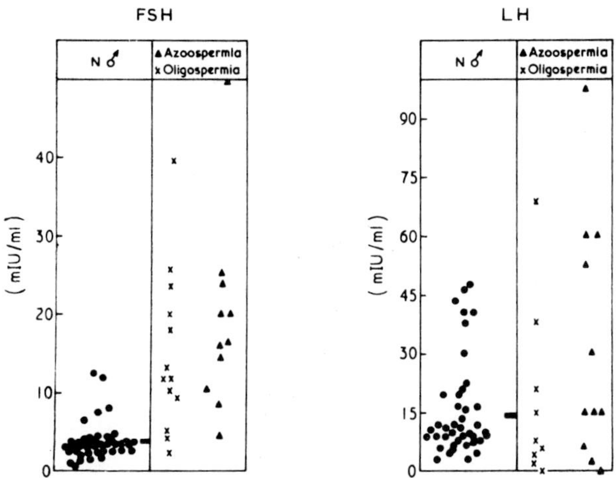

FIG. 12. FSH and LH levels in serum of normal male subjects and in serum of 23 patients with azoospermia or oligospermia.

Figure 12 shows that FSH values are very often higher than normal levels. Therefore this kind of azoospermia and oligospermia is not caused by FSH deficiency. On the contrary these high FSH levels seem to indicate that the primary lesion takes place during spermatogenesis and that there is a secondary FSH secretion.

At the present time we do not know which of the germinal cell types has to be involved before the gonadotropin level will be higher than normal. This finding indicates that exogenous gonadotropin treatment is illogical and explains why it is considered inefficient (Mroueh *et al.*, 1967). In these 23 men the LH levels were often

[5] The sera were kindly provided by Dr. Maquinay, Professor of Urology, University of Liege, and his assistant Dr. Bischops.

normal except in 4 cases of azoospermia and in one case of oligospermia. In all cases with normal LH, the testosterone plasma concentrations ranged within normal value limits.

This pathological condition strongly suggests that the FSH secretion does not depend on androgen production and especially not on testosterone plasma concentration. It seems that the feedback mechanism for FSH arises from spermatogenesis itself. Unfortunately, the substance relaying the information concerning spermatogenesis is not yet known.

V. Conclusions

Gonadotropins can be assayed by radioimmunoassay. In postmenopausal women a high dose of estrogen induces a rapid drop of LH level and a delayed and moderate decrease of FSH. Natural progesterone has no effect, whereas long-acting progestagens such as norethisterone enanthate produce a complete inhibition of LH and a partial inhibition of FSH for more than one month.

In eugonadal women the events observed during the menstrual cycle seem to indicate that: (1) follicle maturation depends on the early FSH rise; (2) ovulation is etiological related to FSH and LH mid-cycle peak; (3) the corpus luteum function seems to be independent of gonadotropin secretion. All contraceptive drugs (sequential, nonsequential, progestagen alone) inhibit the secretion of both gonadotropins, or at least one of them.

In men there is no diurnal variation of FSH; sometimes there is a slight decrease of LH between 12 noon and 4 P.M. Androgens, and especially testosterone, influence LH secretion; i.e., injection of testosterone decreases the serum LH levels and low basal testosterone concentration is accompanied by high serum LH levels in Klinefelter's syndrome. On the contrary, testosterone has little or no effect on FSH levels. When spermatogenesis is altered in azoospermia, oligospermia, or in Klinefelter's syndrome, serum FSH levels are always higher than in normal subjects.

Dexamethasone, hypoglycemia, and synthetic lysine-vasopressin induce no modification of gonadotropin levels.

Posterior pituitary extract stimulates gonadotropin secretion, and synthetic oxytocin increases the FSH levels in serum of young normal men.

ACKNOWLEDGMENTS

Our thanks are due Dr. Ryan and Dr. Hartree, who gave us pure FSH and LH preparations respectively; also to Mrs. Debruche and Miss Gaspar for their excellent technical assistance, and to Dr. Palem and Dr. Buret for performing testosterone, estrogen, and pregnandiol assays. These studies were supported by grant No. 1013 provided by The National Foundation for Medical Research (FRSM) of Belgium.

REFERENCES

Bogdanove, E. M. (1964). *Vitamin Hormones* **22**, 205.

Brostoff, J. V., James, T., and Landon, J. (1968). *J. Clin. Endocrinol.* **28**, 511.

Brown, J. B., Klopper, A., and Loraine, J. A. (1958). *J. Endocrinol.* **17**, 410.

Burger, H. G., Catt, K. J., and Brown, J. B. (1968). *J. Clin. Endocrinol.* **28**, 1508.

Cargille, C. M., Ross, G. T., Howland, L. A., and Rayford, P. L. (1968). *Clin. Res.* **16**, 33.

Cargille, C. M., Ross, G. T., and Yoshimi, T. (1969). *J. Clin. Endocrinol.* **29**, 12.

Catt, K. T. (1970). *Acta Endocrinol.* (Kbh) (In press).

Czarny, D. V., James, T., Landon, L., and Greenwood, F. C. (1968). *Lancet* **II**, 126.

Eddy, R. L. (1968). *J. Clin. Endocrinol.* **28**, 1836.

Faiman, C., and Ryan, R. (1967a). *J. Clin. Endocrinol.* **27**, 444.

Faiman, C., and Ryan, R. (1967b). *Nature* **215**, 857.

Franchimont, P. (1966a). *J. Label. Compounds* **2**, 303.

Franchimont, P. (1966b). " Le Dosage des Hormones Hypophysaires Somatotropes et Gonadotropes et son Application en Clinique." Arscia S. A., Bruxelles.

Franchimont, P. (1966c). *Ann. Endocrinol.* (Paris) **27**, 273.

Franchimont, P. (1968a). In " Protein and Polypeptide Hormones " (M. Margoulies, ed.), pp. 99–116. Excerpta Medica, Amsterdam.

Franchimont, P. (1968b). *Ann. Endocrinol.* (Paris) **29**, 403.

Franchimont, P., and Legros, J. J. (1969). *Ann. Endocrinol.* (Paris) **30**, 125.

Frantz, A. G., and Rabkin, H. (1964). *New England J. Med.* **271**, 1375.

Franchimont, P., Legros, J.J., Cession, G., Ayalon, D., and Motsers, A. (1970). *Obst. Gynec.* (In press).

Fukushima, H., Stevens, V., Gantt, C., and Vorys, N. (1964). *J. Clin. Endocrinol.* **24**, 205.

Gagliardino, J. J., Baily, J. D., and Martin, J. M. (1967). *Lancet* **I**, 1357.

Gans, E., and Van Rees, G. (1962). *Acta Endocrinol.* (Kbh) **39**, 245.

Hearn, W. R., Webel, E. J., Randolph, P. W., and Parks, N. E. (1961). *Proc. Soc. Exp. Biol. Med.* **107**, 515.

Henri, R., and Thevenet, M. (1959). *Bull. Soc. Chim. Biol.* **41**, 1391.

Legros, J. J., and Franchimont, P. (1968). *Lancet* **II**, 735.

Loraine, J. A., and Bell, T. (1968). *In* " Fertility and Contraception in the Human Female " pp. 75–107. E. and S. Livingstone, Edinburgh.

MacArthur, J., Worcester, J., and Ingersoll, F. (1958). *J. Clin. Endocrinol.* **18**, 1186.

MacArthur, J., Antoniades, H., Larson, L., Pennell, R., Ingersoll, F., and Ufelder, H. (1964). *J. Clin. Endocrinol.* **24**, 427.

Mahfouz, E. N. (1958). *J. Pharmacol. Exp. Therap.* **123**, 35.

McCann, S. M., and Ramirez, V. D. (1964). *Recent Progr. Hormone Res.* **20**, 131.

Meyer, V., and Knobil, E. (1966). *Endocrinology* **79**, 1016.

Midgley, A. R. (1967). *J. Clin. Endocrinol.* **27**, 295.

Midgley, A. R., and Jaffe, R. B. (1966). *J. Clin. Endocrinol.* **26**, 1375.

Midgley, A. R., and Jaffe, R. B. (1968). *J. Clin. Endocrinol.* **28**, 1699.

Mroueh, A., Lytton, B., and Kase, N. (1967). *J. Clin. Endocrinol.* **27**, 53.
Nelson, N. (1944). *J. Biol. Chem.* **153**, 375.
Odell, W. D., and Swerdloff, R. S. (1968). *Proc. Nat. Acad. Sci. U. S.* **61**, 529.
Palem, M., Maquinay, A., Margoulies, M., and Coninx, P. (1969). *Rev. Franç. Etudes Clin. Biol.* **14**, 1.
Parlow, A. F. (1964). *Endocrinology* **75**, 1.
Peterson, N. T., Midgley, A. R., and Jaffe, R. B. (1968). *J. Clin. Endocrinol.* **28**, 1473.
Rosenberg, E., and Keller, P. (1965). *J. Clin. Endocrinol.* **25**, 1262.
Ross, G. T., Odell, W. D., and Rayford, P. L. (1966). *Lancet* **II**, 1255.
Rosselin, G., and Dolais, J. (1967). *Presse Med.* **27**, 2027.
Saxena, B. B., Demura, H., Gandy, H., and Peterson, R. E. (1968). *J. Clin. Endocrinol.* **28**, 519.
Schally, A. V., Carter, W. H., Saito, M., Arimura, A., and Bowers, C. Y. (1968). *J. Clin. Endocrinol.* **28**, 1747.
Swerdloff, R. S., and Odell, W. D. (1969). *J. Clin. Endocrinol.* **29**, 157.
Taleisnik, S., and McCann, S. M. (1961). *Endocrinology* **68**, 263.
Vorys, N., Ellery, D., and Stevens, V. (1965). *Am. J. Obst. Gynecol.* **93**, 641.
Yalow, R. S., Varsano-Aharon, N., Ecehmendia, E., and Berson, S. (1969). *Hormone Metab. Res.* **1**, 3.

Control of Prolactin Secretion

J. L. Pasteels

I. Exteroceptive Stimuli

In cyclic female laboratory mammals, or in the male, prolactin secretion is very low. It becomes strongly stimulated at the onset of pregnancy, as indeed it does by the act of coitus itself. This stimulation of prolactin explains the physiological luteotropic activation of the corpus luteum. The effect of coitus was established as early as 1922 (Long and Evans, 1922; Slonacker, 1929; Meyer *et al.*, 1929) by pseudopregnancy experiments in which the luteotropic stimulation was induced by sterile mating or by various mechanical or electrical stimulations of the cervix (Shelesnyak, 1931; Greep and Hisaw, 1938; Rothchild and Dickey, 1960; Zeilmaker, 1965; De Feo, 1966; Hetherington, 1968; Yang, 1968). During lactation the release of prolactin is dependent on the stimuli of suckling (Selye, 1934; Meites and Turner, 1948; Folley, 1952; Grosvenor and Turner, 1958a,b). When the mothers are deprived of their pups for a few hours the prolactin secretory granules are stored within the pituitary prolactin cells. The granules are massively excreted after ½ hr of suckling stimulus (Pasteels, 1963). But when the litter is removed for a longer time, lysosomes destroy the unnecessary hormone in the cells as well as the excess of ribosomes and mitochondria. This accounts for the return of prolactin cells to their resting condition (Smith and Farquhar, 1966). Other exteroceptive stimuli can influence prolactin secretion. Parkes and Bruce (1961) and Bruce (1965) described the inhibition of prolactin release in pregnant mice exposed to the odor of strange males. The work of Grosvenor (1965a,b) and his co-workers (Grosvenor *et al.*, 1968) also appears to indicate that stimuli different from those of suckling (not well identified at the present time) contribute to the normal secretion of prolactin in lactating rats. Recent confirmation of this finding has been reported by Moltz *et al.* (1969) who surgically removed nipples not previously suckled. Non specific stress has been shown to be a powerful stimulus of prolactin secretion (Nicoll *et al.*, 1960; Gavazzi *et al.*, 1961), but the physiological significance of this finding remains unclear. In pigeons, visual stimuli are sufficient to elicit prolactin stimulation of the crop gland (Patel, 1936). All exteroceptive stimuli seem to influence prolactin secretion by way of hypothalamic control. This hypothesis is supported by a considerable body of experimental evidence. It has been found, for example, that the prolactin inhibiting activity of the rat hypothalamus was strongly depleted by the suckling stimulus (Ratner and Meites, 1964; Grosvenor, 1965a,b; Minaguchi and Meites, 1967) or by stress (Grosvenor, 1965a). Everett (1967) has shown recently that coital stimuli could induce delayed pseudopregnancy, suggesting the possibility that such stimuli are " remembered " by the central nervous system. Coitus without subsequent ovulation was obtained in barbiturate-blocked rats, but the next ovulation (5 or 6 days later) in these animals, when removed from new contact with the male, was followed by pseudopregnancy.

II. Prolactin Inhibiting Activity of the Hypothalamus

It is now well known that the chronic control of prolactin secretion by the hypothalamus, in male or cyclic female mammals is inhibitory. Evidence of this negative control was first obtained by grafting the hypophysis away from its neural connections (Desclin, 1950, 1956a; Everett, 1954, 1956; Sanders and Rennels, 1957; Alloiteau, 1958; Boot *et al.*, 1959; Quilligan and Rothchild, 1960; Meites and Hopkins, 1960; Ahren, 1961). In such experiments the portal vessels are obliterated, but an indirect vascular link via the general circulation could allow some residual control of the pituitary by the hypothalamus. More conclusive evidence has come from *in vitro* experiments. In organ culture (Meites *et al.*, 1961) and tissue culture (Pasteels, 1961a), it has been shown that the hypophysis (now in an autonomous state) secreted increased amounts of prolactin. The prolactin-inhibiting activity of the hypothalamus was demonstrated by adding hypothalamic tissue to the cultures (Pasteels, 1961c; Danon *et al.*, 1963; Gala and Reece, 1964) or extracts (Pasteels, 1962a, 1963; Talwalker *et al.*, 1963; Schally *et al.*, 1964; Gala and Reece, 1964). *In vitro* experiments are widely used to assay the hypothalamic prolactin inhibiting activity (Kragt and Meites, 1967). The prolactin inhibiting activity has also been assayed *in vivo* by injecting median eminence extracts into lactating rats to inhibit suckling-induced release of prolactin (Grosvenor *et al.*, 1964, 1965) or to prevent the stress-induced fall in pituitary prolactin concentration (Grosvenor *et al.*, 1965). Schally and co-workers have devised another *in vivo* assay based upon the inhibition of depletion of pituitary prolactin following cervical stimulation (Kuroshima *et al.*, 1966; Arimura *et al.*, 1967). The prolactin inhibiting activity has been reported in the hypothalamus of various mammals: for the rat, by Pasteels (1961c, 1962a, 1963), Danon *et al.* (1963), Talwalker *et al.* (1963), Gala and Reece (1964), Ratner and Meites (1964), Grosvenor (1965a, b), Kragt and Meites (1967), Minaguchi and Meites (1967); for cattle, by Grosvenor *et al.*, (1964) and Schally *et al.* (1965); for sheep, by Schally *et al.* (1965) and Dhariwal *et al.* (1968); for pig, by Schally *et al.* (1965) and Kuroshima *et al.* (1966); and for the human, by Pasteels (1963). In birds, however, the hypothalamic control of prolactin secretion appears to be quite different. Hypothalamic extracts were found to stimulate prolactin secretion in the blackbird (Nicoll, 1965), the pigeon (Kragt and Meites, 1965), and the duck (Gourdji and Tixier-Vidal, 1966). This interesting finding draws new attention to work suggesting that in some circumstances the mammalian hypothalamus can stimulate rather than inhibit prolactin secretion. Meites and associates (1960) observed that rat hypothalamic tissue could initiate mammary secretion in estrogen-primed rats. The hypothalami were taken from postpartum or estrogen-treated female rats and the possible effect of prolactin content was ruled out. Recently, Mishkinsky *et al.* (1968) confirmed this result and showed that oxytocin in various doses did not reproduce the effect. These experiments were performed *in vivo*. Crude hypothalamic extracts contain numerous substances capable of influencing prolactin secretion indirectly via the adrenals or the neural system. However, it should be kept in mind that the dominant prolactin inhibiting activity of the hypothalamus in mammals does not exclude a possible stimulatory control (assuming that several hypothalamic factors influence prolactin secretion). Further attempts to purify the hypothalamic neurohumors should solve this question.

III. About the Existence of a Specific Prolactin Inhibiting Factor

The name Prolactin Inhibiting Factor (PIF) is now widely used. By comparison to the well-known Corticotropin Releasing Factor (CRF), Luteinizing Hormone Releasing Factor (LHRF), Follicle-Stimulating Hormone Releasing Factor (FSHRF), Thyrotropin Releasing Factor (TSHRF), it was proposed to denote PIF as the prolactin inhibiting activity of the hypothalamus. *In vitro*, the prolactin inhibiting activity of hypothalamic extracts could not be duplicated by vasopressin, epinephrine, norepinephrine, acetylcholine, serotonin, histamine, substance P, or bradykinin (Talwalker *et al.*, 1963; Meites and Nicoll, 1966). These negative findings were confirmed by Gala and Reece (1963) for vasopressin, norepinephrine, acetylcholine, and serotonin. Gala and Reece described a slight stimulation of prolactin release by epinephrine, but this effect did not seem to be dose-dependent. The well-known stimulation of prolactin secretion by reserpine and similar substances (Barraclough, 1957; Desclin, 1957; Meites, 1957, 1958; Tuchmann-Duplessis and Mercier-Parot, 1957; Mayer *et al.*, 1958; Pasteels, 1961a) can be explained by its effect on the hypothalamus. This was shown by Kanematsu and Sawyer (1963a): when reserpine was implanted into the hypothalamus it stimulated prolactin release, whereas when implanted into the anterior hypophysis it had no effect. The depletion of monoamines induced by its implantation within the median eminence (Smelik, 1969) " might play a role in the release of hypothalamic neurohumoral transmitters, " as suggested by Müller *et al.* (1967) for the release of growth hormone. Reserpine treatment has been shown to depress the prolactin inhibiting activity of the rat hypothalamus (Ratner *et al.*, 1965). Since no monoamines have been demonstrated to inhibit prolactin release by the isolated hypophysis, their role is probably indirect.

Oxytocin is certainly not the prolactin inhibiting factor. When added to the hypophysis *in vitro* it had no action on prolactin release (Talwalker *et al.*, 1963; Pasteels, 1963.) In response to large amounts, only a slight stimulation could be observed (Gala and Reece, 1963). Oxytocin has been considered as a possible prolactin releasing factor: it stimulates milk production in goats and ewes (Denamur, 1953; Denamur and Martinet, 1961), but since it also prevents mammary involution in postpartum hypophysectomized rats (Meites and Hopkins, 1961) the effect on milk production could be at least partially attributed to its direct action on the mammary gland. Some experiments have demonstrated a true prolactin release in response to the injection of large doses of oxytocin (Benson and Folley, 1956, 1960; Desclin, 1956a, 1962; Stutinsky, 1957). However, this action is inhibited by atropine or dibenamine (Grosvenor and Turner, 1958a). At the doses used, oxytocin had a strong influence on the brain (Haun, 1959) and could act as a nonspecific stress. Thus its role as a prolactin releasing factor remains also to be convincingly demonstrated.

Isolation and purification of a specific PIF from hypothalamic extracts would provide unequivocal evidence of its existence. An important step was performed when Kragt and Meites (1967) described a log-dose relationship between prolactin inhibiting activity of rat hypothalami and prolactin release by pituitary tissue *in vitro*. This strongly suggests that PIF could be a specific hormone. However, to our knowledge, no pure PIF has yet been prepared. The main difficulty seems to be the distinction between PIF and LHRF. McCann *et al.* (1966) could easily separate FSHRF, Growth Hormone Releasing Factor (GHRF), and TSHRF from LHRF and prolactin inhibiting activity, but were unable to obtain PIF preparations without LHRF activity. It is worth noting that, according to McCann, " there was a tendency for the PIF to emerge from the Sephadex columns just prior to the LHRF, which suggests that the application of further techniques of purification may ultimately effect a complete separation of the two activities. . . . A final answer to the questions will require additional purification studies aimed at attaining a complete separation of the two activities " (Dhariwal *et al.*, 1968). Schally and coworkers claim to have obtained preparations of LHRF devoid of prolactin inhibiting activity when tested either *in vitro* (Schally *et al.*, 1964) or *in vivo* (Arimura *et al.*, 1967). But Dhariwal *et al.* (1968) raised the objection that neither worker reported PIF activity in other Sephadex fractions, so the location of PIF on their columns remains unknown. Attempts to separate PIF from LHRF are of special importance, for it is well known that the stimulation of prolactin secretion generally occurs during the time when gonadotropic functions are depressed. The strongest evidence for the existence of a separate PIF is found in reports of recent physiological experiments: Everett and co-workers have been able to influence Luteinizing Hormone (LH) and prolactin secretion separately by selective hypothalamic stimulation (Everett and Quinn, 1966; Quinn and Everett, 1967) and possibly by coital stimuli in ovulation-blocked rats (Everett, 1967).

IV. Endocrine Control of Prolactin Secretion

It has been well demonstrated that corticosteroids, progesterone, and androgens are effective in stimulating prolactin secretion *in vivo* (Meites *et al.*, 1963; Meites and Nicoll, 1966). No direct action of these hormones on pituitary prolactin cells could be demonstrated by experiments *in vitro* (Nicoll and Meites, 1964). Their mode of action, probably indirect, remains to be determined. Thyroid hormones can also stimulate prolactin secretion by the pituitary. Nicoll and Meites (1963) reported a direct action of thyroxine and triiodothyronine on organ cultures of rat hypophysis. Estrogens, powerful stimulators of prolactin secretion, act at two levels: on the pituitary cells themselves and via the hypothalamus. In suitable concentrations, estradiol stimulates prolactin production by the rat adenohypophysis *in vitro* (Nicoll and Meites, 1962). These results are in agreement with the more circumstantial evidence obtained by the study of pituitary grafts in living animals (Desclin, 1950; Desclin and Koulischer, 1960; Cohere and Meunier, 1963; Potvliege, 1965). Furthermore, intrahypophyseal implants of estrogen have been shown to stimulate prolactin secretion in rats (Ramirez *et al.*, 1963) or rabbits (Kanematsu and Sawyer, 1963b).

Tritiated estradiol is fixed easily on pituitary cells and induces hypertrophy of the anterior lobe by direct action (Palka *et al.*, 1966). Its receptor could be the nucleus of the cells (King, 1966). However, it has also been demonstrated that estradiol treatment *in vivo* depletes the prolactin inhibiting activity of the hypothalamus of the rat (Ratner and Meites, 1964). In addition, prolactin secretion is stimulated by estrogen implants in the hypothalamus (Kanematsu and Sawyer,

1963b; Ramirez *et al.*, 1963). The preferential uptake of estradiol by the hypothalamus is well known (Kato and Villee, 1967a,b; Stumpf, 1968).

Prolactin could regulate its own secretion by the " short " feedback mechanism. When implanted into intact rats, prolactin-secreting tumors depressed the prolactin content of the normal hypophysis (Chen *et al.*, 1967; McLeod *et al.*, 1968) and increased the prolactin inhibiting activity of the hypothalamus (Chen *et al.*, 1967). The implantation of approximately 250 μg of prolactin into the median eminence produced the same effect (Clemens and Meites, 1968).

The action of progestogens on prolactin secretion is of special interest. When the corpus luteum is receptive to the luteotropic action of prolactin, the prolactin stimulates progesterone secretion which in turn enhances prolactin release. Thus the pseudopregnancy becomes self-sustained by a positive feedback mechanism. Moreover, this action of the progestogens is linked with their antiovulatory activity and can provide information on the neural pathways involved in the control of both prolactin and gonadotropin secretions. With Ectors, we showed in rats that medroxyprogesterone (Provera) inhibited ovulation by inducing pseudopregnancy (Ectors *et al.*, 1966). It had no action on prolactin secretion by the hypophysis *in vitro* but when administered *in vivo* medroxyprogesterone depleted the rat hypothalamus of prolactin inhibiting activity (Pasteels and Ectors, 1967). In more recent experiments, minute amounts of medroxyprogesterone (3–5 μg) were implanted in the hypothalamus. When in a definite area of the suprachiasmatic region, they induced pseudopregnancy. When the implantation was performed more peripherally (at the borders of the same area) only a significant lengthening of estrus occurred (Pasteels and Ectors, 1967). Similar results were obtained by the use of 5 μg of progesterone (Pasteels and Ectors, 1968) and 1 μg of estradiol (Ectors and Pasteels, unpublished observations). It was found that the same hypothalamic area was influenced by estrogens and progestogens in the same way, i.e., stimulation of prolactin secretion and inhibition of LH and Follicle-Stimulating Hormone (FSH) release. (The lengthening of estrus cannot be ascribed to a stimulation of FSH secretion, for we observed that the implantation of 1 μg estradiol within this hypothalamic area inhibits the hypertrophy of FSH cells in rats ovariectomized 3 weeks previously). This progesterone-and estrogen-sensitive area could explain the link between prolactin secretion and gonadotropic functions. Further details of this work will be described in the following section.

V. Hypothalamic Areas Controlling Prolactin Secretion

Pituitary stalk section or destruction of portal vessels results in a marked stimulation of prolactin secretion. Such experimental results are easily explained by the inhibitory control of the hypothalamus on prolactin synthesis and release (see preceding section), and they contribute to the demonstration of this control mechanism in species where little other experimental evidence is available; such species include the cat (Grosz and Rothballer, 1961), the rabbit (Jacobsohn, 1949; Donovan and van der Werff ten Bosch, 1957), the goat (Cowie *et al.*, 1964), and the ferret (Donovan, 1963). Although no equivalent of a hypophysiotropic area has been described to date for prolactin secretion (in this case it would be, rather, a " hypophysis-inhibiting "), the median eminence-basal tuberal hypothalamus is certainly directly concerned in the regulation of prolactin. This has been clearly shown in the rat by the following experiments: lesion by McCann and Friedman (1959, 1960), Nikitovitch-Winer (1960), Kordon (1966), and De Voe *et al.*, (1966); electric stimulation by Everett and Quinn (1966); estrogen implantation by Ramirez *et al.*, (1963) and Ramirez and McCann, (1964); implantation of reserpine by Kanematsu and Sawyer (1963a) and Smelik (1969); and even implantation of prolactin itself (Clemens and Meites, 1968). Effective lesions were found to reach caudally the mammillary bodies (Nikitovitch-Winer, 1960) or at least the ventromedial-posterior-premammillary complex (Kordon, 1966). In the rabbit the effective hypothalamic area was found in the posterior medial basal tuberal region caudal to the pituitary stalk; prolactin secretion occurred when this area was destroyed (Haun and Sawyer, 1960, 1961) or implanted with estradiol benzoate (Sawyer *et al.*, 1960). Gale (1963) and Gale and Larsson (1963), using coagulation experiments with radio-frequency heating, have shown that in the goat the median eminence is also responsible for the inhibition of prolactin secretion.

More remote areas are also involved in the regulation of prolactin, possibly in an indirect way. Quinn and Everett (1967) have induced immediate or delayed pseudopregnancies by selective stimulation of the dorsomedial-ventromedial hypothalamus in rats, and several reports agree in localizing an inhibitory control of prolactin secretion (in rats) in an area including possibly the paraventricular nuclei, but extending more caudally, dorsally, and laterally (Flerkó and Bardos, 1959; Cook, 1959; Flament-Durand and Desclin, 1964; De Voe *et al.*, 1966). The destruction of this area induces mammary stimulation and pseudopregnancy.

In the anterior hypothalamus, the suprachiasmatic region (close to the roof of the suprachiasmatic recess of the third ventricle) is certainly also involved in the control of prolactin secretion. In collaboration with Ectors, we have been able to induce typical pseudopregnancy (as shown by vaginal smears, uterine deciduomata, and mammary gland stimulation) in virgin rats as a result of the implantation of minute amounts of medroxyprogesterone (Ectors and Pasteels, 1967; Pasteels and Ectors, 1968), progesterone (Pasteels and Ectors, 1968), or estradiol benzoate (unpublished observations) into this area. Controls were blank implantation (without steroid) or implantation of cholesterol (approximately 10 μg) and of prednisolone trimethylacetate (approximately 5 μg); these implants induced no significant changes in the estrous cycles. This similar action of both estrogen and progestogens may seem surprising, but it cannot be ascribed to a non specific steroid effect. Moreover, the same anterior hypothalamic area controls gonadotropic secretion, for we found that progesterone and medroxyprogesterone (Pasteels and Ectors, 1968) and estradiol (unpublished observations) implants at the borders of this area resulted in lengthening of estrus, and that the implantation of 1 μg estradiol benzoate inhibited the hypertrophy of gonadotropic cells in rats ovariectomized 3 weeks previously. Our results should be compared with those of Kato and Villee (1967a,b) and of Stumpf (1968).

Independently and by a different procedure (the administration of tritiated estradiol) Kato and Villee (1967a,b) demonstrated the preferential uptake of estradiol by the anterior hypothalamus. By the use of dry-mount autoradiography, to prevent diffusion and redistribution of labeled material, Stumpf (1968) has described a topography of estradiol-concentrating neurons in the hypothalamus of the rat which is absolutely identical to the localization of our effective stereotaxical implants. We have found (as have other authors) that there are two definite areas that respond to estrogen: one in the median eminence and the other in the anterior hypothalamus.

From the recent work of Halász and Gorski (1967) it appears that the more sensitive area may be the anterior one, for the transection of neural connections between the anterior hypothalamus and the hypophysiotropic area prevents the compensatory hypertrophy of the remaining ovary after unilateral ovariectomy. This steroid-binding anterior hypothalamic area is pluripotential, for Dörner *et al.* (1968a,b) have also demonstrated its role as a male mating center. From our results it appears clear that this area is sensitive to progestogens as well as to estrogen and that it regulates prolactin secretion. Such influence of the anterior hypothalamus might have been inferred from the report of Grosz and Rothballer (1961), who found that a transverse section of the tuber cinereum just behind the optic chiasm reinitiated milk secretion in postpartum cats separated from their litter.

Thus an estrogen-and progestogen-sensitive anterior hypothalamic area controls both prolactin secretion and gonadal function. This is probably the link between the regulation of LH (and possibly FSH) and prolactin. We do not favor the hypothesis of a common control of these functions because differential mechanisms of stimulation of gonadotropin secretion and of inhibition of prolactin release have been well demonstrated. Rather, we believe that common hypothalamic neurons, sensitive to both estrogen and progestogens, are controlling the hypophysiotropic areas for gonadotropins and also the specific neurons inhibiting prolactin secretion.

Obviously the hypothalamic areas controlling prolactin secretion are dependent on other parts of the nervous system, as shown by the influence of exteroceptive stimuli. But little is known about the neural pathways involved in this control. Tindal (1966) and Tindal *et al.* (1967) pointed out the possible role of the amygdaloid complex by demonstrating the influence of estrogen implants in this area on lactogenesis in the rabbit. Beyer and Mena (1965) have also induced milk secretion in the rabbit by removal of the telencephalon.

VI. Neural Control of Prolactin Secretion in the Human

Considerable doubt about the existence of human prolactin was raised by the discovery of the prolactin-like activities of human growth hormone (Chadwick *et al.*, 1961; Ferguson and Wallace, 1961; Kovacic, 1962; Forsyth, 1964; Forsyth *et al.*, 1965; Hartree *et al.*,

1965). However, the existence of human prolactin as a separate entity distinct from growth hormone is satisfactorily demonstrated by the study of its neural control. Contrary to what is known for growth hormone, human prolactin secretion is inhibited by the hypothalamus, a control very similar to that in other mammals. We found that tissue cultures of human anterior hypophysis secreted, quite autonomously, a pigeon crop gland-stimulating substance (Pasteels, 1962b). Comparison was made between this prolactin activity and growth hormone assayed immunologically. In the same cultures, removed from hypothalamic influence, prolactin secretion increased while growth hormone release decreased (Pasteels *et al.*, 1963; Pasteels, 1963; Brauman *et al.*, 1964). Moreover, we observed that the addition of hypothalamic extracts significantly inhibited the autonomous secretion of human prolactin, while it increased twofold the release of growth hormone (Pasteels, 1963). Rabbits were immunized against human prolactin elaborated *in vitro* (Pasteels *et al.*, 1965). The immune sera thus obtained neutralized the prolactin activity of human blood collected during postpartum or in pathological case of amenorrhea-galactorrhea (Pasteels, 1967).

We conclude that human prolactin does exist as a separate hormone distinct from growth hormone and that its hypothalamic control is inhibitory. Numerous clinical findings support this hypothesis. Eckles *et al.* (1958) observed that pituitary stalk section in women bearing a carcinoma of the breast resulted in long-lasting milk secretion in the non tumoral parts of the mammary glands. In these patients an increased sensitivity to insulin suggested that growth hormone secretion was depressed. Galactorrhea, in some cases at least, could not be the result of an oversecretion of growth hormone, since immunoassays clearly demonstrated that growth hormone concentrations in the blood were decreased (Van Cauwenberge *et al.*, 1963). In these cases the galactorrhea is due to excess prolactin secretion and could occur as a consequence of the severance of hypothalamo-hypophyseal connections by tumors (for example, craniopharyngioma (Salus, 1935), or even by a thyrotropic tumor (Herlant *et al.*, 1966). We freely admit that no satisfactory prolactin preparation has been yet extracted from human pituitaries. But cytological examination of the hypophysis explains this very easily. Human prolactin cells can be clearly distinguished (Pasteels, 1963; Herlant and Pasteels, 1967), but they appear in significant amounts only during pregnancy and the postpartum period. These cases are only seldom seen in the necropsy room. To get adequate yield from starting material, human prolactin should be extracted from glands containing significant amounts of the hormone, such as hypophyses removed from deceased pregnant or lactating women, or from mass culture of human hypophysis *in vitro*.

References

Ahrén, K. (1961). *Acta Endocrinol.* (Copenhagen) **38**, 449.

Alloiteau, J. J. (1958). *C. R. Acad. Sci.* (Paris) **247**, 1047.

Arimura, A., Saito, T., Müller, E. E., Bowers, C. Y., Sawano, S., and Schally, A. V. (1967). *Endocrinology* **80**, 972.

Barraclough, C. A. (1957). *Anat. Record* **127**, 262.

Benson, G. K., and Folley, S. J. (1956). *Nature* **177**, 700.

Benson, G. K., and Folley, S. J. (1960). *Acta Endocrinol.* (Copenhagen) *Suppl.* **51**, 1147.

Beyer, C., and Mena, F. (1965). *Am. J. Physiol.* **208**, 289.

Boot, L. M., Röpcke, G., and Kaligis, A. (1959). *Acta Physiol. Pharmacol. Neerl.* **8**, 543.

Brauman, H., Brauman, J., and Pasteels, J. L. (1964). *Nature* **202**, 1116.

Bruce, H. M. (1965). *Excerpta Med. Intern. Congr. Ser.* **83**, 193.

Chadwick, A., Folley, S. J., and Gemzell, C. A. (1961). *Lancet* **2**, 241.

Chen, C. L., Minaguchi, H., and Meites, J. (1967). *Proc. Soc. Exp. Biol. Med.* **126**, 317.

Clemens, J. A., and Meites, J. (1968). *Endocrinology* **82**, 878.

Cohere, G., and Meunier, J. M. (1963). *C. R. Soc. Biol.* (Paris) **157**, 1261.

Cook, A. R. (1959). *Tex. Rep. Biol. Med.* **17**, 512.

Cowie, A. T., Daniel, P. M., Knaggs, G. S., Prichard, M. M. L., and Tindal, J. S. (1964). *J. Endocrinol.* **28**, 253.

Danon, A., Dikstein, S., and Sulman, F. G. (1963). *Proc. Soc. Exp. Biol. Med.* **114**, 366.

De Feo, V. J. (1966). *Endocrinology* **79**, 440.

Denamur, R., (1953). *C. R. Soc. Biol.* (Paris) **147**, 88.

Denamur, R., and Martinet, J. (1961). *Ann. Endocrinol.* (Paris) **22**, 777.

Desclin, L. (1950). *Ann. Endocrinol.* (Paris) **11**, 656.

Desclin, L. (1956a). *Ann. Endocrinol.* (Paris) **17**, 586.

Desclin, L. (1956b). *C. R. Soc. Biol.* (Paris) **150**, 1489.

Desclin, L. (1957). *C. R. Soc. Biol.* (Paris) **151**, 1774.

Desclin, L. (1962). *Proc. Intern. Union Physiol. Sci.* **22**, 715.

Desclin, L., and Koulischer, L. (1960). *C. R. Soc. Biol.* (Paris) **154**, 1515.

De Voe, W. F., Ramirez, V. D., and McCann, S. M. (1966). *Endocrinology* **78**, 158.

Dhariwal, A. P. S., Grosvenor, C. E. Antunes-Rodrigues, J., and McCann, S. M. (1968). *Endocrinology* **82**, 1236.

Donovan, B. T. (1963). *J. Endocrinol.* **26**, 201.

Donovan, B. T., and van der Werff ten Bosch, J. (1957). *J. Physiol.* (London) **137**, 410.

Dörner, G., Döcke, F., and Moustafa, S. (1968a). *J. Reprod. Fertility* **17**, 175.

Dörner, G., Döcke, F., and Moustafa, S. (1968b). *J. Reprod. Fertility* **17**, 583.

Eckles, N. E., Ehni, G., and Kirschbaum, A. (1958). *Anat. Record* **130**, 295.

Ectors, F., and Pasteels, J. L. (1967). *C. R. Acad. Sci.* (Paris) **265**, 758.

Ectors, F., Pasteels, J. L., and Herlant, M. (1966). *C. R. Acad. Sci.* (Paris) **263**, 1988.

Everett, J. W. (1954). *Endocrinology* **54**, 685.

Everett, J. W. (1956). *Endocrinology* **58**, 786.
Everett, J. W. (1967). *Endocrinology* **80**, 145.
Everett, J. W., and Quinn, D. L. (1966). *Endocrinology* **78**, 141.
Ferguson, K. A., and Wallace, A. L. (1961). *Nature* **190**, 632.
Flament-Durand, J., and Desclin, L. (1964). *Endocrinology* **75**, 22.
Flerkó, B., and Bardos, V. (1959). *Acta Neuroveget.* **20**, 248.
Folley, S. J. (1952). *Ciba Found. Colloq. Endocrinol.* **4**, 381.
Forsyth, I. A. (1964). *J. Endocrinol.* **31**, XXX.
Forsyth, I. A., Folley, S. J., and Chadwick, A. (1965). *J. Endocrinol.* **31**, 115.
Gala, R. R., and Reece, R. P. (1963). *Fed. Proc.* **22**, 506.
Gala, R. R., and Reece, R. P. (1964). *Proc. Soc. Exp. Biol. Med.* **117**, 833.
Gale, C. C. (1963). *Acta Physiol. Scand.* **59**, 269.
Gale, C. C., and Larsson, B. (1963). *Acta Physiol. Scand.* **59**, 299.
Gavazzi, G., Giuliani, G., Martini, L., and Pecile, A. (1961). *Ann. Endocrinol.* (Paris) **22**, 788.
Gourdji, D., and Tixier-Vidal, A. (1966). *C. R. Acad. Sci.* (Paris) **263**, 162.
Greep, R. O., and Hisaw, F. L. (1938). *Proc. Soc. Exp. Biol. Med.* **39**, 359.
Grosvenor, C. E. (1965a). *Endocrinology* **76**, 340.
Grosvenor, C. E. (1965b). *Endocrinology* **77**, 1037.
Grosvenor, C. E., and Turner, C. W. (1958a). *Proc. Soc. Exp. Biol. Med.* **97**, 463.
Grosvenor, C. E., and Turner, C. W. (1958b). *Endocrinology* **63**, 535.
Grosvenor, C. E., McCann, S. M., and Nallar, R. (1964). *Program 46th Meeting Endocrine Soc.*, p. 96.
Grosvenor, C. E., McCann, S. M., and Nallar, R. (1965). *Endocrinology* **76**, 883.
Grosvenor, C. E., Mena, F., Schaefgen, D. A., Dhariwal, A. P. S., Antunes-Rodriguez, J., and McCann, S. M. (1968). *In* "Pharmacology of Hormonal Polypeptides and Proteins" (N. Back, L. Martini, and R. Paoletti, eds.), pp. 242–253. Plenum Press, New York.
Grosz, H. J., and Rothballer, A. B. (1961). *Nature* **190**, 349.
Halász, B., and Gorski, R. A. (1967). *Endocrinology* **80**, 608.
Hartree, A. S., Kovacic, N., and Thomas, M. (1965). *J. Endocrinol.* **33**, 249.
Haun, C. K. (1959). *Anat. Record* **133**, 286.
Haun, C. K., and Sawyer, C. H. (1960). *Endocrinology* **67**, 270.
Haun, C. K., and Sawyer, C. H. (1961). *Acta Endocrinol.* (Copenhagen) **38**, 99.
Herlant, M., and Pasteels, J. L. (1967). *Meth. Achievmts Exp. Pathol.* **3**, 250.
Herlant, M., Linquette, M., Laine, E., Fossati, P., May, J. P., and Lefebvre, J. (1966). *Ann. Endocrinol.* (Paris) **27**, 181.
Hetherington, C. M. (1968). *J. Reprod. Fertility* **17**, 391.
Jacobsohn, D. (1949). *Acta Physiol. Scand.* **19**, 10.
Kanematsu, S., and Sawyer, C. H. (1963a). *Proc. Soc. Exp. Biol. Med.* **113**, 967.
Kanematsu, S., and Sawyer, C. H. (1963b). *Endocrinology* **72**, 243.
Kato, J., and Villee, C. A. (1967a). *Endocrinology* **80**, 567.
Kato, J., and Villee, C. A. (1967b). *Endocrinology* **80**, 1133.
King, R. J. B. (1966). *In* "La Physiologie de la Reproduction cher les Mammifères" (A. Jost, ed.), pp. 570-583, CNRS, Paris.
Kordon, C. (1966). Thèse de Sciences, Paris.
Kovacic, N. (1962). *Nature* **195**, 1210.
Kragt, C., and Meites, J. (1965). *Endocrinology* **76**, 1169.
Kragt, C., and Meites, J. (1967). *Endocrinology* **80**, 1170.
Kuroshima, A., Arimura, A., Bowers, C. Y., and Schally, A. V. (1966). *Endocrinology* **78**, 216.
Long, J. A., and Evans, H. M. (1922). " Memoirs of the Univ. Calif.", p. **6**.

McCann, S. M., and Friedman, H. (1959). *Fed. Proc.* **18**, 101.
McCann, S. M., and Friedman, H. (1960). *Endocrinology* **67**, 597.
McCann, S. M., Antunes-Rodriguez, J., Watanabe, S., Ratner, A., and Dhariwal, A. P. S. (1966). *Program Ford Found. Conf. Physiol. Human Reprod.*, p. 50.
McLeod, R. M., De Witt, G. W., and Smith, M. C. (1968). *Endocrinology* **82**, 889.
Mayer, G., Meunier, J. M., and Rouault, J. (1958). *C. R. Acad. Sci.* (Paris) **247**, 524.
Meites, J. (1957). *Proc. Soc. Exp. Biol. Med.* **96**, 728.
Meites, J. (1958). *Proc. Soc. Exp. Biol. Med.* **97**, 742.
Meites, J., and Hopkins, T. F. (1960). *Proc. Soc. Exp. Biol. Med.* **104**, 268.
Meites, J., and Hopkins, T. F. (1961). *J. Endocrinol.* **22**, 207.
Meites, J., and Nicoll, C. S. (1966). *Ann. Rev. Physiol.* **28**, 57.
Meites, J., and Turner, C. W. (1948). *Montana State Coll. Agri. Exp. Sta. Res. Bull.* **415**, 1.
Meites, J., Talwalker, P. K., and Nicoll, C. S. (1960). *Proc. Soc. Exp. Biol. Med.* **103**, 298.
Meites, J., Kahn, R. H., and Nicoll, C. S. (1961). *Proc. Soc. Exp. Biol. Med.* **108**, 440.
Meites, J., Nicoll, C. S., and Talwalker, P. K. (1963). *In* " Advances in Neuroendocrinology " (A.V. Nalbandov, ed.), pp. 238–277. University of Illinois Press, Urbana, Ill.
Meyer, R. K., Leonard, S., and Hisaw, F. L. (1929). *Proc. Soc. Exp. Biol. Med.* **27**, 340.
Minaguchi, H., and Meites, J. (1967). *Endocrinology* **80**, 603.
Mishkinsky, J., Khazen, K., and Sulman, F. G. (1968). *Endocrinology* **82**, 611.
Moltz, H., Levin, R., and Leon, M. (1969). *Science* **163**, 1083.
Müller, E. E., Sawano, S., Arimura, A., and Schally, A. V. (1967). *Endocrinology* **80**, 471.
Nicoll, C. S. (1965). *J. Exp. Zool.* **158**, 203.
Nicoll, C. S., and Meites, J. (1962). *Endocrinology* **70**, 272.
Nicoll, C. S., and Meites, J. (1963). *Endocrinology* **72**, 544.
Nicoll, C. S., and Meites, J. (1964). *Proc. Soc. Exp. Biol. Med.* **117**, 579.
Nicoll, C. S., Talwalker, P. K., and Meites, J. (1960). *Am. J. Physiol.* **198**, 1103.
Nikitovitch-Winer, M. (1960). *Mem. Soc. Endocrinol.* **9**, 70.
Palka, Y. S., Ramirez, V. D., and Sawyer, C. H. (1966). *Endocrinology* **78**, 486.
Parkes, A. S., and Bruce, H. M. (1961). *Science* **134**, 1049.
Pasteels, J. L. (1961a). *Ann. Endocrinol.* (Paris) **22**, 257.
Pasteels, J. L. (1961b). *C. R. Acad. Sci.* (Paris) **253**, 3074.
Pasteels, J. L. (1961c). *C. R. Acad. Sci.* (Paris) **253**, 2140.
Pasteels, J. L. (1962a). *C. R. Acad. Sci.* (Paris) **254**, 2664.
Pasteels, J. L. (1962b). *C. R. Acad. Sci.* (Paris) **254**, 4083.
Pasteels, J. L. (1963). *Arch. Biol.* (Brussels) **74**, 439.
Pasteels, J. L. (1967). *Ann. Endocrinol.* (Paris) **28**, 117.
Pasteels, J. L. and Ectors, F. (1967). *C. R. Acad. Sci.* (Paris) **264**, 106.
Pasteels, J. L., and Ectors, F. (1968). *Ann. Endocrinol.* (Paris) **29**, 663.
Pasteels, J. L., Brauman, H., and Brauman, J. (1963). *C. R. Acad. Sci.* (Paris) **256**, 2031.
Pasteels, J. L., Robyn, C., and Hubinont, P. O. (1965). *C. R. Acad. Sci.* (Paris) **260**, 4381.
Patel, M. D. (1936). *Physiol. Zool.* **9**, 129.
Potvliege, P. (1965). *J. Microscop.* (Paris) **4**, 485.
Quilligan, E. J., and Rothchild, I. (1960). *Endocrinology* **67**, 48.
Quinn, D. L., and Everett, J. W. (1967). *Endocrinology* **80**, 155.
Ramirez, V. D., and McCann, S. M. (1964). *Endocrinology* **75**, 206.
Ramirez, V. D., Abrams, R. M., and McCann, S. M. (1963). *Fed. Proc.* **22**, 2063.
Ratner, A., and Meites, J. (1964). *Endocrinology* **75**, 377.

Ratner, A., Talwalker, P. K., and Meites, J. (1965). *Endocrinology* **77**, 315.
Rothchild, I., and Dickey, R. (1960). *Endocrinology* **67**, 42.
Salus, F. (1935). *Deut. Arch. Klin. Med.* **177**, 614.
Sanders, A., and Rennels, E. G. (1957). *Anat. Record* **127**, 360.
Sawyer, C. H., Haun, C. K., Hilliard, J., and Radford, H. M. (1960). *Acta Endocrinol.* (Copenhagen) *Suppl.* **51**, 1139.
Schally, A. V., Meites, J., Bowers, C. Y., and Ratner, A. (1964). *Proc. Soc. Exp. Biol. Med.* **117**, 252.
Schally, A. V., Kuroshima, A., Ishida, Y., Redding, T. W., and Bowers, C. Y. (1965). *Proc. Soc. Exp. Biol. Med.* **118**, 350.
Selye, H. (1934). *Am. J. Physiol.* **107**, 535.
Shelesnyak, M. C. (1931). *Anat. Record* **49**, 179.
Slonacker, J. R. (1929). *Proc. Soc. Exp. Biol. Med.* **89**, 406.
Smelik, P. G. (1969). *Acta Neurol. Psychiat. Belg.* **69**, 540.
Smith, R. E., and Farquhar, M. G. (1966). *J. Cell Biol.* **31**, 319.
Stumpf, W. E. (1968). *Science* **162**, 1001.
Stutinsky, F. (1957). *C. R. Acad. Sci.* (Paris) **244**, 1537.
Talwalker, P. K., Ratner, A., and Meites, J. (1963). *Am. J. Physiol.* **205**, 213.
Tindal, J. S. (1966). *Program Ford Found. Conf. Physiol. Human Reprod.*, p. 61.
Tindal, J. S., Knaggs, G. S., and Turvey, A. (1967). *J. Endocrinol.* **37**, 279.
Tuchman-Duplessis, H., and Mercier-Parot, L. (1957). *C. R. Soc. Biol.* (Paris) **151**, 656.
Van Cauwenberge, H., Franchimont, P., and Salmon, J. (1963). *Bull. Acad. Roy. Med. Belg.* VIIe Série **3**, 53.
Yang, W. H. (1968). *Endocrinology* **82**, 423.
Zeilmaker, G. H. (1965). *Acta Endocrinol.* (Copenhagen) **49**, 558.

Control of Growth Hormone Secretion

R. M. Bala, R. Burgus, K. A. Ferguson, R. Guillemin,
C. F. Kudo, G. C. Olivier, N. W. Rodger, and J. C. Beck

I. Introduction

Knowledge of the regulatory mechanisms governing the synthesis and release of growth hormone has been accumulated slowly in contrast with the information available on other pituitary hormones. This is in part due to the fact that the growth hormone secretory function appears to be more autonomous than the secretion of the other pituitary hormones and lacks an obvious target organ or tissue. In addition, the relative difficulty, nonspecificity, and insensitivity of the bioassay methods which have of necessity been used in these studies have delayed the acquisition of definitive data. The advent of sensitive immunoassay techniques reported by Hunter and Greenwood (1964a) and Glick and his associates (1963) has contributed enormously to recent studies of the regulation of growth hormone secretion and has clarified many previously obscure points. A comprehensive review of past contributions is not the intention of this presentation, and the interested individual is referred to the excellent article by Reichlin (1966) which has been used as a base upon which to build more recent observations in man and the experimental animal as they apply to the data presented in this communication.

Radioimmunoassay has led to the establishment of normal growth hormone values under standard conditions and has demonstrated that plasma growth hormone, as measured by this technique, is subject to wide and rapid fluctuations. Elevated levels are seen during insulin or tolbutamide-induced hypoglycemia (Roth *et al.*, 1963a,b; Greenwood *et al.*, 1966; Frantz and Rabkin, 1964; Hunter and Greenwood, 1964b); during falling blood glucose levels seen after a glucose load or postprandially; during fasting (Glick *et al.*, 1965; Roth *et al.*, 1964), exercise (Roth *et al.*, 1963b; Hunter *et al.*, 1965a, b; Hartog *et al.*, 1967), intravenous aminoacid infusions (Knopf *et al.*, 1965; Merimee *et al.*, 1967, Rabinowitz *et al.*, 1966), operations (Ketterer *et al.*, 1966; Ross *et al.*, 1966; Roth *et al.*, 1963b, Schalch, 1967), physical trauma, psychic stress, and electroconvulsive therapy (Schalch, 1967); and after the administration of pyrogen (Glick *et al.*, 1965; Greenwood and Landon, 1966; Kohler *et al.*, 1967; Frohman *et al.*, 1967a, b; Roth *et al.*, 1964). The effects of exposure to cold on plasma growth hormone remain unclear, although the bulk of evidence suggests no change when shivering is prevented (Berg *et al.*, 1966; Glick, 1968). Plasma growth hormone peaks are observed most commonly during deep sleep, and smaller peaks are occasionally seen during subsequent sleep phases (Takahashi *et al.*, 1968; Hunter *et al.*, 1966; Quabbe *et al.*, 1966; Glick, 1968; Glick and Goldsmith, 1968; Honda *et al.*, 1967). In the monkey, minor trauma, spontaneous excitement, pain, noise, apprehension, intravenous injection of histamine, epinephrine, pitressin and vasopressin, hemorrhage, and arousal from nembutal anesthesia are associated with abrupt increases in growth hormone secretion. It is of interest that these stimuli are comparable to those which have long been known to increase Adrenocorticotropic Hormone (ACTH) release (Meyer and Knobil, 1966, 1967).

The stimulus to growth hormone secretion induced by insulin is a direct result of hypoglycemia, since insulin in the presence of normal or elevated blood glucose levels is without effect (Roth *et al.*, 1964). The elevation of plasma growth hormone in response to 2-deoxy-D-glucose suggests that the metabolism of glucose within a regulatory center is more important than the level of blood glucose (Wegienka *et al.*, 1967; Roth *et al.*, 1964). These observations considered together with the elevations in plasma growth hormone on fasting and exercise indicate one secretory control mechanism to be dependent upon an unidentified carbohydrate substrate and independent of a stress mechanism. This receives further support from the observations of Burday and associates (1968), who showed that the arginine-induced growth hormone release could be blocked by hyperglycemia. In addition, earlier studies (Blanco *et al.*, 1966; Abrams *et al.*, 1966) show that microinfusion of glucose into the hypothalamus and lesions placed in the median eminence of the squirrel monkey effectively block the rise in growth hormone induced by hypoglycemia.

The precise physiologic role of arginine-induced stimulation of growth hormone secretion is unclear. It has been suggested that it is a mechanism for ensuring nitrogen retention with increased availability of amino acids, although evidence in support of this is meager. It would seem most likely that the effect is mediated through the median eminence which appears to play a regulatory role in growth hormone release.

There is increasing evidence that other hormones modulate the growth hormone response to insulin hypoglycemia and arginine infusion. Estrogens enhance the response to arginine and to exercise significantly (Frantz and Rabkin, 1965; Chakmakjian and Bethune, 1968; Merimee *et al.*, 1966). Androgens greatly enhance the plasma growth hormone response to hypoglycemia and arginine infusion (Martin *et al.*, 1968). Suppression of growth hormone release by corticosteroids (Frantz and Rabkin, 1964; Hartog *et al.*, 1964) is well documented, although ACTH either does not inhibit or is much less effective than long-term corticosteroid administration (Friedman and Stimmler, 1966). Diminished responsiveness is seen in both hypo- and hyper-thyroidism (Burgess *et al.*, 1965, 1966). A potent progestational agent, 6-methyl-17-α-hydroxyprogesterone acetate, suppressed the response to both insulin and arginine (Simon *et al.*, 1967). Large doses of vasopressin and epinephrine, the latter in an unanesthetized monkey, increase plasma growth hormone levels in the absence of other provocative stimuli (Glick, 1968).

The hypothalamus would appear to play an important role in the regulation of growth hormone secretion. Patients with pituitary stalk sections become unresponsive to stimuli promoting growth hormone secretion (Roth *et al.*, 1963a). Specific hypothalamic lesions in the monkey can prevent insulin-induced growth hormone secretion (Abrams *et al.*, 1966). More recently, the effects of certain hypothalamic extracts on the plasma levels of growth hormone have been reported (Garcia and Geschwind, 1966; Machlin *et al.*, 1967; Smith *et al.*, 1968) although major controversy exists as to the specificity of this response (Knobil, 1966; Knobil and Meyer, 1968). These data would be in keeping with earlier observations, summarized and contributed to by Reichlin (1960, 1961), on the effects of experimental brain lesions on growth. He pointed out that to be effective such lesions had to involve the anterior median eminence including a portion of the primary plexus. Additional convincing data from stimulation experiments have recently been reported which clarify this further. It has been shown (Frohman *et al.*, 1968 a, b) in a series of experiments that both destruction and stimulation of the ventromedial hypothalamic nucleus in rats results in the predicted changes in plasma and pituitary growth hormone content as measured by immunoassay.

The influence of catecholamines on growth hormone secretion has been difficult to evaluate. Recently a definite inhibitory effect on growth hormone secretion has been suggested by a suppressed plasma response to insulin during alpha adrenergic blockade with phentolamine (Blackard and Heidingsfelder, 1968). Propranolol (Imura *et al.*, 1968), a beta adrenergic blocking agent, increases plasma growth hormone levels in both normal subjects and patients with pheochromocytoma. It is tempting to consider that the mediation of the hypoglycemic and other stimuli may be via adrenergic hypothalamic mechanisms.

Previous discussion in this brief review of the role of the central nervous system in the regulation of growth hormone secretion has been largely confined to the primate. Since a section of this presentation is concerned with hypothalamic factors controlling growth hormone release in the rat, the following will serve as a summary of some of the previous work. Del Vecchio and associates (1958) first suggested that crude posterior pituitary extracts could increase epiphyseal cartilage width in the rat, an observation which was confirmed by Hiroshige and Itoh (1960). Since hypophysectomy prevented this effect, a Growth Hormone Releasing Factor (GHRF) analogous to Thyrotropin Releasing Factor (TSHRF) was postulated. Controversy still exists over the role of posterior pituitary hormones in growth hormone release. Reichlin's observations (1960, 1961) also suggested that hypothalamic control existed over growth hormone production but correctly raised questions concerning the specificity of the growth retardation. Equivocal evidence for the existence of a hypothalamic neurohormonal factor in the dog regulating growth hormone secretion was reported by Franz and associates (1962). Hypothalamic extract stimulated growth hormone release *in vitro* and repeated administration of the extract widened tibial epiphyseal cartilages and increased body weight in intact but not in hypophysectomized rats. More convincing results were obtained by Deuben and Meites (1964) in *in vitro* studies using tissue cultures of rat anterior pituitaries and rat hypothalamic extracts. These observations received confirmation by Schally and his associates (1965) using bovine and porcine hypothalamic extracts. In search of a relatively simple assay method for GHRF, Pecile and associates (1965) reported that intracarotid or intravenous injection of a crude hypothalamic extract in rats resulted in a profound fall of pituitary growth hormone content compared with injections of control tissues. The growth hormone content of these pituitaries was bioassayed by the tibia test method of Greenspan and associates (1949). Since these observations were reported, an intensive search for growth hormone releasing factors has been carried out. Crude hypothalamic extracts from guinea pig (Müller and Pecile, 1965), rat (Deuben and Meites, 1965), cow (Schally *et al.*, 1965), hog (Ishida *et al.*, 1965), sheep (Krulich *et al.*, 1965), monkey, and man (Müller and Pecile, 1966) have been claimed to deplete pituitary growth hormone content *in vivo* or to release pituitary growth hormone *in vitro*. Attempts at obtaining GHRF in increasingly pure form have been made (Schally *et al.*, 1966, 1968a; Dhariwal *et al.*, 1965, 1966) using gel filtration and ion exchange chromatography of crude preparations of beef, sheep, and pig hypothalami. All the *in vivo* evidence rests on the measurement of growth hormone secretion by the depletion of pituitary growth hormone using the tibia test bioassay. The effect of the more highly purified GHRF preparations are not altered by thioglycollate (Schally *et al.*, 1968b), but are destroyed by proteolytic enzymes (Schally *et al.*, 1968a; Smith *et al.*, 1968), suggesting that the effect observed is not due to posterior lobe hormone contaminants and that the purified material is a polypeptide of extremely high biological activity. Recently, Krulich and associates (1968) have claimed that a factor inhibiting the release of growth hormone has been isolated in the Sephadex G-25 purification of sheep and rat hypothalamic extract. However, it must be pointed out that although claims of high degrees of purification have been made, the structure of GHRF has not been determined nor has its synthesis been reported.

II. Effects of Insulin and Arginine on Serum Growth Hormone Levels in the Human

The need for provocative measures to evaluate pituitary growth hormone reserve under standard conditions has led to the development of procedures in which growth hormone response is measured in association with insulin hypoglycemia (Roth *et al.*, 1963a, b) and arginine infusion (Knopf *et al.*, 1965; Merimee *et al.*, 1965). Comparisons of these two procedures have been made with the aim of determining the more reliable one (Merimee *et al.*, 1967; Parker *et al.*, 1967; Raiti *et al.*, 1968); as a result, arginine infusion has been accepted as a useful provocative test and in Parker's view is the preferred procedure. This is in part based on its innocuousness and the avoidance of the undesirable side effects produced by insulin-induced hypoglycemia, particularly in the young. Others (Burday *et al.*, 1968; Best *et al.*, 1968) have presented evidence that insulin-induced hypoglycemia produces a more potent and consistent stimulus to the release of growth hormone. Best's data cast doubt on arginine as a stimulus to the secretion of growth hormone. In a series of controlled arginine infusions in which 500 ml of 0.15 *M* saline infusion was compared with the standard arginine infusion, they report that no patient responding to arginine failed to respond in a similar way to saline. In view of this controversy, I should like to present some observations made during the Medical Research Council (M.R.C.) of Canada Collaborative Study by Friesen and myself, supplemented by unpublished data obtained by Friesen and Pozsonyi in an investigation of growth disorders in patients with mental disorders.

A. Materials and Methods

Seventy-four young subjects, 56 with well-documented hypopituitarism and 18 controls, were studied as part of an M.R.C. (Canada) sponsored trial into the effect of human growth hormone on growth in pituitary dwarfism. All patients were fasted overnight prior to the provocative procedure. Hypoglycemia was induced by injecting 0.1 unit/kg insulin (or greater amounts if hypoglycemia failed to occur) and blood samples were taken at 0, 20, 40, 60, 90, and 120 min. Growth hormone concentrations were determined using the double antibody method of Schalch and Parker (1964). Arginine infusions were given in a dose of 0.5 g l-arginine monochloride per kilogram body weight over a 30-min period, blood samples being taken at 0, 30, and 60 min.

B. Results

The mean peak values and the maximum differences in response to insulin hypoglycemia and arginine infusion are shown in Table I. It can be seen that the two procedures produced no significant differences. If responses of individual control subjects are examined in detail it is clear that some failed to respond to one or another of these stimuli with the expected rise in growth hormone level.

TABLE I

Human Growth Hormone (HGH) Response to Provocative Stimuli

	HGH Concentration (μg/ml)					
	Insulin			Arginine		
Subjects	No.	Peak	Max. Diff.	No.	Peak	Max. Diff.
Controls	18	23.4±11.8	22.0±12	11	18.1±8.5	16.1±9.1
Hypopituitarism	56	1.4± 1	0.8± 0.8	52	1.6±1.1	0.9±0.8

C. Discussion

It would appear from this study and a series of unpublished observations made by Friesen and Pozsonyi in 66 patients with growth disorders and mental retardation that there are no significant differences between insulin-induced hypoglycemia and arginine stimulation in causing and increase in growth hormone concentrations. These data support the conclusions of Roth *et al.*, (1964) and Parker *et al.* (1967) and do not explain the differences claimed by others. It must be emphasized that some apparently normal individuals may not respond to one or another of these provocative stimuli, and the etiology and significance of this phenomenon remain unclear. A single abnormal response does not permit the conclusion that an individual is growth hormone deficient.

III. Variability of Response in the Bioassay for a Hypothalamic Growth Hormone Releasing Factor

Interest in the factors regulating the biosynthesis, storage, and release of growth hormone led us to join hands with Roger Guillemin and Roger Burgus of Houston, in 1965, and I should now like to present the findings from our two laboratories. We were originally interested in localizing GHRF activity in fractions from the large-scale purification program under way in Houston which was directed primarily at isolating the TSHRF. Our aim was to ascertain the presence of GHRF activity in the various fractions and to follow its further purification. Gel filtrations, devoted entirely to the search for GHRF activity, were also carried out in the hope that adequate amounts of GHRF might be obtained for studies in the human. The bioassay based on " pituitary depletion of growth hormone content " in the rat (Pecile *et al.*, 1965) was used to assess " growth hormone releasing activity " in crude extracts of ovine, porcine, and bovine hypothalamus, and in purified fractions of ovine hypothalamic extracts following gel filtration on Sephadex G-25.

When subjected to strict mathematical criteria, the responses observed appear to occur at random and thus do not permit us to conclude that GHRF activity can be isolated or quantified by the " rat pituitary depletion " method. Schalch carried out radioimmunoassays of pituitary and plasma growth hormone concentrations in tissues and blood of the same animals as those studied in the bioassays and also found no evidence of stimulation of growth hormone release by the purified hypothalamic fractions.

A. Materials and Methods

1. *Starting Materials and Separation Procedures*

a. Extracts. Frozen hypothalamic fragments from unbred mature sheep were lyophilized, defatted in acetone, and extracted in 2 *N* acetic acid (Guillemin *et al.*, 1962, 1965), glacial acetic acid (Schally and Bowers, 1964) or in 0.01 *N* ammonium hydroxide. Similar extracts of ovine cerebral cortex and glacial acetic acid extracts of porcine hypothalamus were also prepared. All crude extracts were lyophilized and stored at room temperature *in vacuo*.

b. Filtration. Gel filtration on Sephadex G-25 in 0.1 *M* pyridine acetate of the acetic acid extract of ovine hypothalamus was performed as part of the main purification program of TSHRF. In addition, gel filtrations designed exclusively for the search and assay for GHRF activity were performed; for each of these experiments 10,000 to 30,000 sheep hypothalamic fragments were processed.

2. *Injection of Hypothalamic Preparations*

The crude extracts were dissolved in 0.9% NaCl, centrifuged at high speed to remove any insoluble material; the supernatant was injected as described below. Lyophilized aliquots of the effluent of the various gel filtrations were dissolved in 0.9% NaCl and injected in 0.5-ml volumes, under ether anesthesia, into the external jugular vein in male rats weighing 120 to 200 g, obtained from the Cheek-Jones Farms or the Quebec Breeding Farms. The rats were decapitated 30 min later. The anterior pituitary glands were weighed immediately (to the nearest 0.1 mg) and the glands from each group of rats given similar treatment were pooled and frozen on dry ice. The stump blood, collected in heparinized beakers, was pooled for each treatment group, and the plasma was separated and frozen for subsequent immunoassay of growth hormone (Schalch and Reichlin, 1966). Hypothalamic preparations were injected in solutions of 0.01 *N* acetic acid in 0.9% saline (Dhariwal *et al.*, 1965; Ishida *et al.*, 1965), in aqueous solutions of 1% bovine serum albumin or in the original eluent from Sephadex G-25 filtrations (0.1 *M* pyridine acetate buffer, pH 5.5, rendered isotonic by addition of saturated NaCl solutions). In some experiments, instead of the above schedule, rats were decapitated 15 min after intracarotid injection or 20 min after intracarotid or intravenous infusion of test materials for 20 min under pentobarbital anesthesia.

3. *Extraction of Anterior Pituitary Tissues for the Tibia Test*

The 3 to 8 frozen pituitaries from the rats in each treatment group were homogenized by hand in a total volume of 0.5 ml 0.9% saline, centrifuged at 650 g for 45 min at 4 °C. The supernatant was diluted with saline or $NaHCO_3$ in 0.9% saline (pH 8) to the concentration required for injection into rats. Aliquots of the extracts were frozen for later radioimmunoassay of growth hormone.

4. *Tibia Test for Growth Hormone*

Female rats, hypophysectomized at 28 days of age, were obtained from Hormone Assay or the Charles River Breeding Laboratories. They were maintained at 78 °F (40–50% humidity) and received pablum, meat in some experiments, or just 5% glucose water. Approximately one week after hypophysectomy, daily subcutaneous injections of L-thyroxine, 0.5 μg per rat, were begun. The following week, four daily injections of pituitary extract (0.5 ml/day/rat) or bovine growth hormone, NIH-GH-B7 or B10 reference standards, were given intraperitoneally. In each experiment the extract of control pituitaries from donor rats which had received saline or control diluents was assayed at two dose levels at dilutions having ratios of 3 : 1 or 6 : 1. Pituitary extracts from rats injected with hypothalamic or cerebral cortex preparations were assayed at a dose equivalent in weight to the higher dose of the control, since we were expecting a fall of growth hormone content, or bracketed between the two doses of the control pituitary extract. To facilitate analysis of data in the tibia test, each treatment group comprised four to eight rats; a group of hypohysectomized rats receiving saline only was always included in each experiment. The weights of all hypophysectomized animals were checked prior to and during the period of injection of samples. Any animal losing more than 2 g of body weight during the period of injection was eliminated; this accounts for the unequal number of replicates per treatment group.

At the end of the experiment, widths of the epiphyseal cartilages were examined as described by Greenspan and associates (1949); no less than ten readings were taken per bone. The arithmetic mean of these measurements was considered the measure of the width of a particular cartilage.

5. *Statistical Analysis of Data*

The following discussion pertains to the analysis of the data obtained in the "tibia test" used to assess growth hormone activity of the various pituitary extracts. All the bioassays and experiments reported were planned as for a random block design with six or eight replicates per point. Most experiments were eventually calculated in a totally randomized design, as unequal numbers of animals per group were the rule at the end of the experiment. All experiments were studied by analysis of variance, using as the variable the width of tibial cartilage expressed in the original reading units of the calibrated eyepiece or transformed into microns (μ). Following the analysis of variance, all experimental means were compared by the multiple comparison test of Dunnett (1955) to the mean obtained for the control group of hypophysectomized rats receiving saline alone; the modification of Kramer (1956) for unequal groups was utilized. This comparison indicated whether biological activity was demonstrable in the pituitary extracts. Once this was established, the experimental means were further compared one against another, using the multiple range test of Duncan (1955). This comparison expresses statistical validity of any difference between the means of the experimental groups concerned, i.e., whether the responses obtained in the bioassay allow valid conclusions concerning differences in the growth hormone content of the extracts. The multiple range test of Duncan was followed by four-point or three-point bioassays calculated by factorial analysis, usually as multiple assays (see Finney, 1952). All these calculations were performed using the program Exbiol (Sakiz, 1964) on an IBM 7094 computer.

6. *Radioimmunoassays of Plasma or Pituitary Growth Hormone*

Schalch and his associates performed these assays using techniques for rat growth hormone previously described by them (Schalch and Reichlin, 1966) on aliquots of plasma and pituitary extracts sampled and stored as described above.

B. Results

Three groups of normal male rats weighing 150 g (Cheek-Jones Farms) were decapitated 30 min after rapid ether anesthesia. Their pituitary glands were removed, extracted, and assayed for growth

TABLE II

MULTIPLE FOUR-POINT ASSAY FOR GROWTH HORMONE (GH) USING THE TIBIA CARTILAGE TEST ON THREE DIFFERENT EXTRACTS OF THE PITUITARY GLANDS OF NORMAL RATS COLLECTED AFTER ETHER ANESTHESIA

Donors of Pituitary		Tibia Test				
Group	*No.*	Dose (a)	*No.*	Mean Tibia Cartilage Width ($\mu \pm$SE)	*M* with 95% CL	μg GH/mg with 95% CL
1	6	1 mg	4	177±5	1.186	35.6
		3 mg	4	226±3	(0.9–1.5)	(28.1–45.3)
2	5	1 mg	4	180±5	1.175	35.2
		3 mg	4	223±8	(0.9–1.5)	(27.9–44.9)
3	5	1 mg	4	160±3	0.913	27.4
		3 mg	4	219±7	(0.7–1.2)	(21.3–34.6)
Saline		2 ml	4	130±2	Index of precision for the	
NIH-GH-B7		30 μg	4	169±3	multiple 4-point assay:	
NIH-GH-B7		90 μg	4	219±4	$\lambda = 0.143$	

No = number of animals in each group; (a) total dose of extract in pituitary equivalent or of growth hormone reference injected over 14 days; *M* = potency ratios in the multiple 4-point assay between each pituitary extract and doses of the reference preparation NIH-GH-B7; CL = confidence limits of the potency ratio.

hormone content by the tibia test. The three pituitary extracts from the three different groups were assayed in one single experiment using three doses of NIH-GH-B7 as a reference standard. The results were calculated as described and the final analysis was that of a multiple four-point bioassay using the lower doses of the reference preparation NIH-GH-B7. The results presented in Table II show that the method for measuring growth hormone activity in standard preparations or pituitary extracts worked well. The three samples of normal pituitary gland extracts gave closely related activity, yielding statistically identical pituitary growth hormone contents or concentrations. The overall index of precision of the multiple assay was $\lambda = 0.143$. These data gave us confidence that we would be able to observe variations of pituitary growth hormone content of experimental pituitary glands should such changes occur under various treatments.

Figure 1 shows the responses obtained when 11 aliquots from the effluent of a gel filtration of 10,000 fragments of sheep hypothalamus were injected in a single experiment. The narrow black line referring to the ordinate on the left shows the optical density in the Goa reaction. The shaded area refers to the right-hand ordinate and depicts the range of tibia cartilage width where no difference from the control response exists by the multiple range test. In order to infer a depleting effect, a decrease from the response of the control pituitaries of over 18 μ is necessary. This is seen in responses to 150 to 200 μg fractions from the first eluted peak. These are equivalent to depletions of 45 to 60% of control potency. The increase of potency prior to the second peak possibly corresponds to activity in ovine hypothalamus described by Krulich and his associates (1968), which inhibits the release *in vitro* of pituitary substances active in the tibia test.

Although evidence of a GHRF has been associated with fractions from the first eluted peak in four of the seven experiments manifesting reduction of potency, one such fraction subjected to repeated

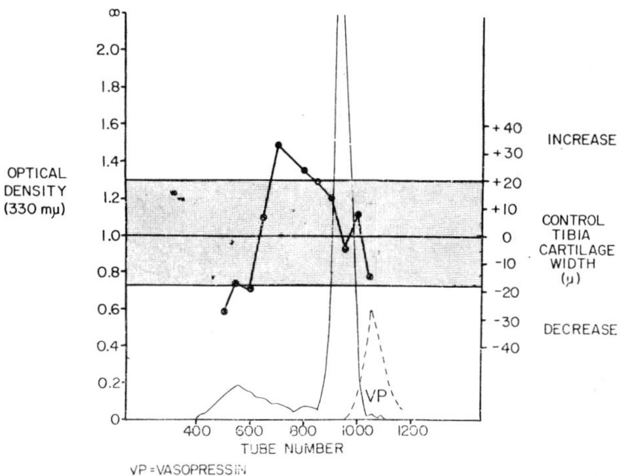

FIG. 1. Bioassay for GHRF of the effluent of a filtration on Sephadex G-25 (column size: 15 × 150 cm; flow rate: 600 ml/hr; $V_0 = 8000$ ml) of a 2 *N* acetic acid extract of ovine hypothalamus (20,000 hypothalamic fragments); volume of each tube = 20 ml. The shaded area represents the 95% confidence limits of the calculated potency (μg growth hormone/mg pituitary) of the control pituitary glands; i.e., any experimental point within this area corresponds to a pituitary growth hormone content or concentration not statistically different from that of the control pituitary glands. Absolute values (in the tibia test) for the two doses of control pituitary extracts were: for 0.66 mg, 215 μ ± 5; for 2.0 mg, 250 μ ± 5; all experimental pituitary extracts were assayed at 2.0 mg/dose. In the same experiment, responses to NIH-GH-B10 were as follows: 30 μg, 215 μ ± 4; 90 μg, 267 μ ± 4.

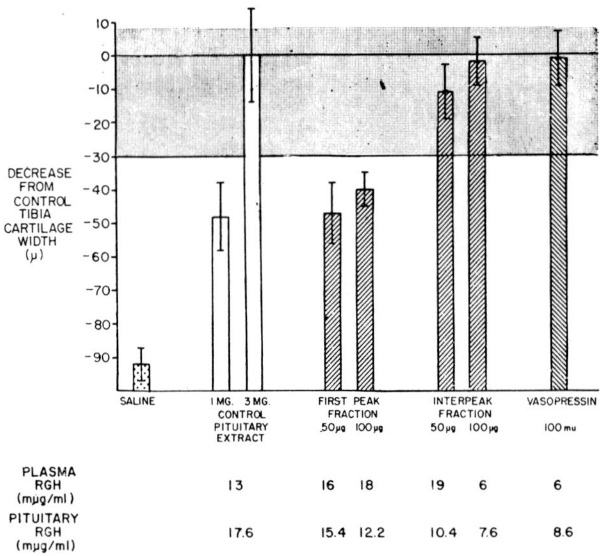

FIG. 2. Bioassay for GHRF of the effluent of a filtration on Sephadex G-25 as described in Fig. 1. Absolute values (in the tibia test) for the two doses of control pituitary extracts were: for 0.66 mg, 220 μ ± 4; for 2.0 mg, 243 μ ± 4; all experimental pituitary extracts were assayed at 2.0 mg/dose. Responses to NIH-GH-B10 in this experiment were as follows: 30 μg, 192 μ ± 8; 90 μg, 217 μ ± 6; 180 μg, 236 μ ± 4.

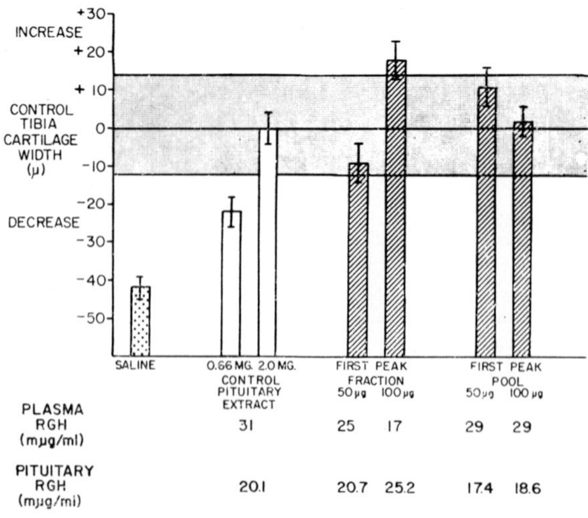

FIG. 3. Bioassay of lyophilized preparations of non-retarded peak. See legend for Fig. 2.

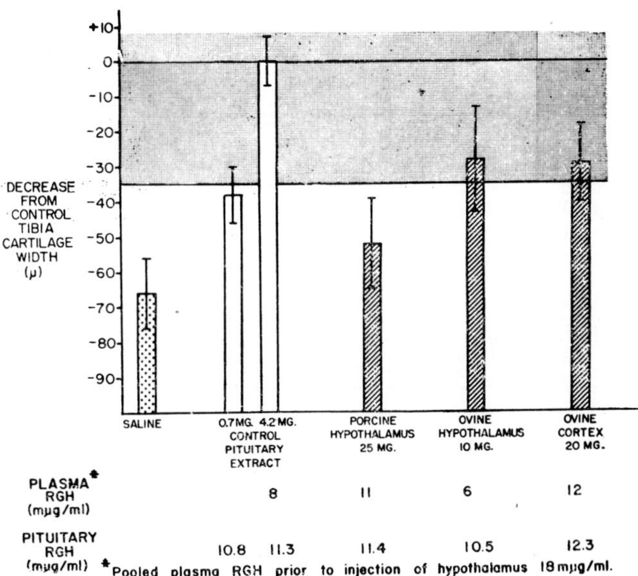

FIG. 4. Bioassay of alkaline extracts of hypothalamus. See legend for Fig. 2.

bioassays at doses of 50 and 100 μg per rat has shown notable variability of response. On some occasions (Fig. 2) a depleting effect was demonstrated in which, with reference to the control extract, decreases of 24 and 30 microns occur. The left-hand bar shows the response to saline alone in tibia test rats. On other occasions (Fig. 3) the same preparation induced an increase of potency at 100 μg, but was inactive at 50 μg. In this latter experiment the assay system was capable of detecting a depletion of 47%, exactly comparable to the one previously described. Repeated bioassays of the retarded fraction have shown equally variable effects.

No consistency of response has been achieved with the addition of albumin to the hypothalamic material, the administration of glucose to the rats prior to intracarotid injection, or the intracarotid infusion of hypothalamic extracts. In some studies, seen in Fig. 4, of the response to an alkaline extract of porcine hypothalamus infused for 20 min into the carotid artery of rats under pentobarbital anesthesia, a depletion was suggested whose 95% confidence limits were from 20 to 100%. Extracts of ovine hypothalamus and cortex were inactive.

TABLE III

SUMMARY OF EFFECTS IN THE BIOASSAY OF THREE SINGLE CHROMATOGRAMS SHOWING 39 TREATMENT GROUPS IN VALID BIOASSAYS [1]

	Non-retarded peak (Tubes 400–650)	Interpeak Zone (Tubes 651–800)	Retarded Peak (Tubes 801–1200)
" Augmentation " of pituitary growth hormone content, $p < 0.05$	1	3	0
No effect	12	6	8
" Depletion " of pituitary growth hormone content, $p < 0.05$	6	1	2

[1] A valid bioassay is defined as one in which the responses (tibia test) to the two doses of the extract of the control pituitary gland are statistically different by the t test of Duncan (1955), thus allowing calculation of a bioassay with $p < 0.05$ for the regression in the factorial analysis.

The summary of the responses obtained in nine bioassays of fractions from the Sephadex G-25 chromatography of ovine hypothalamus is seen in Table III. In all these experiments, responses to the two doses of control pituitary extract in the tibia test were different ($p < 0.05$) by the multiple range test; i.e., a slope of response in the tibia test as the function of the amount of pituitary extract injected was present. These are therefore referred to as valid experiments. The mean detectable decrease in response ($p < 0.05$) from the control values in these experiments was 21 μ, corresponding to a mean depletion of potency of 57%. Of 39 treatment groups injected with hypothalamic fractions from single columns, 9 showed the " depletion " type of response expected from the action of a GHRF, with no consistent localization of this activity in the effluent of the chromatographic separations. Similarly, the " augmentation " effect observed in four treatment groups was not clearly confined to the zone between the optical density peaks of the chromatogram. Twenty-six treatment groups showed no activity (depletion or augmentation), irrespective of their position in the effluent of the chromatogram. In these experiments, an additional six pituitary extracts from donor rats injected with 0.05 to 4 mg of a preparation corresponding to the pooled lyophilized first peak fractions of eight filtrations on Sephadex as above (corresponding to a total of 40,000 sheep hypothalamic fragments, the equivalent of 10 hypothalamic fragments being injected per rat) showed no statistical difference from the control pituitary growth hormone content when assayed simultaneously in the tibia test.

In 12 other experiments considered in retrospect to be invalid (no statistically valid difference between the two doses of extract of pituitary control was observed in the tibia test), 38 treatment groups of rats received one or another of the ovine hypothalamic fractions described above. Significant (multiple comparison test of Duncan, 1955) decreases from the control response (" depletion ") were seen in 4 groups, a significant increase (" augmentation ") in 1, and the remaining 33 groups showed no statistically valid response.

To permit more specific interpretation of these responses, Schalch performed radioimmunoassays of growth hormone in plasma and pituitary extract from rats that received ovine and porcine hypothalamus. Figure 5 shows the radioimmunoassay results of the chromatogram shown previously in which fractions from the non-retarded peak elicited in the bioassay indicated depletions of potency of 45 to 60%, which were consistent with the effects of GHRF. The top-line, referring to the ordinate on the right, shows the absence of the

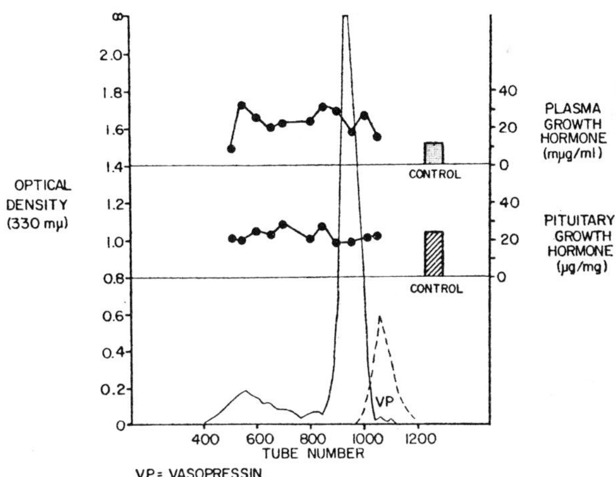

FIG. 5. Plasma and pituitary growth hormone levels by radioimmunoassay performed by Schalch and Reichlin (1966) in rats injected with hypothalamic extracts; comparison with bioassay data seen in Fig. 1. Radioimmunoassays were carried out in duplicate on plasma or pituitary pools. The standard error of individual determinations was ± 10%.

increments of plasma growth hormone that one would expect following the release of such large amounts of pituitary growth hormone. Below, the radioimmunoassay of the pituitary extract from the same rats shows no indication of the decreases or the increases of potency compared with controls that were suggested by responses in the tibia test. Correlation with 11 of our bioassays, including 8 that showed statistically valid positive effects, provides no convincing evidence of growth hormone release.

C. Discussion

The series of bioassays reported here were designed to demonstrate depletion of growth hormone in anterior pituitary tissues as evidence for the action of a GHRF. Strict statistical criteria established different responses to the two dilutions of the control pituitary extracts in less than half of the experiments. In these " valid " bioassays, it was not possible to detect consistent evidence of growth hormone release following the intracarotid injection or the intrajugular injection of acid or alkaline extracts of ovine hypothalamus, or the pooled first peak fraction from the series of gel filtrations performed as the first step in the purification of the TSHRF. In bioassays of individual gel filtrations, inconsistent results have been obtained. The incidence of the " depletion " and the " augmentation " responses relative to the three main fractions of the chromatogram might suggest the presence of GHRF in the first eluted peak and of a factor " augmenting " activity in the tibia test in the adjacent zone. However, the inconsistent pattern of biological activity throughout the effluent and the lack of a dose relation in the responses to these fractions do not permit the conclusion that these responses bear any significance relating them to specific biological activities.

In assessing the validity of these bioassay data, one must consider that decreases from the control response, to attain significance, represent a depletion of growth hormone content of at least 45%, and in 11 valid bioassays, a mean of 62%. If, as it has been suggested (Krulich *et al.*, 1968), a separate factor is capable of augmenting the growth hormone content of the pituitary gland, its presence might be expected to inhibit the response to any GHRF. This was the conclusion proposed by Krulich and associates (1968) following the concurrent injection of a " GHRF " and another factor of ovine origin which was reportedly capable of inhibiting *in vitro* the release of substances active in the tibia test. Although this factor displayed elution characteristics similar to those of the " augmentation " activity reported here, it was not itself active *in vivo* to " augment " pituitary growth hormone content, according to Krulich and associates (1968). If one were to accept that the infrequent responses observed in the assays reported here were due to more than random error of an imprecise technique, the relationship to growth hormone secretion of the differences in tibia test responses would remain to be considered. Conclusions from the bioassay data have been based upon decreases from the control responses of 20 to 40 μ. Li (1953) has indicated that the interplay of pituitary hormones other than somatotropin might account for the degree of difference. It is unlikely that different degrees of stress in various treatment groups caused uncontrolled variation in the content of growth hormone in the pituitary extracts, as bioassay revealed no difference between rats decapitated with and without prior ether anesthesia. Variability in the response could be due to factors in the hypothalamic preparation suspected to be capable of modifying the pituitary secretion of prolactin, Luteinizing Hormone (LH), ACTH and Thyrotropin (TSH). To minimize possible variations due to the effects of TSHRF on the TSH content of pituitary of the donor rats, we administered a minimal dose of thyroxine to the tibia test rats in all experiments. Despite this, one of the nine fractions containing TSHRF activity gave a response in the tibia assay which could be interpreted as the effect of a GHRF.

To demonstrate directly the specificity of response in these bioassays, the growth hormone content of pituitary extracts and plasma from 11 of our experiments was radioimmunoassayed. These studies showed no changes consistent with any acute effect of ovine or porcine hypothalamic preparations on the secretion of growth hormone. Neither the "depletion" nor the "augmentation" responses observed in the bioassay correlated with any change in the radioimmunoassay, even though increments in plasma growth hormone level should have been detected easily had the " depletion " responses corresponded to a true acute release of growth hormone.

These results suggest a need for caution in evaluating conclusions based on the methodology used in this series of experiments. Possible explanations of our results include: (1) ineptitude of our laboratories (although I believe this to be unlikely); (2) species specificity of the releasing factor comparable to that which exists with certain pituitary hormones; and (3) failure of radioimmunoassay of rat growth hormone to measure biologically active growth hormone. Schalch and Reichlin (1966) feel that they have verified the validity of the immunoassay.

We wish to emphasize that we do not consider the results reported here to be evidence against a hypothalamic participation in the control of growth hormone secretion. Earlier reference has been made to the impressive physiological (Reichlin, 1966) evidence in favor of such a mechanism, to which must be added the recent demonstration by Frohman and Bernardis (1968) and Frohman *et al.*, (1968) of an increase in plasma growth hormone levels as measured by radioimmunoassay in rats subjected to electrical stimulation of the ventral hypothalamus. In addition to these data there are several recent reports of the injection of crude extracts of ovine, porcine, or bovine hypothalamus in monkeys (Garcia and Geschwind, 1966; Smith *et al.*, 1968; Knobil *et al.*, 1968) or sheep (Machlin *et al.*, 1967) producing an elevation of plasma growth hormone levels as measured by several types of immunoassays. This effect can reliably be attributed to substances other than vasopressin in these crude extracts (Knobil *et al.*, 1968).

Our intention in this paper is to stress the need for caution in interpreting the results obtained with the method of bioassay which we have described here and which is currently used in many laboratories. Although we employed strict mathematical criteria, it was not possible to draw valid conclusions as to the possible significance of the inconsistent results that the bioassay apparently generated. In this connection, Knobil and associates (1968) have reported that whereas crude hypothalamic extracts do elevate levels of plasma growth hormone in monkeys given pharmacological pretreatment to eliminate non specific effects, highly purified GHRF (prepared by Schally *et al.*) from these same crude extracts, using the rat bioassay of Pecile and associates (1965) to assess the purification, have proved consistently inactive in stimulating secretion of growth hormone in monkeys when measured by the same immunologic methods. Sawano an co-workers (1968) have recently reported that GHRF purified on the basis of the bioassay of Pecile and associates (1965) not only depletes the pituitary growth hormone content but also increases plasma growth hormone levels as measured by bioassay; this would be proof that the pituitary depletion observed corresponds to true release of the hormone. It must be pointed out that the levels of plasma growth hormone reported in this study are hardly comparable to those measured in many laboratories using radioimmunoassays for growth hormone, thus raising the question of the biological significance of one of these two methods. The report by White and associates (1968) that several polyamines found in hypothalamic extracts and extracts of many other tissues produce profound " pituitary depletion " of Follicle-Stimulating Hormone (FSH) in the rat, with no evidence of secretion of FSH, raises further questions as to the significance of the " pituitary depletion " assays as evidence for the secretion of the pituitary growth hormone.

IV. Studies of Growth Hormone Release from Porcine Anterior Pituitary Cell Suspensions and Anterior Pituitary Slices

Our previous experience with the bioassay for GHRF based on rat pituitary growth hormone content led to a search for other models which might be used to study the biosynthesis, storage, and release of growth hormone. Two avenues of approach have been used. The first was an attempt to prepare pituitary cell suspensions and from these to isolate relatively pure populations of specific pituitary cells by a technique which we have used successfully to isolate relatively homogeneous populations of lung parenchymal cells. This approach had only limited success and extensive studies were carried out to pinpoint why this technique seemed successful under some circumstances and failed under others, but no solution was found. The work with mixed pituitary cell suspensions, however, deserves further discussion. The second approach was the measurement of growth hormone release from anterior pituitary slices. These systems will be dealt with sequentially, since much overlap exists between them.

A. Growth Hormone Release from Porcine Anterior Pituitary Cell Suspensions

1. *Materials and Methods*

a. Preparation of cell suspensions. Pituitary glands from 7- to 9-month old hogs were obtained within 20 min of death caused by carbon dioxide narcotization and exsanguination. The whole glands were transported from the abattoir in warm saline (between 25° to 37 °C) in an insulated container, and upon arrival in the laboratory the posterior pituitaries and stalks were separated by blunt dissection from the anterior pituitary. A modification of the collagenase method of Rodbell (1964) was used for the preparation of the cell suspension. The anterior pituitaries were minced with scissors, rinsed with saline to remove excess blood, weighed, and then placed in a siliconized Erlenmeyer flask with Krebs-Ringer Bicarbonate (KRB) buffer, pH 7.4, containing half the usual concentration of calcium, a mixture of the essential amino acids, 1% bovine serum albumin (BSA), glucose at a final concentration of 200 mg%, and 2% collagenase. Alternatively, tissue culture medium 199 containing 1% BSA and 2% collagenase was used. The flask was agitated at 180 cycles/min in an Eberbach metabolic shaker at 37 °C for 60 min and gassed with 95% oxygen and 5% CO_2. Liberation of pituitary cells was manifested by increasing turbidity in the medium, and the remaining tissue fragments were removed by filtering the suspensions through a nylon mesh of 170μ sieve size. The cell suspension was centrifuged at 40 g for 2 min in siliconized tubes, the supernatant was discarded, and the cells were rinsed with the medium containing a small amount of deoxyribonuclease to diminish cell adhesiveness. This procedure was repeated once more and then the cells were resuspended in incubation medium up to the desired volume.

A wet preparation of the cells was examined under the light microscope for intactness of cells, completeness of separation, and the degree of cell agglutination immediately after preparation and occasionally after incubation. A few preparations were also examined after staining with PAS-Orange G to study the nature of the cell population. The time lapse from death of the animal until the cells were ready for incubation was usually about 150 min. All preparatory procedures were carried out at temperatures between 25° and 37 °C.

b. Incubation techniques. In the metabolic studies, 0.5 ml of the cell suspension was placed in 4.5 ml KRB buffer as described above, but with the full complement of calcium and without albumin, and incubated at 37 °C for up to 6 hr, the gas phase being 95% oxygen and 5% CO_2. For the study of glucose oxidation, glucose-1-^{14}C was added to the medium, and for the incorporation of amino acids into proteins, leucine ^{14}C-UL was added.

In later growth hormone release studies, 0.5 ml of the cell suspension was placed in 4.5 ml of KRB buffer containing 1% BSA, or in medium 199 with 1% BSA. The medium was changed every half-hour by centrifuging the tubes at 100 g for 1 min at room temperature, pipetting off the medium, and adding fresh medium that had been warmed to 37 °C. All test materials were dissolved in the incubation medium, pH adjusted, and added to the cells at the appropriate times. The duration of incubation for these studies was usually 2 hr.

c. Analytical procedures.

(1) *$^{14}CO_2$.* Radioactive CO_2 evolved during incubation was trapped in hydroxide of hyamine on filter paper in center well flasks. The filter papers were immersed in toluene-POPOP and counted in a Packard Tricarb liquid scintillation counter.

(2) *Protein.* Cell suspensions after incubation were homogenized in 2 ml of water and an aliquot of this homogenate was used to precipitate protein by the TCA method. The TCA precipitate was dissolved in formic acid, one half was dried and counted in the liquid scintillation counter for determination of ^{14}C labeled protein; the other half was subjected to quantitative measurement of protein by the method of Lowry and associates (1951).

(3) *DNA.* A second aliquot of the cell homogenate was used to determine DNA content by the method of Ceriotti (1952).

(4) *Radioimmunoassay for porcine growth hormone.* In the studies on Porcine Growth Hormone (PGH) release, 0.1 ml of the medium after each medium change and 0.1 ml of the cell homogenate after incubation were taken for immunoassay by the charcoal-dextran separation method of Herbert and co-workers (1965). The usual dilutions required for the samples to fall on the sensitive part of the standard curve were in the order of 1/100 to 1/200 for the media and 1/2.000 to 1/4.000 for the cells after incubation.

d. Preparation of hypothalamic extracts. Crude acid extracts of porcine hypothalami and cerebral cortex were prepared according to the method of Deuben and Meites (1964). Within 30 min of the pigs being killed, median eminences or equivalent weights of cerebral cortex were excised and frozen immediately on dry ice. The hypothalami or cerebral cortices were homogenized in 0.1 *N*

HCl, 50 hypothalami usually requiring 8 to 10 ml of HCl, and centrifuged at 30.000g for 60 min. The supernatant was neutralized with 1 *N* NaOH and recentrifuged at 30.000g for 30 min after which the clear supernatant was decanted and aliquoted for deep-freeze storage until use. All procedures were carried out at 4 °C.

e. Materials. Collagenase was obtained from Worthington Bio-Chemicals. Glucose-1-^{14}C and leucine-^{14}C-UL were purchased from New England Nuclear. Theophylline and dibutyryl cyclic AMP-monosodium (DBcAMP) salt were supplied by Schwartz Bioresearch and the monopotassium salt by Nutritional Biochemicals. Porcine insulin was obtained from Connaught Laboratories, lysine-8-vasopressin from Sandoz (Canada), and L-thyroxine from Abbott Laboratories.

2. *Results*

The cell yield, determined by comparing the DNA content of cell suspensions with that of the tissue minced before collagenase treatment, was usually 5 to 10%. The viability of the cells was tested by incubating them in glucose-1-^{14}C for 2 hr and the evolved $^{14}CO_2$ was measured. Comparison of glucose oxidation in cell suspensions and slices revealed that the cells were indeed viable and more active

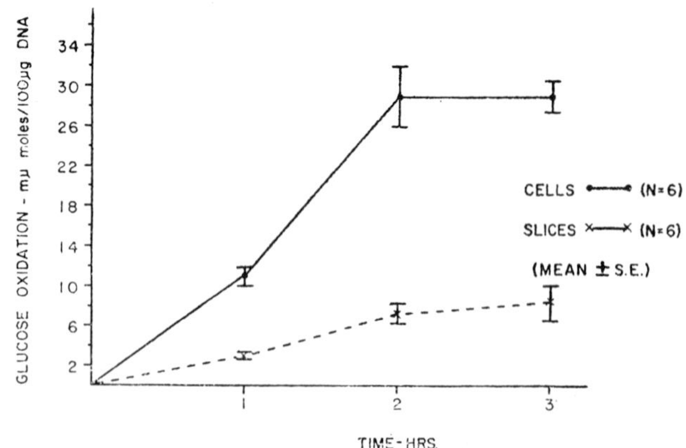

FIG. 6. The production of $^{14}CO_2$ from ^{14}C-labeled glucose in porcine pituitary slices and cell suspension. Details of techniques are seen in the text.

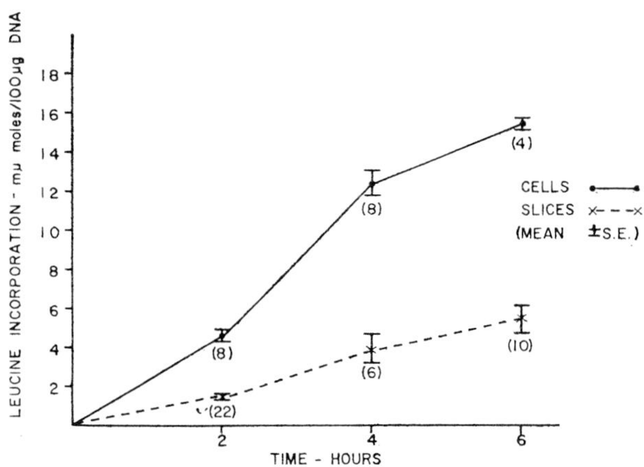

FIG. 7. The comparison of leucine incorporation by porcine pituitary slices and cell suspensions. Details are discussed in the text.

than slices when results were based on DNA content. The data obtained from six experiments are shown in Fig. 6.

The cell suspension also seemed more active than the slices when leucine incorporation into TCA-precipitable protein was studied. Figure 7 compares anterior pituitary slices with cell suspensions over a 6-hr incubation period.

In an attempt to study the possibility of stimulating leucine incorporation into protein, various substances were added to the incubation medium throughout the incubation period. The mean and standard error of four experiments are shown in Fig. 8. These results, subjected to an analysis of variance, failed to show significant stimulation by any of the substances added, although there was a trend suggesting stimulation by DBcAMP and insulin. Because measurement of the total tissue protein was felt to be an insensitive index, attempts were made to isolate growth hormone from the incubation medium by a variety of physicochemical techniques. The amount which could be isolated from the incubation volume was so small as to make this impractical. In view of this, attention was turned to the effect of these substances and of hypothalamic extracts on PGH release by the cell suspensions. The results for DBcAMP, theophylline, insulin, and hypothalamic extracts are shown in Figs. 9 and 10.

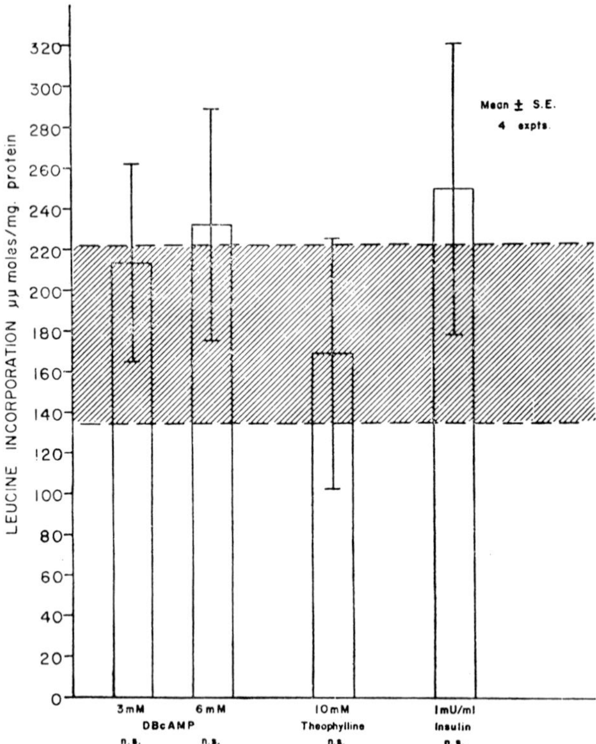

FIG. 8. The effect of DBcAMP (dibutyryl cyclic AMP), theophylline, and insulin on the incorporation of labeled leucine into TCA-precipitable protein by porcine pituitary cell suspensions.

The only significant increase in release occurred with the addition of cyclic AMP from the beginning of incubation, although this effect was short-lived. It may be that the cells are susceptible to stimulation only in this period or that the cyclic AMP is damaging susceptible cells and causing " damage release " of growth hormone. The latter is suggested by the finding that the PGH content of cells after incubation, expressed on the basis of DNA, remains the same for both control and test groups. On the other hand, however, theophylline in large concentrations (10 mM) did not have the same " damaging " effect. Addition of vasopressin (80 mU/ml) to the incubation medium in two experiments and L-thyroxine (1 mM) in one produced no significant change in PGH release when compared with controls.

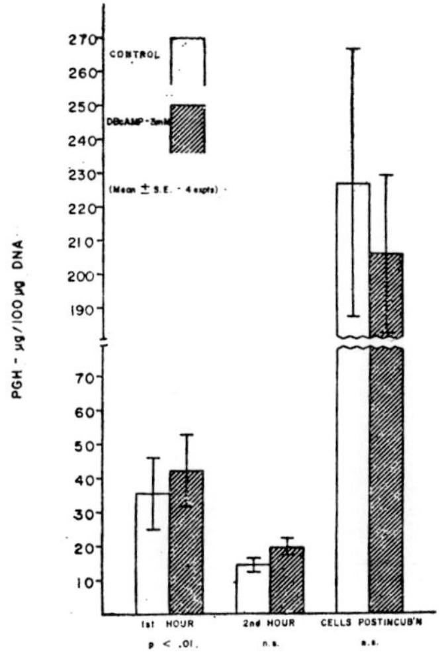

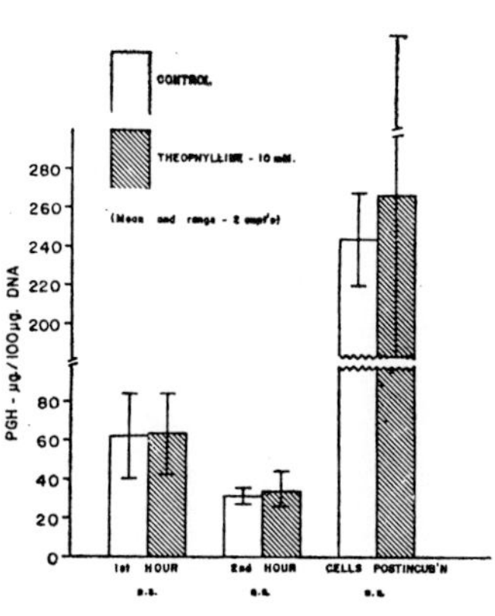

FIG. 9. The effect of DBcAMP and of theophylline on GH release from porcine pituitary cell suspensions. PGH release as measured by immunoassay is expressed in «g/100 μg DNA. The postincubation period represents the amount of PGH remaining in the cells at the end of a 2-hr incubation period. Validity of the data was determined by covariance analysis.

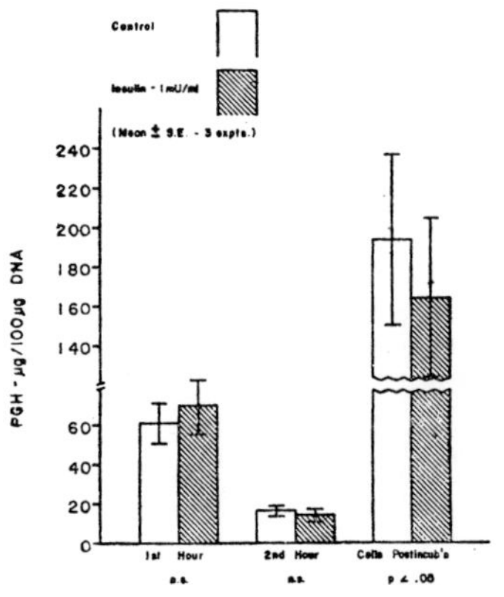

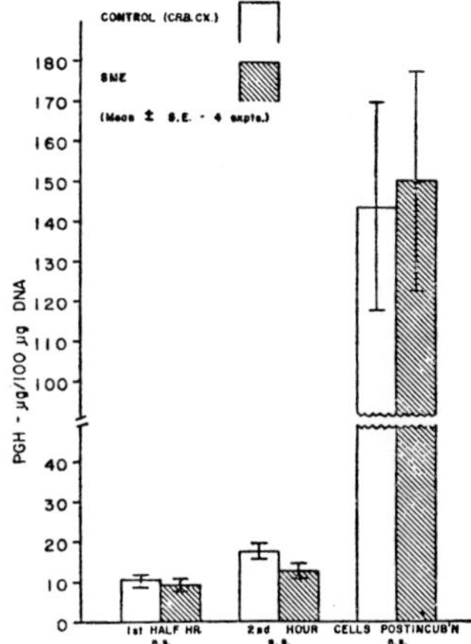

FIG. 10. The effect of insulin and of crude stalk median eminence (SME) extract on GH release from porcine pituitary cell suspensions. See legend for Fig. 9.

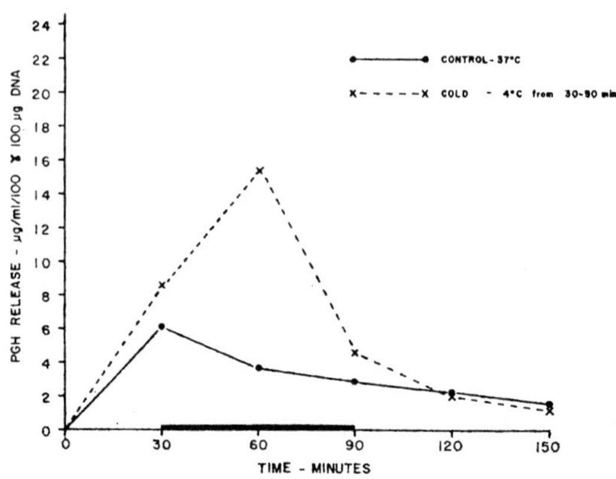

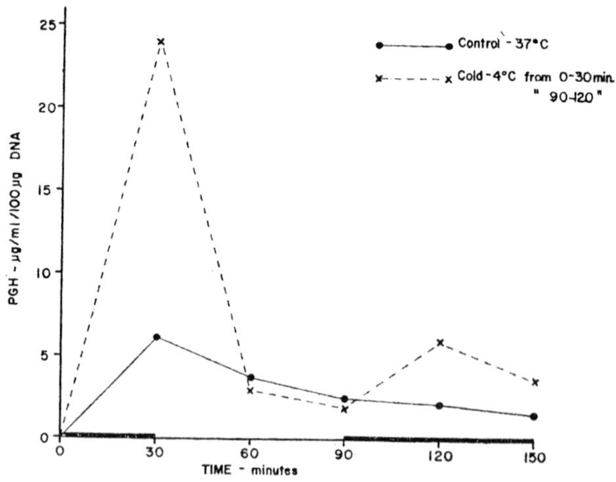

FIG. 11. Effect of temperature on GH release from porcine pituitary cell suspensions. PGH release as determined by immunoassay during the exposure of cells to incubation temperatures of 4 °C. The solid bars along the horizontal axis denote the time period during which incubation took place at 4 °C.

In order to test the hypothesis that these isolated cell suspensions were more susceptible to damage, the cells were subjected to cold (4 °C) during the incubation. In one experiment PGH release from cells incubated at 4 °C for 60 min was compared to controls at 37 °C; Figure 11 summarizes the results.

In a subsequent experiment, also depicted in Fig. 11, alternating temperatures at hourly intervals again precipitated a greater release of PGH in the first half-hour after exposure to cold, as compared with that from the controls. PGH release at 4 °C continued to be greater than that from controls during the second hour, but was substantially less than during the first 30 min. The PGH content of the cells after incubation at 4 °C was the same as that of the control cells, indicating a possible breakdown of the cells with outpouring of PGH content when subjected to sudden cold. This is in keeping with the observations of Schramm and associates (1965) who demonstrated leakage of amylase from zymogen granules in parotid glands exposed to 4 °C and postulated that shrinkage of the lipids in the granule membrane may cause this release.

3. *Discussion*

The viability of isolated cell suspensions has been demonstrated by their ability to oxidize glucose and incorporate leucine into protein. However, the data supporting the use of this model for the study of factors influencing the biosynthesis, storage, and release of growth hormone remain open to further examination. Only DBcAMP has given consistent stimulation of growth hormone release. The apparent stimulation of release with stalk median eminence extracts has been inconsistent for reasons which still elude us. It might be

TABLE IV

EFFECTS OF STALK MEDIAN EMINENCE (SME) ON ISOLATED PORCINE PITUITARY CELLS *in vitro*

Preparation	TSH Release [1] (assay response, %)		GH Release (μg/100 μg DNA)	Adenyl Cyclase (cAMP Produced, cpm)
Control	Dil'n 1 : 10	149	11.5	208
		172	10.7	340
SME	Dil'n 1 : 40	283	9.5	3321
		269	9.3	4471

[1] The details of the methods used for measuring TSH release and adenyl cyclase are found in the article by McKenzie in this volume.

argued on the basis of the data already presented that no evidence exists to prove the isolated cell suspension capable of releasing a pituitary hormone. In cooperation with McKenzie's laboratory, the effect of stalk median eminence on adenyl cyclase, TSH, and PGH release was studied. Results of this experiment are given in Table IV. It can be seen that significant stimulation of adenyl cyclase activity and TSH release was achieved with no change in PGH release.

The medium must be changed at least every half-hour during incubation in order that growth hormone release will continue, and this in itself raises basic issues concerning the handling of the cell suspensions. The fact that the cells require centrifugation every 30 min, with aspiration of the medium at each change, raises the possibility of: (1) further damage to cells during centrifugation, with subsequent " damage release " of growth hormone; and (2) resuspension of cells during aspiration, leading to falsely high " release " values.

B. Growth Hormone Release from Porcine Anterior Pituitary Slices

1. *Materials and Methods*

Only differences from methods outlined in the preceding section will be enumerated.

Pituitary glands were obtained as described previously and transported to the laboratory in cold saline. After dissection of the posterior lobe, the adenohypophysis was divided along the midline and sliced by a Stadie-Riggs microtome. The slices were washed in cold KRB buffer before transfer to 24-ml conical flasks containing 5 ml of the incubation medium, KRB buffer, 200 mg% glucose, 0.5% bovine serum albumin, and amino acid mixture in a final concentration of 1/4 of the concentration of plasma. Each incubation flask contained four to five slices, the total tissue weight being between 80 and 110 mg; slices from one half of the pituitary were used as a control for the other half.

In determing the amounts of immunologically assayable growth hormone present in the pituitaries either before or after incubation, a crude pituitary extract was prepared by homogenization of pituitary slices in cold 0.154 *M* NaCl containing 5 mM NaOH. In the assays of the homogenate, dilution from 1/2.000 to 1/10.000 was necessary so that the aliquots might fall in the sensitive part of the standard curve.

The rates of release of PGH have been calculated and expressed as micrograms PGH per 100 mg wet weight of tissue during the 30-min preincubation and during the first and second hours. The latter were the summation of successive 15 or 30 min periods. The concentration of tissue PGH is usually calculated as the total of the amount remaining in the slices after incubation and the summation of the amounts measured during the incubation time. The percentage of PGH release is calculated from these data.

2. *Results*

Preliminary experiments were done to study the incorporation of L-leucine-1-^{14}C into TCA-precipitable protein as an index of the viability of the tissue slices. The slices were incubated for 2, 4, 6, and 8 hr at 37 °C, and the results (Fig. 7) are expressed in terms of micromoles of L-leucine-1-^{14}C incorporated into TCA-precipitable protein per 100 ml of tissue wet weight. It can be seen

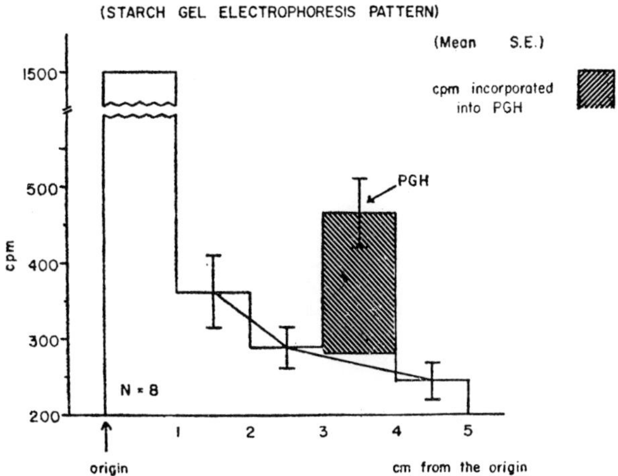

FIG. 12. Incorporation of labeled leucine into a PGH fraction as isolated by starch gel electrophoresis. Details of the methods employed are described in the text. The vertical axis is expressed as cycles per minute and the horizontal axis as centimeters from the origin in a starch gel electrophoretic pattern. The shaded area represents the PGH isolated and confirmed by running standards of "cold" PGH as well as ^{125}I labeled PGH.

that the incorporation of labeled leucine proceeded at a linear rate for at least 6 hr. Approximately 30% of the total isotopic leucine employed was utilized during this period. These experiments did not ensure that the labeled precursor was incorporated into the growth hormone moiety, and an attempt was made to extract the growth hormone and measure the incorporation of the uniformly labeled ^{14}C leucine into it. The pituitaries were extracted and the growth hormone fraction isolated by starch gel electrophoresis, using procedures previously described by Ferguson (1964). Figure 12 depicts

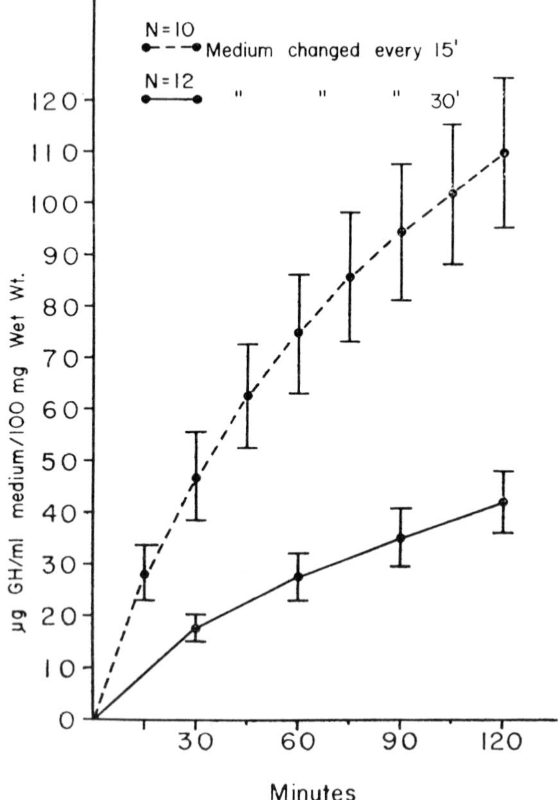

FIG. 13. Cumulative release of GH from porcine anterior pituitary slices. Release of PGH as measured by immunoassay when medium changes carried out every 30 min are compared with those carried out at 15-min intervals. The PGH release is expressed in μg/ml medium/100 mg wet weight of porcine pituitary slices.

a typical experiment carried out with eight porcine pituitaries. Although much of the radioactivity remains at the origin and is unidentified, the fraction made up predominantly of growth hormone contains significant amounts of radioactivity.

In studies of the release of immunologically assayable PGH into the medium, the slices were incubated for a period of 30 min to wash out the cells damaged during the preparation of slices, this being termed the preincubation period. The tissue was then transferred to fresh medium preheated to 37 °C, and the incubation was continued for a 2-hr period with a change of medium every 15 to 30 min. The cumulative release of PGH from slices in which the medium was changed every 15 min is compared in Fig. 13 with that occurring with replacement of the medium at 30 min intervals. It can be seen that the rate of release of PGH is not constant, the output

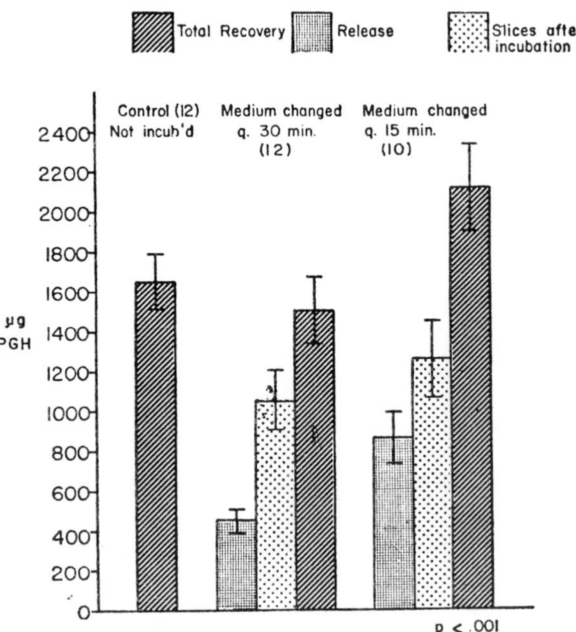

FIG. 14. GH recovery before and after incubation of porcine anterior pituitary slices. The effect of 15-*vs.* 30-min time intervals of medium change on the release of PGH and on the amount of PGH in the slices after incubation. Total recovery is calculated as described in the text. The vertical axis expresses the amount of PGH in micrograms.

during the second hour being significantly less than in the first hour. The PGH output is twice that observed by Schofield (1967) in bovine anterior pituitary slices. It is of interest that the difference between slices obtained from the halves of the same gland is small, although there is great variability in both growth hormone content and release from pituitaries from one animal to another.

As in the studies with the isolated cell suspensions and as reported by Schofield (1967), growth hormone release decreases considerably after 30 min of incubation without change of the medium. This phenomenon is even more clearly demonstrated in Fig. 13 where the release is significantly greater in a system where medium change is carried out every 15 min as compared with that in which the medium is changed at 30-min intervals. Figure 14 shows how the chang-

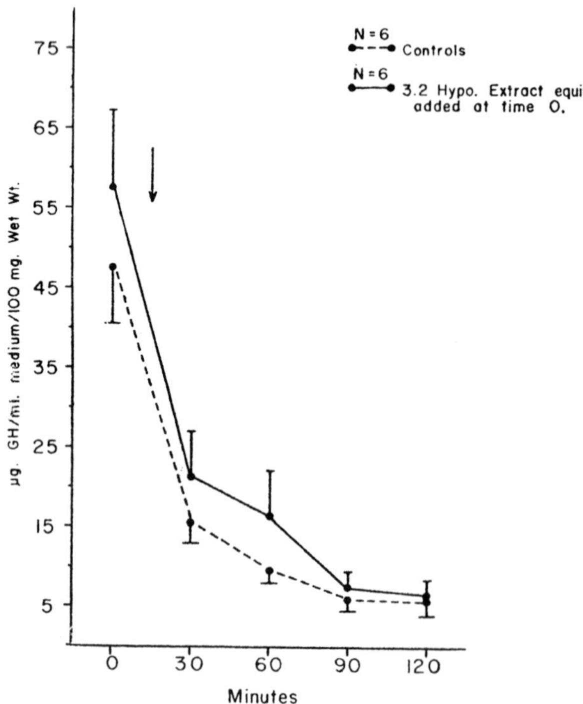

FIG. 15. The effect of hypothalamic extract on PGH release from porcine anterior pituitary slices. The extract equivalent to 3.2 hypothalami was added to all flasks during the total incubation period.

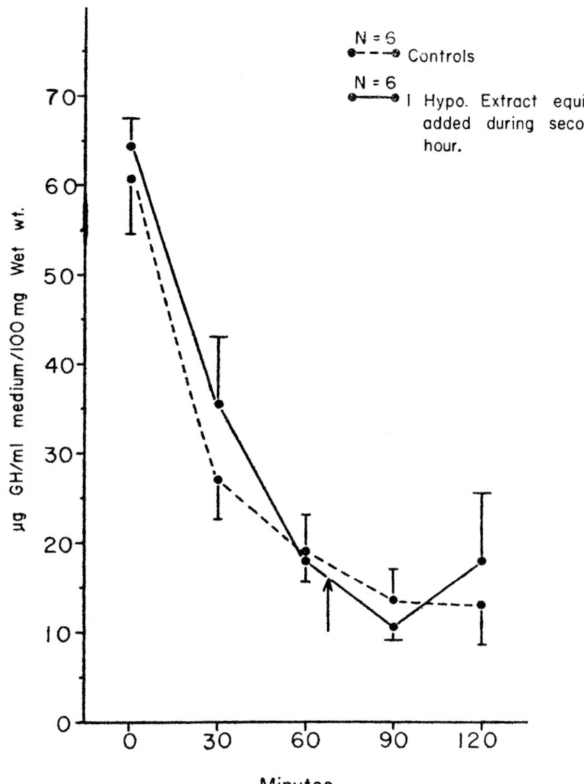

FIG. 16. The effect of hypothalamic extract on PGH release from porcine anterior pituitary slices. The extract equivalent to one hypothalamus was added during the second hour of incubation.

ing of medium at different time intervals influences the amount of growth hormone in the slices at the end of a 2-hr incubation, as well as the release (and by summation, the total recovery) of PGH. The explanation of these observations remains obscure at this time.

The effect of adding stalk median eminence extracts at varying time intervals (i.e., before preincubation, before the first hour, and before the second hour) and at different dose levels, was also studied. Representative data are seen in Figs. 15 and 16. In a large number of experiments the results were variable and only occasionally, as in Table V, was there a significantly different release over the subsequent

TABLE V

EFFECT OF HYPOTHALAMIC EXTRACT ON PORCINE GROWTH HORMONE (PGH) RELEASE [1]

	Total Recovery (μg PGH) [2]	Preincubation	First hour	Second hour
Control	1773	285	257	189
Hypothalamic Extract	2018	385 [3]	414 [3]	483 [3]

[1] Release = μg PGH/100 mg wet weight.

[2] Total recovery = cumulative release in the medium plus the quantity left in the slices after incubation.

[3] The equivalent of 1 hypothalamus was added to the incubation medium.

2½ hr. Figure 15 depicts the results in six experiments, where the extract equivalent to 3.2 hypothalami was added to all flasks during the precincubation period, the first hour and again during the second hour of the incubation. There was no significant difference between the control and treatment groups although cumulative release in the two series suggested a differing trend. The extract equivalent to one hypothalamus was added in six experiments during the second hour of incubation only (Fig. 16). It can be seen that there is no significant change in PGH release during this second hour.

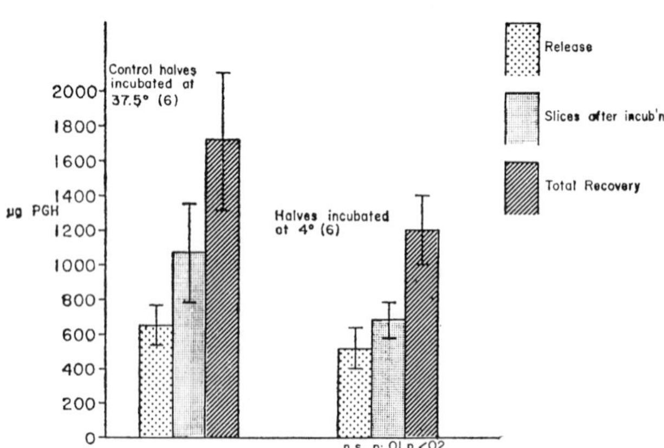

FIG. 17. Effect of temperature on GH release from porcine anterior pituitary slices. The effect of incubation of anterior pituitary slices at 4 °C on PGH release. The amount of PGH remaining in the slices after incubation and total recovery are depicted and expressed as micrograms of PGH.

The effects of incubation temperature on growth hormone were also studied and the results are depicted in Fig. 17. The treatment halves were incubated at either 4° or 37 °C and the PGH concentration in the tissue as well as release in the preincubation period, the first- and second-hour incubation periods was measured. It is seen that the concentration of PGH in the tissue is significantly higher in the slices incubated at 37 °C, suggesting that growth hormone synthesis is occurring at this temperature and that cold arrests it.

3. *Discussion*

In the course of these studies it was shown that no destruction of growth hormone could be detected either during the incubation or the subsequent assay. Our observations showing that 12% of the PGH in anterior pituitary tissue was released in the first hour and 8.1% in the subsequent hour demonstrate release rates in excess of those reported by Schofield (1967) in bovine anterior pituitary slices and correspond more closely to the values for TSH release in halved rat pituitaries described by Solomon and McKenzie (1966). These investigators found that this process stopped after 30 min of incubation *in vitro* and attributed this to inhibition by TSH. Our observation that the amount of growth hormone liberated into the medium and remaining in the pituitary slices is greater with more frequent transfers of medium is in agreement with Schofield's (1967) report. Experiments on the effect of added growth hormone to the two systems described in this presentation remain incomplete. The failure to alter the rate of PGH release in a consistent fashion with hypothalamic extracts remains an enigma and corresponds with results obtained in other laboratories. The response of PGH synthesis and release from the pituitary slices to a lowering of temperature suggests that the output may be dependent upon an enzymic process within the cell and not a result of simple diffusion.

Suwa and Friesen working independently in our department, using squirrel monkey pituitaries and hypothalamic extracts of both monkey and pig origin, have also obtained inconsistent responses. Their observation that DBcAMP is the only substance consistently increasing release corresponds with the observations on the effect of this nucleotide on the isolated cell system described here. As with the cell suspension this effect is a short-lived one.

V. Plasma Growth Hormone-like Activity

In parallel with the studies presented here we embarked on a program designed to isolate and characterize pituitary dependent growth-promoting substance(s) in plasma. We hoped through these investigations to resolve conflicting opinions concerning the identity of plasma Human Growth Hormone (HGH) and HGH as extracted from pituitaries. It was also hoped that insight might be gained into discrepancies between the bioassay and immunoassay data in relation to the search for GHRF, as well as into the inconsistencies which exist concerning the factors controlling growth hormone release *in vitro*.

Figure 18 summarizes some of the reasons for questioning whether plasma and pituitary growth hormones are identical. Plasma biologic and immunoreactive growth hormone-like activity (IRHGH) is increased in acromegaly and decreased in dwarfism, Sheehan's syndrome, and the posthypophysectomy state. In the latter, the levels of activity can be restored to normal by treatment with HGH extracted from pituitaries. These observations are strong arguments in favor of deciding that plasma and pituitary HGH are identical. However,

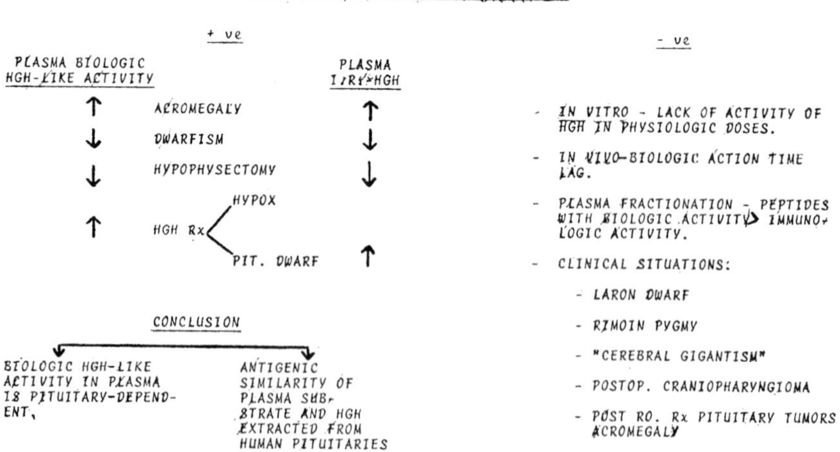

FIG. 18. This figure summarizes the positive and negative factors supporting or questioning the concept of whether circulating growth hormone in plasma is identical with the growth hormone extracted from human pituitaries.

the only firm conclusion that can be made is that the plasma substance with HGH-like activity is pituitary-dependent and has antigenic similarity to HGH extracted from the pituitary gland. The lack of *in vitro* activity by HGH in physiologic amounts and the time lag in certain biological activities when HGH is administered *in vivo* suggest that extracted pitutary HGH may not be the biologically active substance in plasma. The plasma fractions with discrepant biologic and IRHGH-like activities, in favor of the biologic activity, require further clarification. The clinical situations in which inverse relationships exist between plasma IRHGH and stature require further investigation of both plasma HGH-like substance(s) and the target organ response.

If plasma and pituitary HGH are not identical, several other possibilities may be considered. Pituitary HGH may be a large molecule attached to a small biologically active peptide, which may or may not be of pituitary origin. This larger protein may serve as a " transport-storage " system for the active peptide. HGH as extracted from the pituitary may be an inactive form of the plasma hormone which requires modification to an active form after being secreted. It may act by stimulating the production of other biologically active substances or may cause a modification of an intermediate substance to a biologically active one.

To date most investigations of plasma HGH-like substance(s) have involved small samples of plasma which may tax the sensitivity of current procedures. We have attempted to correlate plasma IRHGH and biologic HGH-like activity with more refined physical chemical characterization. The subsequent data are preliminary, and the distribution of plasma IRHGH-like activity from normal and acromegalic subjects is described.

A. Materials and Methods

Plasma samples varying in volume from 80 to 1000 ml from normal, acromegalic subjects and from patients with pituitary dwarfism were used. Some of these samples were fresh; others were aged, frozen, or lyophilized. Beef plasma was used as a control. The plasma samples were fractionated by gel filtration utilizing large columns (i.e., 10 $\times$ 100 cm or larger) with a minimum (Vi) or bed volume of 7000 cm^3 of Sephadex G-75 superfine. Elution was carried out in a reverse flow manner through closed-end columns with 0.15 *M* NH_4HCO_3 and 0.02% $NaNO_3$ (pH 8.2) at 4 °C. The sample volume applied was less than 5% of Vi or the imbibed gel volume. The eluent was monitored continuously through an ultraviolet (UV) absorptiometer at 280 mμ and automatically collected at intervals; the individual fractions were then concentrated by ultrafiltration through Diaflo polyionic membranes at 4 °C with retention of proteins with molecular weight greater than 1.000. The amount of protein was rechecked after concentration.

The IRHGH was determined by a modified solid phase (tube) method as originally described by Catt and Tregear (1967) and Bala and co-workers (1969). The assay was consistently sensitive to less than 0.05 ng HGH and very often to values of 0.025 ng HGH. Bovine plasma fractions, which were processed in a similar manner to the human plasma samples, were used as a control for nonspecific protein effects on the radioimmunoassay. To allow interexperiment comparison, the IRHGH activity was expressed as total activity and activity per milliliter of protein (specific activity) in each fraction, and then calculated per specific distribution coefficient or *Kd* interval. The activity per *Kd* interval was also expressed as a percent of the total activity in each experiment.

B. Results

Figure 19 is a graphic representation of the distribution of protein concentration and the IRHGH activity after fractionation of fresh normal plasma. The apparent secondary protein peak at *Kd* 1.0 reflects the delayed elution of aromatic amino acids from Sephadex, resulting in increasing UV absorption at 280 mμ. The height rather than the area of the hatched bars represents the total IRHGH in each fraction collected, and the height of the solid bars represents the IRHGH activity in nanograms per milligram of protein (specific activity) in that fraction. The distribution coefficient, or *Kd*, of the fractions is shown at the bottom of the slide. HGH has been shown to be eluted from Sephadex G-75 with a *Kd* of approximately 0.25 to 0.35. As may be seen, Sephadex G-75 does not clearly distinguish albumin from other macromolecular proteins. The IRHGH activity per milligram of protein (specific activity = solid bars) is greatest in the *Kd* area corresponding to HGH, but the total amount of

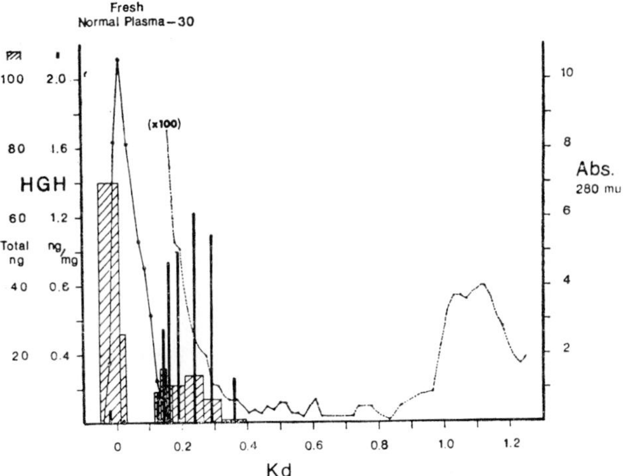

FIG. 19. Distribution of protein concentration and the immunoreactive HGH activity after fractionation of fresh, normal plasma. Detail are described in the text.

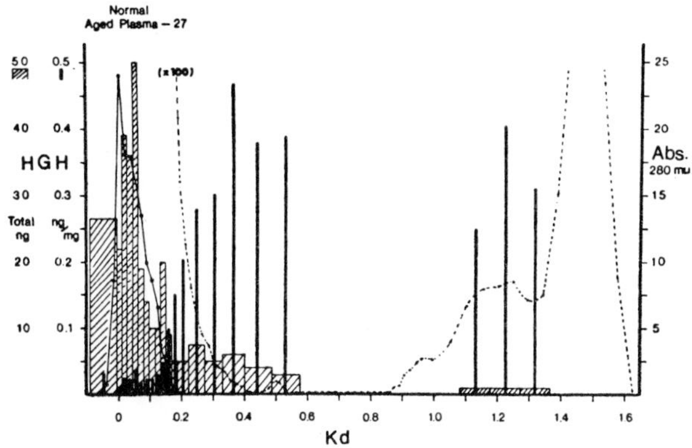

FIG. 20. Distribution of protein concentration and immunoreactive HGH activity after fractionation of an aged plasma sample. For further details see Fig. 19.

IRHGH is greatest in the larger molecular size protein area. As noted, many fractions do not have detectable levels of IRHGH.

Figure 20 is a graphic representation of fractionation of an aged plasma sample. A greater percentage of the total IRHGH as well as more fractions with detectable levels of HGH occur in the *Kd* area corresponding to molecular size greater than HGH.

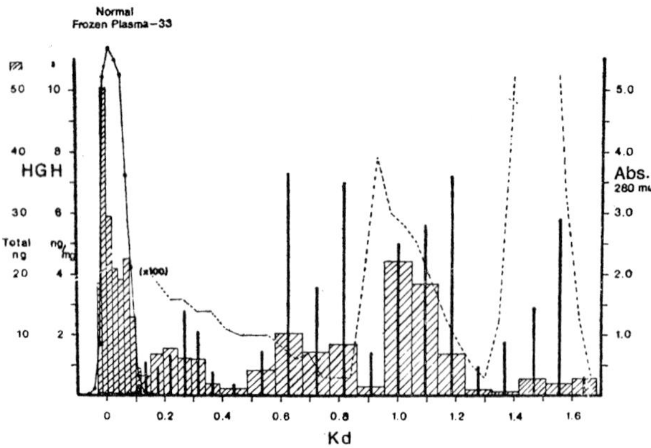

FIG. 21. Distribution of protein concentration and immunoreactive HGH activity after fractionation of an aged, frozen plasma sample from a normal individual.

Figure 21 is a similar representation of a fractionation of an aged frozen plasma sample. A significant amount of IRHGH is detected in the molecular size area smaller than pituitary HGH, as well as in the large protein area. This suggests that small molecules in the area of the aromatic amino acids having antigenic similarity to pituitary HGH have appeared after freezing of the plasma.

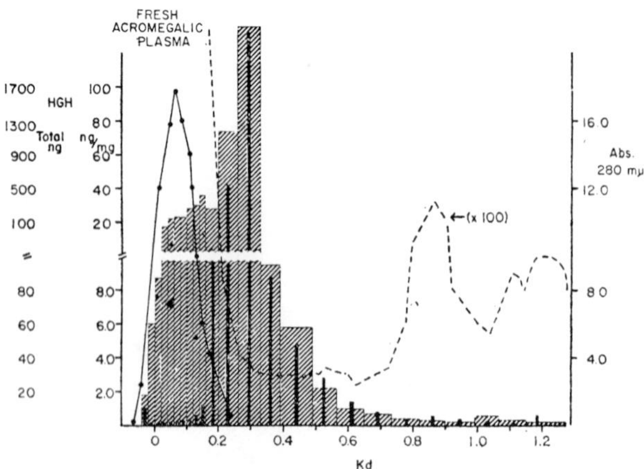

FIG. 22. Distribution of protein concentration and immunoreactive HGH activity after fractionation of a fresh plasma sample from an individual with active acromegaly.

Figure 22 is representative of fractionated fresh acromegalic plasma. The maximum total and specific IRHGH activity occurs in the *Kd* area, corresponding to pituitary HGH, but a greater total amount of low specific activity HGH occurs in the larger protein area.

Figure 23 shows the results of fractionation of a lyophilized acromegalic plasma sample. The results are similar to those from the fresh acromegalic plasma shown in Fig. 22, except that a greater part of the total IRHGH occurs in the larger protein area.

The recoveries in the fractionations ranged from 83 to 129% for protein and 60 to 142% when compared to the starting material or plasma totals.

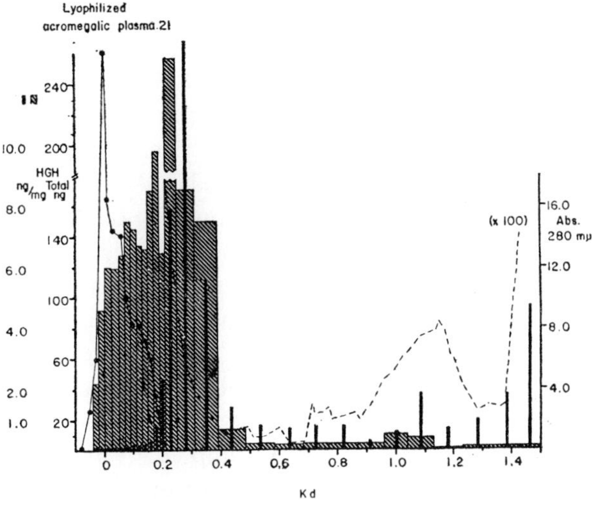

FIG. 23. Distribution of protein concentration and immunoreactive HGH activity after fractionation of a lyophilized plasma sample from a patient with active acromegaly.

C. DISCUSSION

Relatively large volumes of plasma have been fractionated and a comparison of fresh, aged, and frozen normal plasma as well as fresh and lyophilized acromegalic plasmas have been attempted. Consistency of gel column behavior and elimination of nonspecificity of the radioimmunoassay was achieved. The utilization of larger volumes of plasma with consequently increased concentrations has allowed measurement of IRHGH levels which would be undetected if small volumes had been used. The polyionic membrane ultrafiltration method should be less traumatic to proteins than is lyophilization. Adequate column size and use of superfine Sephadex allowed more effective separation of proteins of molecular size of interest in this study. Since the disruptive effect of electrophoresis on protein aggregates or associations is not fully appreciated, we have avoided this technique in primary separations.

In all plasmas, except one fresh acromegalic plasma, more than 50% of the total IRHGH recovered occurred in an area of molecular size greater than that expected for pituitary HGH. This suggests that either IRHGH proteins are aggregated with each other or with other proteins. Aging and freezing of normal plasma was associated with a decreased total and specific activity in the pituitary HGH molecular size proteins. The high total and specific IRHGH activity of acromegalic plasma fraction correspond mainly to a molecular size area of pituitary HGH and may imply an increased relative amount of "free," or "unbound," or "unassociated" IRHGH in acromegaly with increased IRHGH levels. However, a significant percentage of the total IRHGH was still found in the larger molecular size area as well as in the small protein size range.

In conclusion, we feel that HGH extracted from pituitaries and plasma IRHGH have not been conclusively proved to be identical. We believe we have shown that plasma IRHGH activity is present in association with larger proteins as well as existing in a molecular size much smaller than that of extracted pituitary HGH. Further clarification of this problem will be attempted by refractionation of the various molecular size proteins. A correlation of the biologic activity with the physicochemical characteristics of these protein fractions is in progress. This work was carried out by Ferguson and Bala, and was greatly accelerated by Dr. Ferguson's presence in our laboratories under the auspices of the Visiting Scientists Programme, of the MRC of Canada.

VI. Conclusions

These data represent either work already completed or work in progress in our laboratories. In our view GHRF exists and plays an important role in controlling the secretion of growth hormone. However, there are serious and unresolved problems in the isolation, characterization, and assay of this substance, and were these problems overcome the biological significance of GHRF would still remain to be established. This situation contrasts sharply with the progress made with regard to factors controlling the secretion of most of the other pituitary hormones.

Acknowledgments

The research reported here was supported by grants from the Medical Research Council of Canada (MT-631) and the United States Public Health Service (08390-031-04). R. M. Bald, C. Kudo, G. C. Olivier, and N. W. Rodger participated under Medical Research Council (Canada) Fellowship, and K. A. Ferguson as a Visiting Scientist under the auspices of the MRC (Canada).

References

Abrams, R. L., Parker, M. L., Blanco, S., Reichlin, S., and Daughaday, W. H. (1966). *Endocrinology* **78**, 605.

Bala, R. M., Ferguson, K. A., and Beck, J. C. (1969). *Can. J. Physiol.* **47**, 803.

Berg, G. R., Utiger, R. D., Schalch, D. S., and Reichlin, S. (1966). *J. Appl. Physiol.* **21**, 1791.

Best, J., Catt, K. J., and Burger, H. G. (1968). *Lancet* **2**, 124.

Blackard, W. G., and Heidingsfelder, S. A. (1968). *J. Clin. Invest.* **47**, 1407.

Blanco, S., Schalch, D. S., and Reichlin, S. (1966). *Fed. Proc.* **25**, 191.

Burday, S. Z., Find, P. H., and Schalch, D. S. (1968). *J. Lab. Clin. Med.* **71**, 897.

Burgess, J. A., Smith, B. R., and Merimee, T. J. (1965). *Clin. Res.* **13**, 531.

Burgess, J. A., Smith, B. R., and Merimee, T. J. (1966). *J. Clin. Endocrinol. Metab.* **26**, 1257.

Catt, K., and Tregear, G. W. (1967). *Science* **158**, 1570.

Ceriotti, G. (1952). *J. Biol. Chem.* **198**, 297.

Chakmakjian, Z. H., and Bethune, J. E. (1968). *J. Lab. Clin. Med.* **72**, 429.

Del Vecchio, A., Genovese, E., and Martini, L. (1958). *Proc. Soc. Exp. Biol. Med.*. **98**, 641.

Deuben, R. R., and Meites, J. (1964). *Endocrinology* **74**, 408.

Deuben, R. R., and Meites, J. (1965). *Proc. Soc. Exp. Biol. Med.* **118**, 409.

Dhariwal, A.P.S., Krulich, L., Katz, S. H., and McCann, S. M. (1965). *Endocrinology* **77**, 932.

Dhariwal, A. P. S., Antunes-Rodrigues, J., Krulich, L., and McCann, S. M. (1966). *Neuroendocrinology* **1**, 341.

Duncan, D. B. (1955). *Biometrics* **11**, 1.

Dunnett, C. W. (1955). *Am. Stat. Assoc. J.* **50**, 1096.

Ferguson, K. A. (1964). *Metabolism* **13**, 985.

Finney, D. J. (1952). "Statistical Methods in Biological Assay." Hafner Publishing Co., New York.

Frantz, A. G., and Rabkin, M. T. (1964). *New Engl. J. Med.* **271**, 1375.

Frantz, A. G., and Rabkin, M. T. (1965). *J. Clin. Endocrinol. Metab.* **25**, 1470.

Franz, J., Haselbach, C. H., and Libert, O. (1962). *Acta Endocrinol.* **41**, 336.

Friedman, M., and Stimmler, L. (1966). *Lancet* **II**, 944.

Frohman, L. A., and Bernardis, L. L. (1968). *Endocrinology* **82**, 1125.
Frohman, L. A., Aceto, T. Jr., and MacGillivray, M. H. (1967a). *J. Clin. Endocrinol. Metab.* **27**, 1409.
Frohman, L. A., Horton, E. S., and Lebovitz, H. E. (1967b). *Metabolism* **16**, 57.
Frohman, L. A., Bernardis, L. L., and Kant, K. J. (1968). *Science* **162**, 580.
Garcia, J. F., and Geschwind, I. I. (1966). *Nature* **211**, 372.
Glick, S. M. (1968). *Ann. N. Y. Acad. Sci.* **148**, 471.
Glick, S. M., and Goldsmith, S. (1968). *In* "Growth Hormone" (A. Pecile and E. E. Müller, eds.). Excerpta Medica, Amsterdam, pp. 84–88.
Glick, S. M., Roth, J., Yalow, R. S., and Berson, S. A. (1963). *Nature* **199**, 784.
Glick, S. M., Roth, J., Yalow, R. S., and Berson, S. A. (1965). *Recent Progr. Hormone Res.* **21**, 241.
Greenspan, F. S., Li, C. H., Simpson, M. E., and Evans, H. M. (1949). *Endocrinology* **45**, 455.
Greenwood, F. C., and Landon, J. (1966). *Nature* **210**, 540.
Greenwood, F. C., Landon, J., and Stamp, T. C. B. (1966). *J. Clin. Invest.* **45**, 429.
Guillemin, R., Yamazaki, E., Jutisz, M., and Sakiz, E. (1962). *C.R. Acad. Sci.* **255**, 1018.
Guillemin, R., Sakiz, E., and Ward, D. N. (1965). *Proc. Soc. Exp. Biol. Med.* **118**, 1132.
Hartog, M., Gaafar, M. A., and Fraser, R. (1964). *Lancet* **II**, 376.
Hartog, M., Havel, R. J., Copinschi, G., Earll, J., and Ritchie, B. C. (1967). *Quart. J. Exp. Physiol.* **52**, 86.
Herbert, V., Lau, K. S., Gottlieb, C. W., and Bleicher, S. J. (1965). *J. Clin. Endocrinol. Metab.* **25**, 1375.
Hiroshige, T., and Itoh, S. (1960). *Japan. J. Physiol.* **10**, 659.
Honda, Y., Takahashi, S., Takahashi, K., Azumi, K., Shizume, K., Iriye, M., Sakuma, M., and Tsushima, T. (1967). *16th Ann. Meeting Japanese E.E.G. Society.*
Hunter, W. M., and Greenwood, F. C. (1964a). *Biochem. J.* **91**, 43.
Hunter, W. M., and Greenwood, F. C. (1964b). *Brit. Med. J.* **1**, 804.
Hunter, W. M., Fonseka, C., and Passmore, R. (1965a). *Quart. J. Exp. Physiol.* **50**, 406.
Hunter, W. M., Fonseka, C. C., and Passmore, R. (1965b). *Science* **150**, 1051.
Hunter, W. M., Friend, J. A. R., and Strong, J. A. (1966). *J. Endocrinol.* **34**, 139.
Imura, H., Kato, Y., Ikeda, M., Morimoto, M., Yawata, M., and Fukase, M. (1968). *J. Clin. Endocrinol. Metab.* **28**, 1079.
Ishida, Y. A., Kuroshima, A., Bowers, C. Y., and Schally, A. V. (1965). *Endocrinology* **77**, 759.
Ketterer, H., Powell, D., and Unger, R. H. (1966). *Clin. Res.* **14**, 65.
Knobil, E. (1966). *Physiologist* **9**, 25.
Knobil, E., and Meyer, V. (1968). *Ann. N. Y. Acad. Sci.* **148**, 459.
Knobil, E. Meyer, V., and Schally, A. V. (1968). *In* "Growth Hormone" (A. Pecile and E. E. Müller, eds.). Excerpta Medica, Amsterdam, pp. 226–237.
Knopf, R. F., Conn, J. W., Fajans, S. S., Floyd, J. C., Guntsche, E. M., and Rull, J. A. (1965). *J. Clin. Endocrinol. Metab.* **25**, 1140.
Kohler, P. O., O'Malley, B. W., Rayford, P. L., Lipsett, M. B., and Odell, W. D. (1967). *J. Clin. Endocrinol. Metab.* **27**, 219.
Kramer, C. Y. (1956). *Biometrics* **12**, 307.
Krulich, L., Dhariwal, A. P. S., and McCann, S. M. (1965). *Proc. Soc. Exp. Biol. Med.* **120**, 189.
Krulich, L., Dhariwal, A. P. S., and McCann, S, M. (1968). *Endocrinology* **83**, 783.

Li, C. H. (1953). *In* " Bioassay of Anterior Pituitary and Adrenocortical Hormones " (G.E.W. Wolstenholme, ed.). Vol. V. *Ciba Found. Colloq. Endocrinol.* Little, Brown and Co., Boston, Mass., pp. 115-121.

Lowry, O. H., Rosenbrough, N. J., Farr, A. L., and Randall, R. J. (1951). *J. Biol. Chem.* **193**, 265.

Machlin, L. J., Horino, M., Kipnis, D. M., Phillips, S. L., and Gordon, R. S. (1967). *Endocrinology* **80**, 205.

Martin, L. G., Clark, J. W., and Connor, T. B. (1968). *J. Clin. Endocrinol. Metab.* **28**, 425.

Merimee, T. J., Lillicoop, D. A., and Rabinowitz, D. (1965). *Lancet* **II**, 668.

Merimee, T. J., Burgess, J. A., and Rabinowitz, D. (1966). *J. Clin. Endocrinol. Metab.* **26**, 791.

Merimee, T. J., Rabinowitz, D., Riggs, L., Burgess, J. A., Rimoin, D. L., and Mc-Kusick, V. A. (1967). *New Engl. J. Med.* **276**, 434.

Meyer, V., and Knobil, E. (1966). *Endocrinology* **79**, 1016.

Meyer, V., and Knobil, E. (1967). *Endocrinology* **80**, 163.

Müller, E. E., and Pecile, A. (1965). *Proc. Soc. Exp. Biol. Med.* **119**, 1191.

Müller, E. E., and Pecile, A. (1966). *Endocrinology* **79**, 448.

Parker, M. L., Hammond, J. M., and Daughaday, W. H. (1967). *J. Clin. Endocrinol. Metab.* **27**, 1129.

Pecile, A., Müller, E., Falconi, G., and Martini, L. (1965). *Endocrinology* **77**, 241.

Quabbe, H. J., Schilling, E., and Helge, H. (1966). *J. Clin. Endocrinol. Metab.* **26**, 1173.

Rabinowitz, D., Merimee, T. J., Burgess, J. A., and Riggs, L. (1966). *J. Clin. Endocrinol. Metab.* **26**, 1170.

Raiti, S., Blizzard, R. M., Johanson, A., Davis, W. T., and Migeon, C. J. (1968). *Johns Hopkins Med. J.* **122**, 154.

Reichlin, S. (1960). *Endocrinology* **67**, 760.

Reichlin, S. (1961). *Endocrinology* **69**, 225.

Reichlin, S. (1966). *In* " The Pituitary Gland " (G. W. Harris and B. T. Donovan, eds.). Vol. II. Butterworths, London, pp. 270–298.

Rodbell, M. (1964). *J. Biol. Chem.* **239**, 375.

Ross, H., Johnston, I. D. A., Welbor, T. A., and Wright, A. D. (1966). *Lancet* **II**, 563.

Roth, J., Glick, S. M., Yalow, R. S., and Berson, S. A. (1963a). *Metabolism* **12**, 577

Roth, J., Glick, S. M., Yalow, R. S., and Berson, S. A. (1963b). *Science* **140**, 987.

Roth, J., Glick, S. M., Yalow, R. S., and Berson, S. A. (1964). *Diabetes* **13**, 355.

Sakiz, F. (1964). " Statistical analyses in experimental biology with EXBIOL. A simple multiphase program." Common Research Computer Facility, Texas Medical Center, Houston, Texas.

Sawano, S., Arimura, A., Bowers, C. Y., Redding, T. W., and Schally, A. V. (1968). *Proc. Soc. Exp. Biol. Med.* **127**, 1010.

Schalch, D. S. (1967). *J. Lab. Clin. Med.* **69**, 256.

Schalch, D. S., and Parker, M. L. (1964). *Nature* **203**, 1141.

Schalch, D. S., and Reichlin, S. (1966). *Endocrinology* **79**, 275.

Schally, A. V., and Bowers, C. Y. (1964). *Endocrinology* **75**, 608.

Schally, A. V., Steelman, S. L., and Bowers, C. Y. (1965). *Proc. Soc. Exp. Biol. Med.* **119**, 208.

Schally, A. V., Kuroshima, A., Ishida, Y., Arimura, A., Saito, T., Bowers, C. Y., and Steelman, S. L. (1966). *Proc. Soc. Exp. Biol. Med.* **122**, 821.

Schally, A. V., Sawano, S., Müller, E. E., Arimura, A., Bowers, C. Y., Redding, T. W., and Steelman, S. L. (1968a). *In* " Growth Hormone " (A. Pecile and E. E. Müller, eds.). Excerpta Medica, Amsterdam, pp. 185–203.

Schally, A. V., Müller, E. E., Arimura, A., Saito, I., Sawano, S., and Bowers, C. Y., (1968b). *Ann. N. Y. Acad. Sci.* **148**, 372.

Schofield, J. G. (1967). *Biochem. J.* **103**, 331.

Schramm, H., Benzvi, R., and Bdolah, A. (1965). *Biochem. Biophys. Res. Commun.* **18**, 446.

Simon, S., Schiffer, M., Glick, S. M., and Schwartz, E. (1967). *J. Clin. Endocrinol. Metab.* **27**, 1633.

Smith, G. P., Katz, S. H., Root, A. W., Dhariwal, A. P. S., Bongiovanni, A., Eberlein, W., and McCann, S. M. (1968). *Endocrinology* **83**, 25.

Solomon, S. J., and McKenzie, J. M. (1966). *Endocrinology* **78**, 699.

Takahashi, Y., Kipnis, D. M., and Daughaday, W. H. (1968). *J. Clin. Invest.* **47**, 2079.

Wegienka, L. C., Grodsky, G. M., Karam, J. H., Grasso, S. G., and Forsham, P. H. (1967). *Metabolism* **16**, 245.

White, W. F., Cohen, A. I., Rippel, R. H., Story, J. C., and Schally, A. V. (1968). *Endocrinology* **82**, 742.

Preliminary Observations on Prolactin and Growth Hormone Turnover in Rat Adenohyphyses In Vivo

C. S. NICOLL and K. C. SWEARINGEN

I. Introduction

The numerous publications on neural control of prolactin secretion in mammalian and non-mammalian vertebrates have been extensively reviewed by a number of authors in recent years (Desclin, 1962; Meites *et al.*, 1963; Pasteels, 1963; Everett, 1964; Meites and Nicoll, 1966; Meites, 1967; Pasteels, 1967; Grosvenor *et al.*, 1968; Bern and Nicoll, 1968; Nicoll and Fiorindo, 1969; Nicoll *et al.*, 1969, 1970). This subject will not be reviewed again in this article.

Much of the *in vivo* evidence indicating that the hypothalamus exerts inhibitory control over prolactin secretion in mammals is derived from studies in which neural regulation of the anterior pituitary was disrupted by surgical or pharmacologic means. In such circumstances, physiological manifestations of elevated prolactin levels are apparent. These include maintenance of luteal function and stimulation of mammary gland development and secretion. Thus it is frequently stated in the literature that "denervated"[1] anterior pi-

[1] The term "denervated" is used in this article to refer to any procedure which removes or disrupts neural control over the anterior pituitary.

tuitaries in rats and other mammals secrete prolactin in greater than normal quantities or that they hypersecrete the hormone. However, no data are available on the secretion rate of prolactin *in vivo* in any vertebrate species in any physiological or experimental condition.

We have recently been testing a procedure for evaluating the dynamics of prolactin and Growth Hormone (GH) turnover in anterior pituitaries of rats. Aspects of some of our preliminary experiments, which relate to the secretory activity of *in situ* and transplanted adenohypophyses, are reported herein. Estrogen-treated females, untreated males, and females bearing ectopic anterior pituitary transplants were used in these studies. The animals received an intravenous injection of a labeled amino acid and the specific activity of the prolactin and growth hormone in their anterior pituitaries was determined at different times thereafter.

II. Materials and Methods

In the experiments with the estrogen-treated female rats and with the untreated males, six-week-old Sprague-Dawley animals were used. The females had a body weight range of 120 to 170 g and the males of 180 to 250 g. In the study with pituitary transplants, six-week-old rats of the highly inbred Fischer strain were used. These animals had a body weight range of 160 to 220 g.

The estrogen-treated females were given daily injections of 10 μg of ethynyl estradiol in 0.1 ml of corn oil and were killed on the eighth day. The male rats were not treated. In the pituitary transplant experiment, the female recipients received anterior pituitaries from three female donor animals. Pituitary glands were inserted with a trocar under the capsule of the left kidney. This resulted in some fragmentation of the glands. Accordingly, pieces rather than intact lobes were inserted. The anterior pituitaries of the rats bearing the pituitary transplants were left *in situ.* These animals were used for the experiment 50 days after transplantation.

The female rats received an intravenous injection of 2 μc/g body weight of tritium-labeled leucine (40 to 45 C/mM; New Engand Nuclear Company). The leucine was injected in a volume of 0.1 ml/100 g body weight. The male rats were injected intravenously with 3 μc/g body weight of the labeled amino acid. The intravenous injections were done under light ether anethesia and were given in the saphenous vein. In all experiments, the animals were killed at different times after the injection of the labeled amino acid. In the experiments with females, the times of killing ranged from 1 to 12 hr after injection. The males were killed between 2 and 48 hr after the ^{3}H-leucine injection. Six animals were used at each time interval for each experiment.

The anterior pituitaries were rapidly removed from the animal after they were killed by stunning and decapitation. The glands were weighed and frozen on dry ice. The anterior pituitary transplants were carefully removed from the kidney by peeling off the capsule to which the transplanted tissue usually adhered. Excess capsular and kidney tissue was removed, as much as was feasible, and the transplant tissue was weighed. The functional activity of the transplanted anterior pituitaries was assessed by examining the mammary glands of the hosts for evidence of stimulation of development and secretion. A mammotropic effect was seen in all animals. This ranged from moderate stimulation of lobuloalveolar development to extensive stimulation of growth and secretion. These results indicate that in all animals the anterior pituitary transplants had "taken" and were secreting prolactin.

The anterior pituitaries were then processed by disc electrophoretic procedures for the isolation of prolactin and growth hormone. The *in situ* female rat pituitaries and the male glands were run on the electrophoretic columns at 3 mg of anterior pituitary per column. All the transplanted anterior pituitary tissue from the animal was processed on a single electrophoretic column. The gel columns were stained with aniline blue-black and, after destaining, the relative amounts of prolactin and growth hormone in each column were estimated by densitometry, using a standard curve prepared with purified rat prolactin. Details of this procedure have been published previously (Nicoll *et al.*, 1969). The stained prolactin and growth hormone bands were cut from the gel columns and dissolved in 30% hydrogen peroxide in scintillation vials, and then scintillation fluid was added to the vials in accordance with the procedure of Tishler and Epstein (1968). The vials were counted in a Beckman model LS 250 counter.

The specific activities of the prolactin and growth hormone bands from each time interval were plotted on a log scale against time on a linear scale. The data were fitted to both single and double exponential functions by a least squares fit program on a CDC 6400 computer.

III. Results

Data on anterior pituitary weights and their prolactin and GH content and concentrations are shown in Table I. The male glands contained about one-third as much prolactin as the estrogen-treated females and they had about 40% more GH, on a per milligram wet weight basis. The weight of the transplanted glands was about one-fourth, and their prolactin content was about 10%, of that of the estrogen-treated glands. The prolactin concentration of the female transplants was about equal to that of the *in situ* male anterior pituitary and was about 40% of that of the estrogenized female glands. These results with the transplanted glands are in agreement with a previous report that the prolactin content of the transplanted anterior pituitary is lower than that of the *in situ* glands in female rats, as determined by crop-sac assay (Meites *et al.*, 1963). No growth hormone was detectable in the anterior pituitary transplants.

A. Estrogen-Treated Females

The results of the experiment with the estrogen-treated females are shown in Fig. 1. The specific activity of the prolactin band was highest at the one hour collection and declined rapidly by the second hour. Thereafter the decline in specific activity was much slower. These data were resolved by a double exponential equation, as shown

TABLE I

Weights and Prolactin and Growth Hormone Content of Anterior Pituitary (AP)

Experimental Group	AP Wt (mg)	Prolactin (μg/AP)	Prolactin (μg/mg)	Growth Hormone (μg/AP)	Growth Hormone (μg/mg)
Estrogenized Female	7.5 ±0.3	45.0	6.0 ±0.2	173	23.0 ± 0.9
Male	5.5 ±0.2	11.6	2.0 ±0.1	177	32.1 ± 0.9
Transplanted Female	1.9 ±0.1	4.6	2.4 ±0.3	—	< 1

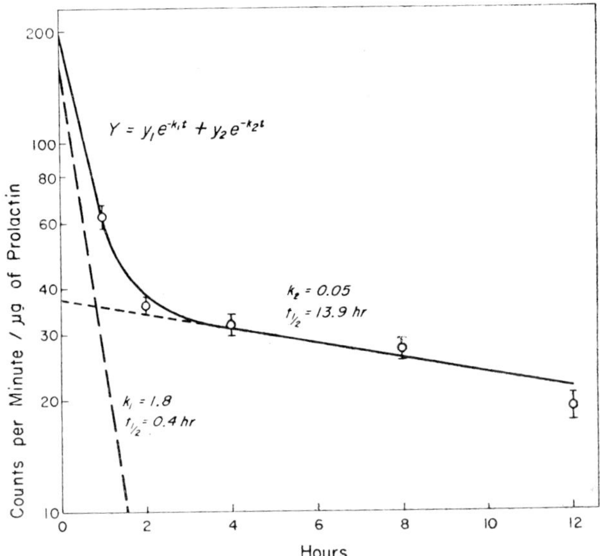

FIG. 1. Temporal changes in the specific activity of prolactin in adenohypophyses of estrogen-treated female rats after intravenous injection of ^{3}H-leucine (2 μc/g body weight).

in Fig. 1. Fast and slow components in the decline in the specific activity of prolactin are apparent from these data. The rate constant of the fast component was 36 times greater than that of the slow component.

Although incorporation of ^{3}H-leucine into the GH band was obtained, the specific activity of the GH did not change significantly through the course of the experiment. Accordingly the rate of turnover of the labeled GH in the estrogen-treated females was too low to be measured by this procedure within 12 hr.

B. UNTREATED MALES

The data from this experiment are shown in Fig. 2. Significant incorporation of the labeled leucine into both prolactin and growth hormone bands in the male anterior pituitary was obtained. However, the specific activity of the male prolactin was substantially lower than that obtained for the prolactin band in the estrogen-treated

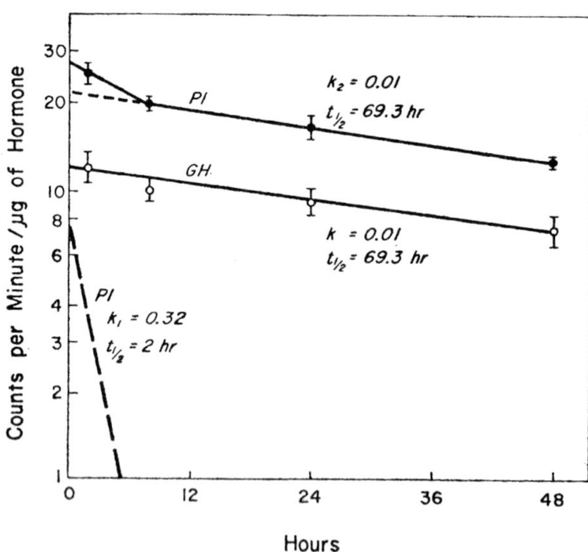

FIG. 2. Temporal changes in the specific activity of prolactin (Pl) and growth hormone (GH) in adenohypophyses of male rats after intravenous injection of 3 μc/g body weight of ³H-leucine.

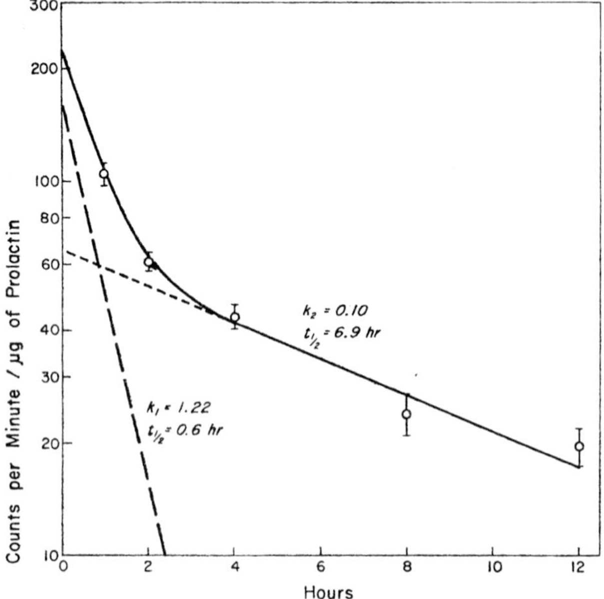

FIG. 3. Temporal changes in the specific activity of prolactin in renal transplants of female adenohypophyses after intravenous injection of 2 μc/g body weight of ³H-leucine.

females, despite the fact that a 50% greater dose of labeled leucine was administered per gram body weight. Although the curve for the decline in the specific activity of the prolactin band was resolved into two components, the curve obtained with growth hormone conformed more closely to a single exponential function. The rate constant for the fast component of the male rat prolactin was about 30 times greater than that of the slow component and was about one fifth that of the fast component of the prolactin in estrogen-treated females. The rate constant of the slow component of the female prolactin curve was about five times greater than that of the male. In the males, the rate constant for the decline in specific activity of growth hormone was identical to that obtained with the slow prolactin component.

C. Adenohypophyseal Transplants

The curve for the decline in the specific activity of the prolactin of the anterior pituitary transplants (Fig. 3) was also resolved by a double exponential function (see Fig. 1 for equation). The fast component had a rate constant about 12 times greater than that of the slow component. The rate constant of the slow component (k_2) was about twice that of the slow component of the estrogen-treated female anterior pituitary and 10 times greater than that of the males. The rate constant of the fast component (k_1) was about 70% of that of the estrogen-treated glands. No growth hormone was detectable in the transplanted glands. Accordingly, no measurement of turnover was obtained.

IV. Discussion

The double exponential functions in the temporal decline in specific activity of the prolactin suggests the existence of fast and slow turnover pools. The absence of a fast component in the GH of males indicates a significant difference in the turnover characteristics of the two hormones in these animals. The significance of the fast turnover component and its quantitative contribution to the total secretion of the hormone by the anterior pituitary are uncertain at this time. These questions are currently being investigated more extensively.

If the relative pool sizes of the fast and slow components of prolactin are estimated (from the counts per minute per milligram of tissue) in the estrogen-treated females, pool sizes of 21 and 24 μg of hormone per anterior pituitary, respectively, are obtained. Assuming that the decline in specific activity of the labeled hormone is a result of its secretion only, multiplying these pool sizes by their respective rate constants [2] gives secretion rates of about 37 and 1 μg of prolactin per milligram of anterior pituitary per hour, for the fast and slow components, respectively. This corresponds to a total secretion rate in the estrogen-treated females of about 38 μg of prolactin per anterior pituitary per hour, or about 900 μg per anterior pituitary per day. At this rate the gland would turn over its entire prolactin content every 1.2 hr, or about 20 times each day. This seems to be an unrealistically high estimate of the prolactin secretion rate in these animals because it represents the secretion of an amount of protein each day equivalent to about 60% of the dry weight of the gland.

Since the significance of the fast component is unknown, we are assuming at this juncture that the slow component is a more accurate reflection of the secretory activity of the gland. The slow component emerges in the later time period of the experiment when the gland is assumed to have reached a steady-state and to be free of possible disturbances which may have occurred as a result of the injection procedure. In addition, it is assumed that intracellular degradation of prolactin and GH (Smith and Farquhar, 1966) did not contribute to the decline in specific activities of the slow prolactin component or the single GH component. The validity of these assumptions may be determined when the nature of the fast component, and other aspects of the turnover of the hormones, are more fully elucidated.

Based on these assumptions, the rate constants of the slow components reflect the time required for the hormone molecules to egress from the gland after they have been synthesized and, presumably, packaged into granules. Comparison of the rate constants of the slow components, as shown in Fig. 4, clearly indicates that the pro-

---

[2] In using these rate constants to calculate secretion rates, it is assumed that they reflect the rate of egress of synthezised hormone from the anterior pituitary. It is also assumed that the labeled hormone is uniformly mixed with unlabeled hormone and that both labeled and unlabeled molecules are secreted at the same rate.

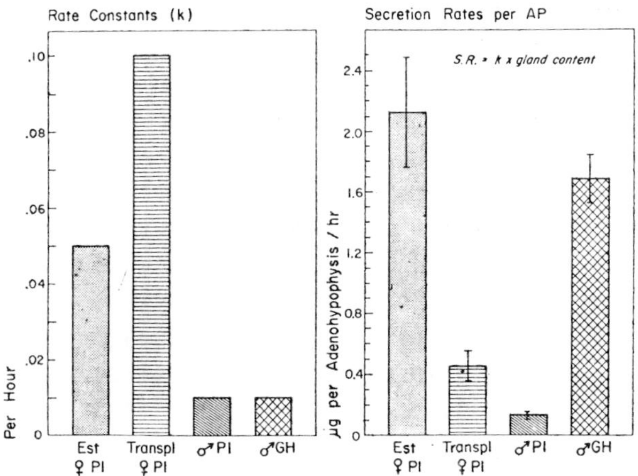

FIG. 4. Rate constants of slow components and estimated secretion rates of prolactin and growth hormone per gland of different rat adenohypophyses (AP). EST, Estrogen; Pl, Prolactin; GH, Growth Hormone.

lactin of the anterior pituitary of the estrogen-treated female rat is egressing five times faster than that of the male gland. The exit rate of prolactin in the anterior pituitary transplants is about twice that of the *in situ* glands of the estrogen-treated animals and about ten times greater than that of the males. The egress rate of growth hormone in the male anterior pituitary is about equal to that of prolactin in the male.

These rate constants were used to calculate secretion rates [3], as shown in Fig. 4. The estrogen-treated female anterior pituitary was estimated to secrete prolactin at a rate of about 2.1 μg per anterior pituitary per hour. This is considerably higher than the secretion rate

[3] An additional assumption inherent in this calculation is that the glands were in a steady-state condition. Accordingly, the prolactin content of the glands should remain constant. The prolactin content of the male anterior pituitary and of the transplanted glands did not show significant changes from one collection time to another, but the content of the estrogen-treated glands did. However, using the counts/minute/mg of anterior pituitary tissue to estimate the rate constant and prolactin secretion rate in the latter group, the rate constant and secretion rate obtained were virtually identical to those obtained with the specific activity data. Accordingly, fluctuations in gland content from one time on another did not influence these estimates in this case.

obtained with the transplanted glands, which showed a rate of about 0.47 μg per anterior pituitary per hour. The male glands secreted prolactin at a rate of 0.13 μg per anterior pituitary per hour and they secreted growth hormone at a rate of 1.7 μg per gland per hour. The observed differences in the secretion rates and rate constants for prolactin of the male and the estrogen-treated female glands are consistent with differences that would be expected from previous studies on the prolactin content of male and female anterior pituitaries (see Meites *et al.*, 1963). The relatively low secretion rate per gland obtained with the transplanted anterior pituitary is largely a result of the small amount of tissue recovered in the transplants (see Table I).

The data on the estimates of prolactin and growth hormone secretion rates per milligram of anterior pituitary tissue are shown in Fig. 5. On this basis the prolactin secretion rate of the transplants is comparable to that of the estrogen-treated glands and the secretion rate of the prolactin by the male glands is about 7% of that of the estrogenized anterior pituitaries. The male glands secreted growth hormone at a rate per milligram of tissue which was about 15 times higher than their rate of prolactin secretion. Our estimate of GH

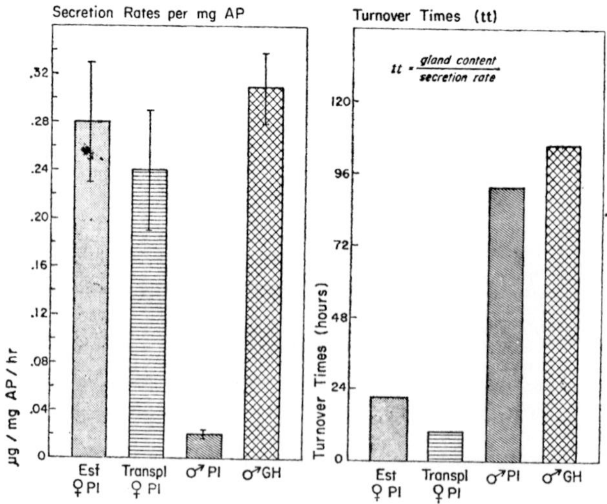

FIG. 5. Estimated secretion rates per milligram of anterior pituitary (AP) and turnover times of prolactin and growth hormone in different rat adenohypophyses (Abbreviations as in Fig. 4).

secretion rate of about 0.3 μg per milligram anterior pituitary per hour is higher than that reported recently by Garcia and Geschwind (1968). They estimated a GH secretion rate of 0.14 μg per milligram per hour in 30-day-old female rats using the plasma levels, distribution volume, and half-life of the hormone to derive their estimate. However, this difference does not represent a gross disparity in the estimated secretion rates and could be due to a number of factors such as the age and sex difference in the animals.

Estimates of the turnover times of the prolactin and growth hormone in these glands have been made on the basis of the estimated secretion rates and gland content of these hormones. The turnover times, shown in Fig. 5, indicate that approximately one day is required to turn over the entire prolactin content of the estrogen-treated female anterior pituitary. The transplanted gland requires only about one-half of a day to turn over its prolactin pool. The male glands require almost 4 days to turn over the entire prolactin content and about 4.5 days are needed to secrete an amount of GH equivalent to its entire content of GH.

V. Conclusions

The estimates of prolactin and GH secretion and turnover rates presented herein must be regarded as tentative at this time. When the significance of the fast turnover component and other aspects of this phenomenon are more fully understood, the validity of assuming that the slow component is a more accurate reflection of the secretion and turnover of prolactin will be determined.

It is of interest to compare our estimates of prolactin secretion and turnover times with those of other investigators who have made similar estimates on other pituitary hormones in rats. We have already pointed out that our estimate for GH secretion rate in the adult male rat compares reasonably well with the estimate of Garcia and Geschwind (1968) on this process in 30-day-old females. The turnover time for growth hormone, calculated from the data of Garcia and Geschwind (1968), is about 146 hr. This also compares reasonably well with our estimate of 105 hr in the adult male rat. Bogdanove and Gay (1968) have estimated secretion rates and turnover times for both Follicle-Stimulating Hormone (FSH) and Luteinizing Hormone (LH) in castrated male rats (in terms of equivalents to NIH ovine hormones). They report that LH is secreted at a rate of about 1.1 μg per anterior pituitary per hour, while FSH is secreted at a rate of about 34 μg per gland per hour. Their estimate of the LH secretion rate is comparable to our estimate for prolactin and GH secretion reported herein. The estimate of FSH secretion seems inordinately high [4]. The turnover time which they estimated for LH was about 120 hr. This compares favorably with our estimate of prolactin and GH turnover times in the male of 91 and 105 hr, respectively. Bogdanove and Gay (1968) also estimated the turnover time for FSH to be about 8 hr [5]. This compares reasonably well with our estimate of turnover times for prolactin in the estrogen-treated female and the transplanted adenohypophysis of 21 and 10 hr, respectively.

Accordingly, our estimates of prolactin and GH secretion rates and turnover times give values which are of the same order ot magnitude as those reported by others for GH and LH, and to some extent for FSH. It is noteworthy in this connection that anterior pituitaries of adult female rats in organ culture secrete prolactin at a rate of about 0.2 to 0.4 μg per milligram wet weight per hour (Nicoll, unpublished observations). This compares very well with our estimate of the *in vivo* secretion rates of the estrogen-treated and transplanted female anterior pituitaries.

Assuming that our data provide accurate estimates of at least the relative rates of prolactin turnover and secretion, it is evident that the male gland secretes much less prolactin than the female gland on either a per anterior pituitary or per milligram basis, because of slower egress rate and lower pool size. The transplanted female anterior pituitary secretes about as much prolactin as the *in situ* gland of estrogen-treated females on a per milligram basis although the egress rate is twice as great as in the transplant. The fact that the amount of prolactin secreted per mg of transplanted tissue is not

---

[4] Bogdanove and Gay (1969) have subsequently reported "corrected" estimates of FSH and LH secretion rates in castrated rats. The corrections were made using estimates of the amount of inert contaminants in the FSH, and LH preparations. Their "corrected" estimates indicate that both hormones are secreted at about 0.3 μg per anterior pituitary per hour.

[5] This estimate of FSH turnover time would not be subject to error from the use of impure FSH standard (Bogdanove and Gay, 1969).

greater than that of the estrogen-treated gland is due in part to the lower prolactin content in the transplant. However, comparisons of the prolactin secretion rate per mg of tissue between transplanted and *in situ* anterior pituitaries are not entirely valid because an unknown amount of the transplant was presumably necrotic or scar tissue and some renal capsular and cortical tissue probably contributed to the weight measurements of these glands. On the other hand, the proportion of prolactin cells in the transplanted anterior pituitary may be relatively high because of loss or atrophy of other cell types.

The prolactin secretion rate of the transplanted anterior pituitary was about 25% of that of the estrogenized gland on a per anterior pituitary basis because of the small size of the transplant and because of the low prolactin content. If the estrogen-treated anterior pituitary is regarded as secreting prolactin at a high " normal " level, it is apparent that the transplanted gland cannot be viewed as " hypersecreting " prolactin. These results suggest that the physiological manifestations of prolactin hypersecretion, seen in animals with " denervated " anterior pituitary, may actually be due to secretion of the hormone at a constant rate (high " normal "?) rather than to its secretion in " greater than normal " quantities. Our studies also indicate that the turnover and secretion rates of GH in male rats is much greater than in the *in situ* glands of estrogen-treated females or transplanted female anterior pituitary in view of the fact that the turnover rates of GH in the anterior pituitary of the latter groups was too low to permit estimation of GH secretion rates.

This method is currently being applied to the evaluation of the turnover of hypophyseal prolactin and GH in female rats in different physiological conditions. It will be of interest to see how the estimates of the turnover and secretion rates in these animals compare with the data obtained with the estrogen-treated and transplanted female glands and with the male anterior pituitary as reported herein. It will also be of interest to learn how our estimates of the secretion rates and turnover times compare with similar estimates obtained using different methods. This procedure should be of considerable value in studying various aspects of the dynamics of prolactin and GH turnover in the anterior pituitaries of rats and other species. Evaluation of the effects of hypophysiotropins on this phenomenon will be an important extension of these studies.

Acknowledgments

The work reported in this article was supported by Institutional Research Grant No. IN-87B from the American Cancer Society and by Grant No. M69.39 from the Population Council.

References

Bern, H. A., and Nicoll, C. S. (1968). *Recent Progr. Hormone Res.* **24**, 681.

Bogdanove, S. M., and Gay, V. L. (1968). *In* " Gonadotropins " (E. Rosemberg, ed.). Geron-X, Los Altos, Calif., pp. 131–136.

Bogdanove, E. M., and Gay, V. L. (1969). *Endocrinology* **84**, 1118.

Desclin, L. (1962). *In* " Proc. XXII Intern. Congr. Intern. Union Physiol. Sci." Vol. **I**, Part. II. Excerpta Medica, Amsterdam, pp. 715–739.

Everett, J. W. (1964). *Physiol. Rev.* **44**, 373.

Garcia, J. F., and Geschwind, I. I. (1968). *In* " Growth Hormone " (A. Pecile and E. E. Müller, eds.). Excerpta Medica, Amsterdam, pp. 267–291·

Grosvenor, C. E., Mena, F., Schaefgen, D. A., Dhariwal, A. P. S., Antunes-Rodriguez, J., and McCann, S. M. (1968). *In* " Pharmacology of Hormonal Polypeptides and Proteins " (N. Back, L. Martini, and R. Paoletti, eds.). Plenum Press, New York, pp. 242–253.

Meites, J. (1967). *Arch. Anat. Microscop. Morphol. Exp.* **56** *Suppl.* 3–4, 516.

Meites, J., and Nicoll, C.S. (1966). *Ann. Rev. Physiol.* **28**, 57.

Meites, J., Nicoll, C. S., and Talwalker, P. K. (1963). *In* " Advances in Neuroendocrinology " (A. V. Nalbandov, ed.). Univ. Illinois Press, Urbana, Ill., pp. 238–277.

Nicoll, C. S., and Fiorindo, R. P. (1969). *Gen. Comp. Endocrinol. Suppl.* **2**, 26.

Nicoll, C. S., Parsons, J., Fiorindo, R. P., and McKennee, C. T. (1970). *In* "Hypophysiotropic Hormones of the Hypothalamus, Assay and Chemistry " (J. Meites, ed.). Williams and Wilkins Co., Baltimore, pp. 115-144.

Nicoll, C. S., Parsons, J. A., Fiorindo, R. P., and Nichols, C. W., Jr. (1969). *J. Endocrinol.* **45**, 183

Pasteels, J. L. (1963). *Arch. Biol.* **74**, 439.

Pasteels, J. L. (1967). *Arch. Anat. Microscop. Morphol. Exp.* **56** *Suppl* 3–4, 530.

Smith, R. E., and Farquhar, M. G. (1966), *J. Cell Biol.* **31**, 319.

Tishler, P. V., and Epstein, C. J. (1968). *Anal. Biochem.* **22**, 89.

The Hypothalamus as the Center of Endocrine Feedback Mechanisms

M. MOTTA, F. PIVA [1], and L. MARTINI

I. Introduction

It has been mentioned in other articles of this volume that the hypothalamus plays an essential role in the control of the secretion of several anterior pituitary hormones (see the articles by Flerkó and by Ganong). The key position of the hypothalamus in neuroendocrine regulation is mainly due to the fact that this region of the brain contains simultaneously the final elements of several afferent pathways (the receptors sensitive to feedback messages) and the first elements of at least six efferent (or executive) pathways (the

[1] Ford Foundation Fellow.

cells synthesizing the releasing and inhibitory factors) (Martini *et al.*, 1969). Because of this the hypothalamus may be considered the major center of the reflex, automatic control of the secretion of anterior pituitary hormones.

Since the precise localization of the nuclei responsible for the synthesis of the releasing factors is discussed in detail in another part of this volume (see the article by Mess and associates), only the evidence regarding the localization of the receptors sensitive to endocrine signals will be reviewed here. From the combined analysis of this article and of that by Mess and associates, it will appear that the areas in which the messages from the periphery are received are different from those where the releasing factors are synthesized; nervous pathways specifically devoted to the transport of the information from the sensory elements to the executive ones must be postulated. Some hypotheses on the physiological significance of these interconnecting systems will be presented as the conclusion of this article.

Table I provides a summary of the different types of feedback mechanisms that have been studied in our laboratory. They can be subdivided into three categories:

1. " Long " feedback mechanisms in which the messages for hypothalamic receptors are provided by hormones synthesized in the peripheral target glands

TABLE I

DIFFERENT TYPES OF ENDOCRINE FEEDBACK MECHANISMS

" *Long* " *feedback mechanisms*
" Negative " feedback effect of corticoids (Corbin *et al.*, 1965)
" Negative " feedback effect of estrogen (Martini *et al.*, 1968 a, b)
" Negative " feedback effect of progesterone (Martini *et al.*, 1968 a, b)
" Negative " feedback effect of testosterone (Martini *et al.*, 1968 a, b)
" Positive " feedback effect of estrogen (Motta *et al.*, 1968)
" Positive " feedback effect of progesterone (Motta *et al.*, 1970)
" *Short* " *feedback mechanisms*
" Negative " feedback effect of ACTH (Motta *et al.*, 1965)
" Negative " feedback effect of LH (Dávid *et al.*, 1966)
" Negative " feedback effect of FSH (Fraschini *et al.*, 1968)
" Positive " feedback effect of TSH (Motta *et al.*, 1969 a)
" *Ultrashort* " *feedback mechanisms*
" Negative " feedback effect of FSHRF (Motta, 1969)

2. " Short " feedback mechanisms in which the signals for the hypothalamus are represented directly by pituitary hormones
3. " Ultrashort " feedback mechanisms in which the information for hypothalamic receptors is apparently provided by the releasing factors themselves.

It appears from Table I that " long " and " short " feedback mechanisms may be of two types, i.e., " negative " (inhibitory) or " positive " (stimulatory). Because of the limitation of space, our discussion will be limited to a few examples selected from the feedback mechanisms that control the secretion of gonadotropins. For more information the reader is referred to the original papers quoted in Table I and to the articles by Mess and Martini (1968) and by Motta *et al.* (1969b).

II. " Long " Feedback Mechanisms

A. " Negative " Feedback Effect of Estrogen on Follicle-Stimulating Hormone Secretion

Intrahypothalamic implants of small quantities of estrogen have been shown to suppress the synthesis as well as the release of Luteinizing Hormone (LH) (Kanematsu and Sawyer, 1963, 1964; McCann and Ramirez, 1964; Ramirez *et al.*, 1964). Since brain implants of estrogen can also induce gonadal atrophy and can block ovarian compensatory hypertrophy, it has been argued that they may also inhibit Follicle-Stimulating Hormone (FSH) secretion (Chowers and McCann, 1967; Davidson and Sawyer, 1961; Davidson, 1969; Fendler and Endröczi, 1966; Flerkó and Szentágothai, 1957; Lisk, 1960, 1965, 1969). However, clear-cut information based on direct measurements of FSH is still lacking. This is mainly due to the fact that the assay methods used for measuring FSH are less sensitive than those used for evaluating LH, so that reliable measurements of plasma FSH cannot be performed. This situation will undoubtedly improve soon with the development of specific and sensitive radioimmunoassay procedures for the measurement of plasma FSH levels also in animal species.

TABLE II

EFFECT OF IMPLANTS OF ESTRADIOL IN THE MEDIAN EMINENCE (ME), AND IN THE ANTERIOR PITUITARY (PIT) ON ORGAN WEIGHTS AND ON PITUITARY FSH LEVELS OF MALE RATS[1]

Groups[2]	Body Wt (g)	Pituitary Wt (mg)	Testes Wt (g)	Prostates Wt (mg)	Seminal Vesicles Wt (mg)	Pituitary FSH (μg/mg[3])	Limits (95%)
ME: sham (60)	231± 4.1	11.0±0.39	2.71±0.53	182.9± 7.8	288.8±16.0	44.83	(38.7–54.1)
						54.31	(47.2–62.1)
ME: estradiol (60)	233± 2.6	13.1±0.45[4]	2.53±0.43[4]	129.1± 7.5[5]	151.8± 9.2[5]	26.73	(19.1–32.3)
						33.33	(27.2–39.5)
Pit: sham (20)	265±11.0	8.9±0.41	2.97±0.70	179.8±13.0	404.0±17.5	40.40	(30.2–48.3)
						36.32	(25.4–40.5)
Pit:estradiol (20)	278± 6.0	12.1±0.60[4]	2.80±0.50	162.2±12.5	346.0±14.8[4]	32.82	(22.8–39.4)
						28.73	(20.3–32.4)

[1] Values are means ± SE.
[2] Number of rats in parentheses.
[3] Microgram equivalents of NIH-FSH-S-3 ovine per mg wet weight of pituitary tissue.
[4] $P \leq 0.01$.
[5] $P \leq 0.001$.

In the experiments to be described here, an attempt has been made to overcome this methodological gap by selecting appropriate experimental conditions. Minute amounts of estradiol have been placed in the median eminence of the hypothalamus or in the anterior pituitary of sexually mature, male rats. Male animals have been selected because they permit obtaining a satisfactory, even if indirect, evaluation of blood levels of FSH and LH; as it is known, the weight of the testes is directly related to the amounts of FSH present in the circulation, and the weights of ventral prostates and of seminal vesicles provide satisfactory indications of the amounts of circulating LH-ICSH (Interstitial Cells-Stimulating Hormone) (Davidson, 1966). Because of the observation (Fraschini and Motta, 1967; Fraschini, 1969; Martini *et al.*, 1968b; see also the article by Fraschini and Martini in this volume) that male rats exhibit a diurnal variation in the concentrations of gonadotropins in their pituitaries, all animals were killed at the same hour of the day, i.e., between 4 and 5 P.M.; at this time the concentration of gonadotropins in the pituitary is particularly elevated. Animals were killed 5 days following implantation.

As shown in Table II, the implantation of estradiol into the median eminence results in a significant decrease in the weights of the testes, of the prostates, and of the seminal vesicles. This suggests that the amounts of FSH and of LH in the general circulation are significantly diminished by the presence of estradiol in this region of the brain. Intrapituitary implants have been completely ineffective on testicular weights, although they have induced a moderate decrease of the weights of the prostates and of the seminal vesicles; only the decrease of the weight of the seminal vesicles is significant. It appears that intrapituitary estrogen does not modify the release of FSH, but probably brings about some reduction of the release of LH; this confirms previous data (Ramirez *et al.*, 1964).

Median eminence implants of estradiol have been shown to induce a significant reduction of the concentration of FSH in the pituitary (as evaluated by the procedure of Steelman and Pohley, 1953) (Table II); the "total" content of FSH per pituitary is also reduced, in spite of the increased pituitary weight found in animals bearing hypothalamic implants of estradiol (see also Chowers and McCann, 1967; Lisk, 1969; Palka *et al.*, 1966). Intrapituitary implants caused a small, statistically insignificant reduction in the concentration of

FSH; however, because of the increase in the pituitary weight found in this group of animals (see also Chowers and McCann, 1967; Lisk, 1969; Palka *et al.*, 1966), the total amount of FSH per pituitary is enhanced. Because of the presence of atrophy of the testes, the decreased pituitary levels of FSH observed following median eminence implants of estrogen cannot be ascribed to a stimulation of the release of the hormone. The data suggest, then, that estrogen, when present in high concentration on their receptors located in the median eminence region, can reduce the synthesis as well as the release of FSH. These data provide the first direct demonstration of the existence in the hypothalamus of estrogen receptors involved in the control of FSH secretion; apparently estrogen receptors for the feedback control of FSH are not present at pituitary level. It may be argued that receptors involved in the feedback control of FSH secretion are different from those that intervene in the regulation of LH secretion; these are apparently present also in the anterior pituitary gland (Kanematsu and Sawyer, 1964; Ramirez *et al.*, 1964).

The participation of the hypothalamus in the feedback effect that estrogen exerts on FSH secretion has been confirmed by using another approach, i.e., by evaluating the effects of systemically administered estradiol on the hypothalamic stores of the Follicle-Stimulating Hormone Releasing Factor (FSHRF). In this group of experiments, estradiol benzoate was administered to castrated rats of both sexes (50 μg/rat per day for 5 days). Animals were killed 24 hr after last treatment. Hypothalamic FSHRF was evaluated according to the procedure of Dávid *et al.* (1965a). Pituitary FSH levels were measured according to the method of Steelman and Pohley (1953). Treatment with estradiol significantly reduced hypothalamic levels of FSHRF in both male and female rats (Fig. 1). Pituitary FSH stores in both groups of animals were also decreased following treatment. The magnitude of the drop of FSH was greater in males, probably because untreated castrated males have higher FSH levels than do females (Martini *et al.*, 1968a,b). If one assumes that the reduction of the stores of FSHRF and of FSH reflects reduced synthesis and release of these two principles, one might conclude that estrogen exerts an inhibitory effect on the secretion of FSH and that this inhibitory effect is exerted via a “ negative ” feedback effect on FSHRF; this interpretation would be in agreement with the results obtained by implanting estrogen in the hypothalamus. The assump-

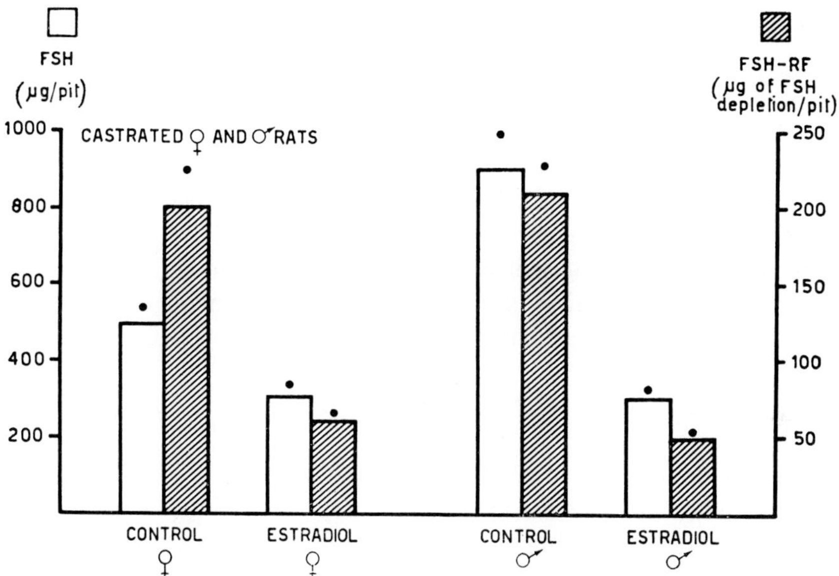

FIG. 1. Effect of systemic administration of estradiol benzoate on pituitary FSH content and on hypothalamic FSHRF stores in castrated female and male rats.

tion that reduced stores mean reduced synthesis and release seems acceptable when one deals with FSH and FSHRF; for these principles, all available evidence shows a very good correlation between their stores in the pituitary and in the hypothalamus and their release. Castration, which increases release of both FSHRF and FSH, also increase their stores (Dávid *et al.*, 1965b; Martini *et al.*, 1968a,b); treatments with sex steroids, which certainly suppress the release of both principles, simultaneously reduce their levels in the hypothalamus and in the pituitary (Dávid *et al.*, 1965b; Martini *et al.*, 1968a,b; Minaguchi and Meites, 1967); in female rats in which hypersecretion of FSHRF is induced by exposure to constant light (Piacsek and Meites, 1967; Piacsek *et al.*, 1969), the storage of this factor in the hypothalamus is also enhanced (Negro-Vilar *et al.*, 1968).

B. "Negative" Feedback Effect of Progesterone on Follicle-Stimulating Hormone Secretion

The idea that progesterone might also inhibit FSH secretion has recently gained a lot of support. The chronic administration of progesterone has been shown to result in the inhibition of ovarian compensatory hypertrophy (a phenomenon due to hypersecretion of FSH) in several species of animals (Benson *et al.*, 1969; Jelinek *et al.*, 1968; Labhsetwar, 1968a,b; Short *et al.*, 1968). In women, plasma FSH levels are particularly low during the second phase of the ovulatory cycle, when progesterone is secreted in high amounts (Odell *et al.*, 1968); in addition, a rapid and significant increase of plasma FSH levels is observed after cessation of corpus luteum function (Faiman and Ryan, 1967; Franchimont, 1968; Igarashi *et al.*, 1967; Midgley and Jaffe, 1968; Swerdloff and Odell, 1969; Taymor *et al.*, 1968).

To determine the site of the inhibitory action of progesterone on FSH secretion, Davidson (1969), Lisk (1969), and Smith *et al.* (1969) have stereotaxically implanted the steroid at various intracranial sites in female rats and have found that intrahypothalamic implants performed in the arcuate or in the median eminence region induce ovarian atrophy. Implants performed in the pituitary gland were completely ineffective. Since ovarian weight is largely dependent on plasma FSH levels, it may be inferred that progesterone receptors for the control of FSH secretion are present in the basal hypothalamus.

It was deemed of interest to evaluate whether the systemic administration of progesterone might result in modification of FSHRF stores at hypothalamic level and of FSH stores at pituitary level. The experimental design was practically the same as that used when estrogen was administered (see Section II, A). Progesterone administered (2 mg/rat per day for 5 days) to adult castrated female rats brought about a reduction of the stores of both FSH and FSHRF. In males, FSHRF stores were not modified by progesterone even if a significant reduction of pituitary FSH content was observed (Fig. 2). Comparable data have been reported in the male by Andjus and Kamberi (1966), who used a less specific technique for the evaluation of FSHRF.

If one accepts again the assumption that the changes in the stores of FSH and of FSHRF reflect changes in the secretory rates of the

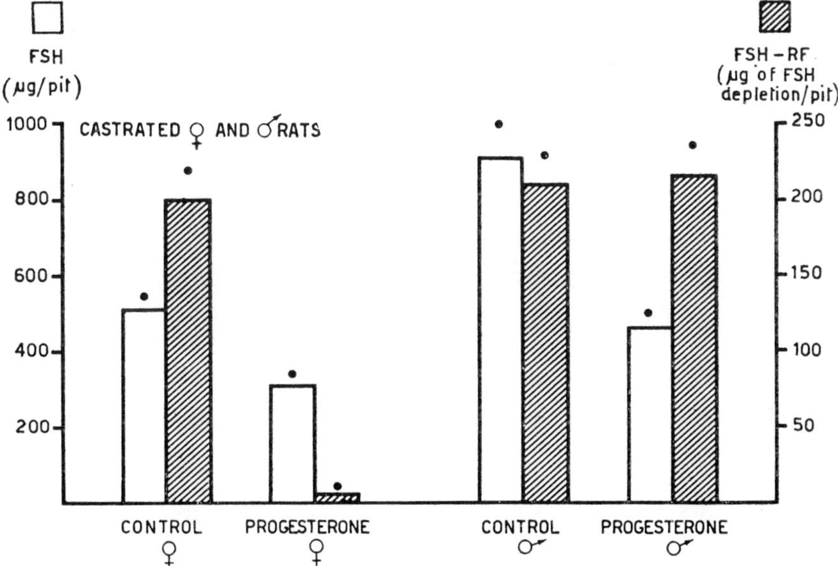

FIG. 2. Effect of systemic administration of progesterone on pituitary FSH content and on hypothalamic FSHRF stores in castrated female and male rats.

two principles, it would appear from the data that progesterone inhibits the secretion of FSH through a transhypothalamic mechanism in females and through an effect at pituitary level in males.

C. THE "NEGATIVE" FEEDBACK EFFECTS OF ESTROGEN AND OF PROGESTERONE ON FOLLICLE-STIMULATING HORMONE SECRETION AND THE INHIBITION OF OVULATION

The question may be asked whether the inhibitory effect estrogen and progesterone exert on the secretion of FSHRF and of FSH (described in Section II, A and B) may play a role in the mode of action of steroid preparations currently used for blocking ovulation in women. In a recent attempt to provide a unitary view of the neuroendocrine events that occur during an ovulatory cycle in women, a prominent role has been assigned to the FSHRF–FSH component (Martini *et al.*, 1970; Piva *et al.*, 1969)[2].

It has been suggested that a new ovulatory cycle begins when a neural stimulus, originating in some unknown extrahypothalamic centers (Halász, 1969), reaches the cells of the paraventricular nuclei in which FSHRF is synthesized (Martini *et al.*, 1968b; Mess *et al.*, 1967; see also the article by Mess and associates in this volume); FSHRF is consequently released into the pituitary portal vessels from the nerve ending of paraventricular neurons in the median eminence region; FSH is then hypersecreted. This results in maturation of the follicle. At a certain stage of its development the follicle becomes competent to respond to the small quantities of LH that are secreted in basal amounts throughout the cycle; consequently the secretion of estrogen can begin. When estrogen reaches a threshold amount in the circulation, a peak of LH secretion is achieved (" positive " feedback effect of estrogen on LH; see Section II, D) and ovulation occurs. If one accepts this interpretation of the ovulatory cycle and the view that FSHRF is essential for transferring the primary stimulus for ovulation down to the anterior pituitary, it is obvious that any treatment able to reduce hypothalamic stores of FSHRF would render the hypothalamus refractory to the transmission of ovulatory stimuli.

Many recent data indicate that estrogenic and progestational steroids used for antifertility purposes may eliminate FSHRF from the hypothalamus and inhibit FSH secretion. Minaguchi and Meites (1967) have reported that the chronic administration of Enovid (a preparation containing the progestational agent norethinodrel and

---

[2] The selection of FSHRF as the " first trigger " of the ovulatory process is based on recent evidence indicating that the release of FSH is increased at the time of ovulation in different species of mammals (rat: Caligaris *et al.*, 1967; Goldman and Mahesh, 1968; McClintock and Schwartz, 1968; pig: Parlow *et al.*, 1964; ewe: Santolucito *et al.*, 1960; sheep: Robertson and Rakha, 1966; cow: Rakha and Robertson, 1965; monkey: Simpson *et al.*, 1956). Evidence obtained in women by using radioimmunological procedures for evaluating plasma levels of gonadotropins also indicates that a peak of FSH precedes the typical mid-cycle peak of LH (Faiman and Ryan, 1967; Franchimont, 1966). The data by Goldman and Mahesh (1969) are also relevant. They have reported that an antiserum prepared against LH and neutralizing both LH and FSH is able to block ovulation in the hamster if injected shortly prior to proestrus; following removal of most of the FSH antibodies by absorption with FSH, the ovulation-blocking potency of the antiserum is greatly reduced.

the estrogenic compound mestranol) decreases hypothalamic stores of FSHRF in both normal and castrated rats. Labhsetwar (1968b) has shown that chlormadinone, administered to intact or to unilaterally spayed rats, results in ovarian atrophy and in the inhibition of ovarian compensatory hypertrophy, respectively. Harris and Sherratt (1969) have observed that rabbits pretreated with chlormadinone show a reduced number of ruptured follicles following electrical stimulation of the median eminence; they have suggested that this is due to the fact that chlormadinone reduces FSH secretion and consequently the number of ripe follicles.

Women receiving nonsequential contraceptive preparations do not exhibit the elevation of plasma and urinary FSH levels, that are typical of the early phases of the cycle, and the mid-cycle FSH peak (Cargille and Ross, 1968; Cargille *et al.*, 1969; Odell *et al.*, 1968; Ross *et al.*, 1966; Schmidt-Elmendorff and Kopera, 1968; Stevens *et al.*, 1965, 1968; Swerdloff and Odell, 1969). Women treated with sequential contraceptives have very low levels of FSH throughout the cycle; surprisingly, preparations of this type seem to stimulate the release of LH (Swerdloff and Odell, 1969) (see Sections II, D and E, for the " positive " feedback of estrogen and progesterone on LH secretion). Long-acting progestational preparations suppress plasma FSH levels in normal, castrated or postmenopausal women (see the article by Franchimont and Legros in this volume).

In the authors' view it appears, then, permissible to suggest that at least some antiovulatory steroids may inhibit ovulation because they prevent the resynthesis and the storage of FSHRF in the hypothalamus after this principle has been secreted into the pituitary portal vessels to initiate an ovulatory cycle. Sawyer, in his article in this volume, has alluded to the fact that the ability to increase hypothalamic " thresholds " is one common feature of antiovulatory steroid (see also Sawyer, 1967); the question may be asked whether the enhancement of hypothalamic " thresholds " might not represent the electrical correlate of the reduction of FSHRF stores.

The suggestion that progestational agents act primarily on the FSHRF-FSH component is compatible with the observation that the electrical stimulation of the median eminence or of the preoptic suprachiasmatic area is able to induce ovulation in progesterone-blocked rats (Moll and Zeilmaker, 1966; Redmond, 1968; Zeilmaker and Moll, 1967) as well as in chlormadinone-treated rabbits

(Exley *et al.*, 1968; Harris and Sherratt, 1969). It is obvious that the stimulation of these two regions will result directly in the liberation of the Luteinizing Hormone Releasing Factor (LHRF); this factor is actually synthesized in neurons whose cell bodies lie exactly in these two areas (Martini *et al.*, 1968b; Mess *et al.*, 1967; see also the article by Mess and associates in this volume).

D. " Positive " Feedback Effect of Estrogen on Luteinizing Hormone Secretion

Our model for studying whether estrogen might exert a " positive " feedback effect on LH secretion has been that of evaluating whether estradiol induces precocious puberty and ovulation in immature female rats following placement in the median eminence region (Motta *et al.*, 1968). Implants were performed at 26 days of age; all animals were killed on day 39. Vagina opened at 36.3 $\pm$ 0.90 days of age in the control group and at 36.0 $\pm$ 0.88 in animals sham-implanted in the median eminence; in animals bearing estrogen implants in this region, vaginal opening was recorded at day 29.0 $\pm$ 0.62 (Table III). In animals with cerebral cortical implants of estrogen the vaginae opened 1 day before the controls; this difference was not significant. Intrapituitary implants of estradiol did not modify the time of appearance of puberty. Pituitary weights showed a considerable increase in all groups of estrogen-implanted animals. Ovarian weight was not altered by intracranial implantations of estrogen. The ovaries of all groups of animals were inspected macroscopically for the presence of fresh corpora lutea; these were found only in the group implanted with estrogen in the median eminence.

The implantation of estradiol in the median eminence was followed by a significant decrease of pituitary LH concentration (as measured by the procedure described by Parlow, 1961). In spite of the enhanced pituitary weight recorded in this group of animals (see also Section II, A) the total LH content per pituitary was also significantly reduced. No modifications of pituitary LH concentrations were induced by estrogen implants performed in the cerebral cortex or directly into the pituitary gland; however, the total LH content in these two groups of animals was higher than in controls, owing to the considerable increase in pituitary weight (see also Section II, A). In animals bearing median eminence implants of estrogen, LH was present

TABLE III

EFFECT OF IMPLANTS OF ESTRADIOL IN THE MEDIAN EMINENCE (ME), THE ANTERIOR PITUITARY (PIT), AND THE CEREBRAL CORTEX (CC) ON ORGAN WEIGHTS, VAGINAL OPENING, AND PITUITARY AND PLASMA LH LEVELS OF IMMATURE FEMALE RATS [1]

Groups [2]	Body Wt (g)	Pituitary Wt (mg)	Ovaries Wt (mg)	Vaginal Opening (days)	Pituitary LH		Plasma LH
					μg/mg [3]	μg/Pit	
Controls (10)	133±2.5	6.4±0.53	49.5±4.2	36.3±0.90	1.24±0.04	7.68	Absent
ME: sham (10)	110±4.1	5.1±0.28	47.3±3.9	36.0±0.88	1.54±0.05	7.85	Absent
ME: estradiol (10)	118±5.9	10.9±0.80 [4]	44.3±3.7	29.0±0.62 [4]	0.28±0.01 [4]	3.05	Present
Pit: estradiol (10)	123±5.2	9.4±0.82 [4]	48.2±4.0	36.4±0.82	1.42±0.04	13.35	Absent
CC: estradiol (10)	117±3.2	8.0±0.45 [4]	46.0±3.7	35.0±0.85	1.36±0.07	10.88	Absent

[1] Values are means ± SE.
[2] Number of rats in parentheses.
[3] Microgram equivalents of NIH–LH-B-1 bovine per mg wet weight of pituitary tissue.
[4] $P \leq 0.001$ *vs.* ME: sham.

in measurable amounts in the blood at time of autopsy (Table III); plasma LH was undetectable by the method of Parlow (1961) in all other groups of animals. The data offer direct evidence for the existence of a stimulatory action of estrogen on LH release (" positive " feedback effect); they also indicate that the crucial site for this stimulatory action is the median eminence. Similar results have been reported by Kannwischer *et al.* (1967) and by Smith and Davidson (1968).

The question may be asked whether such an effect operates also in adult animals and whether it represents an essential step in the process of ovulation. The following data seem to indicate this to be the case. The administration of estrogen at a proper time of the cycle has been shown to induce ovulation in several mammalian species (Everett, 1961, 1964; Greep, 1961; Hammond *et al.*, 1942; Hohlweg, 1934; Piper and Foote, 1968; Ramirez and Sawyer, 1965). Estradiol given to norethisterone-treated rats overcomes the delaying action norethisterone exerts on ovulation (McDonald and Gilmore, 1969); estradiol also overcomes the inhibition of the postcoital elevation in ovarian progestin output induced by norethisterone in the rabbit (Hilliard *et al.*, 1966). Of interest in this connection also are the data recently reported by Ferin *et al.* (1969); they have shown that antisera which block the biological effects of estradiol-17-β inhibit ovulation in Pregnant Mare Serum (PMS)-treated immature rats, and that replacement of the blocked endogenous estradiol with diethylstilbestrol (a synthetic estrogen whose activity is not inhibited by the antiserum) restores ovulation. The fact that estrogen may increase plasma levels of LH in adult animals following either systemic administration or placement into the median eminence has been proved, using both conventional and radioimmunological procedures (Brown *et al.*, 1969; Palka *et al.*, 1966; Callantine *et al.*, 1966; Goding *et al.*, 1969; Radford *et al.*, 1969).

On the other hand, it appears clearly established, in both experimental animals and women, that estrogen levels increase in the peripheral circulation before the ovulatory LH peak is observed (rat: Barnea *et al.*, 1968; Brom and Schwartz, 1968; Hori *et al.*, 1968; Lawton and Sawyer, 1968; Kobayashi *et al.*, 1969; Schwartz and Lawton, 1968; Yoshinaga *et al.*, 1969; monkey: Hopper and Tullner, 1969; women: Burger *et al.*, 1968; Baird and Guevara, 1969; Corker *et al.*, 1969; Korenman *et al.*, 1969).

E. " Positive " Feedback Effect of Progesterone on Luteinizing Hormone Secretion

A large quantity of data obtained in both experimental animals and women suggest that progesterone may facilitate the release of ovulatory amounts of LH. It is not clear, however, whether this can be considered a simple " positive " feedback effect of progesterone (Ramirez, 1969). Data have been presented which indicate that progesterone reduces the " threshold " for the activation of the neural processes that lead to LH release (Beyer and Sawyer, 1969; Sawyer, 1967; Döcke and Dörner, 1969), and that it potentiates the " positive " feedback effect that estrogen exerts on LH secretion (Grayburn and Brown-Grant, 1968; Hagino and Goldzieher, 1968). The evidence available will be considered in some detail because of the theoretical and practical implications involved.

Progesterone, when given at a proper time of the estrous cycle, is able to advance ovulation in rats running either a 4- or a 5-day cycle (Brown-Grant, 1967; Everett, 1948, 1961, 1964; Haller and Barraclough, 1968; Kaasjager, 1969; Zeilmaker, 1966); the steroid, if given on the day of proestrus, is also able to overcome the block of ovulation induced by the administration of pentobarbital, atropine (Redmond, 1968; Zeilmaker, 1966), or norethisterone (McDonald and Gilmore, 1969).

It is known that FSH and PMS, when given in sufficient amounts, may induce ovulation in prepuberal animals. Progesterone will permit non ovulatory doses of these gonadotropins to become fully effective. In addition, when superovulatory doses of PMS are used it will increase the number of animals ovulating and the number of ova per animal (Meyer and McCormack, 1967; Zarrow and Gallo, 1969; Ying and Meyer, 1969).

The ability to potentiate the ovulatory effect of PMS in immature rats is shared by several progestational agents. Table IV, taken from a paper by Seth and Martini (1970), offers an example of this fact. It appears that 15 IU of PMS induce ovulation only in 22% of the animals treated. When subcutaneous progesterone is added, the number of animals ovulating increases to about 70%. An even higher number of animals ovulating is achieved when the retroprogestational agent 1,6-dehydro-6-chloro-retroprogesterone (RO 4-8347) is injected subcutaneously at the place of progesterone. In this test it

TABLE IV

INDUCTION OF OVULATION WITH PROGESTERONE AND PROGESTAGENS IN IMMATURE FEMALE RATS GIVEN A NON OVULATORY DOSE OF PMS

Groups [1]	No. of Rats Treated	No. of Rats Ovulating	Percent of Rats Ovulating	Average of Ova per Rat
PMS	23	5	22	5.7
PMS + Progesterone	19	13	69	11.7
PMS + 1,6-dehydro-6-chloro-retroprogesterone (RO 4-8347)	10	10	100	31.6
PMS + 6-dehydro-retroprogesterone (duphaston)	10	7	70	34.9
PMS + 6-dehydro-6-chloro-17-acetoxy-progesterone (chlormadinone)	8	5	63	25.1

[1] The compounds at a dose of 500 μg/rat were given subcutaneously 50 hr following the injection of 15 IU of PMS. Animals were autopsied 72 hr after the administration of PMS.

is interesting that the activity of two steroids closely related to RO 4-8347 (chlormadinone acetate or 6-dehydro-6-chloro-17-acetoxy-progesterone acetate and duphaston or 6-dehydro-retroprogesterone) is not as pronounced as that of RO 4-8347; the number of animals ovulating is similar to that obtained following the administration of progesterone. If one considers the other parameter presented in Table IV, i.e., the average number of ova found in the tubae, it would appear again that treatment with progesterone increases the effect of PMS and that RO 4-8347 is more effective than progesterone; however based on this parameter, the two other steroids appear as potent as RO 4-8347.

The effectiveness of progesterone in advancing ovulation in normally cycling rats and in potentiating the effect of exogenous PMS appears to be related to the presence of a quota of endogenous estrogen. More direct evidence of the interplay between estrogen and progesterone is provided by the following data. Grayburn and Brown-Grant (1968) have shown that immature animals treated with very low amounts of FSH will ovulate only if treated with a combination of estrogen and progesterone, but not with either steroid alone. The exposure to constant light transforms normally ovulating rats into costant estrous-anovulatory animals (Meyer and McCormack, 1967; Wurtman, 1967); they can be induced to ovulate by treatment with progesterone; a synergistic effect between estrogen and progesterone can be postulated, since animals exposed to constant light have rather high levels of estrogen in the circulation. Rabbits normally ovulate only after exposure to sexual stimulations; however, they may be induced to ovulate " spontaneously " after a treatment with a combination of progesterone plus estrogen; progesterone is ineffective if given to estrogen-deprived rabbits (Sawyer *et al.*, 1950). Immature rats may ovulate following the administration of small doses of estrogen (Hohlweg, 1934; Motta *et al.*, 1968; Ramirez and Sawyer, 1965). Döcke and Dörner, (1966, 1969) have recently reported also that this type of experimentally-induced ovulation can be facilitated by progesterone; it is interesting that both the systemic administration of the steroid (Döcke and Dörner, 1966) and its stereotaxic implantation into the arcuate ventromedial region of the hypothalamus are effective (Döcke and Dörner, 1969).

Caligaris *et al.* (1968) have recently shown that castrated female rats in which plasma levels of LH have been lowered by the chronic administration of low doses of estrogen exhibit an abrupt increase in plasma LH titers following the administration of one single dose of progesterone. Similar data have been reported by Odell and Swerdloff (1968) in postmenopausal women. In these subjects a chronic treatment with estrogen results in a conspicuous reduction of plasma levels of LH; the administration of a single dose of progesterone or of 6-methyl-17-acetoxy-progesterone, after LH has reached its lowest levels, induces an artificial peak of LH secretion that is very similar in shape and magnitude to that physiologically appearing at mid-cycle in normally ovulating women.

A few data recently published in literature seem to indicate that estrogen is not always essential for the effect of progesterone on LH secretion to appear. Ferin *et al.* (1969) have found that exogenous progesterone is able to induce ovulation in animals in which ovulation has been inhibited by treatment with an antiserum that blocks the biological effects of estradiol. Taleisnik *et al.* (1969) have observed that castrated female rats in which plasma levels of LH have been lowered by the chronic administration of testosterone exhibit an abrupt increase in plasma LH titers following the administration of a single dose of progesterone. Progesterone can also induce a comparable increase of plasma LH in castrated male rats pretreated with testosterone.

The question may be asked whether progesterone participates in the physiological release of LH that occurs just prior to spontaneous ovulation. In order to answer this question it would be essential to establish whether progesterone is present in the general circulation before the release of LH occurs. The available information indicates that low amounts of this steroid are present in all animal species before LH release and ovulation (Eto *et al.*, 1962; Telegdy and Endröczi, 1963; Lindner and Zmigrod, 1967; Feder *et al.*, 1968; Hashimoto *et al.*, 1968). There is also some evidence suggesting that progesterone titers might start rising before the ovulatory LH peak (Goldman *et al.*, 1969; Van der Molen and Groen, 1965; Saxena *et al.*, 1968; Neill *et al.*, 1969); however, this evidence has not been confirmed (Neill *et al.*, 1967; Yoshimi and Lipsett, 1968; Cargille *et al.*, 1969; Feder *et al.*, 1969; Goldman and Danhof, 1969; Johansson and Wide, 1969; Chatterton *et al.*, 1968; Roser and Bloch 1969).

III. " Short " Feedback Mechanisms

A. " Short " Feedback Effect of Luteinizing Hormone on Luteinizing Hormone Secretion

The secretion of LH prior to spontaneous ovulation follows a well-defined pattern in many mammalian species; it is peculiar that plasma levels of the hormone increase abruptly and return very rapidly to basal levels (Cargille *et al.*, 1969; Faiman and Ryan, 1967; Neill *et al.*, 1967; Monroe *et al.*, 1969). Surprisingly enough, when LH secretion is induced artificially by treatment with either estrogen (Goding *et al.*, 1969) or progesterone and its derivatives (Caligaris *et al.*, 1968; Odell and Swerdloff, 1968; Taleisnik *et al.*, 1969), plasma LH shows a peak very similar to the one observed during normal ovulation; this is true for all species studied so far.

The rapidity with which the LH peak falls off could be due: (1) to the lack of sufficient stores of LH in the pituitary so that LH hypersecretion cannot be sustained; or (2) to the fact that high plasma levels of LH can stop very rapidly and very efficiently the secretion of further amounts of LH.

TABLE V

EFFECT OF IMPLANTS OF LH IN THE MEDIAN EMINENCE (ME), THE ANTERIOR PITUITARY (PIT), AND THE CEREBRAL CORTEX (CC) ON PITUITARY AND PLASMA LH LEVELS OF CASTRATED FEMALE RATS [1]

Groups [2]	Pituitary Wt (mg)	Pituitary LH (μg/mg [3])	Limits (95%)	Plasma LH
ME: sham (15)	10.6 ± 0.62	3.82	(2.4–5.6)	Present
		4.68	(3.1–7.2)	
ME: LH (14)	10.1 ± 0.45	1.94	(1.2–2.6)	Absent
		1.57	(1.1–2.4)	
Pit: LH (8)	9.8 ± 0.74	4.11	(2.6–6.3)	Present
CC: LH (8)	11.3 ± 0.60	3.06	(1.4–4.1)	Present
		3.44	(2.3–5.4)	
ME: FSH (6)	8.8 ± 0.37	4.36	(2.8–7.2)	Present

[1] Values are means ± SE.
[2] Number of rats in parentheses.
[3] Microgram equivalents of NIH-LH-B-1 bovine per mg wet weight of pituitary tissue.

The first hypothesis does not seem acceptable, since Schwartz and Bartosik (1962) have shown that after the preovulatory discharge of LH, the pituitary still contains rather high amounts of the hormone; in addition, the administration of exogenous progestins can prolong LH hypersecretion after the peak is over (Hilliard *et al.*, 1967).

The possibility that the rapid return of LH to basal levels might be due to the fact that LH is able to shut off its own secretion has been explored by Dávid *et al.* (1966). They have shown (Table V) that the implantation of LH into the median eminence of castrated female rats results in a very significant decrease of pituitary LH stores. In this experiment, castration was performed in order to avoid the possibility that LH absorbed from the implantation site might activate the gonads and operate through the " long " feedback mechanism. The effect of implants of LH on pituitary LH stores is seen only following placement in the median eminence; intrapituitary or cerebral cortical implants are completely ineffective. The effect is also specific for LH, since FSH implants performed in the median eminence region are unable to duplicate it. Table V also shows that median eminence implants of LH reduce the elevated levels of LH found in the plasma of castrated female rats. These data, as well as similar ones reported by Corbin and Cohen (1966) and by Corbin (1966), suggest that the median eminence contains receptors sensitive to changing levels of LH and that when these receptors are activated by the presence of increased levels of the hormone, they operate in such a fashion that they interrupt a further secretion of LH. The data make it likely that the rapid return to basal levels of the LH mid-cycle peak is due to this " short " feedback effect of LH.

IV. " Ultrashort " Feedback Mechanisms

A. " Ultrashort " Feedback Effect of Follicle-Stimulating Hormone Releasing Factor on Follicle-Stimulating Hormone Releasing Factor Secretion

It has been recently decided to explore the possibility that the synthesis, the storage, and the release of the releasing factors might be influenced also by changes of the levels of the releasing factors themselves in the general circulation. The experiments to be reported here were planned in order to study whether the increase of the circulating levels of FSHRF (obtained by chronically administering a crude hypothalamic extract containing FSHRF) might influence the storage of FSHRF at hypothalamic level. In order to avoid the possibility that the administration of exogenous FSHRF might activate the secretion of FSH, and consequently of sex steroids, and might operate via the traditional " short " and " long " feedback mechanisms, the experiments were performed in hypophysectomized-castrated animals. Two groups of male rats, castrated for three weeks and hypophysectomized for one week, were used; one of these groups was injected subcutaneously with an extract of rat hypothalamus (1 median eminence per rat per day for 5 days); the other was treated in a similar way with saline solution. Normal controls were also studied. Animals were killed 3 hr after the last injection. Figure 3 indicates that the treatment with the median eminence extract results in a considerable reduction of hypothalamic stores of FSHRF. As expected, these were particularly elevated in castrated-hypophysectomized controls (Motta, 1969; Motta *et al.*, 1969b). In another set of experiments it was shown that cerebrocortical or liver extracts do not modify FSHRF stores in castrated-hypophysectomized rats (Hyyppa *et al.*, 1970).

Since the hypothalamic extract used was not contaminated with sex steroids or with FSH, the effect observed cannot be due to the

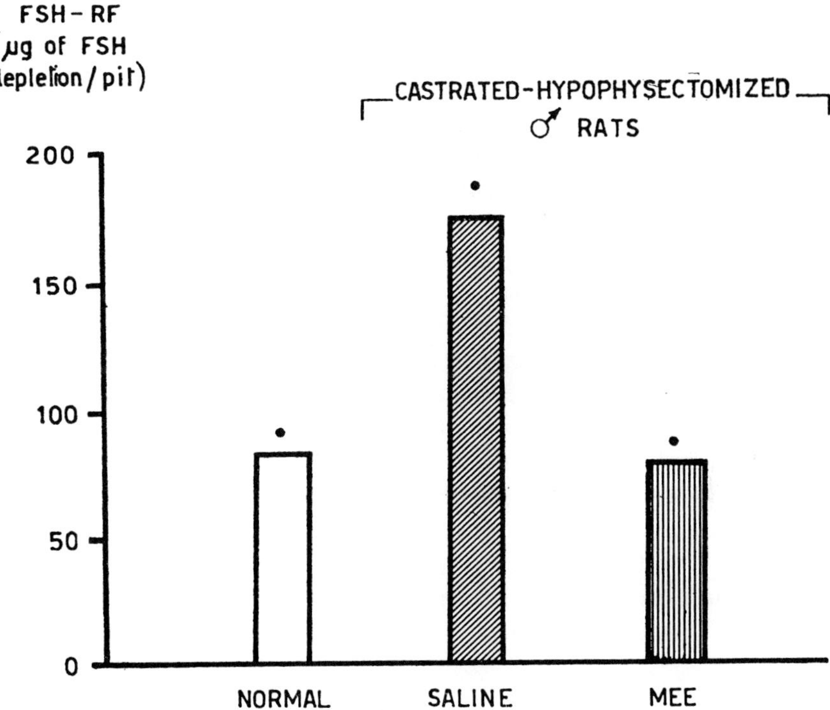

FIG. 3. Effect of systemic administration of median eminence extracts (MEE) on hypothalamic FSHRF stores in castrated-hypophysectomized male rats.

usual " long " and " short " feedback effects of these hormones (see Sections II, A, B and III, A). The data, then, suggest that some unknown factor present in the hypothalamic extract is able to reduce hypothalamic stores of FSHRF. Since the hypothalamic extract injections contained high amounts of FSHRF, it has been tentatively suggested that the administration of exogenous FSHRF is responsible for the effect observed. Even if it is difficult on the basis of this preliminary evidence to establish whether this reduction of stores is due to inhibition of synthesis of FSHRF or to stimulation of its release, it seems reasonable to suggest that the brain contains elements that are sensitive to changing levels of releasing factors. The possibility that some hypothalamic products might influence hypothalamic function is supported by data in the literature. Two well-known polypeptides of hypothalamic origin, oxytocin and vasopressin, apparently exert such effects (see Motta *et al.*, 1969b, for references).

V. Conclusions

The data presented in this article clearly indicate that the hypothalamus participates in several types of endocrine feedback mechanisms. The data discussed suggest that the " long, " the " short, " and the " ultrashort " feedback systems all operate through receptors located in the hypothalamus. The experiments involving localized implants have also indicated that feedback receptors are maximally concentrated in the basal medial hypothalamus and particularly in the median eminence region. On the other hand, the results discussed in the article by Mess and associates in this volume have validated the hypothesis that the releasing factors are synthesized in regions located far from the median eminence. It is, then, apparent that the areas in which the releasing factors are synthesized do not overlap with those where the information from the periphery is received. Nervous interconnections between the afferent component (median eminence receptor area) and the efferent-executive component (nuclei synthesizing the releasing factors) must be postulated.

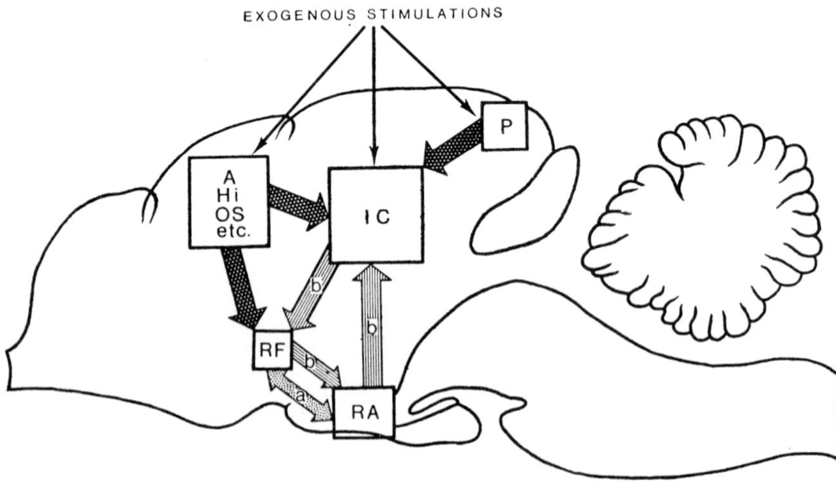

FIG. 4. Schematic representation of the pathways that might participate in transmitting the information received from feedback receptors to the sites where the releasing factors are synthesized. See text for more details on circuits (a) and (b). Abbreviations: A, amygdala; Hi, hippocampus; OS, olfactory structures; P, pineal gland; RA, receptor area; RF, neurons synthesizing the releasing factors; IC, integrative center.

Theoretically speaking, two types of circuits (Fig. 4) may be anticipated: (a) a direct one from median eminence receptors (and probably from other receptor sites located outside the hypothalamus) (Mangili *et al.*, 1966; Mess and Martini, 1968) to the cell bodies where the releasing factors are synthesized; or (b) indirect ones from the median eminence to the releasing factors producing cells via intermediate centers of integration. These centers (located either in the hypothalamus or outside the hypothalamus) might be responsible for integrating the information received from feedback receptors with that originating in other brain structures (amygdala, hippocampus, pineal gland, etc.) or provided by exogenous influences (light, temperature, etc.) (Fig. 4). According to the latter hypothesis, endocrine feedback systems might be considered like multisynaptic reflex loops. The elucidation of all pathways participating in these nervous circuits appears to be of primary importance and will be a very fruitful field of research in the coming few years.

Acknowledgments

The experimental work performed in the authors' laboratory and described here was supported by funds of the Department of Pharmacology of the University of Milan and by the following grants: R/00088 of the World Health Organization, Geneva, Switzerland; AM-10119-01-02-03 and AM-11783-01-02-03 of the National Institute of Health, Bethesda, Maryland; M 64-64, M 65-118; M 66-121 of the Population Council, New York; and 67-530 of the Ford Foundation, New York. Gifts of FSH, and LH were made by the National Institute of Health, Bethesda, Maryland. Ro 4-8347, Duphaston, and Chlormadinone were supplied by F. Hoffman La Roche and Co., Basle, Switzerland.

All such support is gratefully acknowledged.

References

Andjus, R. K., and Kamberi, I. (1966). *Excerpta Med. Intern. Congr. Ser.* **111**, 359.

Baird, D. T., and Guevara, A. (1969). *J. Clin. Endocrinol.* **29**, 149.

Barnea, A., Gershonowitz, T., and Shelesnyak, M. C. (1968). *J. Endocrinol.* **41**, 281.

Benson, B., Sorrentino, S., and Evans, J. S. (1969). *Endocrinology* **84**, 369.

Beyer, C., and Sawyer, C. H. (1969). *In* " Frontiers in Neuroendocrinology 1969 " (W. F. Ganong and L. Martini, eds.), pp. 255–287. Oxford University Press, New York.

Brom, G. M., and Schwartz, N. B. (1968). *Neuroendocrinology* **3**, 366.

Brown, J. M., Catt, K. J., Cumming, I. A., Goding, J. R., Kaltenbach, C. C., and Mole, B. J. (1969). *J. Physiol.* (London) **201**, 98 P.

Brown-Grant, K. (1967). *J. Physiol.* (London) **190**, 101.

Burger, H. G., Catt, K. J., and Brown, J. B. (1968). *J. Clin. Endocrinol.* **28**, 1508.

Caligaris, L., Astrada, J. J., and Taleisnik, S. (1967). *Endocrinology* **81**, 1261.

Caligaris, L. Astrada, J. J., and Taleisnik, S. (1968). *Acta Endocrinol.* **59**, 177.
Callantine, M. R., Humphrey, R. R., and Nesset, B. L. (1966). *Endocrinology* **79**, 455.
Cargille, C. M., and Ross, G. T. (1968). *Lancet* **I**, 924.
Cargille, C. M., Ross, G. T., and Yoshimi, T. (1969). *J. Clin. Endocrinol.* **29**, 12.
Chatterton, R. T., Jr., MacDonald, G. J., and Greep, R. O. (1968). *Endocrinology* **83**, 1.
Chowers, I., and McCann, S. M. (1967). *Proc. Soc. Exp. Biol. Med.* **124**, 260.
Corbin, A. (1966). *Endocrinology* **77**, 893.
Corbin, A., and Cohen, A. I. (1966). *Endocrinology* **78**, 41.
Corbin, A., Mangili, G., Motta, M., and Martini, L. (1965). *Endocrinology* **76**, 811.
Corker, C. S., Naftolin, F., and Exley, D. (1969). *Nature* **222**, 1063.
Dávid, M. A., Fraschini, F., and Martini, L. (1965a). *Experientia* **21**, 483.
Dávid, M. A., Fraschini, F., and Martini, L. (1965b). *C. R. Acad. Sci.* (Paris) **261**, 2249.
Dávid, M. A., Fraschini, F., and Martini, L. (1966). *Endocrinology* **78**, 55.
Davidson, J. M. (1966). *In* " Neuroendocrinology " (L. Martini and W. F. Ganong, eds.), pp. 565–611. Academic Press, New York.
Davidson, J. M. (1969). *In* " Frontiers in Neuroendocrinology 1969 " (W. F. Ganong and L. Martini, eds.), pp. 343–388. Oxford University Press, New York.
Davidson, J. M., and Sawyer, C. H. (1961). *Acta Endocrinol.* **37**, 385.
Döcke, F., and Dörner, G. (1966). *J. Endocrinol.* **36**, 209.
Döcke, F., and Dörner, G. (1969). *Neuroendocrinology* **4**, 139.
Eto, J., Masuda, H., Suzuki, Y., and Hosi, T. (1962). *Jap. J. Anim. Reprod.* **8**, 34.
Everett, J. W. (1948). *Endocrinology* **43**, 389.
Everett, J. W. (1961). *In* " Sex and Internal Secretions " (W. C. Young, ed.). Vol. I, pp. 497–555. Williams and Wilkins, Baltimore.
Everett, J. W. (1964). *Physiol. Rev.* **44**, 373.
Exley, D., Gellert, R. J., Harris, G. W., and Nadler, R. D. (1968). *J. Physiol.* (London) **195**, 697.
Faiman, C., and Ryan, R. J. (1967). *J. Clin. Endocrinol.* **27**, 1711.
Feder, H. H., Resko, J. A., and Goy, R. W. (1968). *J. Endocrinol.* **40**, 505.
Feder, H. H., Brown-Grant, K., Corker, C. S., and Exley D. (1969). *J. Endocrinol.* **43**, XXIX.
Fendler, K., and Endröczi, E. (1966). *Neuroendocrinology* **1**, 129.
Ferin, M., Zimmering, P. E., and Vande Wiele, R. L. (1969). *Endocrinology* **84**, 893.
Flerkó, B., and Szentágothai, J. (1957). *Acta Endocrinol.* **26**, 121.
Franchimont, P. (1966). *J. Label. Compounds* **2**, 303.
Franchimont, P. (1968). *Ann. Endocrinol.* (Paris) **29**, 403.
Fraschini, F. (1969). *In* " Progress in Endocrinology " (C. Gual, ed.), pp. 637–644. Excerpta Medica, Amsterdam.
Fraschini, F., and Motta, M. (1967). *Program 49th Meeting Endocrine Soc.*, p. 128.
Fraschini, F., Motta, M., and Martini, L. (1968). *Experientia* **24**, 230.
Goding, J. R., Catt, K. J., Brown, J. M., Kaltenbach, C. C., Cumming, I. A., and Mole, B. J. (1969). *Endocrinology* **85**, 133.
Goldman, B. D., and Danhof, I. E. (1969). *Fed. Proc.* **28**, 771.
Goldman, B. D., and Mahesh, V. B. (1968). *Endocrinology* **83**, 97.
Goldman, B. D., and Mahesh, V. B. (1969). *Endocrinology* **84**, 236.
Goldman, B. D., Kamberi, I. A., Siiteri, P. K., and Porter, J. C. (1969). *Endocrinology* **85**, 1137.
Grayburn, J. A., and Brown-Grant, K. (1968). *J. Endocrinol.* **42**, 409.
Greep, R. O. (1961). *In* " Sex and Internal Secretions " (W. C. Young, ed.). Vol. I, pp. 240–301. Williams and Wilkins, Baltimore.

Hagino, N., and Goldzieher, J. W. (1968). *In* " Proc. 24th Intern. Congr. International Union of Physiological Sciences ". Vol. VII, p. 175. Federation of American Societies for Experimental Biology, Washington.
Halász, B. (1969). *In* " Frontiers in Neuroendocrinology 1969 " (W. F. Ganong and L. Martini, eds.), pp. 307-342. Oxford University Press, New York.
Haller, E. W., and Barraclough, C. A. (1968). *Proc. Soc. Exp. Biol. Med.* **129,** 291.
Hammond, J., Jr., Hammond, J., and Parkes, A. S. (1942). *J. Agr. Sci.* **32,** 308.
Harris, G. W., and Sherratt, R. M. (1969). *J. Physiol.* (London) **203,** 59.
Hashimoto, I., Henricks, D. M., Anderson, L. L., and Melampy, R. M. (1968). *Endocrinology* **82,** 333.
Hilliard, J., Hayward, J. N., Croxatto, H. B., and Sawyer, C. H. (1966). *Endocrinology* **78,** 151.
Hilliard, J., Penardi, R., and Sawyer, C. H. (1967). *Endocrinology* **80,** 901.
Hohlweg, W. (1934). *Klin. Wochenschr.* **13,** 92.
Hopper, B. R., and Tullner, W. W. (1969). *Fed. Proc.* **28,** 771.
Hori, T., Ide, M., and Miyake, T. (1968). *Endocrinol. Japon.* **15,** 215.
Hyyppa, M., Motta, M., and Martini, L. (1970). *Neuroendocrinology* (In press).
Igarashi, M., Kamioka, J., Ehara, Y., and Matsumoto, S. (1967). *Fertil. Steril.* **18,** 672.
Jelinek, J. M., Seda, M., and Marhan, O. (1968). *Steroids* **11,** 565.
Johansson, E. D. B., and Wide, L. (1969). *Acta Endocrinol.* **62,** 82.
Kaasjager, W. A. (1969). *J. Endocrinol.* **43,** XIX.
Kanematsu, S., and Sawyer, C. H. (1963). *Am. J. Physiol.* **205,** 1073.
Kanematsu, S., and Sawyer, C. H. (1964). *Endocrinology* **75,** 579.
Kannwischer, R., Wagner, J., and Critchlow, V. (1967). *Anat. Record* **157,** 268.
Kobayashi, F., Hara, K., and Miyake, T. (1969). *Endocrinol. Japon.* **16,** 251.
Korenman, S., Perrin, L., and Rao, B. R. (1969). *Program. 51st Meeting Endocrine Soc.*, p. 116.
Labhsetwar, A. P. (1968a). *Anat. Record* **160,** 380.
Labhsetwar, A. P. (1968b). *J. Reprod. Fertil.* **17,** 101.
Lawton, I. E., and Sawyer, C. H. (1968). *Endocrinology* **82,** 831.
Lindner, H. R., and Zmigrod, A. (1967). *Acta Endocrinol.* **56,** 16.
Lisk, R. D. (1960). *J. Exp. Zool.* **145,** 197.
Lisk, R. D. (1965). *Acta Endocrinol.* **48,** 209.
Lisk, R. D. (1969). *Neuroendocrinology* **4,** 368.
Mangili, G., Motta, M., and Martini, L. (1966). *In* " Neuroendocrinology " (L. Martini and W. F. Ganong, eds.). Vol. I, pp. 297–370. Academic Press, New York.
Martini, L., Fraschini, F., and Motta, M. (1968a). *In* " Endocrinology and Human Behaviour " (R. P. Michael, ed.), pp. 175–187. Oxford University Press, London.
Martini, L., Fraschini, F., and Motta, M. (1968b). *Recent Progr. Hormone Res.* **24,** 439.
Martini, L., Fraschini, F., and Motta, M. (1969). *In* " Advances in the Biosciences " (G. Raspé, ed.), pp. 201–212. Pergamon Press-Vieweg, Braunschweig.
Martini, L., Piva, F., and Motta, M. (1970). *In* " Ovo Implantation, Human Gonadotropins and Prolactin " (P. O. Hubinont, ed.), pp. 170–180. S. Karger, Basel.
McCann, S. M., and Ramirez, V. D. (1964). *Recent Progr. Hormone Res.* **20,** 131.
McClintock, J. A., and Schwartz, N. B. (1968). *Endocrinology* **83,** 433.
McDonald, P. G., and Gilmore, D. P. (1969). *J. Endocrinol.* **45,** 51.
Mess, B., and Martini, L. (1968). *In* " Recent Advances in Endocrinology " (V. H. T. James, ed.), pp. 1–49. J. and A. Churchill, London.
Mess, B., Fraschini, F., Motta, M., and Martini, L. (1967). *In* " Hormonal Steroids " (L. Martini, F. Fraschini and M. Motta, eds.), pp. 1004-1013. Excerpta Medica, Amsterdam.
Meyer, R. K., and McCormack, C. E. (1967). *J. Endocrinol.* **38,** 187.

Midgley, A. R., Jr., and Jaffe, R. B. (1968). *J. Clin. Endocrinol.* **28**, 1699.
Minaguchi, H., and Meites, J. (1967). *Endocrinology* **81**, 826.
Moll, J., and Zeilmaker, G. H. (1966). *Acta Endocrinol.* **51**, 281.
Monroe, S. E., Rebar, R. W., Gay, V. L., and Midgley, A. R., Jr. (1969). *Endocrinology* **85**, 720.
Motta, M. (1969). *In* "Progress in Endocrinology" (C. Gual, ed.), pp. 523–531. Excerpta Medica, Amsterdam.
Motta, M., Mangili, G., and Martini, L. (1965). *Endocrinology* **77**, 392.
Motta, M., Fraschini, F., Giuliani, G., and Martini, L. (1968). *Endocrinology* **83**, 1101.
Motta, M., Sterescu, N., Piva, F., and Martini, L. (1969a). *Acta Neurol. Belg.* **69**, 501.
Motta, M., Fraschini, F., and Martini, L. (1969b). *In* "Frontiers in Neuroendocrinology 1969" (W. F. Ganong and L. Martini, eds.), pp. 211–253. Oxford University Press, New York.
Motta, M., Piva, F., and Martini, L. (1970). *Bull. schweiz. Akad. med. Wiss.* (In press).
Negro-Vilar, A., Dickerman, E., and Meites, J. (1968). *Proc. Soc. Exp. Biol. Med.* **127**, 751.
Neill, J. D., Johansson, E. D. B., Datta, J. K., and Knobil, E. (1967). *J. Clin. Endocrinol.*, **27**, 1167.
Neill, J. D., Johansson, E. D. B., and Knobil, E. (1969). *Endocrinology* **84**, 45.
Odell, W. D., and Swerdloff, R. S. (1968). *Proc. Natn. Acad. Sci. U.S.* **61**, 529.
Odell, W. D., Parlow, A. F., Cargille, C. M., and Ross, G. T. (1968). *J. Clin. Invest.* **47**, 2551.
Palka, Y. S., Ramirez, V. D., and Sawyer, C. H. (1966). *Endocrinology* **78**, 487.
Parlow, A. F. (1961). *In* "Human Pituitary Gonadotropins" (A. Albert, ed.), pp. 300–326. C. C. Thomas, Springfield, Ill.
Parlow, A. F., Anderson, L. L., and Melampy, R. M. (1964). *Endocrinology* **75**, 365.
Piacsek, B. E., and Meites, J. (1967). *Neuroendocrinology* **2**, 129.
Piacsek, B. E., Armstrong, D. T., and Greep, R. O. (1969). *Endocrinology* **84**, 1184.
Piper, E. L., and Foote, W. C. (1968). *J. Reprod. Fertil.* **16**, 253.
Piva, F., Sterescu, N., Zanisi, M., and Martini, L. (1969). *Bull. World Health Org.* **41**, 275.
Radford, H. M., Wheatley, I. S., and Wallace, A. L. C. (1969). *J. Endocrinol.* **44**, 135.
Rakha, A. M., and Robertson, H. A. (1965). *J. Endocrinol.* **31**, 245.
Ramirez, V. D. (1969). *In* "Progress in Endocrinology" (C. Gual, ed.), pp. 532–541. Excerpta Medica, Amsterdam.
Ramirez, V. D., and Sawyer, C. H. (1965). *Endocrinology* **76**, 1158.
Ramirez, V. D., Abrams, R. M., and McCann, S. M. (1964). *Endocrinology* **75**, 243.
Redmond, W. C. (1968). *Endocrinology* **83**, 1013.
Robertson, H. A., and Rakha, A. M. (1966). *J. Endocrinol.* **35**, 177.
Roser, S., and Bloch, R. B. (1969). *C. R. Acad. Sci.* (Paris) **268**, 1318.
Ross, G. T., Odell, W. D., and Rayford, P. L. (1966). *Lancet* **II**, 1255.
Santolucito, J. A., Clegg, M. T., and Cole, H. H. (1960). *Endocrinology* **66**, 273.
Sawyer, C. H. (1967). *In* "Hormonal Steroids" (L. Martini, F. Fraschini and M., Motta, eds.), pp. 123–135. Excerpta Medica, Amsterdam.
Sawyer, C. H., Everett, J. W., and Markee, J. E. (1950). *Proc. Soc. Exp. Biol. Med.* **74**, 185.
Saxena, B. B., Demura, H., Gandy, H. M., and Peterson, R. E. (1968). *J. Clin. Endocrinol.* **28**, 519.
Schmidt-Eldmendorff, H., and Kopera, H. (1968). *Lancet* **I**, 1194.
Schwartz, N. B., and Bartosik, D. (1962). *Endocrinology* **71**, 756.
Schwartz, N. B., and Lawton, I. E. (1968). *Neuroendocrinology* **3**, 9.
Seth, P., and Martini, L. (1970). *Neuroendocrinology*. (In press).

Short, R. E., Peters, J. B., First, N. L., and Casida, L. E. (1968). *J. Animal Sci.* **27**, 705.
Simpson, M. E., Van Wagenen, G., and Carter, F. (1956). *Proc. Soc. Exp. Biol. Med.* **91**, 6.
Smith, E. R., and Davidson, J. M. (1968). *Endocrinology* **82**, 100.
Smith, E. R., Weick, R. F., and Davidson, J. M. (1969). *Endocrinology* **85**, 1129.
Steelman, S. L., and Pohley, F. M. (1953). *Endocrinology* **53**, 604.
Stevens, V. C., Vorys, N., Besch, P. K., and Barry, R. D. (1965). *Metabolism* **14**, 327.
Stevens, V. C., Goldzieher, J. W., and Vorys, N. (1968). *Am. J. Obst. Gynecol.* **102**, 95.
Swerdloff, R. S., and Odell, W. D. (1969). *J. Clin. Endocrinol.* **29**, 157.
Taleisnik, S., Caligaris, L., and Astrada, J. J. (1969). *J. Endocrinol.* **44**, 313.
Taymor, M. L., Aono, T., and Pheteplace, C. (1968). *Acta Endocrinol.* **59**, 298.
Telegdy, G., and Endröczi, E. (1963). *Steroids* **2**, 119.
Van der Molen, H. J., and Groen, D. (1965). *J. Clin. Endocrinol.* **25**, 1625.
Wurtman, R. J. (1967). *In* " Neuroendocrinology " (L. Martini and W. F. Ganong, eds.), pp. 19–59. Academic Press, New York.
Ying, S. Y., and Meyer, R. K. (1969). *Endocrinology* **84**, 1466.
Yoshimi, T., and Lipsett, M. B. (1968). *Steroids* **11**, 527.
Yoshinaga, K., Hawkins, R. A., and Stocker, J. F. (1969). *Endocrinology* **85**, 103.
Zarrow, M. X., and Gallo, R. V. (1969). *Endocrinology* **84**, 1274.
Zeilmaker, G. H. (1966). *Acta Endocrinol.* **51**, 461.
Zeilmaker, G. H., and Moll, J. (1967). *Acta Endocrinol.* **55**, 378.

Integrated Hypothalamic Responses to Stress

P. G. SMELIK

I. Introduction

The response to noxious or threatening stimuli is one of the basic adaptive mechanisms of animals and man. In spite of the fact that the nervous system and the endocrine system have a fundamentally different mode of operation, it has long been recognized that both are involved in adaptation or defense mechanisms. It has become increasingly clear that the organism responds to stimuli that would disturb the homeostatic equilibrium with an integrated neuroendocrine reaction; it is interesting to see how research and concepts starting from different points of view have converged in the field of neuroendocrinology.

II. Concepts of Stress

The studies of Cannon (1914, 1939) have emphasized the role of the sympathetic nervous system in the defense reaction of the body. His work has shown that the adrenal medullary hormones are secreted by a sympathetic reflex response to emotional stimuli that prepare the animal for "fight or flight." The emergency function of this mechanism, which has a clearly neuroendocrine character, was thought to subserve a goal-directed activity of the organism which required mobilization and expenditure of energy.

Of course we know now that this is only part of the story, but the mechanism is a very essential and immediate phase in the adaptive pattern of reaction to noxious stimuli. Nevertheless it seems that in the history of concepts, this part has been somewhat neglected for some time. This is mainly due to the emphasis that has been laid upon the role of the adrenal cortex since the discovery of the function of the adrenocortical steroids and their regulation.

The concept of " stress, " advocated by Selye (1953), focused mainly on the idea that a great variety of harmful stimuli are capable of eliciting a fixed and stereotypical response in which hyperactivity of the adrenal cortex is the central phenomenon. The pattern of responses belonging to the General Adaptation Syndrome can be described in terms of an hourglass model in which a wide spectrum of stimuli converge into a narrow common element and then give rise to a great number of adaptive reactions that can be modified by particular conditions or demands.

The idea of associating the stress response with a stereotypical pituitary-adrenal activation has been enormously fruitful. Nevertheless, at present it is somewhat surprising to realize that as late as 1953 Selye was still puzzled by the ignorance about the nature of the so-called " first mediator, " the stimulus that should promote Adrenocorticotropic Hormone (ACTH) secretion. The essential role of the hypothalamus in the control of the pituitary-adrenal axis apparently was not yet recognized.

At that time, however, it became rapidly clear that the " first mediator " was to be sought in the hypothalamus (which appeared to control the pituitary-adrenal axis), and subsequently an impressive amount of work has been done on the identity of the hypothalamic ACTH-regulating factor. Although vasopressin has been a serious candidate for many years, there is now general agreement that the Corticotropin Releasing Factor (CRF) is not identical to vasopressin. Nevertheless, many workers remain impressed by the high potency and specificity of vasopressin as an ACTH releaser and by the fact that every type of stress readily releases vasopressin in addition to ACTH. More will be said later about this point when evaluating the exceptional position of vasopressin, which clearly cannot be classified as a nonspecific stressful agent.

Meanwhile, the question about the " first mediator " has been shifted from ACTH to CRF. If CRF releases ACTH, what then releases CRF?

III. Possible Role of Sympathetic Nervous System

Based upon Cannon's concept of adrenomedullary hormone discharge during the early phase of the sympathetic emergency reaction, several investigators have assumed that possibly it is the reflex release of catecholamines which in turn activates the pituitary-adrenocortical system. Long and co-workers for instance, presented a concept in which medullary adrenaline was thought to be released during the so-called "autonomic phase". Adrenaline would set off the ACTH hypersecretion and subsequently an increased production of adrenocortical steroids, introducing the "metabolic phase" (McDermott *et al.*, 1950). It must be admitted that this concept nicely linked up adrenomedullary and adrenocortical activation and also promoted adrenaline as a very potent ACTH releaser. In our experiments (Smelik, 1960) as little as 0.6 μg of adrenaline administered over 1 hr as an intramuscular infusion was capable of inducing maximal adrenal ascorbic acid depletion. Curiously, this effect was completely abolished in rats in which the posterior lobe had been removed (Smelik, 1960).

Unfortunately for this unifying concept it soon turned out that neither removal of the adrenal medulla (Vogt, 1951) nor adrenergic blockade (Guillemin, 1955) could prevent the stress-induced ACTH secretion. Therefore it can be concluded that peripheral mechanisms are not essential for the control of ACTH release.

IV. The Hypothalamus

Nevertheless, despite later experimental results, the original idea of an involvement of adrenergic activity in the regulation of the pituitary-adrenal secretions remained alive. Work by Ranson and Magoun (1939) and by Hess (1949) had shown that electrical stimulation of the hypothalamus resulted in two types of reaction patterns, one of which could be elicited by activation of the posterior and medial part of the hypothalamus (characterized by a sympathetic outflow) and the other of which, being of parasympathetic character, resulted from stimulation of the anterior and lateral hypothalamus.

Since it was known that stimulation of the posterior hypothalamus caused an adrenomedullary discharge of catecholamines, it had been thought (Porter, 1954) that this part of the diencephalon would at the same time exert a stimulatory control over ACTH secretion. However, as further experiments progressed, it appeared that the controlling centers for the pituitary-adrenal axis were situated in the anterior part of the hypothalamus. Moreover, later evidence suggested that an inhibitory influence on ACTH release may be exerted by the posterior hypothalamus. It has now been found that lesions in the posterior hypothalamus actually may facilitate ACTH secretion (Smelik, 1959; Knigge *et al.*, 1959).

A neural influence on CRF-producing neurons is generally accepted at present. Analysis of the nature of such inputs cannot be done solely with hypothalamic lesions, since we now know that a number of neural networks are intermingled within the hypothalamus. These networks are not characterized by a distinct and strict localization but by a different type of neurotransmission. Since cholinergic, noradrenergic, and other monoaminergic fiber systems are present in the hypothalamus, it is obvious that specific pharmacological blockade (rather than stimulation!) of a certain type of neurotransmission could give us an indication about the nature of the nervous elements that may be involved in the activation or inhibition of CRF production.

In view of these considerations, we tried to obtain more information about the role of hypothalamic fiber systems in the control of ACTH secretion. Depletion of the monoamines in the median eminence by reserpine implants failed to effect subsequent stimulation of the pituitary-adrenal axis (Smelik, 1967). In contrast, implantation of atropine into the anterior part of the hypothalamus resulted in a prompt and dramatic inhibition of stress-induced ACTH release (Hedge and Smelik, 1968), suggesting a cholinergic link in the activity of the CRF-producing neurons. This would fit into the concept of a balance between the " trophotropic " anterior part and the " ergotropic " posterior part of the hypothalamus, a division that would suggest stimulation (by cholinergic transmission) of CRF production by the anterior part, whereas the posterior portion would activate the release of peripheral noradrenaline and adrenaline. If activation of one part of the hypothalamus would result in suppression of the other part, as has been repeatedly suggested by Gellhorn (1957, 1967), then an inhibitory influence of the posterior part on the controlling center for ACTH secretion would not be inconceivable.

Hypothalamic cholinergic transmission from the supraoptic neurosecretory cells, resulting in vasopressin release, has also been proposed by several workers (Abrahams and Pickford, 1956; Walker, 1957). This suggests that activation of cholinergic impulses in the anterior hypothalamus may result in the concomitant release of vasopressin as well as CRF. This brings us back to the intimate relationship between CRF and vasopressin.

V. Role of Vasopressin

Recently we came across an interesting phenomenon concerning the ACTH-releasing capacity of vasopressin. When the pituitary-adrenal response to stressful stimuli had been blocked by pretreatment with dexamethasone, only two stimuli were still capable of causing a short-lasting release of ACTH; namely mechanical stimulation of the anterior hypothalamus by an electrode, and the injection of vasopressin. This suggested that these stimuli would act at a site distal to the level of corticoid blockade and would be capable of releasing directly from the CRF neurons a small amount of CRF still present in these neurons, in which the synthesis of new CRF had been impaired by the dexamethasone. When such animals so treated were challenged for a second time with vasopressin no further adrenocortical response was obtained. This indicated that the stores of releasable CRF were depleted. In such conditions the effect of a CRF preparation was unchanged (Hedge and Smelik, 1969).

These results suggest that vasopressin may act on the CRF stores in the nerve endings and may deplete them, perhaps in a way similar to the action of indirect sympathicomimetics on adrenergic nerve endings. This might explain why it has always been so difficult to dissociate the effects of vasopressin and CRF.

VI. Integrated Response to Stress

So far, this article has concentrated on hypothalamic mechanisms involved in the regulation of the response to stress. It should be clear that the hypothalamus represents the controlling structure for the integrated neuroendocrine response to stress. It is well to emphasize here that the word " stress " is somewhat misleading, since hypothalamic regulation via the autonomic nervous system and the endocrine system is continuously functioning in the readjustment of the organism to the flux of surrounding conditions. It is the minor and predominantly emotional disturbance of everyday life which requires a continuous adaptation of the body to environmental changes. The very extreme conditions that we apply as a " stressful stimulus " can serve only as an experimental model for a permanently changing situation.

Moreover, pituitary-adrenal activation is but a part of the defense mechanism and one of the corollaries of the stress response. The initial phase is the arousal and activation of the " ergotropic " posterior part of the hypothalamus, resulting in a number of autonomic reactions, e.g., increased heart rate, muscle vasodilatation, adrenomedullary secretion, glucose and fat mobilization, pupillary dilatation, constriction of the capillary bed of the skin, piloerection, and increased respiration (Abrahams *et al.*, 1960). The second phase is an activation of the anterior hypothalamic area (causing the release of vasopressin and ACTH), the production of the protecting glucocorticoids, retention of water and salt, and a number of parasympathetic reactions.

This chain of events can be considered as the basic pattern of the integrated response to stress. If we realize that the hypothalamus in its turn is controlled by higher structures, then it will be obvious that this pattern can be influenced and modified in many respects. Since the hypothalamus is intimately connected with the limbic structures, which subserve many functions connected with emotional and instinctive behavior, it can be expected that both aversive and rewarding stimuli inducing behavioral reactions also cause concomitant neuroendocrine reactions.

Experimental research is beginning to explore this field. There are some indications that limbic structures exert stimulatory as well as inhibitory influences on the pituitary-adrenal axis (Mason, 1958), and it has been reported that, whereas negative reinforcement will activate the adrenocortical system, a positive reinforcement or rewarding stimulus will actually inhibit the system (Mason, 1959; Kling, 1963). The import of such findings to a better understanding of the adaptive mechanisms that respond to changes in the physical and social environment, and especially of maladjustments in psychosomatic disorders, should be recognized at present. A concept that adopts the central position of the hypothalamus as the integrating and controlling structure for both autonomic and endocrine adaptive regulations, would stimulate and facilitate such a development.

References

Abrahams, V. C., and Pickford, M. (1956). *J. Physiol.* (London) **133**, 330.

Abrahams, V. C., Hilton, S. M., and Zbrozyna, A. (1960). *J. Physiol.* (London) **154**, 491.

Cannon, W. B. (1914). *Am. J. Physiol.* **33**, 356.

Cannon, W. B. (1939). "The Wisdom of the Body." Norton, New York.

Gellhorn, E. (1957). "Autonomic Imbalance and the Hypothalamus." University of Minnesota Press, Minneapolis.

Gellhorn, E. (1967). "Principles of Autonomic-Somatic Integrations." University of Minnesota Press, Minneapolis.

Guillemin, R. (1955). *Endocrinology* **56**, 248.

Hedge, G. A., and Smelik, P. G. (1968). *Science* **159**, 891.

Hedge, G. A., and Smelik, P. G. (1969). *Neuroendocrinology* **4**, 242.

Hess, W. R. (1949). "Das Zwischenhirn: Syndrome, Lokalisationen, Funktionen." Schwabe, Basel.

Kling, A. (1963). *Psychosomat. Med.* **25**, 489.

Knigge, K. M., Penrod, C. H., and Schindler, W. J. (1959). *Am. J. Physiol.* **196**, 579.

Mason, J. W. (1958). *In* "Reticular Formation of the Brain." (H. H. Jaspers *et al.*, eds.), pp. 645–661. Little Brown and Co., Boston.

Mason, J. W. (1959). *Recent Progr. Brain Res.* **15**, 345.

McDermott, W. V., Fry, E. G., Brobeck, J. R., and Long, C. N. H. (1950). *Yale J. Biol. Med.* **23**, 52.

Porter, R. W. (1954). *Recent Progr. Brain Res.* **10**, 1.

Ranson, S. W., and Magoun, H. W. (1939). *Ergeb. Physiol.* **41**, 56.

Selye, H. (1953). "The Story of the Adaptation Syndrome." Acta, Montreal.

Smelik, P. G. (1959). "Autonomic Nervous Involvement in Stress-Induced ACTH Secretion." Born, Assen.

Smelik, P. G. (1960). *Acta Endocrinol.* (Copenhagen) **33**, 437.

Smelik, P. G. (1967). *Neuroendocrinology* **2**, 247.

Vogt, M. (1951). *J. Physiol.* **114**, 465.

Walker, J. M. (1957). *In* "The Neurohypophysis" (H. Heller, ed.), pp. 221–229. Butterworths, London.

Control of Secretion of the Pars Nervosa of the Pituitary

M. Pickford

I. Introduction

Before beginning to discuss the control of secretion from the neurohypophysis it is as well to state briefly one or two facts relating to the hormones released, namely, vasopressin and oxytocin. They can be extracted from the whole extent of the hypothalamico-hypophyseal system which consists of two pairs of grouped nerve cells lying in the hypothalamus and their extensions into the pars nervosa. The cells concerned are in the supraoptic and paraventricular nuclei. Pituicytes are abundant in the pars nervosa, though absent from the hypothalamus. It is likely that these cells exert some kind of influence, but since such influence has not been identified, nothing more will be said about the pituicytes.

What will be discussed is the nature of those stimuli which alter the quantity of hormones in the gland or circulating blood, or which lead to a change in function or activity of the target organs. It will then be necessary to consider how it is that these stimuli affect release of the active substance.

II. Factors which Stimulate the Pars Nervosa System

A. Plasma Osmotic Pressure, Antidiuretic Hormone and the Kidneys

The condition of diabetes insipidus draws attention to the pituitary and related hypothalamus. We no longer believe that the anterior lobe makes a diuretic hormone or that vasopressin exerts a diuretic action except when used in large doses. There is ample reason for thinking that small doses of vasopressin induce antidiuresis by a direct action on the kidney and that therefore it may rightly be named the Antidiuretic Hormone, or ADH. Following loss of the neurohypophyseal system, there is abolition of facultative water reabsorption by the kidney, with the result that the 24-hr urine volume greatly increases.

Verney (1926) clearly showed that ADH exerts a moment-to-moment control over the volume of water excreted by the kidneys. First he noted that the isolated perfused kidney excreted a large volume of dilute urine which was able to concentrate in a smaller volume if ADH was added to the perfusate or if the perfusing blood was allowed to pass through an isolated head containing a pituitary gland. No antidiuretic principle was picked up when the perfusate passed through a head from which the pituitary had been removed. Having made this clear he next examined the effect on conscious dogs of orally administered water and noted particularly the time relationship between the peak water load and the maximum of diuresis. There was a time difference of about 10 to 15 min between the two peaks. Verney (1926) suggested that this was due to the cessation or reduction in the rate of release of ADH combined with the rate of disappearance of hormone already in the blood. It was appreciated that water was a good diuretic and normal saline a poor one. This suggests that a change in the osmotic pressure of the plasma

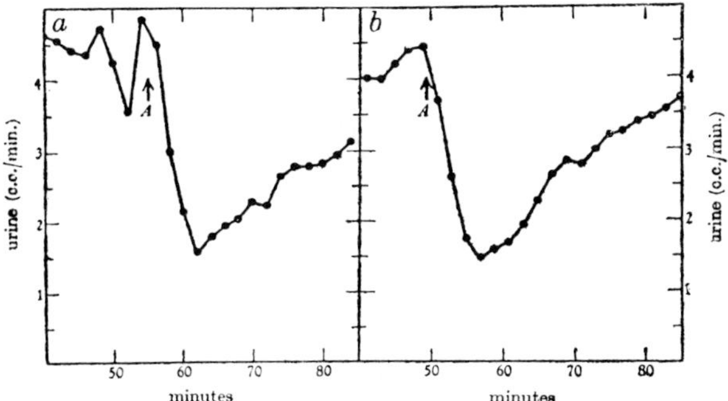

FIG. 1. Inhibition of water diuresis in a dog. In *a* at A, injection of 0.428 *M* NaCl into right carotid artery. In *b* at A, injection of 1 mU of posterior pituitary extract intravenously (From Verney, 1947).

may influence the output of ADH. Verney (1947) tested this idea with great care. He trained dogs for use in water diuresis experiments and prepared them with exteriorized carotid arteries. He could then test the effect on urine output of intracarotid injections of a variety of solutions. Verney (1947) matched any antidiuresis seen with injections of ADH and then assayed in the same animal the quantity of hormone released by the solution injected (Fig. 1). He found that NaCl was a higly effective antidiuretic agent (also Na_2SO_4), that sucrose exerted a moderate effect, glucose very little, and urea none at all. Thus he showed that those substances which exert an osmotic pressure and do not easily pass through cell membranes act as releasing agents for ADH. It was immaterial whether the kidneys were innervated or denervated. The rise in effective concentration of NaCl lay within physiological limits; it was necessary to increase the osmotic pressure of the blood by only 2%. By ligating the external carotid artery and a number of intracranial branches at operation, it was possible later to demonstrate that NaCl solution injected into the exteriorized common carotid artery was still active and must therefore affect a site in the anterior hypothalamus in the region where the supraoptic and paraventricular nuclei are found (Jewell and Verney, 1957).

Verney's work led to the use of the term " osmoreceptor " for those cells in the hypothalamus which are sensitive to changes in plasma

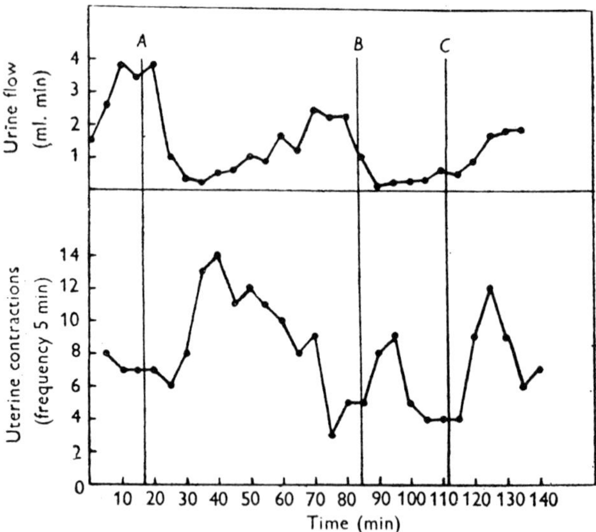

FIG. 2. Uterine motility (below) and inhibition of water diuresis (above) in a dog. At A, 2.4 ml 1.7 *M* NaCl solution injected into a carotid artery. At B, 5 mU pitressin injected intravenously. At C, 5 mU pitocin injected intravenously (From Abrahams and Pickford, 1954).

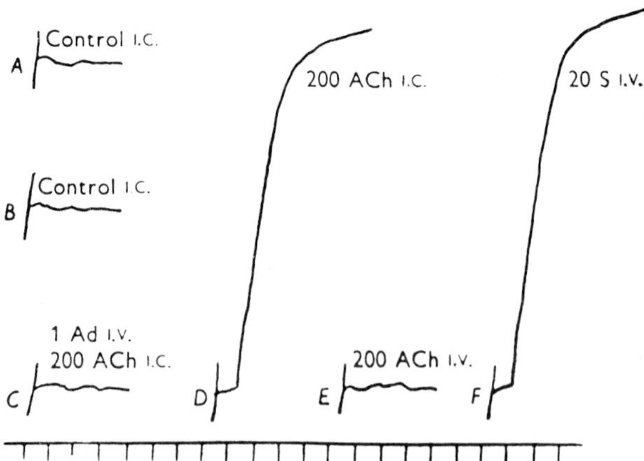

FIG. 3. Record of milk ejection in a conscious dog. All injections into a carotid artery (ic) except where otherwise stated. At A and B, 1 ml 0.9% NaCl solution injected. At C, 1 μg adrenaline intravenously (iv), and 10 sec later, 200 μg acetylcholine (ACh) into carotid artery. At D, 200 μg ACh. At E, 200 μg ACh intravenously. At F, 20 mU oxytocin intravenously (From Pickford, 1960).

osmotic pressure. In several species the intracarotid injection of small volumes of hypertonic NaCl solution brings about the release of oxytocin as well as ADH (Fig. 2) (Abrahams and Pickford, 1954; Holland *et al.*, 1959; Pickford, 1960; also consult Heller and Ginsburg, 1966). A number of drugs also induce hormone release (Fig. 3) (Walker, 1957; Dyball, 1968a).

B. Electrical Stimulation of the Neurohypophyseal Tract

Haterius (1940) showed in anesthetized rabbits that stimulation of the pituitary stalk caused antidiuresis and Harris (1947) demonstrated by remote control in conscious rabbits that electrical stimulation of the neurohypophyseal system brought about the release of both antidiuretic and oxytocic hormones, as judged by a reduction in rate of urine flow and an increase in uterine motility. Cross and Harris (1952) obtained milk ejection in rabbits by stimulation of the neurohypophyseal tract and Andersson (1951) found that sheep and goats responded similarly to hypothalamic stimulation. Where an assay was made it seemed that stimulation caused a greater release of oxytocin than of vasopressin. This point will be discussed later. None of the events described followed stimulation in animals with diabetes insipidus (Peeters and Coussens, 1950; Cross, 1951; Kallialla and Karvonen, 1951).

C. Reflex Stimuli

There is a reflex release of the hormones in both man and animals. Suckling leads to milk ejection and also on occasion to uterine contractions. Less often, antidiuresis also occurs (Harris, 1955). Coitus may induce uterine contractions, milk ejection, and antidiuresis (Eränko *et al.*, 1953; Harris and Pickles, 1953).

It may be accepted, then, that reflex, osmotic, chemical, and electrical stimulation induce release of pars nervosa hormones. The reflex stimuli, whatever their route, are ultimately transmitted to the cells in the supraoptic and paraventricular nuclei. Hypertonic solutions and certain drugs probably act directly on these same nuclei. There is no doubt that these cells are similar to other neurons in receiving and transmitting information in addition to being secretory cells whose products enter the blood stream in order to be carried to a distant organ. Their neuronal character has been demonstrated by a number of workers (Morita *et al.*, 1961; Yagi *et al.*, 1966; Ishikawa *et al.*, 1966; Dyball and Koizumi, 1969).

D. Chemotransmission to Supraoptic and Paraventricular Cells

Since the supraoptic and paraventricular cells are neurons, they must respond to a chemotransmitter and this is probably acetylcholine (ACh) (Pickford, 1939; also consult Heller and Ginsburg, 1966). Recent work supports this earlier suggestion. Unit recordings have been obtained from cells in the relevant nuclei after ascertaining by means of antidromic stimulation from the pars nervosa that the cells impaled did indeed send processes into the gland; it was found that intracarotid injections of ACh excited the cells and that the degree of excitation corresponded approximately to the amount of hormone released by the stimulus (Brooks *et al.*, 1966; Dyball and Koizumi, 1969). These cells are also excited by hypertonic NaCl solution injected into the carotid artery and by those reflex stimuli which are known to bring about the release of the two hormones (Fig. 4).

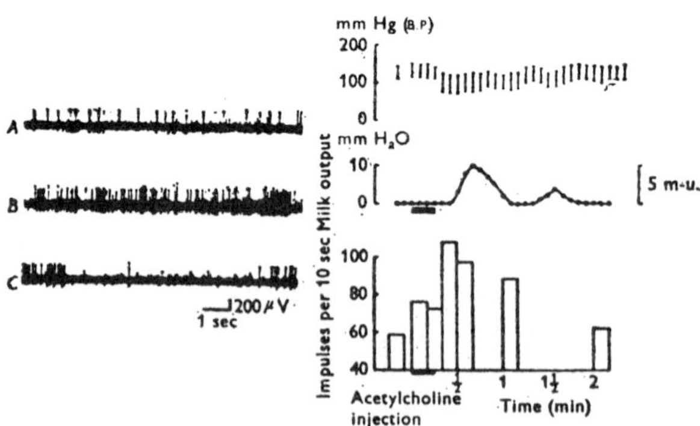

FIG. 4. Effect of acetylcholine on neuron activity in paraventricular cell, on milk ejection, and blood pressure. Time of intracarotid injection of 40 μg ACh marked by bar in lower right-hand graph. Cellular discharge shown on left; A, control; B, 20 sec after injection; C, 5 min after injection. Rates of discharge plotted in bar graph at bottom right. Mammary duct pressure shown at center right. Calibration sign designates maximum pressure change caused by 5 mU oxytocin intravenously. Blood pressure record at top right (From Brooks *et al.*, 1966).

III. Formation and Release of Hormones

A. Neurosecretory Material and its Site of Formation

Histochemical techniques have advanced our knowledge of neurohypophyseal events. Bargmann (1949) found that the supraoptic and paraventricular cells and their processes showed a particular affinity for the chrome alum hematoxylin stain and it seemed possible that the stainable neurosecretory material (NSM) was the carrier substance to which the hormones were attached. If this were the case, then the quantity of NSM should vary with demand and with the amount of active substance in the neurons and their processes. In some instances this assumption appears to hold good. Thus Ortmann (1951) and Hild and Zetler (1953a, b) found that the NSM decreased as a result of dehydration. If animals were allowed free water after a period of dehydration, then depletion of NSM was not apparent. NSM was also seen to decrease following administration of hypertonic NaCl solutions (Ortmann, 1951; Leveque and Scharrer, 1953). Nevertheless the situation is not so simple as this. For instance, the disappearence of NSM following painful stimuli (Rothballer, 1953) occurs within minutes. When the painful stimulus ceases, repletion is also rapid and it is difficult to believe that the whole system can empty and refill at such a speed. Moreover, reduction of NSM without a parallel loss of hormones can occur (Moses *et al.*, 1963). When NSM vanishes from view it disappears from the whole neuron, not from just a part of it. Does this mean that the active and the carrier substances are made throughout the neuron; or does it mean that they are made in the cells, move down the fibers to the storage area and are lost from there? The rate at which repletion occurs refutes the latter supposition.

It has been suggested that NSM can be made throughout the neuron with an activity gradient towards the distal end (Green and Maxwell, 1959). Others have reached the same conclusion (Sachs, 1959, 1961). Sloper (1966) looked to see which regions accumulated ^{35}S soon after the injection of labeled methionine and inferred that these areas were the site of active amino acid synthesis. He concluded that the cellular part of the system was the main site of amino acid formation. That movement can take place in the axoplasm seems likely from the observation that NSM accumulates on the proximal side of a transected stalk and that streaming occurs in tissue cultures from the hypothalamus (Hild, 1954). Another fact suggests that the whole neuron takes some part in hormone formation: the ratio of vasopressin to oxytocin is not the same throughout the system. In dogs, for example, the hypothalamus contains only about 7% as much oxytocin as vasopressin, and in the neurohypophysis the two substances are present in equal amounts (Vogt, 1953; Van Dyke *et al.*, 1955).

B. Electron Microscopic Findings

More recently electron microscopy has provided some interesting information on the formation and release of hormones. So-called elementary granules are found throughout the hypothalamico-hypophyseal system and these electron-dense structures are believed to correspond to NSM for the following reasons. By means of differential centrifugation of neural lobe homogenates in sucrose solutions, it is possible to separate the granules from the other cell structures; bioassay has demonstrated clearly that the active substances are to be found in the granular layer (Lederis and Heller, 1960; Heller and Lederis, 1962; Weinstein *et al.*, 1961). Following dehydration or administration of hypertonic saline solutions the granules decrease in number and the vesicles appear. Simultaneous observations made on electron microscopic findings and assay for hormone content have shown that loss of density does not necessarily mean loss of hormones (Lederis, 1963, 1966). Some rats were anesthetized with ether and others anesthetized and also bled. Both procedures are known to release vasopressin.

Electron microscopic examination showed that, compared with controls, the stimulated rats had suffered a loss of dense material from the elementary granules in the neurohypophyseal nerve fibers, their swellings, and terminations. The decrease in hormone content of the glands was variable and, relative to the controls, was greater in the animals which were bled as well as anesthetized. However, there was certainly not the extreme loss of vasopressin that might have been anticipated from the paucity of dense granules and the large number of apparently empty vesicles. Therefore it was suggested that the appearance of depletion is due to some undefined internal change resulting from stimulation (Daniel and Lederis, 1966).

C. Release of Hormones from Nerve Terminals and Importance of Ca

As already mentioned, the greater part of the hormones is bound in the elementary granules and only a small proportion is " free ". The question then arises as to how the hormones leave the nerve fibers in order to reach the blood stream. Douglas and Poisner (1964a, b) studied the isolated pars nervosa and found that it failed to release its active substances when the osmotic pressure of the bathing fluid was raised but that it did respond to a rise in the external K concentration and to electrical stimulation. They suggested that nerve impulses reaching the terminals acted in the same way as K and induced depolarization. Secretion could take place in the absence of Na, so the point of depolarization was not to permit the entry of Na into the terminals. Further, in the presence of excess K (concentrations above 30 mM) there was an increased uptake of Ca, in the absence of which K was unable to evoke secretion. Ca was itself a stimulant to secretion. It was concluded that the entry of Ca was essential to the process of release of hormone in the neurohypophysis, just as it is in the adrenal medulla. Further discussion of this aspect of the subject con be found in Heller and Ginsburg (1966). It is of interest that *in vitro* hormone release from neurophysin is affected by Ca in concentrations within the range found in the cells (Ginsburg *et al.*, 1966).

Earlier it was stated that ACh is the chemotransmitter to the supraoptic and paraventricular neurons. It has been suggested that ACh is also involved in the process of hormone release from the nerve terminals. This view was based on the finding that structures resembling synaptic vesicles are present in the nerve ends and that the neurohypophyseal system contains cholinesterase (see Douglas and Poisner, 1964a). However, Douglas and Poisner (1964a) found that the isolated neurohypophysis failed to respond to ACh, which it should have done if the latter view was correct. Further, Daniel and Lederis (1966), using an intact isolated hypothalamico-hypophyseal system *in vitro* (i.e., supraoptic nerve cells were present in the preparation), found that ACh could be shown to stimulate hormone release. These findings indicate that the role of ACh is limited to the cell bodies and not to the terminals.

D. Differential and Simultaneous Release of Antidiuretic Hormone and Oxytocin

The next problem to consider is whether there is a differential release of vasopressin and oxytocin and if so, how it is controlled. Certain stimuli such as ACh and hypertonic NaCl solutions were found to bring about the release of both hormones. But even here a difficulty soon became apparent in that the two substances were released in unequal quantities, more oxytocin than vasopressin being liberated on any one occasion. In rabbits, electrical stimulation liberated four times more oxytocin than vasopressin (Harris, 1955) and suckling gave 100 times more oxytocin (Harris, 1960). In dogs, when hypertonic NaCl was injected into the carotid artery, the ratio of oxytocin/vasopressin was about 30 (Pickford, 1960). On the other hand, hemorrhage and a number of drugs induce the release of vasopressin alone (Ginsburg and Brown, 1957; Walker, 1957; Beleslin *et al.*, 1967; Bisset *et al.*, 1967; Dyball, 1968b).

To explain these findings it is necessary to suppose that there are separate nerve tracts relating to the release of the two hormones which must be produced in different cells. Information about the central nerve paths to the supraoptic and paraventricular cells is still meager but Bisset *et al.* (1967) have succeeded in tracing independent pathways in the tuberal region of the hypothalamus itself and have shown by assay techniques that one route releases vasopressin and the other oxytocin. Part of the route concerned in milk ejection has also been traced by Cleverley *et al.* (1968) who found that stimulation of two sites in the ventromedial nucleus was able to yield a considerable output of oxytocin. With regard to the cells, Lederis (1962) showed that the ratio of vasopressin/oxytocin is higher in the supraoptic than in the paraventricular nuclei in both sheep and dogs; moreover the ratio differed in the two species. Olivecrona (1957) destroyed the paraventricular nuclei electrolytically in rats and found that the operation led to a marked reduction in the oxytocin content of the neurohypophysis. Finally there is now evidence that vasopressin and oxytocin are contained in different particles and in separate nerve fibers (Bindler *et al.*, 1967). It may be concluded that vasopressin and oxytocin can be differentially released, that the formation of oxytocin is related more to the paraventricular cells than to the supraoptic, and that the pathways concerned with the liberation of the two hormones are beginning to be discovered.

E. Findings from Unit Recordings

Evidence of another kind has come from the study of the behavior of individual nerve cells in the hypothalamus. Cross and Green (1959) recorded from single units in the paraventricular nucleus and noted that not all the cells responded similarly to a given stimulus; thus hypertonic NaCl solution excited a far greater number of cells than did solutions of glucose. Also, some cells increased in activity and others were inhibited. No assays were made of the quantity of hormones released during the periods of stimulation. Ishikawa *et al.* (1966) also recorded from paraventricular cells and in addition related the activity of the cells to the output of oxytocin as assayed on the milk ejection response in lactating cats. They found that somewhat more than half of the cells tested were stimulated by intracarotid injections of strong NaCl solutions. The remaining cells were either unaffected or inhibited. However, when less concentrated NaCl solutions were used and given by slow infusion no depression was observed (Fig. 5). It is possible, then, that the inhibition seen by Cross and Green (1959) may have depended on indirect sensory impulses and not on a direct depression of the paraventricular cells themselves. Intracarotid injections of ACh also increased the rate of discharge of the neurons and was accompanied by a milk ejection response (Fig. 4). Supraoptic cells, too, were responsive to both

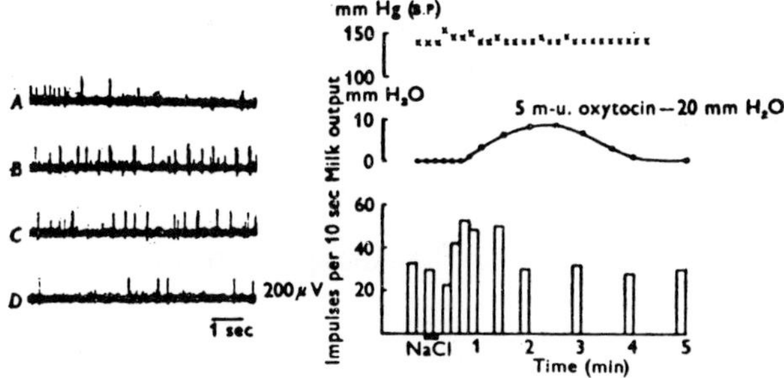

FIG. 5. Effect of osmotic stimulus on paraventricular discharge, milk ejection, and blood pressure in cat. Time of intracarotid injection of 1 ml 1 *M* NaCl solution marked by bar in lower right-hand graph. Cellular discharge shown on left; A, control; B, 30 sec after injection; C, 2 min, and D, 5 min after osmotic stimulus. Rates of discharge plotted on bar graph at bottom right. Mammary duct pressure shown in center right. Calibrating dose of 5 mU oxytocin intravenously resulted in pressure change of 20 mm H_2O. Top right, blood pressure record (From Brooks *et al.*, 1966).

hypertonic NaCl solution and ACh. Paraventricular cells were only moderately responsive to stimulation of the skin or afferent nerves or electrical stimulation of the nipples; but they were easily activated by gentle intermittent suction of the nipples and by uterine distension. Any change in frequency of neuronal discharge was closely related to the degree of alteration in mammary duct pressure. Thus the supraoptic cells react to the same stimuli as the paraventricular ones but not necessarily to the same extent. The supraoptic cells were the most sensitive to osmotic stimulation and also to various influences from a number of central regions.

A further comparison of the behavior of the supraoptic and paraventricular cells was made in which the units were identified as being part of the neurohypophyseal system by means of antidromic stimulation from the pars nervosa (Dyball and Koizumi, 1969). Again it was noticed that the supraoptic cells were more easily excited than the paraventricular by strong NaCl solutions and that they also reacted more strongly to vagal stimulation, $CaCl_2$ solution, and possibly also to ACh. This pattern of sensitivity of the supraoptic cells agrees reasonably well with the results obtained by Dyball (1968a) who measured the amount of vasopressin released in response to vagal stimulation, $CaCl_2$ and ACh, and found that oxytocin and vasopressin were released in different amounts by the three forms of stimulation. These findings again support the idea that the paraventricular cells are more concerned with the release of oxytocin than of vasopressin. Probably there is some overlap in function of the two groups of neurons, though in the rat (Olivecrona, 1957) and in the cat (Nibbelink, 1961) the separation of function seems to be fairly clear-cut.

IV. Some Remaining Problems

It is not surprising that electrical stimulation should induce the liberation of both ADH and ACh, since all nerve fibers in the area will be stimulated indiscriminately. Since ACh is the chemotransmitter, it too would affect both groups of cells, though here there appears to be a difference in the sensitivity of the neurons. The same situation holds with hypertonic solutions. Both groups of cells respond but the supraoptic are more sensitive than the paraventricular.

Problems still remain in connection with differential control. Why does a given adequate stimulus release more oxytocin than vasopressin when the content of the system is greater in vasopressin than oxytocin and when the supraoptic cells (with their particular relationship to vasopressin) seem to be more sensitive to the effective stimuli? Is it useful or merely incidental that an osmotic stimulus releases both hormones? In answer to the first question the only existing hint is that when the rat neurohypophysis was used for the purpose of hormone extraction it was found that oxytocin was preferentially soluble in the watery acetone commonly used in the process. Oxytocin was even more easily dissolved out by acetone from the glands of lactating rats (Heller, 1959). There may, then, be a difference in the physical nature of the membrane of the elementary granules. The answer to the second question is unknown but the suggestion was made that the liberation of oxytocin in response to hypertonicity might be related to the need for a vasodilator substance which would counteract the tendency of ADH to reduce both glomerular filtration rate and renal plasma flow by its vasoconstrictor action (Brooks and Pickford, 1958; Ezrin *et al.*, 1962).

V. Inhibitors of Hormone Release

So far there has been no mention of factors which inhibit release of the neurohypophyseal hormones. From what has already been said it might be anticipated that a reduction in plasma osmotic pressure and overhydration should be inhibitory. Andersson and Jewell (1957) found that after prolonged hydration the supraoptic cells of dogs stained very poorly with chrome alum hematoxylin, as though they were depleted of NSM, and at the same time the nerve fibers were loaded. They interpreted this as cessation of activity on the part of the cells because of lack of demand and an accumulation in the fibers of unwanted active substance. Certain drugs can also inhibit the release of the hormones (Dyball, 1968b).

It has been assumed for a long time that alcohol inhibits release of vasopressin and oxytocin; hence the use of alcohol-anesthetized rats for the assay of ADH. The most extensive observations have been made on the effect of emotion and adrenaline on hormone release. Rydin and Verney (1938) noticed that mild stress and excitement could prevent the appearance of an expected antidiuresis in diuretic dogs. O'Connor and Verney (1942) showed that emotion-induced antidiuresis depended on the presence of the posterior pituitary gland. Later they showed that sympathetic activity could inhibit the liberation of ADH and that the active agent was apparently adrenaline, since an intravenous injection of this substance shortly before the emotional stimulus prevented the antidiuresis usually seen (O'Connor and Verney, 1945). Further, denervation of the adrenal glands prevented the spontaneous inhibition of antidiuresis which occurred in some dogs. Somewhat later, Abrahams and Pickford (1956) showed by means of intracarotid injections in dogs that adrenaline could prevent the response to a following injection of ACh and suggested that the blocking effect of adrenaline was one on the supraoptic and paraventricular cells directly. That adrenaline can inhibit milk ejection by a central action is seen in Fig. 3.

There is still a great deal more that could be said about the control of neurohypophyseal hormone secretion; the foregoing discussion is no more than a summary of some of the main points. Much remains to be discovered, above all why oxytocin is released in any circumstance other than parturition and suckling.

References

Abrahams, V. C., and Pickford, M. (1954). *J. Physiol.* (London) **126**, 329.

Abrahams, V. C., and Pickford, M. (1956). *J. Physiol.* (London) **131**, 712.

Andersson, B. (1951). *Acta Physiol. Scand.* **23**, 8.

Andersson, P., and Jewell, P. A. (1957). *J. Endocrinol.* **15**, 232.

Bargmann, W. (1949). *Klin. Wochschr.* **37**, 617.

Beleslin, D., Bisset, G. W., Haldar, J., and Polak, R. L. (1967). *Proc. Roy. Soc.* **166**, 443.

Bindler, E., La Bella, F. S., and Sanwall, M. (1967). *J. Cell Biol.* **34**, 185.

Bisset, G. W., Hilton, S. M., and Poisner, A. M. (1967). *Proc. Roy. Soc.* **166**, 422.

Brooks, C. McC., Ishikawa, T., and Koizumi, K. (1966). *J. Physiol.* (London) **182**, 217.

Brooks, F. P., and Pickford, M. (1958). *J. Physiol.* (London) **142**, 468.

Cleverley, J. D., Knaggs, G. S., Tindal, J. S., and Turvey, A. (1968). *J. Endocrinol.* **42**, 609.

Cross, B. A. (1951). *J. Physiol.* (London) **114**, 447.

Cross, B. A., and Green, J. D. (1959). *J. Physiol.* (London) **148**, 544.

Cross, B. A., and Harris, G. W. (1952). *J. Endocrinol.* **8**, 148.

Daniel, A. R., and Lederis, K. (1966). *J. Endocrinol.* **34**, 91.

Douglas, W. W., and Poisner, A. M. (1964a). *J. Physiol.* (London) **172**, 1.

Douglas, W. W., and Poisner, A. M. (1964b). *J. Physiol.* (London) **172**, 19.

Dyball, R. E. J. (1968a). *Brit. J. Pharmacol. Chemother.* **33**, 319.

Dyball, R. E. J. (1968b). *Brit. J. Pharmacol. Chemother.* **33**, 329.

Dyball, R. E. J., and Koizumi, K. (1969). *J. Physiol.* (London) **201**, 711.

Eränko, O., Friberg, O., and Karvonen, M. J. (1953). *Acta Endocrinol.* (Copenhagen) **12**, 197.

Ezrin, C., Loach, L. W., and Nicholson, T. F. (1962). *Can. Med. Assoc. J.* **87**, 673.

Ginsburg, M., and Brown, L. M. (1957). *Brit. J. Pharmacol. Chemother.* **11**, 236.

Ginsburg, M., Jayasena, K., and Thomas, P. J. (1966). *J. Physiol.* (London) **184**, 387-

Green, J. D., and Maxwell, D. S. (1959). *In* "Comparative Endocrinology" (A. Gorb. man, ed.). J. Wiley, New York, pp. 368-384.

Harris, G. W. (1947). *Phil. Trans. Roy. Soc.* (London) **232**, 385.

Harris, G. W. (1955). "Neural Control of the Pituitary Gland." Arnold, London.

Harris, G. W. (1960). *In* "Handbook of Physiology" (J. Field, ed.). Sect. 1, Vol. II, American Physiological Society, Washington, pp. 1007–1038.

Harris, G. W., and Pickles, V. R. (1953). *Nature* **172**, 1049.

Haterius, H. O. (1940). *Am. J. Physiol.* **128**, 506.

Heller, H. (1959). *In* "Recent Progress in Endocrinology of Reproduction" (C. W. Lloyd, ed.). Academic Press, New York, pp. 365-385.

Heller, H., and Ginsburg, M. (1966). *In* "The Pituitary Gland" (G. W. Harris and B. T. Donovan, eds.). Vol. III. Butterworths, London, pp. 330–373.

Heller, H., and Lederis, K. (1962). *J. Physiol.* (London) **158**, 27P.

Hild, W. (1954). *Z. Zellforsch.* **40**, 257.

Hild, W., and Zetler, G. (1953a). *Pfluegers Arch. Ges. Physiol.* **257**, 169.

Hild, W., and Zetler, G. (1953b). *Z. Ges. Exp. Med.* **120**, 236.

Holland, R. C., Cross, B. A., and Sawyer, C. H. (1959). *J. Physiol.* (London) **196**, 791.

Ishikawa, K., Koizumi, K., and Brooks, C. McC. (1966). *Neurology* **16**, 101.

Jewell, P. A., and Verney, E. B. (1957). *Phil. Trans. Roy. Soc.* (London) **240**, 197.

Kallialla, H., and Karvonen, M. J. (1951). *Ann. Med. Exp. Biol. Fenniae* (Helsinki) **29**, 233.

Lederis, K. (1962). *Mem. Soc. Endocrinol.* **12**, 227.

Lederis, K. (1963). *Gen. Comp. Endocrinol.* **3**, 714.

Lederis, K. (1966). *J. Endocrinol.* **37**, XXIV.

Lederis, K., and Heller, H. (1960). *In* "Proc. I Intern. Congr. Endocrinol." (F. Fuchs, ed.). Periodica, Copenhagen, p. 115.

Leveque, T. F., and Scharrer, E. (1953). *Endocrinology* **65**, 909.

Morita, H., Ishibashi, T., and Yamashita, S. (1961). *Nature* **191**, 183.

Moses, A. M., Leveque, T. F., Giambattista, M., and Lloyd, C. W. (1963). *J. Endocrinol.* **26**, 273.

Nibbelink, D. W. (1961). *Am. J. Physiol.* **200**, 1229.

O'Connor, W. J., and Verney, E. B. (1942). *Quart. J. Exp. Physiol.* **31**, 393.

O'Connor, W. J., and Verney, E. B. (1945). *Quart. J. Exp. Physiol.* **33**, 77.

Olivecrona, H. (1957). *Acta Physiol. Scand., Suppl.* **40**, 136.

Ortmann, R. (1951). *Z. Zellforsch. Mikroskop. Anat. Abt. Histoche.* **36**, 92.

Peeters, G., and Coussens, R. (1950). *Arch. Int. Pharmacodyn.* **84**, 209.

Pickford, M. (1939). *J. Physiol.* (London) **95**, 226.

Pickford, M. (1960). *J. Physiol.* (London) **152**, 515.

Rothballer, A. B. (1953). *Anat. Record* **115**, 21.

Rydin, H., and Verney, E. B. (1938). *Quart. J. Exp. Physiol.* **27**, 343.

Sachs, H. (1959). *Biochem. Biophys. Acta* **34**, 572.

Sachs, H. (1961). *In* "Regional Neurochemistry" (S. Kety and J. Elkes, eds.). Pergamon Press, Oxford, pp. 264–273.

Sloper, J. C. (1966). *In* "The Pituitary Gland" (G. W. Harris and B. T. Donovan, eds.), Vol. III. Butterworths, London, pp. 131–139.

Van Dyke, H. B., Adamsons, K., and Engel, S. L. (1955). *Recent Progr. Hormone Res.* **11**, 1.

Verney, E. B. (1926). *Proc. Roy. Soc.* **99**, 487.

Verney, E. B. (1947). *Proc. Roy. Soc.* **135**, 25.

Vogt, M. (1953). *Brit. J. Pharmacol.* **8**, 193.

Walker, J. M. (1957). *In* "The Neurohypophysis" (H. Heller, ed.). Butterworths, London, pp. 221–229.

Weinstein, H., Malamud, A., and Sachs, H. (1961). *Biochim. Biophys. Acta* **50**, 386.

Yagi, K., Azuma, T., and Matsuda, K. (1966). *Science* **154**, 778.

Control of Rhythmic Secretion of Gonadotropins

N. B. Schwartz

I. Introduction

A salient feature of reproductive processes in the vertebrate organism is the many aspects which demonstrate a rhythmic or cyclic character (Asdell, 1964; Marshall, 1956; Van Tienhoven, 1968). This article is concerned with the factors which contribute to the rhythmic control in secretion rates of gonadotropic hormones in particular and with the issue of cyclic reproduction in general.

In the first part of the article, the general aspects of the differences among various species in cycle lengths will be considered, and the events and processes subsumed under a " cycle " will be briefly reviewed. The problem of cyclicity in reproduction will then be viewed in detail in three species (rat, mouse, and hamster) in order to show how the interplay between internal and external events determines the differences among them in the control of cycle regularity. The possible adaptive significance of rhythmic regulation of reproductive processes will then be summarized.

II. Reproductive Cyclicity and Gonadotropic Hormone Rhythms

A restricted definition of "gonadotropic hormone cycle" would include only periodicities in levels of gonadotropic hormones in blood or pituitary gland. Such a definition, however, would be too restrictive to be really useful because so few of such measurements have been made. Accordingly, our discussion will be expanded to include rhythmic changes in gonadal function where these are presumed to result from alterations in secretion rates of gonadotropic hormones, and rhythmic changes in accessory sex tissues and secondary sex characteristics where these are known to be secondary or tertiary consequences of gonadotropic hormonal-gonadal rhythms. Thus we will actually discuss "reproductive cycles."

A. Elements and Processes Common among Species

There is a common sequence of processes occurring in reproduction in different species, but with different sets of these processes being included within a defined "reproductive cycle": ovarian quiescence; follicular maturation and estrogen (and progesterone) secretion; growth and secretion of accessory sex tissue and changes in secondary sex characters; coitus; ovulation; corpus luteum maintenance; fertilization; implantation; gestation; parturition; lactation; maternal behavior; weaning (Van Tienhoven, 1968). Generally it is considered that some combination of Luteinizing Hormone (LH) and Follicle-Stimulating Hormone (FSH) is required for follicular maturation, both morphological and secretory. Either FSH or LH alone may cause ovulation, or some combination may be required. Corpus luteum maintenance may require prolactin from the pituitary gland, alone or in combination with other hormones, or may require placental secretion of prolactin-like hormones. Implantation is probably ultimately dependent on pituitary secretion of at least LH, but other pituitary hormones may also be involved. Lactation is dependent on a host of pituitary hormones. Behavior (mating, courtship, etc.) and accessory sex tissue function are indirectly dependent on ovarian steroid function as a result of pituitary gonadotropic hormone secretion (Van Tienhoven, 1968).

B. Some Differences among Species in Reproductive Cycles

Table I summarizes ten different cycle types. Perusal of the Table reveals that animals can be classified broadly as follows: by whether their breeding periods are seasonal or (relatively) nonseasonal; by whether there are cycles within the seasonal breeding period; by whether there are long delays between successive events; by whether ovulation is spontaneous or copulation-induced; by whether the luteal stage (nonpregnant) follows automatically after ovulation

TABLE I

Differences among Species in Reproductive Cycles

Overall Breeding Period	Cycles within Breeding Period	Delays	Ovulation	Luteal Stage	Gestation Period	Example
Seasonal (Jan.-Feb.)	Monoestrous (few days)	None	Spont.	Follows Ovul.	52	Fox
Seasonal (Fall)	Polyestrous (17 days)	None	Spont.	Follows Ovul.	145	Sheep
Seasonal (Spring to Summer)	Continuous	None	Induced	Follows Ovul.	42	Ferret
Seasonal (Spring and Fall)	Polyestrous (2 weeks)	None	Induced	Follows Ovul.	62	Cat
Seasonal (March)	Polyestrous (8–9 days)	Implantation	Induced	Follows Ovul.	48	Mink
Seasonal (Cop. in Fall)	Monoestrous (?)	Fertilization	Spont. (Spring)	Follows Ovul.	55	Bat
Seasonal	Polyestrous	Implant. (Dec.-Jan.)	Spont. (Summer)	Follows Ovul.	42	Badger
Continuous	Continuous	None	Induced	Follows Ovul.	31	Rabbit
Continuous	Polyestrous (16 days)	None	Spont.	Follows Ovul.	68	Guinea Pig
Continuous	Polyestrous (4–6 days)	None	Spont.	Copulation Induced	22	Rat

This table is summarized from information in: Asdell, 1964; Van Tienhoven, 1968; Marshall, 1956; Enders, 1963.

(whether spontaneous or induced) or has to follow copulation in spontaneously ovulating species. A " cycle " may be defined as the events between successive manifestations of behavioral estrus (cat) or between ovulations (rat) or between successive luteal phases (guinea pig), etc.

C. " Nonseasonal, Polyestrous, Spontaneously Ovulating, Non-luteal Phase Rodents ": Rat, Mouse, Hamster

The cycle of the rat as described in the last line of Table I is similar to that of the laboratory mouse and hamster (Van Tienhoven, 1968; Asdell, 1964). All three species show cycle lengths of 4, 5, or 6 days. LH release from the pituitary occurs on the afternoon of the day called " proestrus," as does mating behavior, and spontaneous ovulation (with no mating required) takes place after midnight. Copulation or cervical stimulation is required in all three species to produce corpora luteal secretion of progesterone; pseudopregnancy following sterile mating takes 13 days in rat and mouse and 8 or 9 days in the hamster. Pregnancy lasts 22 days in rats, 19 or 20 days in mice, and 16 days in hamsters. An excellent review of luteal function in these three species has recently been published (Greenwald and Rothchild, 1968). All three species require alternating conditions of light and dark in the environment in order for normal cyclicity to persist, and all three show reversal of ovulation and mating when the light-dark environment is reversed.

Thus these species are generally classified together as having a similar kind of cycle. We will examine this thesis closely, focusing on the differences which have been elucidated among them in the manner of control of their reproductive cycles. It will then be of interest to return to our broader spectrum of species (Table I) with the information gained from the three species viewed in detail.

III. Differences in Control of Rhythmic Gonadotropin Secretion in the Rat, Mouse, and Hamster

A. The Estrous Cycle

The following description of the estrous cycle of the laboratory rat was taken from Everett (1961) and Schwartz (1969). When maintained under an alternating light-dark environment (L/D 14:10;

midnight is the midpoint of the dark period) ovulation occurs at about 1 A.M. every four or five days, between the days of proestrus and estrus, as the result of a four- or five-day cycle in gonadotropic hormone release. In four-day cycles, on the morning of estrus, all rats show cornified cells in the vagina, ova " in cumulus " in the oviducts, and new corpora lutea in the ovaries. A new set of small follicles is visible in the ovary. On the next day (metestrus) follicles continue to grow and there is little or no ovarian secretion of estrogen; the vaginal smear is leukocytic. On the following day (diestrus) the vaginal mucosa is still predominantly leukocytic, but the ovarian follicles have increased in size. On this day there is gonadotropic hormone secretion by the pituitary before 3 P.M. (as seen by effects of hypophysectomy or injection of anti-LH serum: Lawton and Sawyer, 1968; Schwartz and Ely, 1969a) which causes ovarian estrogen secretion by 7 P.M. (see Schwartz, 1969, for documentation). This estrogen is responsible for the maximal uterine wet and dry weight, for the accumulation of intralumenal fluid seen by the morning of proestrus, and for the vaginal cornification seen on the morning of estrus. The estrogen is also responsible for the discharge of the ovulatory surge of LH and FSH seen after 2 P.M. and before 5 P.M. of the day of proestrus. Estrogen release on the morning of proestrus between 6 A.M. and 8 A.M. is responsible, when followed by progesterone secretion after 2 P.M., for the mating behavior manifested starting (average) at 5 P.M.. On the morning of proestrus pituitary FSH and LH contents are high and minimal values are seen by the next morning. Correspondingly, LH and FSH levels in the plasma show a significant increase on the afternoon of proestrus (Schwartz, 1969), as does prolactin (Kwa and Verhofstad, 1967). Following release of LH and FSH, progesterone secretion starts and estrogen secretion stops (Schwartz, 1969), and preovulatory swelling of the follicles is initiated, with ovulation occurring after midnight.

In the five-day cycle the events described above are similar, but the estrogen secretion appears to be delayed by some fraction of a 24-hr period. Since there is a 24-hr periodicity in facilitation for the LH surge of proestrus, this event is delayed by a full 24 hr (see Schwartz, 1969). A given rat may run predominantly cycles of a given length, but may also abruptly switch from one cycle length to another.

The golden hamster, unlike the rat, exhibits only four-day estrous cycles (Orsini, 1961; Kent, 1968). The cycle can be followed more reliably if the vaginal discharge rather than the vaginal smears are examined; there is a so-called postestrous discharge seen in the morning, when fresh ova are found "in cumulus" in the oviducts (Orsini, 1961). The changes in pituitary LH content and uterine weight between proestrus and estrus are like those seen in the rat: the values are very high on the morning of proestrus and drop significantly by the morning of estrus (Orsini and Schwartz, 1966; Brom and Schwartz, 1968). Uterine ballooning is seen at proestrus and disappears by the morning of estrus. Estrogen secretion on the day before proestrus is responsible for the above pituitary, vaginal, and uterine events (Brom and Schwartz, 1968). Pituitary FSH content appears to drop a day earlier in the hamster than in the rat, between diestrus and proestrus (Keever and Greenwald, 1967). Mating behavior in the hamster occurs during the afternoon of proestrus; as in the rat, estrogen secretion during the morning of proestrus, followed by progesterone after 2 P.M., is responsible for the behavior (Brom and Schwartz, 1968). A striking difference in ovarian morphology in the hamster, which distinguishes this species from the rat (and mouse) is that only one set instead of several sets of corpora lutea is visible at a time in the cyclic ovary.

When mice are housed under the conditions used for studying rats and hamsters (alternating light-dark; single mouse per cage) 50% show no cycles or cycles which are very irregular; the other 50% show four-, five-, or six-day cycles (Bingel and Schwartz, 1969a; Whitten, 1966). Under these conditions, in mice showing five-day cycles (Bingel and Schwartz, 1969a, b), pituitary LH content is maximal on the second day of diestrus rather than the next day (proestrus), but drops to minimal values at estrus when fresh ova are found in the oviduct. This is also true in the deermouse, where plasma LH values have been shown to be maximal at proestrus (Eleftheriou and Zolovick, 1967). Uterine wet weight and intralumenal fluid content are maximal on the morning of proestrus and drop significantly by the next morning. The vaginal smear is most useful when examined twice daily, revealing that the vagina is still cornified on the morning of metestrus (day after estrus) but not in the afternoon, thus distinguishing this day readily from estrus, when both smears show cornified cells. Mating behavior occurs late in the day of proestrus

(Whitten, 1966). Ovulation occurs late in the dark period or within a few hours after the lights go on (Bingel and Schwartz, 1969a).

If the individual female mouse is caged near a male mouse, the cycle is more regular and shortens to a four-day length (Whitten, 1966). Conversely, if female mice are caged in groups of ten, they do not run cycles at all; placing a male in the cage induces a cycle, with normal days of proestrus and estrus (Whitten, 1966; Bingel and Schwartz, 1969a). Pregnancy can also be interrupted by a "strange" male, the Bruce effect. Rats do not show this suppressive effect of grouped females on cycles, nor the inductive, facilitatory effects of males (Whitten, 1966).

B. Timing of the Critical Period for Luteinizing Hormone Release

On the day of proestrus in the adult cyclic four-day rat, there is a surge of LH (and FSH) released, starting after 2 P.M. (Everett, 1961). Hypophysectomy at 2 P.M., but not at 4 P.M., prevents the ovulation expected by the morning of estrus. Central nervous system depressants such as pentobarbital, also block the ovulation (and LH release; see Schwartz, 1969) when administered at 2 P.M. but not when injected at 4 P.M. In the case of the 2 P.M. injection, ovulation is delayed by 24 hr but can again be blocked on the afternoon of the next day. These data, plus others (Everett, 1961) suggest that there is a 24-hr periodicity, starting at 2 P.M., of the facilitation of release of the ovulatory surge of LH; however, the surge is released only on the day of proestrus (Schwartz, 1969). The spread of the critical period is wider in the five-day rat (Hoffmann and Schwartz, 1965). In either the four-day or five-day cycle, anti-LH serum administered just before the critical period on the day of proestrus will block ovulation; the absorption of the anti-LH serum with sheep tissue and FSH will not prevent it from blocking (Schwartz and Ely, unpublished observations).

The timing of the critical period for the release of LH for postpartum ovulation has also been investigated under the same lighting conditions as those described above. The results indicate that: (1) rats which deliver between 4 P.M. and 8 A.M. do not release LH until the afternoon of that day and do not ovulate until some time during that night; (2) rats delivering between 8 A.M. and 4 P.M. may release LH on the same afternoon or wait until the following afternoon;

(3) the critical period for the LH release on a given afternoon is broader than for cyclic ovulation (Hoffmann and Schwartz, 1965).

The critical period for the release of gonadotropic hormones in the cyclic hamster shows a timing similar to that of the four-day rat. Hypophysectomy at 2 P.M. but not at 4 P.M. blocks the expected ovulation; pentobarbital in anesthetic doses at 2 P.M. does not block ovulation, but phenylisopropylhydrazine or phenobarbital at 2 P.M. do block (Alleva and Umberger, 1966; Goldman and Mahesh, 1969). Anti-ovine-LH serum before the critical period blocks ovulation, but if the antiserum is absorbed with FSH the blockade is lost except at high doses (Goldman and Mahesh, 1969). There is no postpartum ovulation in the hamster.

In the mouse, barbital (a long-acting barbiturate) blocks ovulation at a fairly high dose whether injected at 2 P.M. or as late as 5 P.M.; injection at 9 P.M. no longer completely blocks ovulation (Bingel and Schwartz, 1969b). The later time of LH release in the mouse correlates with the later time of ovulation in this species. The postpartum ovulation in the mouse, under L/D 14:10, as in the rat (Hoffmann and Schwartz, 1965), is also associated with a longer critical period for LH release than is the cyclic ovulation. However, unlike the rat, mice delivering between 5 P.M. and 9 P.M. can still release LH that night and ovulate at some time during the following morning (Bingel and Schwartz, 1969c). Thus the mouse, whether cyclic or postpartum, can release LH later than the rat.

The 24-hr periodicity implicit in the timing of the critical period for gonadotropin release in these three species has also been seen in noncyclic rats under specified conditions. The immature rat treated with Pregnant Mare Serum Gonadotropin (PMSG) or PMSG plus Human Chorionic Gonadotropin (HCG), shows a 24-hr periodicity in LH release (Strauss and Meyer, 1962; Lawton and Schwartz, 1965) as does the ovariectomized adult rat (Lawton and Schwartz, 1968). In intact adult rats not separated by cycle stages there is also a 24-hr periodicity in pituitary prolactin content (Clark and Baker, 1964).

C. Timing of Ovarian Secretions during the Estrous Cycle

In the rat, ovariectomy at 10 A.M. on the day before proestrus blocks the increase in uterine growth, the pituitary LH discharge (judged by " estrous " pituitary LH content), and the mating behavior of proestrus, as well as the vaginal cornification expected at " estrus. " Similar results are obtained with injection of MER-25 (an estrogen antagonist), by hypophysectomy, or by injection of anti-ovine-LH serum (Schwartz and Ely, 1969a, b; Lawton and Sawyer, 1968; Schwartz, 1969). The later in the day any of these treatments are performed, the less drastic the results. Ovariectomy on the day of proestrus at 10 A.M. does not alter the low pituitary LH contents, vaginal cornification, or uterine weights at " estrus," indicating that the necessary estrogen for these events has been secreted by this time. Ovariectomy at 10 A.M. provokes a premature mating response (Lorenzen-Nequin and Schwartz, 1968; Schwartz, 1969).

Sham ovariectomy at 10 A.M. on the day before proestrus can block the events of proestrus and estrus in a significant number of rats, but when performed later in the day it does not. Sham ovariectomy on the morning of proestrus does not alter any event except to precipitate a premature mating response (see Schwartz, 1969). The response to either sham ovariectomy or ovariectomy can be blocked by adrenalectomy (Lorenzen-Nequin and Schwartz, 1968).

In the hamster, ovariectomy at 10 A.M. on the day before proestrus, as in the rat, blocks the uterine response of proestrus and the vaginal " postestrous " discharge of estrus. Unlike the rat, after this treatment the hamster does not show the normal accumulation of pituitary LH on the morning of proestrus; content is low instead of high and remains so on the day of " estrus." Ovariectomy at 10 A.M. on the day of proestrus, does not alter proestrous LH discharge nor does it prevent the vaginal " postestrous " discharge the next morning; however, it does prevent mating behavior that afternoon. Sham ovariectomy has no effect whatsoever at any time on any variable, the cycle proceeding completely normally (Brom and Schwartz, 1968).

The timing of ovarian secretion during the mouse cycle has not been investigated. However, the concurrence of maximal uterine size with the day of release of the ovulatory surge of LH, and of vaginal cornification with the day when fresh ova are seen in the oviduct, tends to suggest that the same qualitative changes in ovarian secretion are occurring. The late critical period in the mouse may cause a prolongation of estrogen secretion (see Schwartz, 1969), which could be responsible for the persistence of the vaginal cornification until the morning of metestrus (Bingel and Schwartz, 1969a).

D. Effects of Changes in Lighting and Temperature Environments

In the rat, continuous light initially leads to asynchrony of pituitary and ovarian events (with an alteration of timing of the critical period), but eventually proceeds into a persistent estrus, anovulatory syndrome (Lawton and Schwartz, 1967; Jochle, 1966; Dempsey and Searles, 1943). Constant dark, on the other hand, does not disrupt the cycles of the majority of rats (although the time of the critical period shifts), nor does ovarian atrophy generally follow (Hoffmann, 1967; Reiter, 1967). However, there may be some strain differences in response to dark (Reiter, 1967). Superior cervical ganglionectomy, blinding, or melatonin injections can suppress the estrus-stimulating effects of continuous light in rats (Wurtman *et al.*, 1968), but pinealectomy does not, suggesting that continuous light may " functionally pinealectomize " the rat. Pinealectomy can prevent the effects of blinding in reducing ovarian weight in the prepuberal rat; however, even without pinealectomy the ovarian weight recovers as the blinded rat becomes older (Reiter, 1967). A cold environment, as does continuous light, tends to throw the rat into a persistent estrous syndrome (Denison and Zarrow, 1955); in either case, copulation-induced ovulation followed by pregnancy can occur (Dempsey and Searles, 1943).

The hamster differs considerably from the rat in its responses to changes in lighting and temperature environment. Continuous light does not lead to persistent estrus or a nonovulatory state; the only apparent effect is to change the timing of the critical period for release of the ovulating hormones (Alleva *et al.*, 1968). Blinding hamsters leads to a severe gonadal atrophy, which can be prevented by pinealectomy or superior cervical ganglionectomy (Reiter, 1967; Wurtman *et al.*, 1968). Thus there appears to be a difference between the rat and hamster in the innervation of the pineal: in the rat, removal of the superior cervical ganglion is equivalent to either blinding or the administration of melatonin in overcoming the effects of L/L; in the hamster it is the reverse and equivalent to pinealectomy. Exposure to cold will eventually lead to severe gonadal atrophy and acyclicity, not prevented by pinealectomy (Grindeland and Folk, 1962; Reiter, 1967). Seasonal factors may also block pituitary-ovarian function in the hamster under controlled animal-room temperatures and lighting conditions (Hoffman, 1968; Orsini and Schwartz, 1966).

Constant light for the mouse tends to cause a persistent estrus syndrome, although strains differ in suceptibility (Jochle, 1966). Reproduction is possible in the mouse under extreme cold conditions, but since the cycle was not followed it is possible that ovulation might have been induced (Barnett and Manly, 1954). Neither blindness nor pinealectomy affected gonadal size in mice (Reiter, 1967).

E. Conclusions

The fundamental regulation of the gonadotropic hormone rhythm known as the "estrous cycle" in rats, mice, and hamsters is probably quite similar. The sequence of changes in the variables is the same: the light-dark alternation sets the timing of the release of the ovulatory surge of LH and thus sets the timing of ovulation and probably mating behavior as well (because of LH-stimulated progesterone secretion from the ovary). Ovarian feedback signals (probably estrogen) set the length of the cycle by determining the day which is "proestrus." The mouse apparently can release LH later in the day, which may result in prolongation of estrogen secretion and might contribute to the relative irregularity of the mouse cycle. Since the four-day rat releases estrogen earlier than the five-day rat, it is possible that the seemingly inevitable hamster four-day cycle might result partially from a fast response on the part of the ovary so that estrogen thresholds for triggering the ovulatory surge during the critical period of proestrus are always achieved on the third day after the last ovulation in this species. A theoretical model of the rat estrous cycle, which probably can also represent the mouse and hamster cycle, has recently been published (Schwartz, 1969).

The species appear to differ drastically in the nature of environmental changes which are received as inputs by the hypothalamic-pituitary-gonadal axis and by the type of response of the latter system to these inputs. For the rat, cessation or disruption of the estrous cycle can result from continuous light and/or cold. However, the disruption which occurs is a prolongation of LH and FSH release for maintenance of estrogen secretion, with no release of the ovulatory surge of LH. Reproduction is still possible because follicles are present and the rat then becomes an induced ovulator in the presence of the male. The effects of acute stress in disrupting the cycle are probably the result of triggering of adrenal secretion of progesterone, which can either facilitate events of the cycle or block them, depending on the time of secretion (or injection) (Schwartz, 1969). The response of the mouse to continuous light or cold appears to be similar to that of the rat. The major difference between the two species is in the prolonged anestrus, seen in grouped female mice, which can be broken within three days by the presence of the male. The observations suggest that a pheromone produced by females suppresses the secretion of FSH and LH necessary for follicular growth and estrogen production, but that male pheromones can induce such gonadotropic hormone secretion very quickly (Whitten, 1966).

In the hamster, continuous light does not have much effect, but darkness and/or cold can cause drastic gonadal atrophy and acyclicity. These two environmental stimuli can also induce hibernation in the hamster, particularly at certain seasons (Hoffman, 1968). The hamster is a " permissive " hibernator, and it seems likely that the effects of cold, dark, and season are related to the gonadal quiescence which accompanies hibernation. The fact that FSH and LH can reverse the effects of blindness on the hamster ovary (Reiter, 1967) suggests that pituitary secretion is simply " switched off " under the influence of these environmental changes.

The lack of effect of sham ovariectomy on the hamster cycle is interesting, since progesterone in this species, as in the rat, is necessary for mating and can block the cycle if injected (Kent, 1968; Keever and Greenwald, 1967). It may be that the hamster adrenal does not secrete progesterone under conditions of stress-induced Adrenocorticotropic Hormone (ACTH) release, as seems to be true in the rat (Lorenzen-Nequin and Schwartz, 1968). This could be the result of differences in adrenal biosynthetic pathways, since cortisol is the principal glucocorticoid secreted by the hamster (Frenkel *et al.*, 1965), in contrast to corticosterone in the rat (Glenister and Yates, 1961). Another difference between the two species is the reversed sexual dimorphism of adrenal weight, the adrenals being larger in female rats than in males but larger in male hamsters than in females. Glenister and Yates (1961) postulated that the adrenal weight difference in rats was due to differences in liver metabolism in the two sexes, resulting in a greater turnover rate of corticosterone in the female and thus a greater ACTH stimulation of the female adrenal. Liver metabolism of cortisol in the hamster has not been investigated, but the half-life of cortisol does not appear to be different in male and female hamsters (based on a small sample; Frenkel *et al.*, 1965).

IV. Adaptive Significance of the Rhythmic Secretion of Gonadotropins

" A requirement for successful internal fertilization is that one of the following conditions be met: the ova or the sperm should retain their capability for a long time to become fertilized or to fertilize, there must be a rather accurate synchronization between copulation and ovulation, or copulations must occur frequently throughout the breeding season to ensure the presence of viable sperm at the time of ovulation " (Van Tienhoven, 1968). Ovum life span, once ovulated, is very short; if one adds to this the necessity of also synchronizing reproductive events with geophysical periodicities in order that optimum external conditions be present for breeding, the necessity of a dual control of a cyclic gonadotropic release becomes obvious. Internal events are signaled by ovarian hormone feedbacks and external events are signaled by various inputs, dictated by the ecological niche of the animal (Table I; Asdell, 1966; Amoroso and Marshall, 1960). It is apparent that the rat, mouse, and hamster are all " good " breeders; the persistent estrous syndromes of the rat do not preclude breeding in the presence of the male just as the anestrus of grouped female mice disappears in the presence of a male. For the hamster, the consideration of protection against cold may temporarily override (with darkness) other considerations so that hamsters do not breed under unfavorable conditions. But when " seasonal " conditions are favorable the persistence of the regular hamster cycle is remarkable, perhaps the result of relative inpenetrability of other external events to the pituitary-ovarian system.

Acknowledgments

Much of the work mentioned as coming from this laboratory was supported by PHS Grant HD-0440. I would like to thank Miss H. P. Bartlett, a graduate student, for her participation in stimulating discussions of the differences in the three species described in this article.

References

Alleva, J. J., and Umberger, E. J. (1966). *Endocrinology* **78**, 1125.

Alleva, J. J., Waleski, M. W., Alleva, F. R., and Umberger, E. J. (1968). *Endocrinology* **82**, 1227.

Amoroso, E. C., and Marshall, F. H. A. (1960). *In* " Marshall's Physiology of Reproduction." (A. S. Parkes, ed.). Vol. II, Part 2. Longmans and Green, London pp. 707–831.

Asdell, S. A. (1964). " Patterns of Mammalian Reproduction. " 2nd ed. Cornell Univ. Press, Ithaca.

Asdell, S. A. (1966). *In* " Comparative Biology of Reproduction in Mammals " (I. W.

Rowlands, ed.). Symp. Zool. Soc. London, No. 15. Academic Press, New York, pp. 1–13.
Barnett, S. A., and Manly, B. M. (1954). *Nature* **173**, 355.
Bingel, A. S., and Schwartz, N. B. (1969a). *J. Reprod. Fertility* **19**, 215.
Bingel, A. S., and Schwartz, N. B. (1969b). *J. Reprod. Fertility* **19**, 223.
Bingel, A. S., and Schwartz, N. B. (1969c). *J. Reprod. Fertility* **19**, 231.
Brom, G. M., and Schwartz, N. B. (1968). *Neuroendocrinology* **3**, 366.
Clark, R. H., and Baker, B. L. (1964). *Science* **143**, 375.
Dempsey, E. W., and Searles, H. F. (1943). *Endocrinology* **32**, 119.
Denison, M. E., and Zarrow, M. X. (1955)., *Proc. Soc. Exp. Biol. Med.* **89**, 632.
Eleftheriou, B. E., and Zolovick, A. J. (1967). *J. Reprod. Fertility* **14**, 33.
Enders, A. C. (1963). " Delayed Implantation." The Univ. of Chicago Press, Chicago.
Everett, J. W. (1961). *In* " Sex and Internal Secretions " (W. C. Young, ed.), 3rd ed., Vol. I. Williams and Wilkins, Baltimore, pp. 497–555.
Frenkel, J. K., Cook, K., Grady, H. J., and Pendleton, S. K. (1965). *Lab. Invest.* **14**, 142.
Glenister, D. W., and Yates, F. E. (1961). *Endocrinology* **68**, 747.
Goldman, B. D., and Mahesh, V. B. (1969). *Endocrinology* **84**, 236.
Greenwald, G. S., and Rothchild, I. (1968). *J. Animal Sci.* **27**, *Suppl.* **1**, 139.
Grindeland, R. E., and Folk, G. E. Jr. (1962). *J. Reprod. Fertility* **4**, 1.
Hoffmann, J. C. (1967). *Neuroendocrinology* **2**, 1.
Hoffmann, J. C., and Schwartz, N. B. (1965). *Endocrinology* **76**, 620.
Hoffman, R. A. (1968). *In* " The Golden Hamster " (R. A. Hoffman, P. F. Robinson and H. Magalhaes, eds.). Iowa State Univ. Press, Ames, pp. 25–39.
Jochle, W. (1966). *In* " Reproduction in the Female Mammal " (G. E. Lamming and E. C. Amoroso, eds.). Plenum Press, New York, pp. 267–281.
Keever, J. E., and Greenwald, G. S. (1967). *Acta Endocrinol.* (Copenhagen) **56**, 244.
Kent, G. C. Jr., (1968). *In* " The Golden Hamster " (R. A. Hoffman, P. F. Robinson and H. Magalhaes, eds.). Iowa State Univ. Press, Ames, pp. 119–138.
Kwa, H. G., and Verhofstad, F. (1967). *J. Endocrinol.* **39**, 455.
Lawton, I. E., and Sawyer, C. H. (1968). *Endocrinology* **82**, 831.
Lawton, I. E., and Schwartz, N. B. (1965). *Endocrinology* **76**, 276.
Lawton, I. E., and Schwartz, N. B. (1967). *Endocrinology* **84**, 497.
Lawton, I. E., and Schwartz, N. B. (1968). *Am. J. Physiol.* **214**, 213.
Lorenzen-Nequin, L., and Schwartz, N. B. (1968). *Excerpta Med. Intern. Congr. Ser.* **157**, 28.
Marshall, F. H. A. (1956). *In* " Marshall's Physiology of Reproduction " (A. S. Parkes, ed.). Vol. I, Part. 1. Longmans and Green, London, pp. 1–42.
Orsini, M. W. (1961). *Proc. Anim. Care Panel* **11**, 193.
Orsini, M. W., and Schwartz, N. B. (1966). *Endocrinology* **78**, 34.
Reiter, R. J. (1967). " The Pineal Gland: A Report of Some Recent Physiological Studies." Edgewood Arsenal Technical Report, EATR 4110.
Schwartz, N. B. (1969). *Recent Progr. Hormone Res.* **25**. 1.
Schwartz, N. B., and Ely, C. A. (1969a). *In* " Progress in Endocrinology " (C. Gual, ed.). Excerpta Medica, Amsterdam, pp. 1005–1014.
Schwartz, N. B., and Ely, C. A. (1969b). *Fed. Proc.* **28**, 381.
Strauss, W. F., and Meyer, R. K., (1962). *Science* **137**, 860.
Van Tienhoven, A. (1968). " Reproductive Physiology of Vertebrates." Saunders, Philadelpia.
Whitten, W. K. (1966). *In* " Advances in Reproductive Physiology " (A. McLaren, ed.). Vol. I. Academic Press, New York, pp. 155–177.
Wurtman, R. J., Axelrod, J., and Kelly, D. E. (1968). " The Pineal." Academic Press, New York.

Rhythmic Phenomena and Pineal Principles

F. FRASCHINI AND L. MARTINI

I. Introduction

It has recently been suggested that the mammalian pineal gland may act as a " biological clock " and may intervene in the control of the rhythmicity of several biological processes (Wurtman, 1967). This suggestion is based on the observations: (1) that the biosynthesis of pineal principles and particularly of pineal methoxyindoles (e.g. 5-methoxy-N-acetyltryptamine or melatonin, 5-methoxytryptophol, etc.), is cyclic in nature, and is dependent on environmental light conditions; and (2) that exogenously administered pineal principles may modify several cyclic endocrine and non endocrine phenomena occurring in the body. The formation of pineal indoles, and the effects of different light-dark schedules on their biosynthesis, have been dealt with in considerable detail in the article by Wurtman in this volume. Consequently, this paper will be devoted mainly to the description of some biological effects exerted by pineal principles. It will appear that the physiological role (s) of the pineal gland may be established using the two approaches which have permitted to elucidate the secretory functions of other endocrine structures, i.e. surgical extirpation of the gland (pinealectomy) and administration of exogenous pineal principles. A peculiarity of the pineal body, not shared by other endocrine glands, is that its complete denervation, obtained through the surgical elimination of the two superior cervical ganglia, brings about effects perfectly similar to those which follow pinealectomy; for the reasons stated in the article by Wurtman in this volume, the animal submitted to a bilateral superior cervical ganglionectomy may be considered " functionally " pinealectomized. For additional information the reader is referred to the chapters by Wurtman (1967) and Kitay (1967), to the book by Wurtman *et al.* (1968) and to the recent reviews by Fraschini (1969) and Reiter and Fraschini (1969).

II. The Pineal Gland and the Secretion of Pituitary Gonadotropins

A. Effects of Pinealectomy

In female rodents, pinealectomy causes acceleration of vaginal opening (precocious puberty), ovarian hypertrophy and an increased incidence of cornified cells in vaginal smears (Wurtman *et al.*, 1959, 1963). The operation also enhances the ovulatory response that follows the administration of exogenous gonadotropins to immature animals (Dunaway, 1966; Dunaway and O'Steen, 1966, 1967). Both pinealectomy and superior cervical ganglionectomy abolish the negative effect of light deprivation on ovarian and uterine weights in neonatally-androgenized female rats (Reiter, 1968a; Reiter *et al.*, 1968a); the two operations also prevent the stimulating effect exposure to additional light exerts upon the onset of estrus in ferrets (Herbert, 1969).

In male rodents, pinealectomy and superior cervical ganglionectomy have been reported to induce hypertrophy of the testes, of the prostates and of the seminal vesicles (Carnicelli *et al.*, 1963; Houssay and Pazo, 1966; Izawa, 1926; Kincl and Benagiano, 1967) and to prevent the involution of the gonads and of the accessory sexual organs that follows light deprivation (Czyba *et al.*, 1964; Hoffman and Reiter, 1965; Reiter, 1967a, b, 1968b; Reiter *et al.*, 1968b; Reiter and Hester, 1966).

All the data quoted so far are compatible with the hypothesis that the pineal gland normally exerts an inhibitory influence on the secretion of pituitary gonadotropins (Kitay, 1967; Wurtman, 1967), and that the inhibitory influence of the pineal body is increased when the biosynthesis of methoxyindoles in the gland is activated by depriving the animals of sufficient amounts of light (Wurtman, 1967; Anton-Tay and Wurtman, 1968). However, the evidence supporting such hypothesis is absolutely indirect. This evidence includes the demonstration that pinealectomy causes hypertrophy of the pituitary gland, increases the number of basophils, enhances the mitotic activity of adenohypophyseal cells (Bindoni and Raffaele, 1968; Clementi *et al.*, 1969; Moszkowska, 1965; Thieblot, 1965), and induces the development of massive pituitary adenomata (Quay, 1967).

The experiments here to be described were planned in order to assess, in a more direct fashion, whether the pineal gland really inhibits the secretion of pituitary gonadotropins. Experiments were performed in adult male rats. Male animals were chosen since it is still necessary to rely on indirect criteria in order to have some information on the amounts and the type of gonadotropins present in the general circulation. Modifications of testicular weight represent a reliable index of plasma levels of Follicle-Stimulating Hormone (FSH), while changes of the weights of the prostates and of the seminal vesicles are strictly related to the amounts of circulating Luteinizing Hormone (Interstitial Cells-Stimulating Hormone, LH-ICSH) (Davidson, 1966, 1969). These parameters, although not ideal, are certainly more specific and sensitive than those provided by the study of the histological picture of the ovary (Martini *et al.*, 1968). Table I shows that, in sexually mature male rats, testicular weight

TABLE I

EFFECT OF PINEALECTOMY ON THE WEIGHT OF ENDOCRINE STRUCTURES IN MALE RATS [1]

Groups [2]	Final body Wt, g	Pituitary Wt, mg	Testes Wt, g	Seminal Vesicles Wt, mg	Prostates Wt, mg
Sham-operated (20)	263.7±7.2	7.7±.34	2.82±.21	290.5±11.3	184.8±15.2
Pinealectomized (15)	258.3±5.4	8.1±.49	3.24±.19	530.7±18.1[3]	274.1±17.0[3]

[1] Value are means ± SE.

[2] No. of rats in parentheses.

[3] $P \leq 0.001$.

TABLE II

EFFECTS OF PINEALECTOMY AND OF CASTRATION ON PITUITARY LH AND FSH LEVELS IN MALE RATS [1]

Groups [2]		Pituitary LH [3] µg/Pit	Pituitary FSH [3] µg/Pit
Controls	(20)	9.05±0.75 (4)	64.35± 2.09 (3)
Pinealectomized	(15)	21.47±2.10 (4) [4]	216.38±16.70 (3) [4]
Castrated	(18)	25.63±3.10 (6) [4]	372.98±15.30 (3) [4]

[1] Values are means ± SE.
[2] No. of rats in parentheses.
[3] Microgram equivalents of NIH-LH-B-1 bovine or NIH-FSH-S-3 ovine per pituitary. No. of assays in parentheses.
[4] $P \leq 0.005$ *vs.* controls.

is enhanced 12 days following pinealectomy; a very significant increase of the weights of the prostates and of the seminal vesicles is also observed (Motta *et al.*, 1967). These data, which confirm some of the reports quoted above, indicate that following pinealectomy, plasma levels of both FSH (increase in testis weight) and LH-ICSH (increase in prostate and seminal vesicle weights) are enhanced. Pinealectomy (Table II) results also in conspicuous and highly significant increase in the pituitary content of both gonadotropins (Fraschini, 1969). The amount of LH stored in the pituitary 12 days following pinealectomy is more than twice the amount of the hormone present in controls; the increase of pituitary FSH content induced by pinealectomy is even greater. The levels of LH and FSH found in the pituitary of adult male rats 12 days after gonadectomy have been included in Table II for comparison; it is evident that the increase in the stores of both gonadotropins observed after pinealectomy is similar to that induced by castration; the small difference may be ascribed to the fact that castration normally increases pituitary weight, while pinealectomy does not (Table I) (Motta *et al.*, 1967).

The results of these experiments indicate that pinealectomy stimulates the synthesis as well as the release of the two gonadotropins, and provide strong evidence in support of the thesis that the pineal gland normally keeps the anterior pituitary under an inhibitory control. It is apparent from the data that this inhibitory tone is usually rather strong; indeed, the stimulus provided to the pituitary by the elimination of the pineal gland is at least as great as that provided by castration.

B. Effects of Administration of Pineal Principles

1. *Systemic Administration*

Pineal indoles and methoxyindoles have been reported to counteract several effects of pinealectomy and to mimic the activity of crude pineal extracts (see Kitay and Altschule, 1954, for information on the effects of such extracts). Melatonin and 5-methoxytryptophol have been shown to depress ovarian weight (Farrell *et al.*, 1966; Wurtman *et al.*, 1963; Motta *et al.*, 1967) and to reduce the incidence of estrous smears in normally cycling rats and in animals exhibiting a constant estrous syndrome (Chu *et al.*, 1964); both compounds exert also an inhibitory effect on copulation-induced ovulation in rabbits (Farrell *et al.*, 1966, 1968). As stated in Section II, A, it is difficult to assess in a quantitative fashion in female animals, how much of these effects can be ascribed to suppression of LH release, and how much is rather due to inhibition of FSH secretion. Consequently, it was decided to carry on a study of the endocrine effect of exogenous melatonin in male rats.

Table III demonstrates that subcutaneous treatment of 30-day old male rats with melatonin (200 μg/100 g b.w./day, for 21 days) does not modify pituitary and testicular weights; however, prostates and seminal vesicles are significantly atrophied following treatment. These data, as well as the similar ones recently reported

TABLE III

Effect of Systemic Administration of Melatonin on the Weight of Endocrine Structures in Male Rats [1]

Groups [2]		Body Wt, g	Pituitary Wt, mg	Testes Wt, g	Seminal Vesicles Wt, mg	Prostates Wt, mg
Controls (12)						
	initial	96.3±4.2				
	final	223.5±8.4	7.3±0.50	2.53±0.29	240.3±10.6	167±11.0
Melatonin (12)						
200 μg/100 g						
s.c.	initial	97.1±4.3				
	final	227.6±7.9	6.9±0.42	2.69±0.31	190.1±15.0[3]	134±10.0[3]

[1] Values are means ± SE.
[2] No. of rats in parentheses.
[3] $P \leq 0.02$.

by Debeljuk (1969), indicate that melatonin significantly reduces the secretion of LH, but does not modify the secretion of FSH. Because of the observation previously reported (see Section II, A) that the pineal gland normally inhibits the secretion of both gonadotropins, one must postulate that the pineal body influences the secretion of FSH through principles different from melatonin (Motta *et al.*, 1967).

2. *Brain Implants*

The negative result obtained with melatonin on FSH secretion suggested to study whether other pineal principle(s) might be involved in the control of the secretion of this gonadotropin. As it has been reported in the article by Wurtman in this volume, the pineal gland synthesizes, in addition to melatonin, several other indole and methoxyindole derivatives (e.g., 5-hydroxytryptamine or serotonin, 5-methoxytryptophol, 5-hydroxytryptophol, etc.). In the experiments here to be reported pineal indoles and methoxyindoles were not administered systemically; they were placed directly into the brain, using a stereotaxic apparatus. The reasons for selecting this route of administration were: (1) the possibility of clarifying with one single experiment whether pineal principles which inhibit the anterior pituitary act directly on the gland, or whether they need the mediation of the nervous structures which control the secretion of pituitary gonadotropins; and (2) the elimination of the danger to classify as ineffective compounds which might give negative results following systemic administration because they do not cross the blood-brain barrier (Wurtman, 1967).

a. Median eminence implants. Figure 1 summarizes the effects of median eminence implants of indole and methoxyindole derivatives on pituitary FSH stores. Implants have been performed in adult male rats castrated three weeks before; they have been left *in situ* for 5 days. It appears that two principles, 5-methoxytryptophol and serotonin, are effective in reducing pituitary FSH stores; the other two, melatonin and 5-hydroxytryptophol, do not exert any effect at all. Figure 2 shows that 5-methoxytryptophol implanted directly into the pituitary does not reduce pituitary FSH; serotonin seems to increase FSH stores following direct placement in the pituitary. The experiments here reported indicate that two pineal principles, 5-methoxytryptophol and serotonin, may intervene in the control of

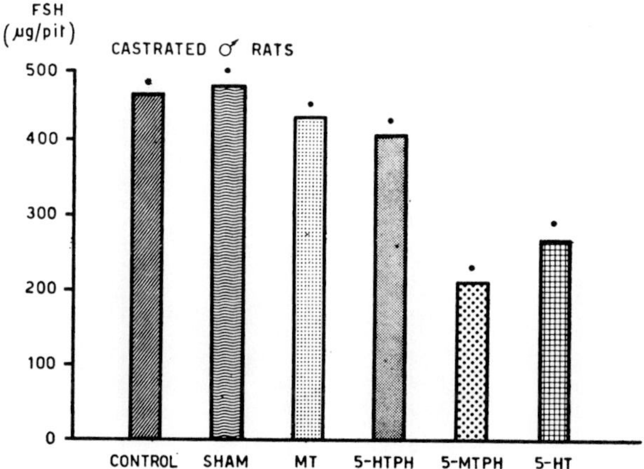

FIG. 1. Effect of implants of indole and methoxyindole derivatives in the median eminence on pituitary FSH content of castrated male rats. Abbreviations: MT, melatonin; 5-HTPH, 5-hydroxytryptophol; 5-MTPH, 5-methoxytryptophol; 5-HT, 5-hydroxytryptamine (serotonin). Each column represents the mean (with the standard error) of three assays performed on three different pools of pituitary glands (nine or ten pituitaries per pool). Results are expressed as FSH content per pituitary since there were no significant variations in pituitary weights in the different groups of animals (From Fraschini, 1969).

FSH secretion; apparently, they do so by activating special receptors localized in the median eminence of the hypothalamus, and not through a direct action on the pituitary gland. The magnitude of the decrease of pituitary FSH stores induced by the implantation of small amounts of the two compounds suggests that median eminence receptors sensitive to indole principles may play a relevant role in the control of the secretion of FSH. These data confirm, on the basis of direct measurements of FSH, the conclusion derived from the experiments in which the systemic injection of melatonin has been used (see section II, B, 1), i.e. that melatonin is not directly involved in the control of the secretion of this gonadotropin. They also explain why the administration of pineal extracts, but not that of synthetic melatonin, completely counteracts the effects of pinealectomy (Thieblot and Blaise, 1963); it is possible that crude pineal extracts contain in addition to melatonin, the two principles which specifically inhibit FSH secretion, namely 5-methoxytryptophol and serotonin. It is interesting to recall here that Corbin and Schottelius (1961) have reported that the injection of serotonin into the third

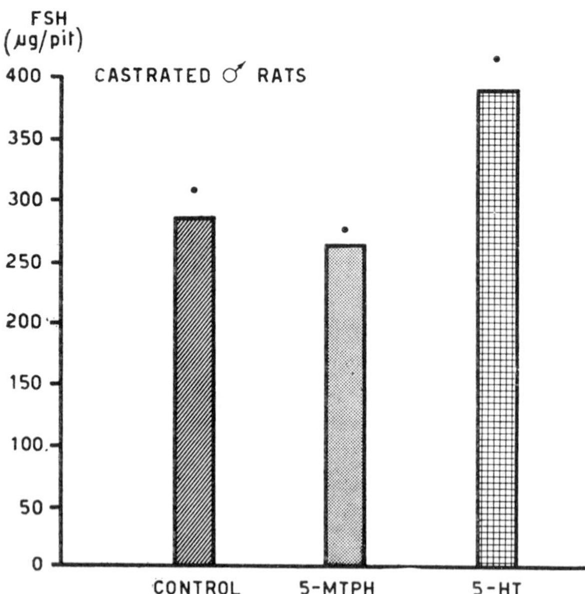

FIG. 2. Effect of implants of 5-methoxytryptophol and 5-hydroxytryptamine (serotonin) in the anterior pituitary on pituitary FSH content of castrated male rats. Abbreviations: 5-MTPH, 5-methoxytryptophol; 5-HT, 5-hydroxytryptamine (serotonin). Each column represents the mean (with the standard error) of three assays performed on three different pools of pituitary glands (nine or ten pituitaries per pool). Results are expressed as FSH content per pituitary since there were no significant variations in pituitary weights in the different groups of animals.

ventricle of immature female rats significantly postpones the onset of puberty; the need for a hypersecretion of FSH at time of puberty has been recently underlined (Corbin and Daniels, 1967, 1969).

Figure 3 summarizes the results of experiments in which pituitary LH levels have been measured following placement of pineal indoles and methoxyindoles in the median eminence of castrated male rats. As in the experiments previously described implants have been left in the brain for 5 days. Median eminence implants of serotonin and of 5-methoxytryptophol do not reduce pituitary stores of LH; on the contrary, melatonin and 5-hydroxytryptophol induce very significant decreases in the pituitary content of LH. Melatonin is unable to modify the pituitary reserve of LH when placed in control brain structures (i.e., in the cerebral cortex) or in the anterior pituitary gland itself (Fig. 4). The reduced levels of LH found in the pituitary

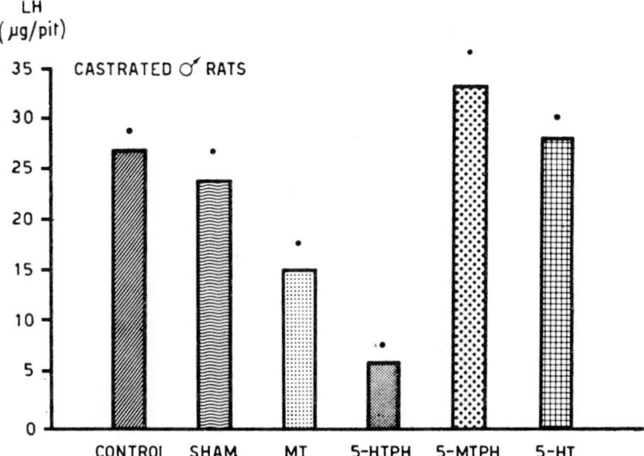

FIG. 3. Effect of implants of indole and methoxyindole derivatives in the median eminence on pituitary LH content of castrated male rats. Abbreviations: MT, melatonin; 5-HTPH, 5-hydroxytryptophol; 5-MTPH, 5-methoxytryptophol; 5-HT, 5-hydroxytryptamine (serotonin). Each column represents the mean (with the standard error) of three assays performed on three different pools of pituitary glands (four or five pituitaries per pool). Results are expressed as LH content per pituitary since there were no significant variations in pituitary weights in the different groups of animals (From Fraschini *et al.*, 1968a).

of animals having median eminence implants of melatonin or of 5-hydroxytryptophol might indicate either that release of LH has been activated, or that the synthesis of the hormone has been inhibited. The data summarized in Table IV show that plasma levels of LH are reduced in animals bearing median eminence implants of melatonin; this indicates that release of LH is not activated and suggests that brain implants of melatonin (and probably also those of 5-hy-

TABLE IV

EFFECT OF IMPLANTS OF MELATONIN INTO THE MEDIAN EMINENCE (ME) AND THE CEREBRAL CORTEX (CC) ON PLASMA LH LEVELS OF CASTRATED MALE RATS [1]

Groups [2]		% Ovarian ascorbic acid depletion	Plasma LH
ME-Sham	(7)	21.4 ± 1.7	present
ME-Melatonin	(12)	5.1 ± 0.9	absent
CC-Melatonin	(10)	24.7 ± 2.8	present

[1] Values are means ± SE.
[2] No. of rats in parentheses.

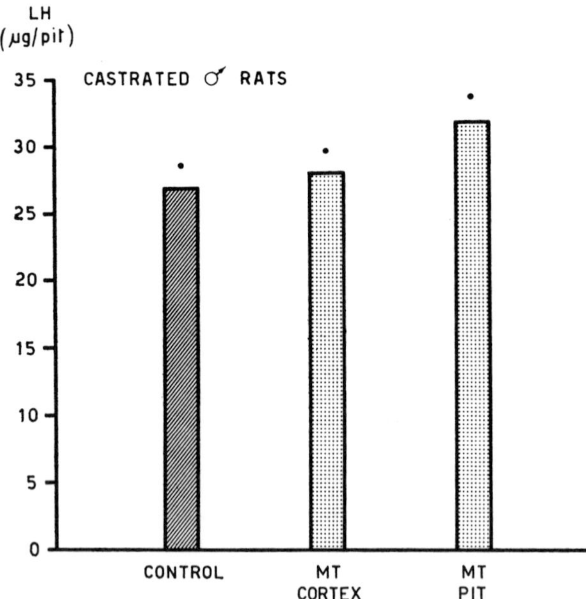

FIG. 4. Effect of implants of melatonin in the cerebral cortex and in the anterior pituitary on pituitary LH content of castrated male rats. Abbreviations: PIT, pituitary; MT, melatonin. Each column represents the mean (with the standard error) of three assays performed on three different pools of pituitary glands (four or five pituitaries per pool). Results are expressed as LH content per pituitary since there were no significant variations in pituitary weights in the different groups of animals.

droxytryptophol) inhibit the synthesis of LH. These data support then the statement (see Section II, B, 1) that the most important action of melatonin is that of inhibiting LH secretion; they show in addition that another pineal principle, 5-hydroxytryptophol, may intervene in the control of LH secretion with a mechanism similar to that of melatonin. Farrell and co-workers reported that 5-hydroxytryptophol had no effect on endocrine phenomena when given systemically (Farrell *et al.*, 1966); it is quite possible that the activity of this compound was not discovered in Farrell's experiments because it does not cross the blood brain barrier; obviously this factor is not involved when the compound is implanted directly in the brain. It is interesting to point out that also the two indole derivatives which control LH secretion, operate at the median eminence level. This could suggest that the median eminence has two types of receptors sensitive to pineal compounds; the first type, sensitive to changing levels of melatonin and of 5-hydroxytryptophol, appears to be involved in the control of the secretion of LH; the second type seems to be sensitive to changing levels of 5-methoxytryptophol and of serotonin and apparently regulates FSH secretion.

The general conclusions which may be derived from these experiments are: (1) the pineal gland controls the secretion of pituitary gonadotropins through two different humoral channels: melatonin and 5-hydroxytryptophol function as messengers for inhibiting LH secretion, while serotonin and 5-methoxytryptophol are the humoral principles which reduce FSH secretion; (2) the effects of pineal indoles and methoxyindoles on the secretion of gonadotropins are exerted on specific brain receptors rather than directly on pituitary cells; (3) since methoxyindoles in the mammalian organism are formed only in the pineal gland (see the article by Wurtman in this volume), 5-methoxytryptophol and melatonin are probably more important than serotonin and 5-hydroxytryptophol as regulators of pituitary activity. It is however surprising that in the control of each gonadotropin apparently intervene one indole and one methoxyindole of pineal origin.

b. Midbrain implants. In another set of experiments it has been shown that implantation of melatonin in the midbrain of castrated male rats results in a significant decrease of pituitary and plasma levels of LH (Fraschini *et al.*, 1968a, b). From these data it has been inferred that receptors sensitive to melatonin are localized also in the midbrain. These observations appear particularly relevant after the demonstration: (1) that melatonin, administered intravenously or placed in the lateral cerebral ventricles, is selectively concentrated within the midbrain (Anton-Tay and Wurtman, 1969); and (2) that the intraperitoneal administration of melatonin is followed by a significant increase of serotonin in the midbrain (Anton-Tay *et al.*, 1968). Since serotoninergic axons arising from cell bodies in the midbrain provide a major input to the medial forebrain bundle and to the hypothalamus (Dahlström and Fuxe, 1964), it is possible that part af the endocrine effects of melatonin are linked to a modification of the activity of this serotoninergic pathway (see the article by Wurtman in this volume for additional information). Unfortunately the data regarding the participation of the midbrain in the control of gonadotropin secretion are still rather scanty (Appeltauer *et al.*, 1966; Mess and Martini, 1968; Sas *et al.*, 1965). The evidence summarized in this paragraph indicates, however, that the clarification of midbrain-gonadotropin relationships will represent a very fruitful area for future work.

3. *Intraventricular Administration*

Recent data from our laboratory indicate that intraventricular injections of melatonin significantly decrease pituitary LH stores in normal adult male rats; such a decrease is already apparent 90 min after the intraventricular administration of the pineal principle. Pituitary FSH stores have been found to be unmodified after treatment (Fraschini and Piva, unpublished observations) (see Section III, A, for explanations on the choice of the intraventricular route of administration). It is noteworthy that in these experiments the effect of melatonin was found to be limited to suppression of LH secretion, exactly like in the experiments in which systemic administration and brain implants have been used (see Section II, B, 1 and 2). Since Anton-Tay and Wurtman (1969) have recently demonstrated that, following intraventricular injections, melatonin is distributed to the hypothalamus and the midbrain, one might conclude that using this route of administration the compound reaches simultaneously its receptors in these two regions of the brain. However, an effect through an activation of the midbrain-hypothalamic serotoninergic pathway cannot be disregarded (see Section II, B, 2, b). The second hypothesis would be supported by the observation by Kordon *et al.* (1968) that treatments with drugs which increase brain serotonin levels suppress gonadotropin-induced superovulation in immature female rats.

C. The Pineal Gland and Diurnal Gonadotropin Rhythms

The cyclic nature of gonadotropin secretion in females has been established since several years (Everett, 1961, 1964; Flerkó, 1966; Szentágothai *et al.*, 1968). More recently it has been recognized that a certain measure of cyclicity is present also in male mammals (Davidson, 1966; Martini *et al.*, 1968). The cycle seems to have a diurnal periodicity in the male; in females, the situation is more complex since a diurnal cycle, when it exists, appears to be superimposed on the ovulatory cycle, which has different lengths in different species.

In male mammals (including man) diurnal changes have been described in the nuclear volume of Leydig cells (Kovács, 1959); in the plasma levels of testosterone and of its precursors (Crafts *et al.*, 1968; Dray *et al.*, 1965; Kniewald and Martini, unpublished observations; Laatikainen and Vihko, 1968; McKenna and Rippon, 1965; Migeon *et al.*, 1957; Resko and Eik-Nes, 1966; Southren *et al.*, 1965, 1967) as well as in plasma levels of FSH (Faiman and Ryan, 1967; Saxena *et al.*, 1968, 1969).

With regard to diurnal rhythms of gonadotropin secretion in female mammals, the following evidence may be quoted. Ovarian cholesterol levels have been found to exhibit a diurnal rhythm both in mature and in immature female rats (Clark and Zarrow, 1967; Zarrow *et al.*, 1969). Similar diurnal variations have been found in the ascorbic acid and progesterone contents of the ovaries of pseudopregnant immature rats (De Groot, 1967; Stevens *et al.*, 1964). In long-term ovariectomized rats plasma levels of LH have been reported to be higher in the morning than in the afternoon (Lawton and Schwartz, 1968). Diurnal rhythms in the concentrations of both LH and FSH have been described to occur during the follicular phase of the human menstrual cycle (Midgley and Jaffe, 1968). A daily neural facilitatory period for the release of the ovulatory surge of LH is a well documented phenomenon (Barraclough, 1967; Everett, 1961, 1964; Hoffmann and Schwartz, 1965; Schwartz, 1964; Schwartz and Bartosik, 1962; Schwartz and Caldarelli, 1965). A diurnal fluctuation has been also found in the responses of the hypothalamic-pituitary complex to ovarian steroids: Caligaris *et al.* (1968) have reported that the administration of progesterone to castrated estrogen-treated female rats is followed in 4 hrs by an increase of plasma LH levels when the steroid is administered at 12 noon, but not when the steroid is given at 8 A.M.; Terasawa and Sawyer (personal communication) were able to record a diurnal variation in the electrical responsiveness to progesterone of neurons of the arcuate nucleus and of the median eminence (see the article by Sawyer in this volume for additional information).

In 1967, Fraschini and Motta reported that in normal male rats caged under standard light conditions (14 hours of light beginning at 6.30 A.M., 10 hours of dark) there is a diurnal cycle in the levels of gonadotropins stored in the anterior pituitary gland; a peak of FSH and LH content was observed to occur in the afternoon, between 4 P.M. and 6 P.M. (Fig. 5) (see Martini *et al.*, 1968 for additional information). More recently, similar diurnal cycles of pituitary FSH and LH have been found also in adult male rats a few weeks following castration. They also appear in female rats, when the ovulatory cycle has been obliterated by ovariectomy performed 3 weeks before measurement (Fraschini, unpublished observations). Since pineal principles, and particularly melatonin and 5-methoxytryptophol, have been shown in previous sections of this paper to modify the secretion of LH and of FSH (see Section II, B, 2), the question may be asked whether the diurnal changes in the synthesis, storage and release of gonadotropins might be connected with the diurnal modifications which are observed in the biosynthesis of pineal methoxyindoles. It has been clearly established (Wurtman, 1967; Wurman *et al.*, 1968; see also the article by Wurtman in this volume) that the synthesis of melatonin and of 5-methoxytryptophol is cyclic in nature, and is strictly regulated by the light-dark schedule to which the animal is exposed. The biosynthesis of pineal methoxyindoles is inhibited during the day, and activated during the night,

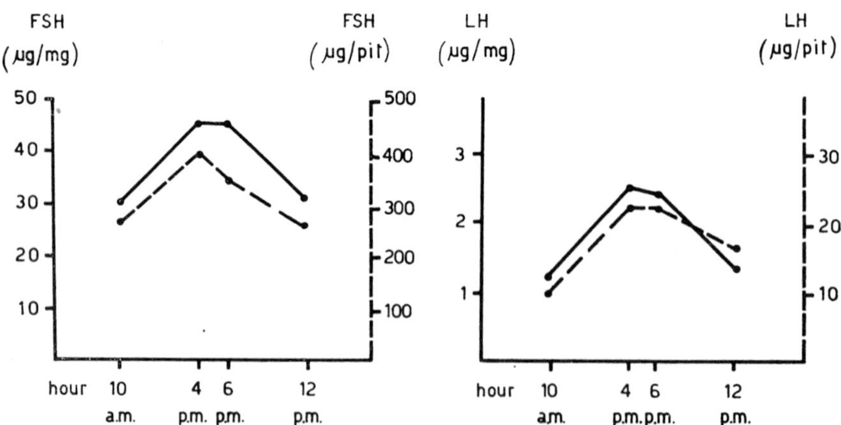

FIG. 5. Diurnal rhythm of pituitary FSH and LH content in normal male rats. Each point represents the mean of three assays performed on three different pools of pituitary glands (nine or ten pituitaries per pool for FSH; four or five pituitaries per pool for LH). Results are expressed both as FSH and LH content per pituitary (dotted line) and as FSH and LH concentration (µg/mg of pituitary tissue) (solid line).

because light inhibits the activity of the enzyme hydroxyindole-o-methyltransferase (HIOMT) which is essential for introducing the methoxy-group on the indole molecule. These data make it logical to suggest that, during the light period, the pituitary gland is released from the usual pineal inhibition so that LH and FSH can be synthesized at a higher rate and accumulated in the pituitary gland; conversely, during the dark period, when the pineal gland is activated, the pituitary gland is submitted to an increased inhibitory tone, which prevents further accumulation of gonadotropins. This hypothesis has been confirmed in a preliminary fashion by the observation that the removal of the superior cervical ganglia (which prevents the pineal gland to be influenced by the light stimulus, and which brings about a sort of " functional " pinealectomy) abolishes the decrease of pituitary FSH and LH concentrations which occur in normal animals at midnight, following the exposure to a few hours of dark (Fig. 5) (Fraschini, unpublished observations).

III. The Pineal Gland and Brain Activity

A. Effect of Pineal Principles on Barbiturate-Induced Sleep

Observations are accumulating which indicate that the pineal gland and its principles may influence also non endocrinologically oriented neural functions. Particularly relevant appear the data regarding the possibility that pineal principles might intervene in the control of motor activity and of alertness. Pinealectomized rats show increased motor activity and an activated electroencephalogram (EEG) (Nir *et al.*, 1969; Reiss *et al.*, 1963). Conversely, the administration of pineal extracts or of melatonin decreases the spontaneous motor activity in the rat (Reiss *et al.*, 1963; Wong and Whiteside, 1968). In addition, the intravenous injection of melatonin in newborn chicken (Barchas *et al.*, 1967), or the implantation of small amounts of melatonin into the preoptic area of unrestrained cats, produces sleep (Marczynski *et al.*, 1964). Melatonin-induced sleep is characterized by slow EEG activity of high voltage and by a decrease of " paradoxical sleep " periods (Hishikawa *et al.*, 1969; Lerner and Case, 1960; Marczynski *et al.*, 1964; Supniewski *et al.*, 1961). Serotonin and 5-hydroxytryptophol have been reported to modify optic-evoked potentials following intravenous administration in rabbits (Sabelli, 1970).

Because of these data, it was decided to study whether the sedative effect of melatonin might be demonstrated also by using a traditional pharmacological test, i.e. the prolongation of barbiturate-induced sleep. In the experiments here to be described melatonin was injected into the lateral cerebral ventricles of adult male rats, immediately after the intraperitoneal injection of a standard dose of pentobarbital (3 mg/100 g b.w.) (Fioretti *et al.*, 1968). It was felt that if it could be demonstrated that melatonin influences a biological parameter after being directly added to the cerebrospinal fluid (CSF), this might be taken as good evidence in favor of the possibility that pineal methoxyindoles are physiologically released into the CSF rather than in the blood. Since melatonin has never been recovered in the general circulation, it has been suggested that the CSF might be responsible for transporting and distributing this compound to its receptor sites (Sheridan *et al.*, 1969; Wurtman, 1967; Wurtman *et al.*, 1968; see also the article by Wurtman in this volume). This suggestion appears particularly valid because of the observations:

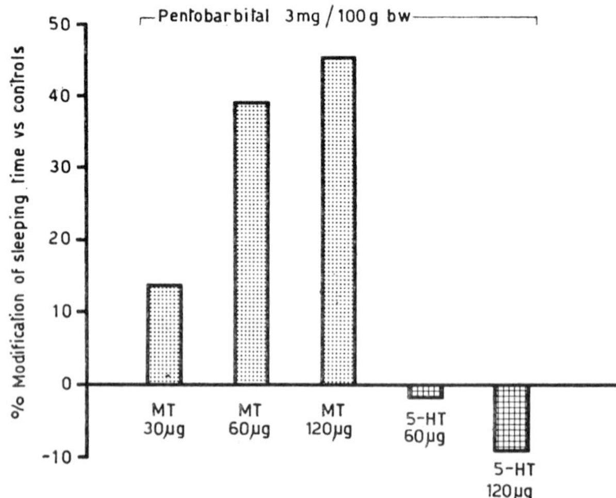

FIG. 6. Effect of intraventricular injections of melatonin on pentobarbital-induced sleep in male rats. Controls have been injected with 3 mg/100 g b.w. of pentobarbital intraperitoneally and with saline solution intraventricularly. Experimental groups have been injected with 3 mg/100 g b.w. of pentobarbital intraperitoneally and with melatonin (MT) or 5-hydroxytryptamine (serotonin, 5-HT) intraventricularly.

(1) that practically all receptors for melatonin-controlled effects are located in the brain (see Section II, B, 2); and (2) that more than 100 times as large a percentage of the administered dose of melatonin is retained in the brain after its intraventricular administration than after its intravenous injection.

As shown in Fig. 6, intraventricular administrations of melatonin prolong the sleeping effect of pentobarbital. The potentiation of the barbiturate effect appears to be related to the dose of melatonin injected (Fioretti *et al.*, 1968). This effect, which is similar to the one reported by Barchas (1968) and by Arutyunyan *et al.* (1963) in mice following intraperitoneal administration of much larger doses of melatonin, seems specific for melatonin since comparable amounts of serotonin injected intraventricularly in conjunction with the usual dose of barbiturate are followed by a small decline in barbiturate-induced sleeping time. These results unambiguously suggest that melatonin facilitates sleep. The mechanism through which this effect is exerted is not clear at the moment. On the basis of the observation that serotonin and its precursor 5-hydroxytryptophan induce hypnosis (Heliman *et al.*, 1961; Rothballer, 1967) and that p-chlorophenylalanine (a blocker of serotonin synthesis) normally induces insomnia (Jouvet, 1967; Koella *et al.*, 1968; Weitzman *et al.*, 1968) it has been suggested that serotonin might be the " sleep hormone ". Since the administration of melatonin increases serotonin stores in several regions of the brain (Anton-Tay *et al.*, 1968), one might suggest that melatonin facilitates sleep through the activation of serotoninergic pathways. However, the results obtained with intraventricular injections of serotonin do not support this mode of action; they even cast some doubt on the possibility that serotonin might be considered the physiological inducer of sleep.

B. Effect of Pineal Principles on Body Temperature

In another set of experiments it has been shown that intraventricularly injected melatonin exerts a hyperthermic effect in rats (Fioretti and Martini, 1968). This result stresses once more the species differences which exist in the mechanisms controlling body temperature; it had been previously reported that melatonin depresses body temperature in rabbits (Supniewski *et al.*, 1961). As in the case of the prolongation of barbiturate-induced sleeping time, the effect of melatonin on body temperature seems rather specific. In agreement with previous evidence (Feldberg and Lotti, 1967; see also the article by Feldberg in this volume) serotonin has been found to depress body temperature in rats (Fioretti and Martini, 1968). This result indicates that melatonin probably operates on thermoregulatory mechanisms in a direct fashion and not by modifying the serotoninergic tone.

IV. Conclusions

The data summarized in the previous sections of this paper indicate that the pineal gland and its indole and methoxyindole derivatives may influence several endocrinological and non endocrinological processes.

It has been clearly established that the pineal gland exerts a strong inhibitory influence on the secretion of pituitary gonadotropins. The chemical nature of the messengers used for suppressing the release of the two gonadotropic factors has also been ascertained: it appears that serotonin and 5-methoxytryptophol are involved in the control of the secretion of FSH, while melatonin and 5-hydroxytryptophol regulate the secretion of LH. It has also been clarified that the inhibitory effect of pineal principles is mainly exerted on specific indole- and methoxyindole-sensitive receptors, localized in the hypothalamus and in the midbrain. The hypothesis has been put forward that the diurnal cyclicity in the secretion of LH and FSH, which has been recently described in normal and in castrated male rats and in ovariectomized females, might be connected with the diurnal rhythm which exists in the biosynthesis of pineal methoxyindoles. Much more work is obviously needed in this area before firm conclusions can be reached; however, it appears possible to suggest, at least as a working hypothesis, that also the daily facilitatory period for the release of the ovulatory surge of LH which operates in several mammalian species, might be regulated by the pineal gland; it seems quite possible that the release of LH might be facilitated during the light hours, when the negative effect exerted by pineal principles on the neural structures which control the secretion of gonadotropins is eliminated.

Melatonin has also been shown to depress nervous activity (as indicated by the prolongation of barbiturate-induced sleeping time) and to exert an hyperthermic effect.

It seems important that melatonin might depress pituitary LH stores, modify sleeping patterns and increase body temperature after direct administration into the cerebrospinal fluid of the lateral ventricles. These results prove that melatonin is easily distributed and transported to its receptors when the compound is present in the cerebrospinal fluid; they provide strong evidence in support to the hypothesis that melatonin and other pineal principles might physiologically be released into the cerebrospinal fluid rather than into the blood.

Acknowledgments

The work here presented has been supported by the funds of the Department of Pharmacology of the University of Milan and by the following Grants: 67-530 of the Ford Foundation, New York; M65-118-M66-121 of the Population Council, New York; R-00088 of the World Health Organization, Geneva, Switzerland. Gifts of FSH and LH have been supplied by the National Institutes of Health, Bethesda, Maryland. Bibliographic assistance was received from the UCLA brain information service.

References

Anton-Tay, F., and Wurtman, R. J. (1968). *Endocrinology* **82**, 1245.
Anton-Tay, F., and Wurtman, R. J. (1969). *Nature* **221**, 474.
Anton-Tay, F., Chou, C., Anton, S., and Wurtman, R. J. (1968). *Science* **162**, 277.
Appeltauer, L. C., Reissenweber, N. J., Dominguez, R., Griño, E., Sas, J., and Benedetti, W. (1966). *Acta Neuroveget.* **29**, 75.
Arutyunyan, G. S., Mashkovskii, M. D., and Roshchina, L. F. (1963). *Fed. Proc.* **23**, T 1330.
Barchas, J. (1968). *Proc. Western Pharmacol. Soc.* **11**, 22.
Barchas, J., Da Costa, F., and Spector, S. (1967). *Nature* **214**, 919.
Barraclough, C. A. (1967). *In* " Neuroendocrinology " (L. Martini and W. F. Ganong, eds.). Vol. II, pp. 61–99. Academic Press, New York.
Bindoni, M., and Raffaele, R. (1968). *J. Endocrinol.* **41**, 451.
Caligaris, L., Astrada, J. J., and Taleisnik, S. (1968). *Acta Endocrinol.* **59**, 177.
Carnicelli, A., Saba, P., Cella, P. L., and Marescotti, V. (1963). *Folia Endocrinol.* **16**, 229.
Chu, E. W., Wurtman, R. J., and Axelrod, J. (1964). *Endocrinology* **75**, 238.
Clark, J. H., and Zarrow, M. X. (1967). *Acta Endocrinol.* **56**, 445.
Clementi, F., De Virgiliis, G., and Mess, B. (1969). *J. Endocrinol.* **44**, 241.
Corbin, A., and Daniels, E. L. (1967). *Neuroendocrinology* **2**, 304.
Corbin, A., and Daniels, E. L. (1969). *Neuroendocrinology* **4**, 65.
Corbin, A., and Schottelius, B. A. (1961). *Am. J. Physiol.* **201**, 1176.
Crafts, R., Llerena, L. A., Guevara, A., Lobotsky, J., and Lloyd, C. W. (1968). *Steroids* **12**, 151.
Czyba, J. C., Girod, C., and Durand, N. (1964). *C. R. Soc. Biol.* **158**, 742.
Dahlström, A., and Fuxe, K. (1964). *Acta Physiol. Scand. Suppl. 232*, 1.
Davidson, J. M. (1966). *In* " Neuroendocrinology " (L. Martini and W. F. Ganong, eds.). Vol. I, pp. 565–611. Academic Press, New York.

Davidson, J. M. (1969). *In* " Frontiers in Neuroendocrinology 1969 " (W. F. Ganong and L. Martini, eds.), pp. 343–388. Oxford Univ. Press, New York.

Debeljuk, L. (1969). *Endocrinology* **84**, 937.

De Groot, C. A. (1967). *Acta Endocrinol.* **56**, 459.

Dray, F., Reinberg, A., and Sebaoun, J. (1965). *C. R. Acad. Sci.* **261**, 573.

Dunaway, J. F. (1966). *Anat. Record* **154**, 340.

Dunaway, J. F., and O'Steen, W. K. (1966). *Texas Rep. Biol. Med.* **24**, 503.

Dunaway, J. E., and O'Steen, W. K. (1967). *Texas. Rep. Biol. Med.* **25**, 525.

Everett, J. W. (1961). *In* " Sex and Internal Secretions " (W. C. Young, ed.). Vol. I, pp. 497–555. Williams and Wilkins Co., Baltimore.

Everett, J. W. (1964). *Physiol. Rev.* **44**, 373.

Faiman, C., and Ryan, R. J. (1967). *Nature* **215**, 857.

Farrell, G., McIsaac, W. M., and Powers, D. (1966). *Program 48th Meeting Endocrine Soc.*, p. 98.

Farrell, G., Powers, D., and Otani, T. (1968). *Endocrinology* **83**, 599.

Feldberg, W., and Lotti, V. J. (1967). *Brit. J. Pharmacol.* **31**, 152.

Fioretti, M. C., and Martini, L. (1968). *Ann. Accad. Med. Perugia* **60**, 426.

Fioretti, M. C., Barzi, F., Bececco, D., and Fraschini, F. (1968). *Ann. Accad. Med. Perugia*, **59**, 318.

Flerkó, B. (1966). *In* " Neuroendocrinology " (L. Martini and W. F. Ganong, eds.). Vol. I, pp. 612–668. Academic Press, New York.

Fraschini, F. (1969). *In* " Progress in Endocrinology " (C. Gual, ed.), pp. 637–644. Excerpta Medica, Amsterdam.

Fraschini, F., and Motta, M. (1967). *Program 49th Meeting Endocrine Soc.*, p. 128.

Fraschini, F., Mess, B., Piva, F., and Martini, L. (1968a). *Science* **159**, 1104.

Fraschini, F., Mess, B., and Martini, L. (1968b). *Endocrinology* **82**, 919.

Heliman, K. N., Vonderhae, A. R., and Peters, J. J. (1961). *Neurology* **11**, 1011.

Herbert, J. (1969). *J. Endocrinol.* **43**, 625.

Hishikawa, Y., Cramer, H., and Kuhlo, W. (1969). *Exp. Brain Res.* 7, 84.

Hoffman, R. A., and Reiter, R. J. (1965). *Science* **148**, 1609.

Hoffmann, J. C., and Schwartz, N. B. (1965). *Endocrinology* **76**, 626.

Houssay, A. B., and Pazo, J. H. (1966). *In* " Abstracts Book of the III Intern. Pharmacol. Congress (Saõ Paulo) ", p. 147.

Izawa, Y. (1926). *J. Physiol.* **77**, 126.

Jouvet, M. (1967). *Sci. Am.* **216**, 62.

Kincl, F. A., and Benagiano, G. (1967). *Acta Endocrinol.* **54**, 189.

Kitay, J. I. (1967). *In* " Neuroendocrinology " (L. Martini and W. F. Ganong, eds.). Vol. II, pp. 641–664. Academic Press, New York.

Kitay, J. I., and Altschule, M. D. (1954). " The Pineal Gland. " Harvard Univ. Press, Cambridge, Mass.

Koella, W. P., Feldstein, A., and Czicman, J. S. (1968). *Electroencephalog. Clin. Neurophysiol.* **25**, 481.

Kordon, C., Javoy, F., Vassent, G., and Glowinski, J. (1968). *Europ. J. Pharmacol.* **4**, 169.

Kovács, J. (1959). *Acta Biol. Acad. Sci. Hung.* **10**, 69.

Laatikainen, T., and Vihko, R. (1968). *J. Clin. Endocrinol. Metab.* **28**, 1356.

Lawton, I. E., and Schwartz, N. B. (1968). *Am. J. Physiol.* **214**, 213.

Lerner, A. B., and Case, H. D. (1960). *Fed. Proc.* **19**, 590.

Marczynski, T. J., Yamaguchi, N., Ling, G. M., and Grodzinska, L. (1964). *Experientia* **20**, 435.

Martini, L., Fraschini, F., and Motta, M., (1968). *Recent Progr. Hormone Res.* **24**, 439.

McKenna, J., and Rippon, A. E. (1965). *Biochem. J.* **95**, 107.

Mess, B., and Martini, L. (1968). *In* " Recent Advances in Endocrinology " (V. H. T. James, ed.), pp. 1–49. J. and A. Churchill, London.

Midgley, A. R., Jr., and Jaffe, R. B. (1968). *J. Clin. Endocrinol. Metab.* **28**, 1699.

Migeon, C. J., Keller, A. R., Lawrence, B., and Shepard, T. H. (1957). *J. Clin. Endocrinol. Metab.* **17**, 1051.

Moszkowska, A. (1965). *Progr. Brain Res.* **10**, 564.

Motta, M., Fraschini, F., and Martini, L. (1967). *Proc. Soc. Exp. Biol. Med.* **126**, 431.

Nir, I., Behroozi, K., Assael, M., Ivriani, I., and Sulman, F. G. (1969). *Neuroendocrinology* **4**, 122.

Quay, W. B. (1967). *Experientia* **23**, 129.

Reiss, M., Davis, R. H., Sideman, M. B., and Plichta, E. S. (1963). *J. Endocrinol.* **28**, 127.

Reiter, R. J. (1967a). *J. Endocrinol.* **38**, 199.

Reiter, R. J. (1967b). *Neuroendocrinology* **2**, 138.

Reiter, R. J. (1968a). *Fed. Proc.* **27**, 319.

Reiter, R. J. (1968b). *Fertil. Steril.* **19**, 1009.

Reiter, R. J., and Fraschini, F. (1969). *Neuroendocrinology* **5**, 219.

Reiter, R. J., and Hester, R. J. (1966). *Endocrinology* **79**, 1168.

Reiter, R. J., Rubin, P. H., and Reichert, J. R. (1968a). *Life Sci.* **7**, 299.

Reiter, R. J., Hoffmann, J. C., and Rubin, P. H. (1968b). *Science* **160**, 420.

Resko, J. A., and Eik-Nes, K. B. (1966). *J. Clin. Endocrinol. Metab.* **26**, 573.

Rothballer, A. B. (1967). *Neurophysiology* **4**, 409.

Sabelli, H. C. (1970). *Experientia* **26**, 58.

Sas, J., Griño, E., Benedetti, W. L., Appeltauer, L. C., and Dominguez, R. (1965). *Acta Morphol. Acad. Sci. Hung.* **13**, 377.

Saxena, B. B., Demura, H. M., Gandy, H., and Peterson, R. E. (1968). *J. Clin. Endocrinol. Metab.* **28**, 519.

Saxena, B. B., Leyendecker, G., Chen, W., Gandy, H. M., and Peterson, R. E., (1969). *Acta Endocrinol. Suppl.* **142**, 185.

Schwartz, N. B. (1964). *Am. J. Physiol.* **207**, 1251.

Schwartz, N. B., and Bartosik, D. (1962). *Endocrinology* **71**, 756.

Schwartz, N. B., and Caldarelli, D. (1965). *Proc. Soc. Exp. Biol. Med.* **119**, 16.

Sheridan, M. N., Reiter, R. J., and Jacobs, J. J. (1969). *J. Endocrinol.* **45**, 131.

Southren, A. L., Tochimoto, S., Carmody, N. C., and Isurugi, K. (1965). *J. Clin. Endocrinol. Metab.* **25**, 1441.

Southren, A. L., Gordon, G. G., Tochimoto, S., Pinron, G., Lane, D. R., and Stypulkowski, W. (1967). *J. Clin. Endocrinol. Metab.* **27**, 686.

Stevens, V. C., Owen, L., Fukushima, M., and Vorys, N. (1964). *Endocrinology* **74**, 493.

Supniewski, J., Misztal, S., and Marczynski, T. J. (1961). *Diss. Pharm.* (Warsz) **13**, 205.

Szentágothai, J., Flerkó, B., Mess, B., and Halász, B. (1968). " Hypothalamic Control of the Anterior Pituitary ". Akadémiai Kiadó, Budapest.

Thieblot, L. (1965). *Progr. Brain. Res.* **10**, 479.

Thieblot, L., and Blaise, S. (1963). *Ann. Endocrinol.* **24**, 270.

Weitzman, E. D., Rapport, M. M., McGregor, P., and Jacoby, J. (1968). *Science* **160**, 1361.

Wong, R., and Whiteside, C. B. C. (1968). *J. Endocrinol.* **40**, 383.

Wurtman, R. J. (1967). *In* " Neuroendocrinology " (L. Martini and W. F. Ganong, eds.). Vol. II, pp. 19–59. Academic Press, New York.

Wurtman, R. J., Altschule, M. D., and Holmgren, U. (1959). *Am. J. Physiol.* **197**, 108.

Wurtman, R. J., Axelrod, J., and Chu, E. W. (1963). *Science* **141**, 277.

Wurtman, R. J., Axelrod, J., and Kelly, D. E. (1968). " The Pineal ". Academic Press, New York.

Zarrow, M. X., Clark, J. H., and Denenberg, V. H. (1969). *Neuroendocrinology* **4**, 270.

Control of Circadian Periodicities in Pituitary Function

K. RETIENE

I. Introduction

A. HISTORICAL DEVELOPMENT OF RHYTHM RESEARCH

Mankind has observed for centuries that plants and animals perform certain activities at fairly fixed times of the day. Complete recorded information about rhythmic phenomena in the animated world dates to the middle of the eighteenth century when the French astronomer De Mairan (1729) described rhythmic daily movements of plant leaves, as recorded by a crude chymograph. During the next 250 years the literature contained numerous studies describing biological rhythms in every measurable biological process from plant and animal unicellular organisms up to whole organ systems, including man. It is important to note that these observed rhythms possess a striking correlation with external geophysical events on earth, such as solar or lunar cycles with a period length of 12 or 24 hr.

As a matter of fact, most rhythmic biological functions seemed so precisely synchronized to primary day-night cycles that they have been termed " diurnal rhythms."

All these classic publications have been generally restricted to the description of rhythmic phenomena in plants and animals, with little or no attempt to interpret the underlying mechanisms generating such rhythmic behavior. The most logical explanation from a teleologic standpoint was that these rhythmic events were apparently triggered and entrained by the external day-night cycle.

B. General Rules of Circadian Rhythms

Some recent observations, however, contradict the conclusion of the day-night cycle as a totally valid hypothesis. With newly established methods of time-series analysis, as well as statistical computations, it has been found that such rhythms are seen to persist even after day-night cycles are abolished and do exist in blind individuals. What might be even more important is that rhythms under constant conditions do not possess a precise 24-hr timing, rather these cycles are somewhat shorter or longer than a 24-hr period. In fact, because of this latter observation, the new term " circadian " (which means " about a day ") has entered the literature (Halberg, 1959). These findings have led the majority of researchers to believe that some form of endogenous clock mechanism must exist which integrates external signals such as light, temperature, and humidity. Of these signals light is the most important, the so-called " Zeitgeber " (timekeeper) that acts on all circadian rhythms in a predictable way as follows: under constant light and increasing light intensity the period lenghth of circadian rhythms in day-active animals becomes shorter; in night-active animals (like the rat) it becomes longer. Under constant darkness the opposite behavior is found (Aschoff, 1958, 1963) (for detailed literature see " Cold Spring Harbor Symposium ", 1960).

II. Unsolved Problems

In contrast to these briefly summarized general rules of biological rhythm research, very little is known about the nature, control, and interrelationship of circadian periodicities of the endocrine system in the mammal. The regular circadian variation in adrenocortical activity, first claimed by Pincus (1943), is now one of the best known and most studied biological rhythms that are present in all mammals (Brown *et al.*, 1957; Nugent *et al.*, 1960; Ungar, 1967). However, perhaps due to the difficulty of bioassay procedures our present knowledge is not definite with regard to the circadian rhythmicity of the pituitary hormones and the activating neurohumoral substances of the hypothalamus. Neither are we certain that the circadian rhytmicity of hormones is necessary for the overall homeostasis of the endocrine system or that a hierarchy exists in the endocrine rhythms themselves. For instance we do not know whether or not rhythmic secretion of pituitary Adrenocorticotropic Hormone (ACTH) is critical and the rhythmic secretion of Thyrotropin (TSH) or vasopressin is rather unimportant for adequate function of the whole endocrine system. Furthermore, how do external stress factors or exogenous hormone administration act on these rhythms? Answers to these major questions are necessary for complete understanding of endocrine physiology.

In recent years we have been expending efforts to find answers to these questions. The results of our investigations are presented in this article.

III. Experimental Conditions

For all animal experiments we have been using male and female Sprague-Dawley rats. The animals were placed in individual cages for at least 10 days to become acclimatized to a controlled lighting schedule of 14 hr of light per day, from 04.00 to 18.00, alternating with 10 hr of darkness. Constant temperature and humidity were maintained in the animal room.

To obtain resting steroid blood levels, groups of 5 to 8 rats were alternately decapitated within 15 to 30 sec and trunk blood was collected in heparinized beakers for fluorometric determination of corticosterone by the method of Silber *et al.* (1958). Values were expressed as micrograms corticosterone per 100 ml of plasma or per gram of adrenal tissue, respectively.

A. Adrenocorticotropic Hormone

Pituitary glands were removed, and anterior and neural lobes were separated and frozen on dry ice and weighed. Pooled anterior and posterior lobes from each group were ground in 2 ml of 0.1 *N* HCl in saline and stored for ACTH determinations. Fresh solutions were prepared so that doses of 0.2 ml represented the extract of 0.1 mg of pituitary tissue. An *in vivo* assay measuring corticosterone increase in peripheral blood of hypophysectomized rats was used (Guillemin *et al.*, 1958). Extracts were administered in a single dose level to seven animals at each period in a randomized block design. Mean values and standard errors were expressed as microgram corticosterone per 100 ml of plasma or milliunits ACTH per milligram of pituitary tissue. ACTH determinations in plasma were done by a modification (Retiene *et al.*, 1962) of the method of Lipscomb and Nelson (1959) which measures the corticosterone increase in the adrenal venous blood of hypophysectomized rats. An amount of 5 ml of plasma from patients or pooled plasma from each group of rats were injected into the assay animals. Previous studies had shown a linear response between 0.01 mU and 1.0 mU USP Reference Standard (2nd International Standard).

B. Thyrotropin

Extracts of pituitary anterior lobes were also used for the determination of TSH by a modification of the McKenzie assay (McKenzie, 1958). The extracts were assayed at two dosage levels, and the TSH activity was determined by comparison with USP Reference Thyrotropin Standards at the same dilution and at two dosage levels. Five to eight mice were used for each dose and the activity of each extract was determined in two to four assays on separate days.

C. Vasopressin

Extracts of posterior lobes contained 0.01 mg of tissue per 0.2 ml of solvent. For the assay of vasopressin, ether-anesthesized rats were pithed through the inner canthus of the eye and immediately put on artificial respiration. A carotid artery was cannulated and blood pressure recorded continuously by means of a sensitive transducer. The test substances were injected into a cannulated femoral vein. Increases in arterial blood pressure were expressed as changes in millimeter of Hg. Previously, 1.0 to 5.0 mU of USP Reference Standard vasopressin showed a linear dose-response curve.

D. Corticotropin Releasing Factor

To obtain a kind of Corticotropin Releasing Factor (CRF) material, a circular plug of median eminence tissue was punched out with a sharp 4-mm trephine. These plugs of similar size contained the hypothalamic tissue situated behind the optic chiasma and rostral to the mammillary bodies, extending to the third ventricle. The fragments were pooled and ground in 4 ml of 0.1 *N* HCl saline. Solutions in 0.01 *N* HCl saline were made so that the extract of one hypothalamus could be injected for both ACTH and CRF determinations. Assay animals were blocked with chlorpromazine, morphine, and nembutal for the evaluation of CRF (Arimura *et al.*, 1967) and hypophysectomized for the evaluation of ACTH. Chemically blocked rats as well as hypophysectomized animals showed the same dose-response curve to standard ACTH. Unknowns were injected at random into different assay animals.

IV. A Comparative Study of Endocrine Rhythms in Rats

Circadian rhythms of different hormones in one experimental group of rats are summarized in Fig. 1. The values for adrenal and plasma corticosterone throughout the 24-hr course resemble the well-known diurnal variation, with a peak before the onset of darkness at 17.00, i.e., before the period of activity in nocturnal animals, and with lowest values in the early morning at 05.00 (Critchlow *et al.*, 1963; Hauss and Halberg, 1960). The absolute values are comparable with the results reported in previous publications, which demonstrated a typical sex difference in that the peaks of female rats exceeded those of male rats by more than 100% (Retiene *et al.*, 1968).

Changes in ACTH, TSH, and vasopressin content of the pituitary were also comparable with those of corticosterone. Significant peaks at 17.00 are well in phase with the adrenal rhythm in the same animals. A parallel fluctuation of blood ACTH and adrenal steroid levels has previously been demonstrated in rats and man (Retiene *et al.*, 1965). However, a circadian rhythm of vasopressin (a hypothalamic hormone stored only in the posterior lobe of the pituitary) has not been published elswhere. In contrast to earlier publications of Bakke and Lawrence (1965) and Schindler *et al.* (1965), but actually

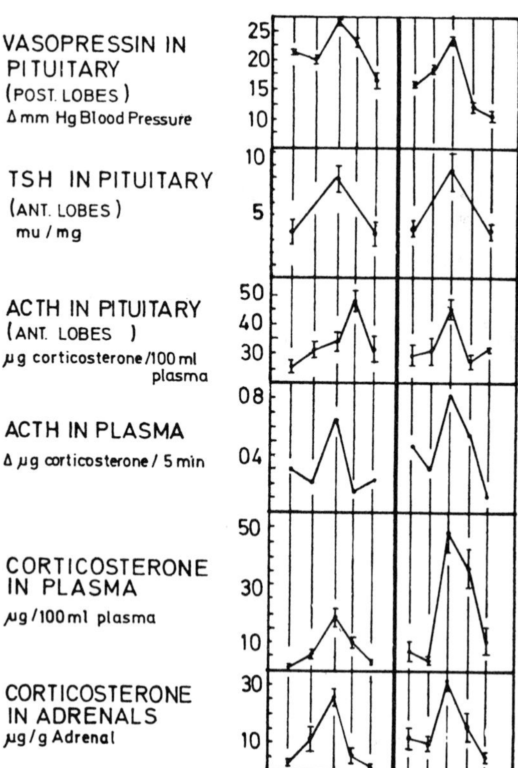

FIG. 1. Comparative study of endocrine rhythms in rats. The vertical lines represent standard errors of the mean. Graphs on the left are males; on the right are females.

not very surprising, are our findings of a parallel circadian rhythm of the TSH content in the pituitary. These other authors observed an opposite behavior of TSH content in male rats, with lowest values in the late afternoon, and no variation in female rats. We think that the discrepancy between these results and our findings can be explained by the fact that we were dealing exclusively with rats under absolute resting conditions. We have observed an opposite behavior of TSH and ACTH content in the pituitary of stressed animals (see below). In addition to our findings, recent papers have described similar circadian periodicities for Luteinizing Hormone (LH) (Lawton and Schwartz, 1964), oxytocin (Ezrin *et al.*, 1962) and melatonin in the pineal gland (Wurtman and Axelrod, 1965).

V. Circadian Rhythms under a Randomized Lighting Schedule

The precise synchrony in the rhythmic behavior of all these animals was achieved with photoperiods of 14 hr of light alternating with 10 hr of darkness. Studies of circadian rhythms have in general involved observations and analysis of physiological rhythms under constant environmental conditions. However, constancy of environment is only one of the possible modifications that can be used to study circadian rhythms. Another method, which to our knowledge has not been investigated, is a randomization of lighting conditions in which periods of light and darkness follow each other at irregular intervals. This experimental design provides control and experimental animals with identical amounts of light per day and yet prevents the acquisition of rhythmic light information by the experimental group. Figures 2 and 3 demonstrate the results of a 17-day study under these conditions. The graph of the corticosterone values for the control animals demonstrates the usual diurnal cycle found in normal rats, with a distinct peak just prior to the onset of darkness. The peak value at 17.00 is significantly different from all other values. The graph of adrenal and plasma steroids for the experimental animals has some resemblance to that of the controls; however, the

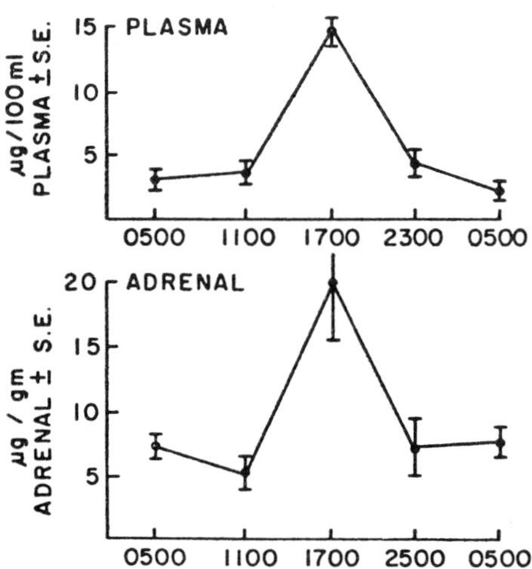

FIG. 2. Diurnal variation of plasma and adrenal corticosterone in rats under normal lighting (14 hr light, 10 hr darkness).

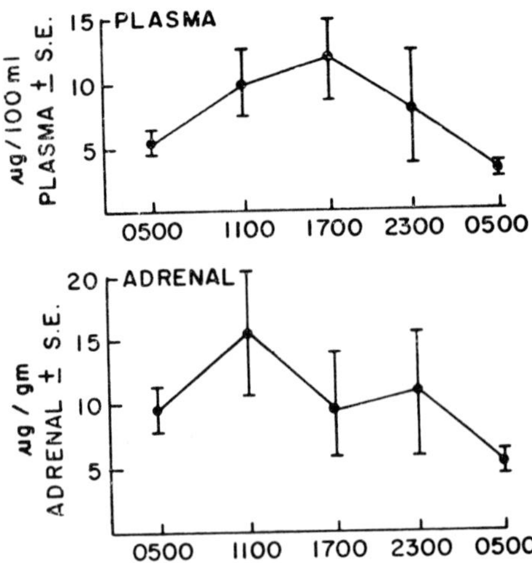

FIG. 3. Diurnal variation of plasma and adrenal corticosterone in rats under random light.

standard errors for these values are so large that none of the means is statistically significant. The only explanation for this phenomenon under a random light pattern seems to be that individual circadian rhythms, both longer and shorter than 24 hr in period, appear to have replaced the synchronized rhythmicity of the whole group. Simple removal of the temporal aspect of the lighting environment, without alteration of the net photoperiod, therefore appears to be a method of modifying and even nullifying the important effect of light on biological rhythms in mammals (Holmquest *et al.*, 1966).

It can be stated from these experiments that rhythmic fluctuation of synthesis and release of hormones is a characteristic phenomenon of the whole endocrine system in mammals. The fact that the peak values of all hormones are observed before the period of activity (i.e., that the animals can measure time and can control their physiological processes accordingly) as well as the fact that these rhythms are maintained under constant or randomized lighting conditions, support the hypothesis of the endogenous nature of these rhythms. There seems to be only one exception, namely, the striking fluctuation of melatonin, which depends exclusively on intact neural pathways connecting the retina with the pineal gland, as stated by Axelrod *et al.* (1965).

VI. Effect of Exogenous Steroid Administration on Circadian Rhythmicity of the Pituitary-Adrenal System

In our group we have not studied the rhythmicity of melatonin *per se*; however, we did not observe any alteration of adrenal rhythms in pinealectomized rats. Thus we found no positive evidence regarding the occasionally raised question of whether or not the pineal gland plays an important role in maintaining circadian periodicity of other hormones. It is generally acknowledged that synthesis and release of all pituitary hormones is controlled by the hypothalamus. Specific releasing and/or inhibiting factors for each of the hypophyseal hormones have been isolated (Schally and Bowers, 1964; Guillemin, 1965). The hypothalamus itself has been mapped for the location of centers synthesizing the releasing factors and for the receptor areas for the feedback effect of peripheral hormones (see Martini *et al.*, 1968; De Wied *et al.* 1965; Endröczi *et al.* 1959). However, the central nervous control mechanisms of the striking circadian rhythms of the pituitary hormones are still problematic. The rhythmicity can be viewed by a cyclic resetting of the feedback mechanisms, but may also depend on a rhythmic stimulation by a certain brain area.

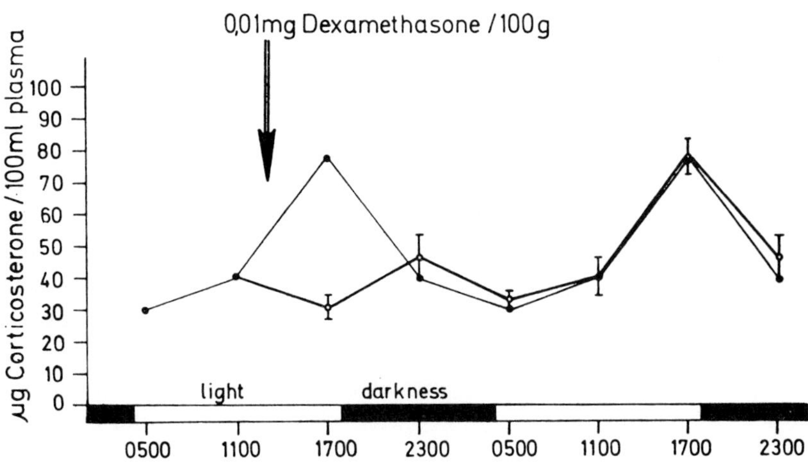

FIG. 4. Circadian rhythmicity of plasma costicosterone in female rats after a single dexamethasone injection. Closed circles = controls; open circles = treated animals.

For elucidation of these problems it seemed to be important to study the effects of isolated hormone administration and of isolated external stress stimuli. Figure 4 demonstrates the effect of a single injection of 0.01 mg of dexamethasone per 100 g body weight, administered just before the diurnal peak in the late afternoon. The circadian rise is completely abolished; however, the next peak does not occur at a random time thereafter. Rather it is seen at exactly the right time of the next day. Thus we have suppressed a circadian rise of corticosterone, but probably have not altered the hypothetical biological clock in the central nervous system.

VII. Effect of Ether Stress on the Circadian Rhythmicity of the Pituitary-Adrenal System

We have summarized in Fig. 5 the results obtained in an experiment that studied the effect of a standardized ether stress of exactly 90 sec. The absolute rise in the steroid level was approximately the same throughout the day and was added to the different basal hormone values. As Fig. 5 shows, the physiological peak at 17.00 was completely abolished by a single injection of 0.01 mg of dexamethasone per 100 g body weight administered at 13.00. However, the stress-induced steroid increase was about the same as in the untreated animals. These findings are in full agreement with experiments of Critchlow and Zimmermann (1966) and show that there are different mechanisms responsible for the release of ACTH under resting and stress conditions, which can be differentiated by the sensitivy to peripheral corticosteroids.

Since the stress-induced reactions of the hypothalamus-pituitary system seemed to be independent of the time of day as well as the actual steroid blood level, we have only studied pituitary ACTH and TSH content at 17.00. The light bars on Fig. 6 demonstrate the resting corticosterone blood level and ACTH content as well as the TSH content; the striped bars indicate the stress-induced change of the same parameters. In the untreated animals the ACTH content is increasing even though ACTH was released to provoke the acute corticosterone increase in the peripheral blood. At the same time, TSH content of the pituitary has decreased to less than half that of the control animals. Thus we find an opposite behavior of ACTH

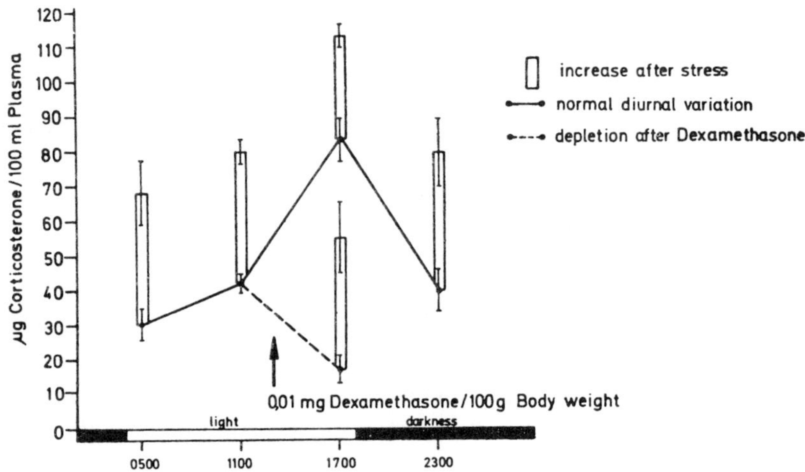

FIG. 5. Stress-induced corticosterone increase in normal and dexamethasone-blocked rats at different times of the day.

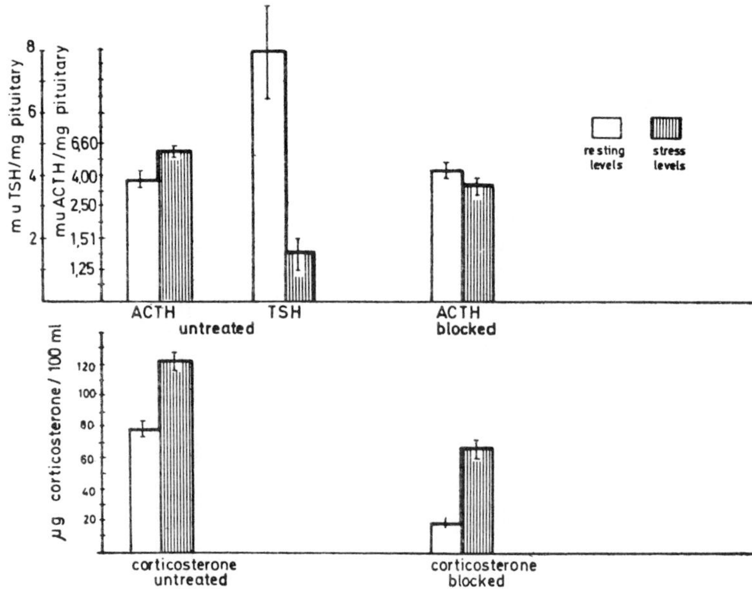

FIG. 6. ACTH and TSH content of the pituitary gland and corticosterone blood level in normal and dexamethasone-blocked rats at 17.00 before and after stress (see text).

and TSH content in the pituitary under stress conditions, whereas under resting conditions both hormones seemed to be governed by a similar control systems. Ether stress therefore triggered the pituitary to release ACTH and at the same time to synthesize an excess of the hormone. This may happen to the detriment of TSH content, as demonstrated by *in vitro* experiments with CRF and Thyrotropin Releasing Factor (TSHRF) preparations by Sakiz and Guillemin (1965). After a single dexamethasone injection just prior to the diurnal peak, corticosterone was markedly decreased, but the ACTH content of the pituitary in these animals was not altered in comparison to that of the controls. The ether stress provoked the same absolute corticosterone increase and a slight decrease of the ACTH content. We must therefore suggest that corticosteroids inhibit the circadian rhythmicity and the synthesis of ACTH, but do not influence hypothalamic structures being employed for the release of ACTH under stress conditions.

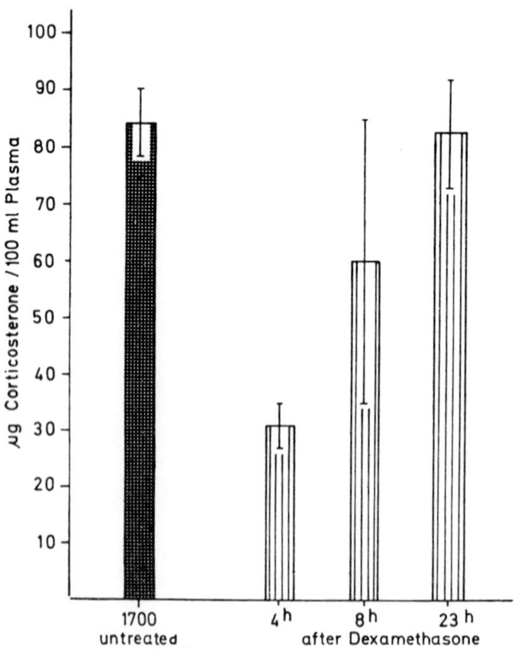

FIG. 7. Corticosterone blood level at 17.00; 4 hr, 8 hr, and 22 hr after dexamethasone.

VIII. Circadian Rhythmicity of the Pituitary-Adrenal System after Long-Term Treatment with Dexamethasone

Further studies revealed that the corticosterone blood level at 17.00 was less altered or remained unchanged in comparison to that of controls when dexamethasone was administered 8 or 22 hr before the peak, respectively (Fig. 7). In other words, dexamethasone injected shortly after the diurnal peak of the ACTH secretion had no influence on the pituitary-adrenal circadian function (Nichols *et al.*, 1965; Retiene *et al.*, 1967a). Stimulated by these preliminary studies we performed a long-term experiment by treating rats at different times of day, but keeping the animals under the standardized environmental conditions with a regular day-night cycle of 14 hr light and 10 hr darkness. One group of rats was daily injected with 0.01 mg of dexamethasone per 100 g body weight at 13.00 and a second group at 18.00, i.e., just after the diurnal peak of pituitary-adrenal function.

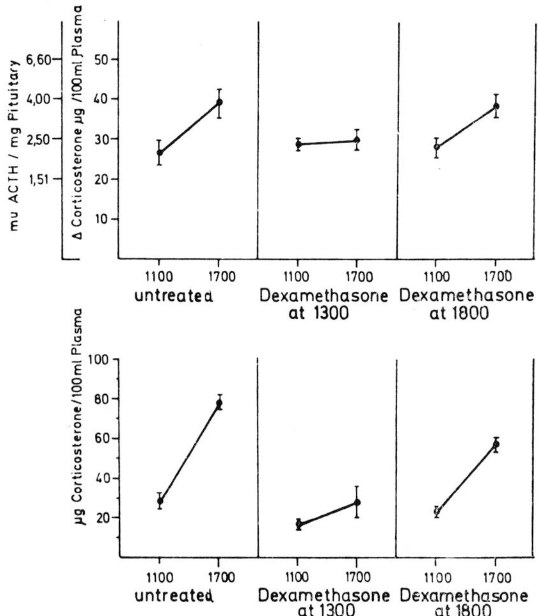

FIG. 8. Circadian rhythmicity of pituitary ACTH and plasma corticosterone after long-term treatment with dexamethasone at different times of the day (see text).

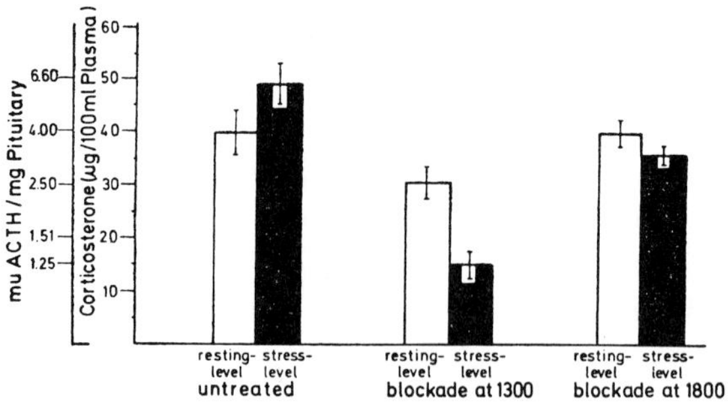

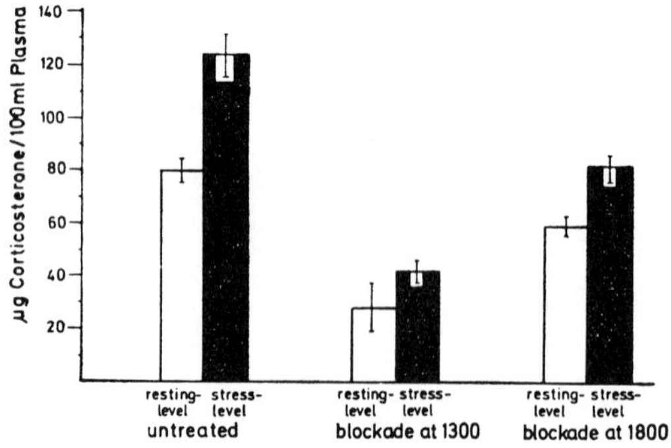

FIG. 9. Stress-induced change of pituitary ACTH and plasma corticosterone after long-term treatment with dexamethasone at different times of the day (see text).

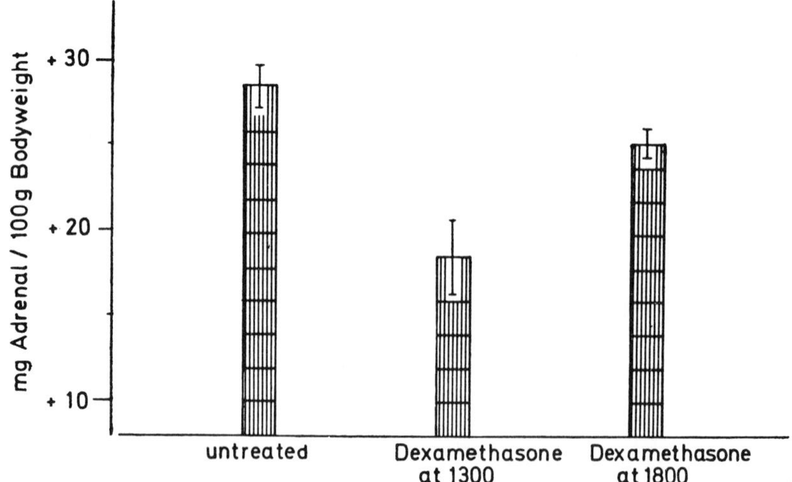

FIG. 10. Adrenal weight of rats after long-term treatment with dexamethasone at different times of the day (see text).

As can be seen in Fig. 8, the circadian rise of pituitary ACTH content and plasma corticosterone is very much the same in control animals and in rats that were treated in the evening. Abolishment of circadian fluctuation was observed in rats that were treated in the rest period around noon.

Ether-induced changes of these parameters are demonstrated in Fig. 9. The animals blocked at 18.00 had a normal resting ACTH content in the pituitary, with no significant change after ether stress. The absolute corticosterone increase in the peripheral blood was only slightly different from that of the controls. Animals blocked at 13.00 had a decreased ACTH content and corticosterone blood level under resting conditions. A very slight (statistically not significant) increase of the corticosterone was observed with a marked depletion of the ACTH content after stress. Whereas in the latter group the synthesis of ACTH seems to be completely inhibited, stress-induced mechanisms to release ACTH are not impaired. In contrast, the animals treated at 18.00 seemed to maintain some ability to synthesize and release ACTH under stress conditions. Fig. 10 demonstrates the adrenal weight in milligrams per 100 g body weight in the same groups of rats. Whereas the adrenal weight of the controls and the experimental group treated at 18.00 are not statistically different from each other, the group of rats treated at 13.00 has the lowest adrenal weight as a sign of adrenal atrophy in this chronically treated group (Retiene *et al.*, 1967b).

IX. Conclusions

At this point the question comes up as to where the " controller " of rhythmicity is situated in the endocrine chain of regulation and how it acts. No doubt the median eminence and the releasing factors serve as the link between the central nervous system and the peripheral hormones. It therefore seems reasonable to postulate a circadian rhythmicity of the hypothalamic releasing factors. Our attempt to measure CRF in hypothalamic extracts, however, revealed very low concentrations, with no fluctuations at all throughout the day in untreated animals, but with a twofold increase after stress. We are convinced that there are many reasons for this failure, particularly methodological. On the other hand, no such rhythm of any releasing factor has been published elsewhere. In spite of this, all the evidence from the literature and from our experiments supports the idea of a special " rhythm center " in the central nervous system. Slusher (1962) as well as Halász and Pupp (1965) described experiments in which they made cuts to separate the most anterior part of the brain from the basal ventral medial part of the hypothalamus. In these rats, ACTH secretion was slightly elevated, but at the same time the circadian fluctuation of ACTH secretion was completely abolished. These structures of the brain, therefore, might have an essential effect on the hypothalamus in the maintenance of diurnal rhythms. However, experiments by Endröczi and Lissák (1960), Knigge and Hays (1964), and Bohus and De Wied (1967) give evidence that this same area exerts inhibitory influences on ACTH release.

The following working hypothesis is proposed and we believe it to fit best with most experimental data at hand. A special " rhythm center, " probably situated in the anterior hypothalamus, acts rhythmically but in an inhibitory fashion on the synthesis of the releasing factors. Thus, circadian peaks of ACTH under resting conditions represent a passive overflow of the hormone pool in the gland, whereas stress stimuli mainly provoke hormone release, which is not related to the " rhythm center." The feedback effect of corticosteroids at the hypothalamic level is similar to the inhibitory effect of the " rhythm center." If both influences appear at the same time of day, normal synthesis of hormones is not impaired. Alternating inhibition by the " rhythm center " and by exogenous steroids leads to a depression of hormone synthesis, and stress mechanisms become less effective.

At the moment we are only beginning to understand the mechanisms controlling and maintaining circadian rhythmicity of endocrine functions. But we do agree that this aspect of endocrine physiology is not merely a subject of academic interest. The experiments to date have revealed important implications regarding the choice of the proper time of day for bioassays, and standardization and quantitation of the concentrations of blood levels of hormones. The clinical investigator has to take into account these new data, since he is now in a situation in which he can modify the action of a hormone by altering not only the dosage but also the time of day at which it is administered.

Addendum: After closing the manuscript we have been able to demonstrate a circadian rhythm of hypothalamic CRF in rats. After stress the absolute hormone content decreased in the morning and increased in the evening, thus giving strong evidence for the discussed hypothesis of a "controller" which rhythmically inhibits hormone synthesis (Retiene and co-workers, unpublished observations).

Acknowlegdment

The experiments of this study were partially supported by the "Deutsche Forschungs-Gemeinschaft."

Some of the experiments were done in collaboration with Dr. H. S. Lipscomb, Dr. W. J. Schindler, Dr. D. L. Holmquest, and Dr. E. Zimmermann, Department of Physiology, Baylor University, Houston, Texas.

References

Arimura, A., Saito, T., and Schally, A. V. (1967). *Endocrinology* **81**, 235.

Aschoff, J. (1958). *Z. Tierpsychol.* **15**, 1.

Aschoff, J. (1963). *Deut. Med. Wochschr.* **88**, 1930.

Axelrod, J., and Wurtman, R. J. (1964). *Abstracts 6th Intern. Congr. Biochemistry.* New York.

Axelrod, J., Wurtman, R. J., and Snyder, S. H. (1965). *J. Biol. Chem.* **240**, 949.

Bakke, I. J., and Lawrence, N. (1965). *Metabolism* **14**, 841.

Bohus, B., and De Wied, D. (1967). *In* "Proc. 1st Intern. Symposium on Biorhythms in Clinical and Experimental Endocrinology" (G. Sayers and A. Lunedei, eds.), pp. 71–83. Università degli Studi, Firenze.

Brown, H., Englert, E., Wallach, S., and Simons, E. L., (1957). *J. Clin. Endocrinol.* **17**, 1191.

Cold Spring Harbor Symposium, 1960, Vol. 25.The Biological Laboratory, Cold Spring Harbor, L. I., New York.

Critchlow, V., and Zimmermann, E. (1966). *Excerpta Med. Intern. Congr. Ser.* **111**, 209.

Critchlow, V., Liebelt, R. A., Bar-Sela, M., Mountcastle, W., and Lipscomb, H. S. (1963). *Am. J. Physiol.* **205**, 807.

De Mairan (1729). "Histoire de l'Academie Royale des Sciences." p. 34.

De Wied, D., Lammers, J. G., and Smelik, P. G. (1965). *Proceedings of the 23rd Int. Congr. Physiol.*, Excerpta Medica, Amsterdam, p. 36.

Endröczi, E., and Lissák, K. (1960). *Acta Physiol. Acad. Sci. Hung* **17**, 39.

Endröczi, E., Lissák, K., Bohus, B., and Kovács, S. (1959). *Acta Physiol. Acad. Sci. Hung.* **16**, 17.

Ezrin, C., Loach, L. W., and Nicholson, T. F. (1962). *Canad. Med. Ass. J.* **87**, 673.

Guillemin, R. (1965). *Verhandl. Deut. Ges. Inn. Med.* **71**, 61.

Guillemin, R., Clayton, G. W., Smith, J. D., and Lipscomb, H. S. (1958). *Endocrinology* **63**, 359.

Halász, B., and Pupp, L. (1965). *Endocrinology* **77**, 553.

Halberg, F. (1959). *Z. Vitamin- Hormon- Fermentforsch.* **10**, 225.
Hauss, E., and Halberg, F. (1960). *Acta Endocrinol.* (Kbh) *Suppl.* **51**, 219.
Holmquest, D. L., Retiene, K., and Lipscomb, H. S. (1966). *Science* **152**, 662.
Knigge, K. M., and Hays, M. (1964). *Proc. Soc. Exp. Biol. Med.* **114**, 67.
Lawton, J. E., and Schwartz, N. B. (1964). *Physiologist* **7**, 3.
Lipscomb, H. S., and Nelson, D. (1959). *Fed. Proc.* **18**, 95.
Martini, L., Fraschini, F., and Motta, M. (1968). *Recent Progr. Hormone Res.* **24**, 439.
McKenzie, I. M. (1958). *Endocrinology* **63**, 372.
Nichols, Th., Nugent, Ch. A., and Tyler, F. H. (1965). *J. Clin. Endocrinol.* **25**, 343.
Nugent, Ch. A., Eik-Nes, K., Kent, H. S., Samuels, L. T., and Tyler, F. H. (1960). *J. Clin. Endocrinol.* **20**, 1259.
Pincus, G. (1943). *J. Clin. Endocrinol.* **3**, 195.
Retiene, K., Ditschuneit, H., Fischer, M., Kopp, K., and Pfeiffer, E. F. (1962). *Acta Endocrinol.* (Kbh) **41**, 211.
Retiene, K., Espinoza, A., Marx, K. H., and Pfeiffer, E. F. (1965). *Klin. Wschr.* **43**, 205.
Retiene, K., Schulz, F., and Marco, J. (1967a). *In* " Proc. 1st Intern. Symposium on Biorhythms in Clinical and Experimental Endocrinology " (G. Sayers and A. Lunedei, eds.), pp. 217–221. Università degli Studi, Firenze.
Retiene, K., Frohns, T., and Schulz, F. (1967b). *Verhandl. Deut. Ges. Inn. Med.* **73**, 990.
Retiene, K. Zimmermann, E., Schindler, W. J., Neuenschwander, J., and Lipscomb, H. S. (1968). *Acta Endocrinol.* (Kbh) **57**, 615.
Sakiz, E., and Guillemin, R. (1965). *Endocrinology* **77**, 797.
Schally, A. V., and Bowers, C. Y. (1964). *Metabolism* **13**, 1190.
Schindler, W. J., Critchlow, V., Krause, D. M., and McHorse, T. S. (1965). *Excerpta Med. Intern. Congr. Ser.* **99**, 65.
Silber, R. H., Bush, R. O., and Oslapas, R. (1958). *Clin. Chem.* **4**, 278.
Slusher, M. A. (1962). *Fed. Proc.* **21**, 196.
Ungar, F. (1967). *In* " Proc. 1st Intern. Symposium on Biorhythms in Clinical and Experimental Endocrinology " (G. Sayers and A. Lunedei, eds.), pp. 57–70. Università degli Studi, Firenze.
Wurtman, R. J., and Axelrod, J. (1965). *Sci. Am.* **213**, 50.

Hormones and Brain Differentiation

F. NEUMANN, H. STEINBECK, and J. D. HAHN

I. Differentiation of Gonadotropin Secretion Patterns

A. PFEIFFER'S STUDIES

Since the classical studies of Pfeiffer (1936) it has become known that male sex hormones play an important role in the differentiation of the sexual centers. Pfeiffer spayed newborn female rats during the first few postnatal days, and upon subsequent implantation of ovaries, normal vaginal cycles in adulthood accompanied by the development of corpora lutea were retained. However, after castration and implantation of testicular tissue into newborn female rats, no corpora lutea were seen to develop in implanted ovaries in adulthood. The animals remained in a state of permanent estrus. Even when the female ovaries were left intact, the output of testosterone from the testicular implant prevented normal functioning of the ovaries after puberty. On the other hand, following castration of newborn male rats and subsequent ovarian implantation, Pfeiffer observed normal ovarian development with corpus luteum formation. Reimplantation of testicular tissue in these animals immediately after castration, however, resulted in the later implanted ovaries remaining small, with follicles only but no corpora lutea development.

The occurrence of corpora lutea is considered proof of a preceding ovulation, signifying cyclic secretion of those tropic hormones that regulate gonadal function.

Pfeiffer, who believed the pituitary to be the target organ of testicular hormones, came to the following conclusions: male and female rats are born with an undifferentiated pituitary that has the latent capacity for the cyclic secretion of gonadotropins. The sex-specific function of the pituitary is not genetically determined. In male animals it is determined secondarily by an action of the testicular androgens during an appropriate phase of development.

At the same time it had already been postulated by Hohlweg and Junkmann (1932) that the hypothalamus controls gonadotropin secretion. Indeed, it could be demonstrated in a number of experiments that the pituitary remains pluripotent.

B. Effect of Androgens, Castration, Antiandrogens, and Estrogens

1. *Site of Action*

Pituitaries of male, female, or neonatally-androgenized female rats have the capacity to maintain ovulation, vaginal cycles, and pregnancies when transplanted beneath the hypothalamus in hypophysectomized but otherwise normal female rats (Harris, 1964; Harris and Jacobsohn, 1952; Martinez and Bittner, 1956; Segal and Johnson, 1959).

The differentiation of the ovaries is also not impaired by neonatal androgen treatment. Even polycystic ovaries of androgenized rats respond with ovulation to a normal gonadotropin secretion pattern when transplanted into castrated but otherwise normal female rats or neonatally-castrated males (Barraclough, 1967a, b; Bradbury, 1941; Harris, 1964; Yazaki, 1959, 1960). It was also possible to induce ovulation in such ovaries by injection of Luteinizing Hormone (LH), Human Chorionic Gonadotropin (HCG), Pregnant Mare Serum (PMS), and even of Luteinizing Hormone Releasing Factor (LHRF) (Arai, 1963; Barraclough, 1967b; Courrier *et al.*, 1961; Harris, 1964; Schuetz and Meyer, 1963; Segal and Johnson, 1959).

The studies cited above proved that androgens impress a male functional pattern upon distinct regulatory centers of the hypothalamus. Before continuing with this discussion, the timing factor of these differentiating processes should be mentioned.

2. *Timing*

In rats and mice the critical period for the differentiation of the hypothalamic control centers of gonadotropin secretion extends from the first to the tenth day of life. Treatment of female rats and mice with testosterone up to the tenth day of life will lead to permanent sterility; however, the effective dosage has to be increased with advancing age (Arai and Gorski, 1968; Barraclough, 1955, 1961; Barraclough and Leathem, 1954; Boyd and Johnson, 1968; Bradbury, 1941; Gorski, 1966, 1968; Gorski and Barraclough, 1963; Gorski and Wagner, 1965; Harris, 1964; Harris and Levine, 1962; Lloyd and Weisz, 1967; Mazer and Mazer, 1939; Selye, 1940; Shay *et al.*, 1939; Swanson and Van der Werff ten Bosch, 1964a,b; Wilson *et al.*, 1941). After the tenth day, androgen treatment no longer influences the capacity for cyclic gonadotropin secretion, which is the prerequisite for ovulation (Barraclough, 1961; Barraclough and Leathem, 1954; Schuetz and Meyer, 1963; Wilson *et al.*, 1941).

On the first day of life, 5 to 10 μg of testosterone propionate administered subcutaneously or 1.25 μg of testosterone propionate intracerebrally implanted are sufficient to cause permanent sterility (Table I). Following treatment with smaller doses, ovulation and vaginal cycles are maintained quite normally over a limited period after puberty, in some cases before the early androgen syndrome becomes established (Gorski, 1968; Swanson and Van der Werff ten Bosch, 1964a, b). The larger the dose of testosterone, the fewer normal cycles will be maintained; hence the sooner the early androgen syndrome is fully developed (Fig. 1; see also Fig. 6). The cause of this delayed anovulatory syndrome is certainly not a continued differentiation of the central nervous system but rather the result of a subsequent and independently developing modification of an incompletely differentiated system (Gorski, 1968). With the progress of time, however, this capacity is lost. Experimental data provide no proof that the differentiation of hypothalamic centers which regulate gonadotropin secretion is initiated in rats before birth. Most results are contradictory to this.

TABLE I

INFLUENCE OF VARIOUS DOSES OF TESTOSTERONE PROPIONATE ADMINISTERED AT DIFFERENT AGES TO FEMALE RATS ON THE FERTILITY OF THE ADULT ANIMAL (10 TO 14 WEEKS OF AGE)

Dose [1]	Days after Birth (= d 1)									
(μg/animal)	1	2	3	4	5	6	8	10	20	Reference
1250	—	100[2]	—	—	100	—	—	40	0	Barraclough (1961)
	—	—	—	—	100	—	—	—	—	Gorski (1966)
	—	—	—	—	100	—	—	—	—	Gorski and Barraclough (1963)
500	—	—	—	100	—	—	—	—	—	Harris (1964)
	—	—	—	100	—	—	—	—	—	Harris and Levine (1962)
250	—	—	—	—	100	—	—	—	—	Boyd and Johnson (1968)
100	—	—	—	—	100	—	—	—	—	Gorski and Barraclough (1963)
	—	—	—	—	67	—	—	—	—	Swanson and Van der Werff ten Bosch (1964a)
50	—	—	—	—	75	—	—	—	—	Swanson and Van der Werff ten Bosch (1964a)
30	—	—	—	—	92	—	—	—	—	Arai and Gorski (1968)
10	100	91	83	100	100	100	17	0	—	Gorski (1968)
	—	—	—	—	71	—	—	—	—	Gorski (1966) Gorski and Barraclough (1963)
	—	—	—	—	50	—	—	—	—	Gorski and Wagner (1965)
	—	—	—	—	30	—	—	—	—	Swanson and Van der Werff ten Bosch (1964a)
	—	100	—	—	14	—	—	—	—	Swanson and Van der Werff ten Bosch (1964b)
5	—	—	—	—	44	—	—	—	—	Gorski (1966) Gorski and Barraclough (1963)
	—	60	—	—	0	—	—	—	—	Swanson and Van der Werff ten Bosch (1964a)
	—	—	—	—	0	—	—	—	—	Swanson and Van der Werff ten Bosch (1964a)
1	—	—	—	—	30	—	—	—	—	Gorski and Barraclough (1963)
	—	—	—	—	0	—	—	—	—	Swanson and Van der Werff ten Bosch (1964a)
0.5	—	—	—	—	0	—	—	—	—	Swanson and Van der Werff ten Bosch (1964a)

[1] All doses given as single injection.
[2] Numbers: % infertile (anovulatory) rats.

On the other hand, it is not possible to establish a cyclic gonadotropin secretion pattern in male rats by castration beyond the third day of life (Gorski, 1966; Gorski and Wagner, 1965; Harris, 1964; Kawashima, 1960; Yazaki, 1960). Obviously, androgens produced by the testes in newborn rats during the first three days of life are sufficient to imprint permanently the central nervous system during that short exposure time (Table II).

TABLE II

OCCURRENCE OF CORPORA LUTEA IN OVARIAN TRANSPLANTS PERFORMED IN MALE RATS AND ITS DEPENDENCE ON THE TIME OF POSTNATAL CASTRATION

Castration Days after Birth (d = 1)										Ovarian Transplants with Corpora Lutea at 10–20 Weeks of Age	Reference
1	2	3	4	5	6	7	10	11	20 and more		
×										+	Pfeiffer (1936)
										+	Kawashima (1960)
										+	Yazaki (1960)
										+	Takewaki (1962)
										+	Harris (1964)
	×									+	Kawashima (1960)
										+	Gorski (1966)
										±	Harris (1964)
		×								±	Yazaki (1960)
										±	Kawashima (1960)
										±	Gorski and Wagner (1965)
										±	Gorski (1966)
										—	Harris (1964)
			×							—	Harris (1964)
				×						—	Yazaki (1960)
										—	Harris (1964)
										—	Gorski and Wagner (1965)
						×				—	Yazaki (1960)
										—	Harris (1964)
							×			—	Yazaki (1960)
										—	Kawashima (1960)
										—	Gorski and Wagner (1965)
								×		—	Gorski (1966)
									×	—	Kawashima (1960)
										—	Yazaki (1960)
										—	Gorski and Wagner (1965)

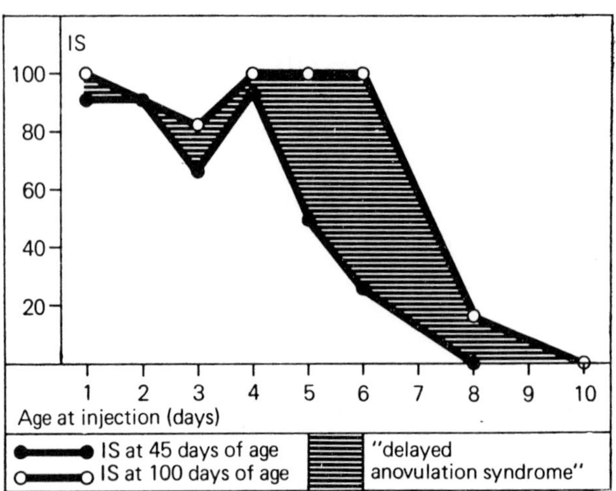

FIG. 1. Influence of animal age at injection of 10 μg of testosterone propionate and at observation of the incidence of sterility (IS) (From Gorski, 1968).

3. *Antiandrogens*

Our method of abolishing androgen effects at any desired time during fetal or neonatal life has been the use of the antiandrogen cyproterone acetate (6-chloro-17-acetoxy-1α, 2α-methylene-4,6-pregna-diene-3,20-dione) (Neumann and Elger, 1965a, b, 1966; Neumann and Kramer, 1967; Neumann *et al.*, 1966a). By treatment of pregnant animals, an extensive female differentiation of the somatic sex characteristics of the fetuses was achieved; e.g., these genetic male animals were born with a vagina. After castration and ovarian implantation in adulthood, we were able to detect quite easily the onset of cyclic changes in vaginal smears. One week after ovarian implantation the smears indicated an estrous condition that generally lasted for several days. Cyclic changes could be demonstrated in some of the animals but not in all. The cycles in feminized male rats are not so regular as in female controls and they are frequently incomplete. The precise state of diestrus or proestrus is seldom attained. With regard to the duration of the cycles, only a few were observed to last 4 to 6 days as in normal female control animals. Frequently the estrous phase was prolonged so that the cycle continued for 6 to 12 days or more. Thus there is a tendency toward continuous estrus (Fig. 2). Corpora lutea were also found in the implanted ovaries of animals with a pronounced vaginal cycle, thus proving ovulation (Figs. 3, 4).

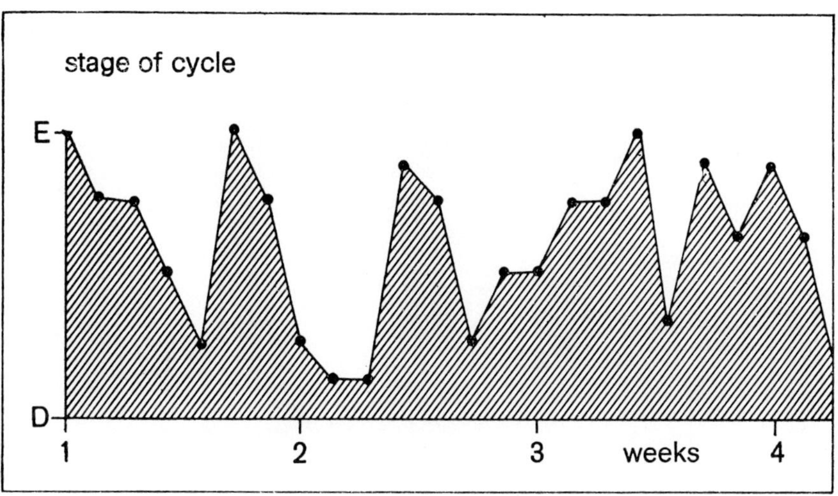

FIG. 2. Cyclic changes in the vaginal smear of a feminized male rat, castrated and ovary-implanted in adulthood. Mother treated with 10 mg/animal per day cyproterone acetate from day 13 to 22 of pregnancy; newborn treated for the first 3 weeks of life with 0.3 mg/animal per day cyproterone acetate s.c. D, diestrus; E, estrus.

In other experiments, only the newborn males were treated with the antiandrogen during the first three weeks of life (Neumann, *et al.*, 1967). The result was essentially the same as after treatment of both pregnant mothers and newborn. Following castration and ovary implantation in adulthood, these animals had also a great number of corpora lutea in the implants, as verified by gross inspection and histological examination.

In summary, this seems to be evidence (at least in rats) that the differentiation of hypothalamic centers which regulate gonadotropin secretion takes place only after birth.

4. *Estrogens*

It has been shown that the anovulatory syndrome can also be evoked by neonatal estrogen injection (Barraclough, 1967b).

C. Prevention of Androgen Effects

The early androgen syndrome can be postponed by progesterone given simultaneously with testosterone; this means that progesterone exerts a protective effect (see Fig. 5).

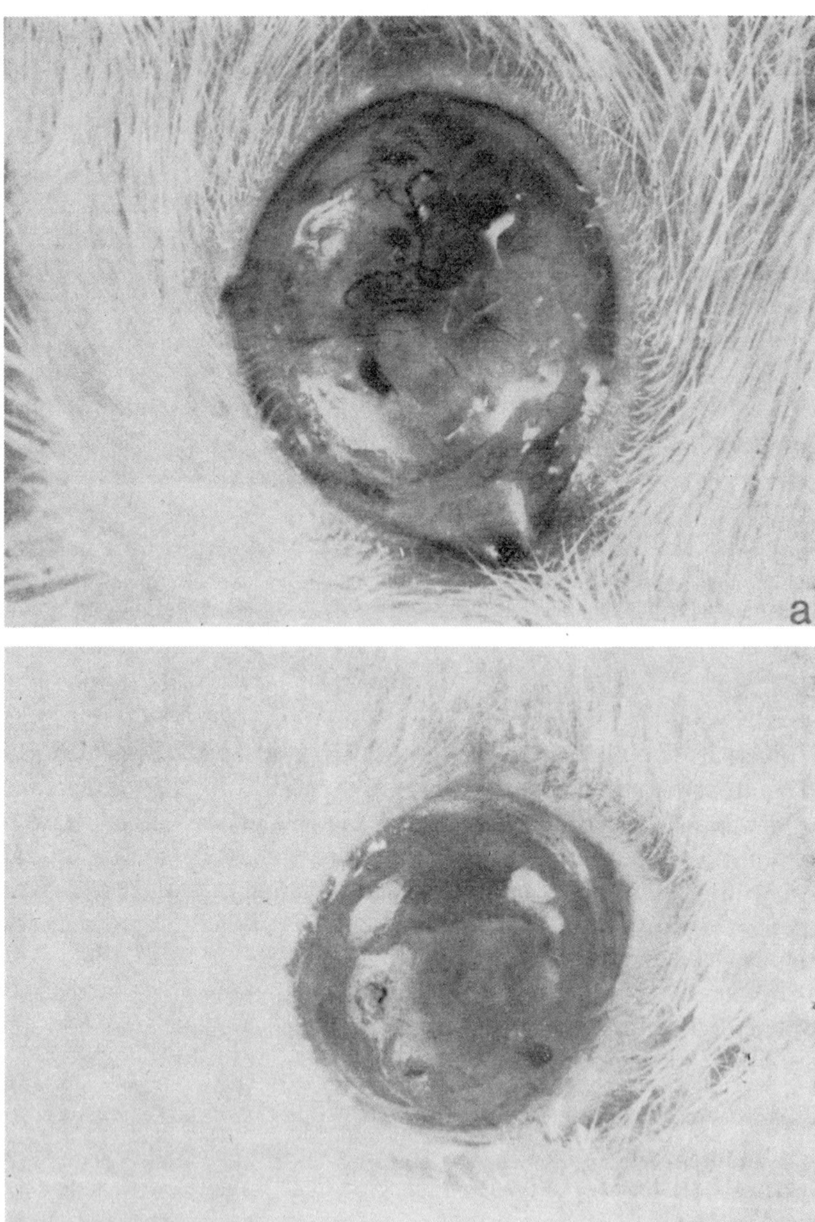

FIG. 3. Ovarian grafts in the anterior eye chamber of rats 8 weeks after castration and implantation in adulthood. (a) Normal male; (b) feminized male. Mother treated with 30 mg/animal per day cyproterone acetate from day 13 to 22 of pregnancy; newborn treated for the first 10 days of life with 2 mg/animal per day cyproterone acetate s.c.

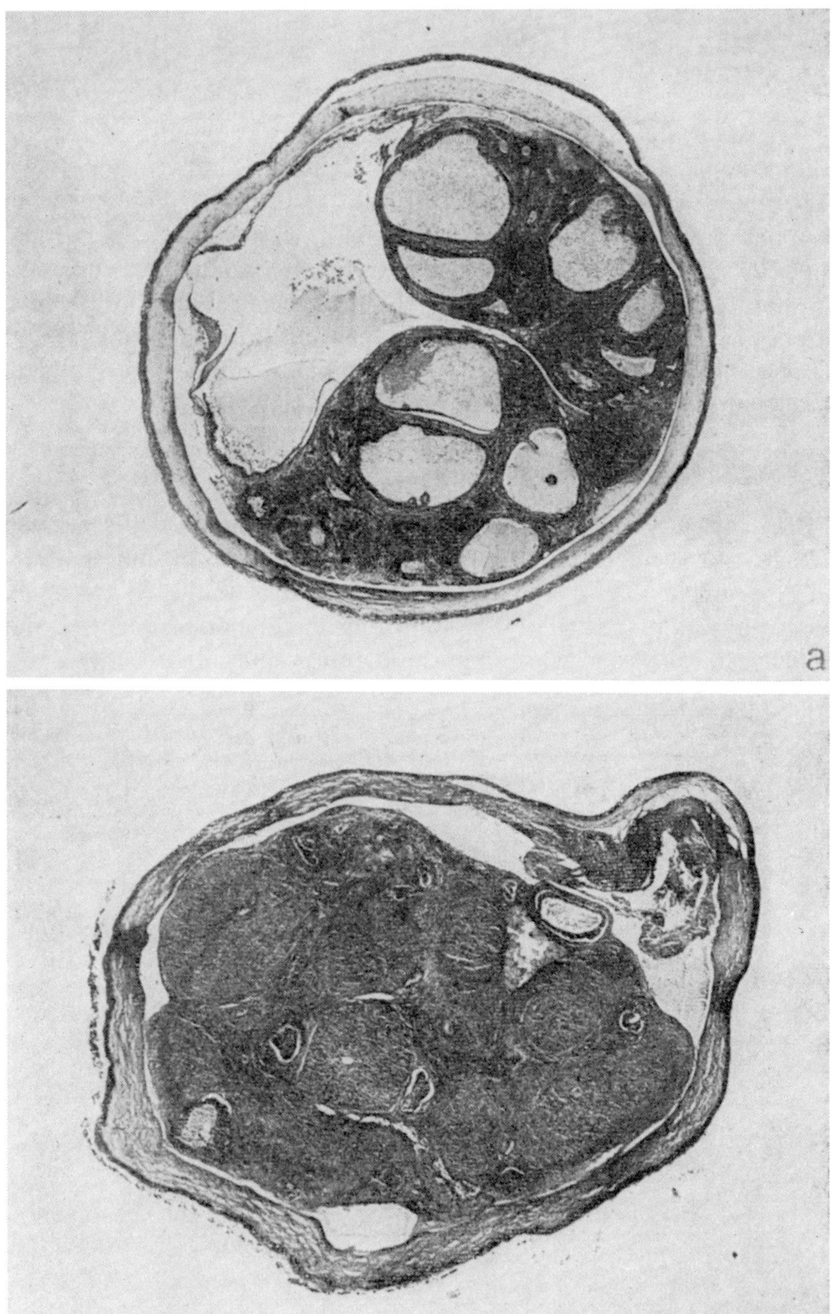

FIG. 4. Histological sections of the ovarian grafts shown in Fig. 3. (a) Normal male; (b) feminized male.

We were able to abolish the effect of androgens also by concomitant application of an antiandrogen (Neumann and Kramer, 1967). Treatment of female rats with 10 μg of testosterone propionate on the first day of life resulted in polycystic ovaries without any corpora lutea in adulthood. In contrast, normal ovaries containing numerous corpora lutea were found when, in addition to testosterone propionate, 100 μg of the antiandrogen cyproterone were injected neonatally.

Our findings have been confirmed, among others, by Wollman and Hamilton (1967) and Arai and Gorski (1968).

D. Mechanism of Action

The primary defect of gonadotropin secretion in androgenized female rats seems to be an impairment of the LH secretion pattern. It is assumed that those centers which control tonic LH secretion and which are located in the ventromedial hypothalamus are not influenced. However, those higher preoptic centers that catalyze the

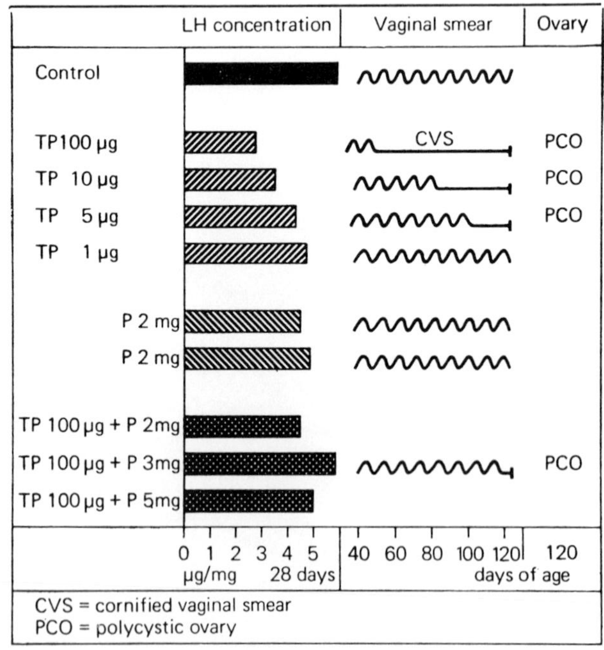

Fig. 5. Effect of testosterone propionate (TP) with or without progesterone (P) on pituitary LH, vaginal cycles, and ovarian histology of the rat (From Lloyd and Weisz, 1967).

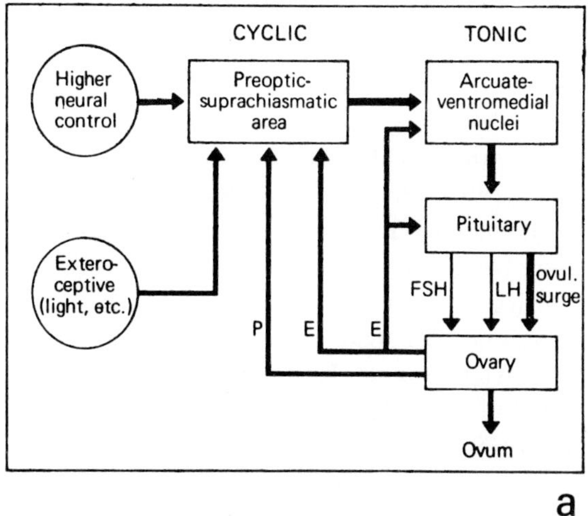

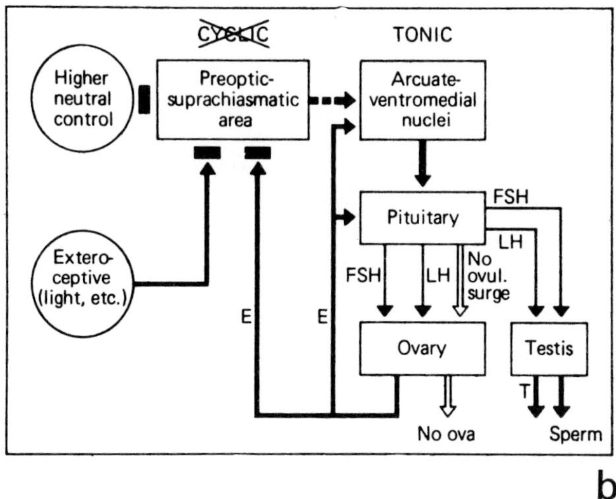

FIG. 6. (a) Diagrammatic representation of events at the hypothalamic, pituitary, and gonadal levels, which result in steroid secretion and ovulation in normal females. When proper estrogen (E) to progesterone (P) ratios are reached, the preoptic area becomes responsive to exteroceptive and enteroceptive influences, is activated, and in turn activates the arcuate-ventromedial nuclear area. Sufficient LHRF is released to cause the ovulatory discharge of gonadotropin and ovulation. (b) Diagrammatic representation of events at the hypothalamic, pituitary, and gonadal levels in normal males or in females after destruction of the preoptic area or neonatal androgen treatment. In the absence of the cyclic control of the ovulatory discharge of gonadotropins, only tonic hypothalamic influences on adenohypophyseal function can be manifested. While sufficient FSH and LH are released to cause follicular development and estrogen (E) secretion, ovulation does not occur and the persistent estrous syndrome ensues. In the male, this control is adequate for maintenance of spermatogenesis and androgen (T) production (From Barraclough, 1966a).

cyclic LH surge necessary for ovulation (Flerkó and Szentágothai, 1957; Flerkó *et al.*, 1967) are rendered permanently refractory to stimulation by early androgen treatment (Gorski and Wagner, 1965; Petrusz and Flerkó, 1965; Petrusz and Nagy, 1967). This results in normal, basal LH secretion, but the ovulatory surge of gonadotropin is absent and therefore permanent sterility ensues.

By electrolytic lesions of the preoptic area the anovulatory, persistent estrous syndrome can also be induced in adult female rats (Alloiteau, 1954; D'Angelo and Kravatz, 1959; Barraclough and Gorski, 1961; Barraclough *et al.*, 1964; Van Dyke *et al.*, 1957; Greer, 1953; Halász *et al.*, 1965; Hillarp, 1949; Van Rees *et al.*, 1962; Takewaki, 1962; Taleisnik and McCann, 1961). Conversely, by electrical stimulation of this region, ovulation can be triggered (Critchlow, 1958; Everett and Radford, 1961). According to Barraclough and Gorski (1961), the arcuate-ventromedial nuclear region of the hypothalamus is responsible for tonic LH secretion, whereas nuclei of the preoptic area modify it and thus determine cyclicity or acyclicity of secretion (Fig. 6).

Halász and Pupp (1965) have, by isolating distinct hypothalamic areas surgically, confirmed this opinion in an impressive series of experiments.

In rats, the preoptic area is still bipotential at birth. Its function is rendered acyclic by the influence of neonatal androgen or estrogen.

Unfortunately there is not space enough to keynote sufficiently all investigations dealing with this mechanism. Therefore some of the important data available are summarized here. In androgenized animals the hypophyseal LH concentration is lower than in normal female rats (Fig. 7; see also Fig. 6) (Barraclough, 1966a, 1967a,b; Gorski and Barraclough, 1962a; Lloyd and Weisz, 1967; Wolthuis *et al.*, 1962).

There are minor or no changes in Follicle-Stimulating Hormone (FSH) concentration (Barraclough, 1967b; Gorski and Barraclough, 1962b; Swanson and Van der Werff ten Bosch, 1964a). The rate of LH secretion and probably also that of FSH secretion is not enhanced in androgen-sterilized female rats but is, rather, lower than in normal females (Kurcz and Gerhardt, 1968; Lloyd and Weisz, 1967; Wolthuis *et al.*, 1962).

These findings are most likely secondary effects mediated by impairment of the feedback mechanism, owing to abnormal steroid

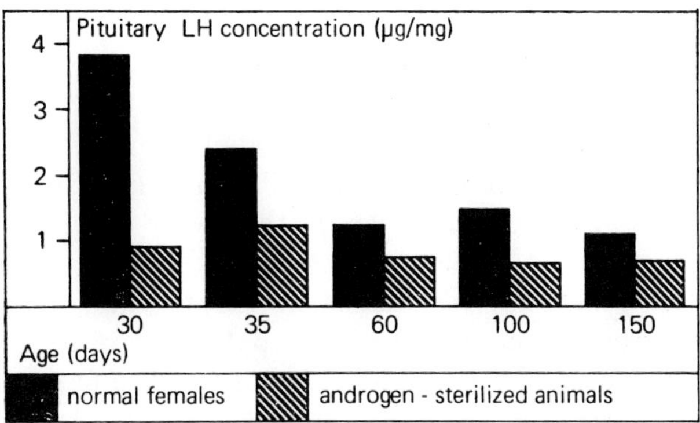

FIG. 7. Comparison of hypophyseal LH concentration at different ages in normal and androgen-sterilized female rats (1.25 mg at 5 days of age) (From Barraclough, 1966b).

secretion from the polycystic ovaries (Goldzieher and Axelrod, 1963; Lloyd and Weisz, 1967; Weisz and Lloyd, 1965). Secondary changes in the circuit may reach such intensity that the primary defect (namely, impaired hypothalamic differentiation) is completely masked.

E. PROLACTIN

In rats, pseudopregnancy can be readily induced by adequate progesterone and estradiol treatment. This is accompanied by a considerable increase of mammary gland tissue, as occurs in normal pregnancy. We applied this treatment to castrated male and female rats and also to castrated virilized female offspring of androgen-treated mothers as well as castrated feminized male rats whose mothers had been treated with an antiandrogen during pregnancy.

Similar to male controls, the growth of mammary gland tissue in virilized female animals is far below that of female controls in response to estradiol–progesterone treatment. Conversely, mammary gland weights in feminized males far exceeded those of male controls. Our first interpretation was that, under androgen influence, less glandular tissue is formed (Neumann *et al.*, 1969).

However, later on we found that neither at birth nor at 30 days of age do any quantitative differences of mammary gland development exist between female, male, virilized female, or feminized male rats. Therefore our first hypothesis could not be correct.

Thereafter we assumed that the different mammary gland growth of our various groups could be due to differences in prolactin secretion. A Hungarian research team (Kurcz *et al.*, 1967, 1968a,b) has found that the hypophyseal prolactin content of androgenized female animals and males is far below that of females (see Table III).

It should be mentioned, however, that with advancing age the hypophyseal prolactin content and prolactin secretion of androgen-sterilized female rats exceeds that of normal females, in response to prolonged estrogen secretion from the polycystic ovaries (Kurcz *et al.*, 1968a; Lloyd and Weisz, 1967). Since a positive feedback mechanism exists between estrogen and prolactin, the pituitaries enlarge; this may be evidence of increased prolactin secretion.

Other results should be mentioned. Harris and Levine (1965) treated androgenized female rats with 200 μg of estradiol benzoate daily for 6 to 8 weeks. The pituitary enlargement was considerably smaller in androgenized animals than in female controls. Conversely, we found pituitaries of feminized male rats to be significantly larger than pituitaries of normal males following a 3-week estradiol treatment. Therefore it is fairly certain that those structures which regulate prolactin secretion are also determined by androgens in male fashion to react less sensitively with reduction of the prolactin inhibiting factor in response to estrogens in adulthood.

It has not been ascertained yet whether this process takes place before or after birth. Although only rats have been used so far, we believe that the process also occurs in other species, including man.

TABLE III

EFFECT OF ANDROGENIZATION ON THE PROLACTIN CONTENT OF THE ADENOHYPOPHYSIS

Group	Sex	Number of Animals	Number of Determinations	Number of Pigeons	Prolactin Content: IU/100 mg Adenohypophysis	Prolactin Content: IU/Adenohypophysis
Control	F	29	7	28	7.8 ± 1.5	0.67 ± 0.30
Androgenized [1]	F	53	14	56	4.2 ± 0.4	0.34 ± 0.03
Control	M	10	3	10	3.9 ± 0.3	0.28 ± 0.01

(From Kurz *et al.*, 1967).

[1] 500 μg of testosterone phenylpropionate on the 1st or 2nd day of life.

Some indication of this is the generally poor development of breasts in women suffering from congenital adrenogenital syndrome (AGS) or the excellent breast development in most cases of testicular feminization, i.e., in chromosomal male individuals. In the first case, AGS, androgens become effective at an early time, but they do not have any effect in the second case, testicular feminization.

F. Growth Hormone

Androgenized rats grow more quickly than female controls (Barraclough, 1961; Harris 1964; Swanson and Van der Werff ten Bosch, 1963). Therefore it can be tentatively concluded that more growth hormone is synthesized in androgenized female animals than in normal females. The Hungarian team of Kurcz (1968a,b) found that the hypophyseal growth hormone content of androgenized female rats was higher than that of female controls. With reservation, it can be assumed that androgens also have a determining influence with regard to growth hormone. However, in this case more detailed investigations are needed.

G. Onset of Puberty and Feedback Mechanism

In female rats that have been treated neonatally with either testosterone propionate or estradiol the onset of puberty is advanced (as measured by the time of vaginal opening; Barraclough, 1967b). The feedback mechanism is also operative in androgenized animals, i.e., after ovariectomy the hypophyseal castration cells appear in androgenized animals in a manner similar to castrated normal controls (Harris and Levine, 1965).

II. Differentiation of Behavioral Patterns

Almost all vertebrate species exhibit distinct behavioral patterns that essentially serve to maintain the individual or the species. Most patterns of behavior are more or less closely correlated with the reproductive cycle; they differ in male and female individuals. Most evident are sex differences in sexual behavior. However, there are also sex differences in other behavioral patterns that serve to protect or to nourish the offspring. Maternal instinct, nest building instinct, and aggressive and retrieve behavior are examples of this. Finally, sex differences also exist in less specific behavioral patterns; e.g., spontaneous activity or defecation frequency. Similar to the mode of gonadotropin secretion, a good number of the mentioned behavioral patterns are determined at early stages of development, in the course of which androgens play an important role.

A. Sex Behavior

Of all behavioral patterns, sex behavior is the best to measure objectively. As early as 1938, Dantchakoff had observed that the female offspring of guinea pigs which had been treated with testosterone during pregnancy exhibited male sex behavior when treated with testosterone in adulthood. In spite of this observation, intensive study in this area was not reported until the second decade after Dantchakoff's disclosure. In the main, rats, mice, and guinea pigs have been employed and other species to a lesser degree.

1. *Rats*

Here again space does not permit all studies in this field to be cited. Moreover, not all results or interpretations thereof are in agreement. Everyone who has carried out this kind of behavioral testing knows how complex the patterns of sex behavior are and how difficult concrete statements can be. Female rats who were treated with an androgen within the first 3 days of life are infertile during adulthood, as already mentioned. Moreover, the capacity to display feminine behavior is greatly reduced (Fig. 8a) (Barraclough and Gorski, 1962; Feder, 1967; Feder *et al.*, 1966; Goy *et al.*, 1962; Harris and Levine, 1965; Meyerson, 1968). One could assume disorders of steroid biosynthesis in the polycystic ovaries to be the cause of the impaired female sex behavior; i.e., simply inadequate hormonal conditions. Obviously, however, this is not the case because, after castration in adulthood and appropriate replacement therapy with estrogens and progesterone, androgenized females show little or no feminine sex behavior (Harris and Levine, 1965; Levine and Mullins, 1967). In normal female animals this replacement therapy is always successful (Barraclough and Gorski, 1962; Beach, 1945).

Androgenized female animals also show more intromission patterns than do control females after adequate replacement therapy with testosterone propionate (Harris and Levine, 1965) (see Fig. 8b).

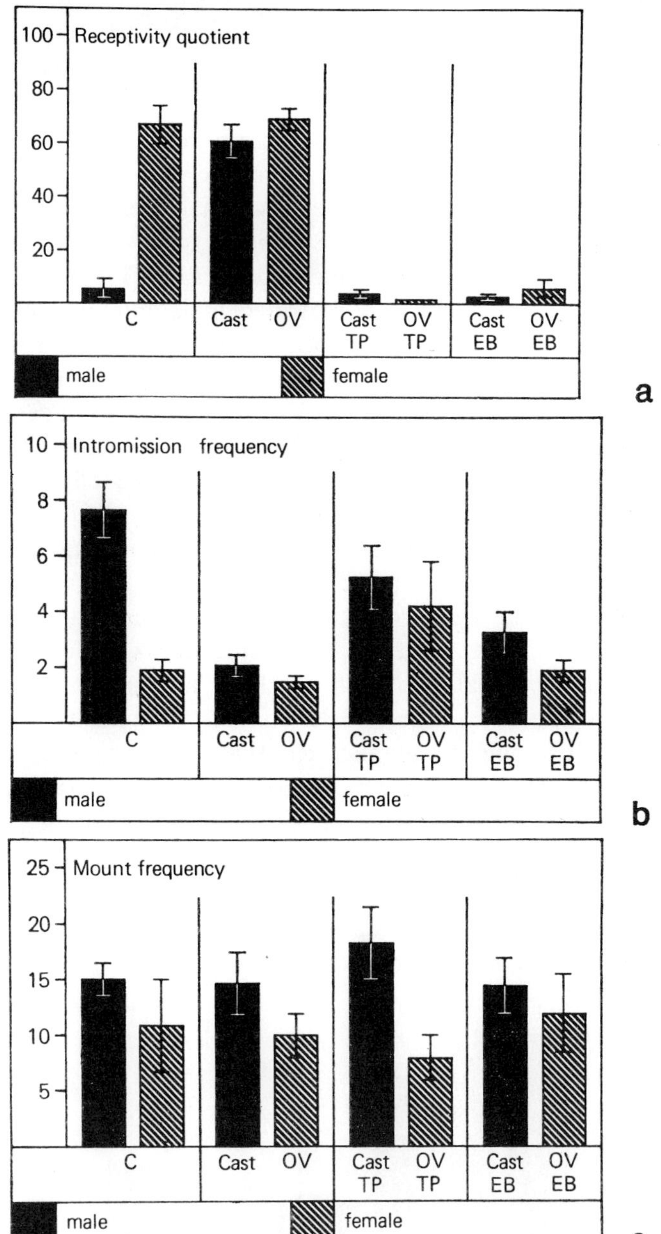

FIG. 8. Mount frequency, intromission frequency, and receptivity quotient of male and female rats treated with testosterone in adulthood following hormonal manipulation at birth. Abbreviations: C, controls; Cast, castrated at birth; Cast TP, castration + testosterone treatment at birth; Cast EB, castration + estrogen treatment at birth; OV, ovariectomy at birth; OV TP, ovariectomy + testosterone treatment at birth; OV EB, ovariectomy + estrogen treatment at birth (From Whalen and Edwards, 1967).

However, mounting frequency does not seem to be influenced (Whalen and Edwards, 1967) (see Fig. 8c). Conversely, neonatally castrated male rats later display male as well as female behavioral patterns. They are bisexual (compare Figs. 8a and 8c). This is also indicated by a graph from a paper by Levine and Mullins (1967) (see Fig. 9).

It can be seen that neonatally-castrated male animals will in time attain a mean lordosis/mount quotient similar to female controls. Neonatal administration of testosterone propionate to female animals leads to dose-dependent lowering of this quotient toward zero. These animals still accept normal males, but they no longer display the lordosis reaction. Yet the testosterone propionate effect is clearly more marked in males than in females (Levine and Mullins, 1967).

We employed the antiandrogen model in our tests, again working with cyproterone acetate. In one experiment the pregnant mothers received the antiandrogen from the 13th day of pregnancy to term; the newborn was treated for the first three weeks of life. We compared the adult sexual behavior of these animals toward estrous females with that of normal male animals (Neumann and Elger, 1966, 1969). It should be emphasized that these animals possessed testes in spite of complete feminization of the external genitals. Therefore it can be assumed that androgens were available, whereas

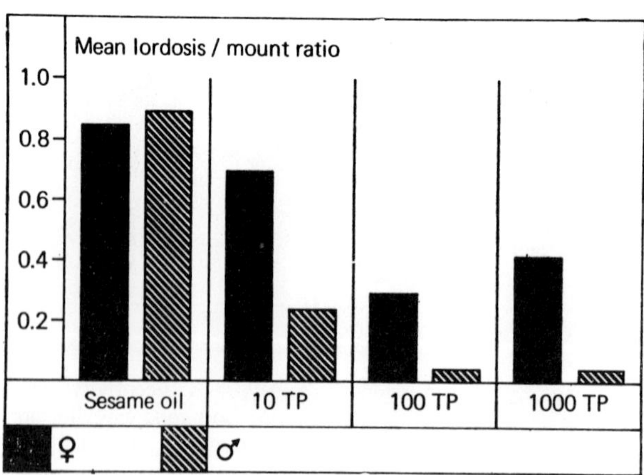

FIG. 9. Feminine sexual behavior in female and neonatally castrated male rats injected with androgen after birth and given estrogen–progesterone in adulthood. Neonatal testosterone propionate (TP) dose given in micrograms (From Levine and Mullins, 1967).

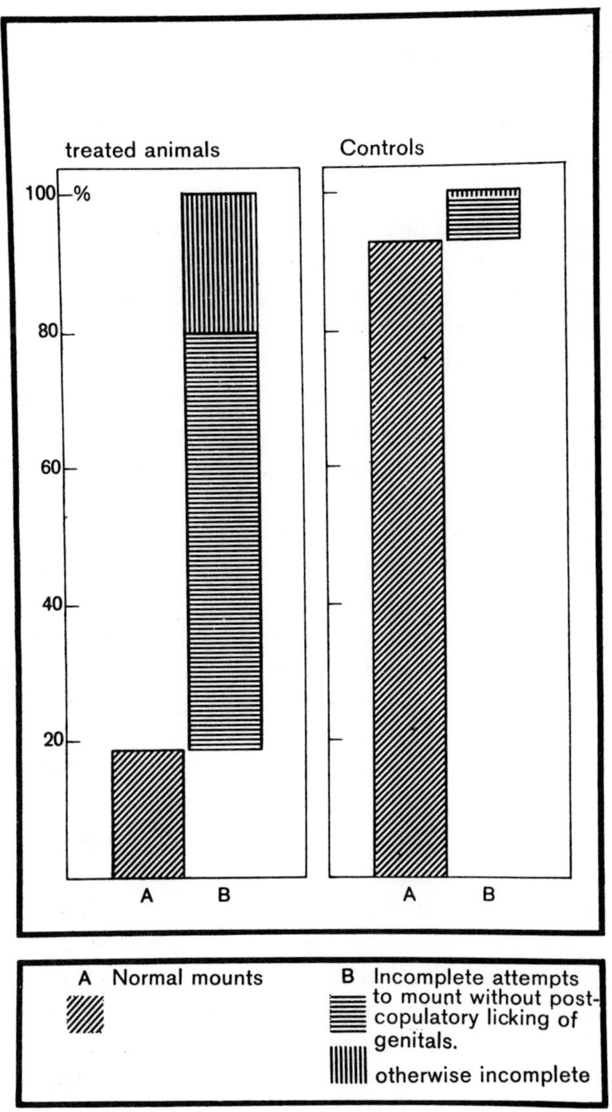

FIG. 10. Sexual behavior of male rats treated neonatally with cyproterone acetate toward females in estrus.

estrogens and progesterone were not. Nevertheless, these animals behaved bisexually. Rats in estrus were pursued, the genitals were sniffed and licked, and mounting took place, although with less purposeful behavior than that shown by normal male animals.

When the same animals, however, were exposed to normal male rats, they behaved like females and were considered as such by the normal male rats. The mounting attempts, however, were not, or only rarely, answered with lordosis. Figure 10 illustrates the male sexual behavior of male rats given the antiandrogen (2 mg per animal daily) during the first 14 days of life (Neumann *et al.*, 1967).

As compared to normal male animals placed with estrous females in an 8-min test, the number of mounting attempts of the treated animals was significantly reduced. Only a relatively limited percentage of the mounting attempts was regarded as normal. More than 80% of the mounting attempts were incomplete. Incomplete mounting was characterized by the absence of postcopulatory licking of the genitals. Other deviations from the normal copulatory behavior were mounting attempts from the side, over the head of the female and the indecisive manner of placing the paws on the back of estrous females.

However, when we castrated such animals in adulthood and implanted ovaries they displayed marked feminine behavior, even though the pregnant mothers and the young or the newborn only received the antiandrogen.

In the first case the external genitals were also differentiated in the female way, as already mentioned. Thereby we were able to follow up the cycle by means of vaginal smears (Neumann and Elger, 1965b, 1966; Neumann and Kramer, 1967). In the animals with marked cyclic changes in the vaginal smear, a correlation between the cyclic state determined by the vaginal smear and the readiness to mate was noted. These animals evoked the interest of normal males much more strongly than those with only slight or no cyclic changes. During estrus, they were mounted much more frequently than during other cyclic phases, occasionally five to ten times within a few minutes. Frequently the feminized animals responded with lordosis to mounting attempts of the males. When these animals were in diestrus or late metestrus, the defense reactions were much more marked, and the mounting attempts of the males were not often answered with lordosis.

It can be generally stated that neonatally castrated male rats will later have the capacity for male as well as for female sex behavior (Beach and Holz, 1946; Grady *et al.*, 1965; Whalen and Edwards, 1966; Young, 1961). It seems, therefore, that endogenous testicular hormones are not necessary after birth to organize the neural systems that mediate sexual performance. The primary role of testicular hormones in the male would seem to be the suppression of those neural systems which mediate feminine sexual behavior (Beach, 1942; Feder, 1967; Feder *et al.*, 1966; Whalen and Edwards, 1967). If the male rat is castrated at birth and administered the appropriate hormones in adulthood, he will display both lordosis, when given estrogen and progesterone, and mounting, when given testosterone (Beach and Holz, 1946; Levine, 1966; Whalen and Edwards, 1966).

In spite of the fact that male animals have been generally tested for feminine sex behavior only after appropriate replacement therapy with estrogens and progesterone, this seems not to be necessary, as recent investigations have shown (Dörner, 1968, 1969; Dörner and Hinz, 1967; Palka and Sawyer, 1966). Dörner (1968) reached the following conclusions: " In adulthood (activation period), androgens stimulated predominantly female sexual behavior in a female organized hypothalamus, but predominantly male behavior in a male organized hypothalamus. " Figure 11 illustrates one of Dörner's experiments (Dörner, 1968). It shows the sexual behavior of female rats neonatally treated with testosterone propionate alone or with an antiandrogen simultaneously after castration and androgen administration in adult life.

In those female animals treated neonatally with testosterone propionate only, male sex behavior prevails, i.e., the number of mounts was greater than the number of lordosis reactions. Quite the opposite situation was created when an antiandrogen was given concomitantly. Here, the number of lordosis reactions was greater as compared with the number of mounts.

According to investigations of Dörner and co-workers, the male mating center is located in the anterior hypothalamus, whereas the female mating center is located in the central hypothalamus (Dörner and Staudt, 1968; Dörner *et al.*, 1968a, b; Dörner, 1969). Testosterone implants in the anterior hypothalamus of spayed, adult female rats led predominantly to male behavior. Testosterone implants in the central hypothalamus, however, caused predominantly female beha-

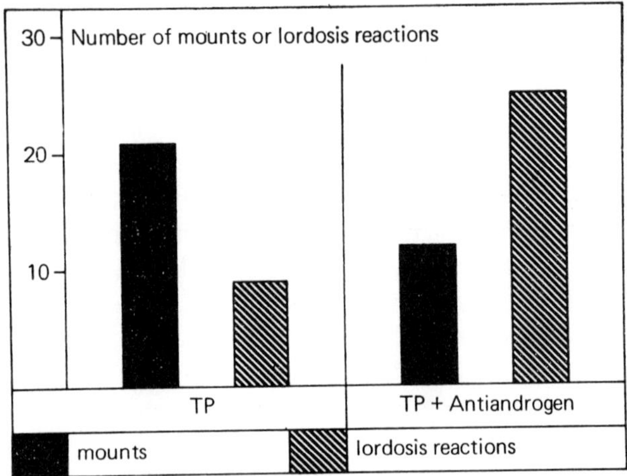

FIG. 11. Sexual behavior of female rats neonatally treated with testosterone propionate (TP) alone (0.02 mg TP on day 3 of life) or concomitantly with an antiandrogen (0.02 mg TP on day 3 of life + 0.1 mg cyproterone acetate on 3rd, 4th, and 5th day) after castration and androgen administration in adult life (1 mg TP/day for 10 days and 0.25 mg TP/day for the following 18 days). In each test the rat was exposed for 5 min to an estrous female or to a male, respectively (From Dörner, 1968).

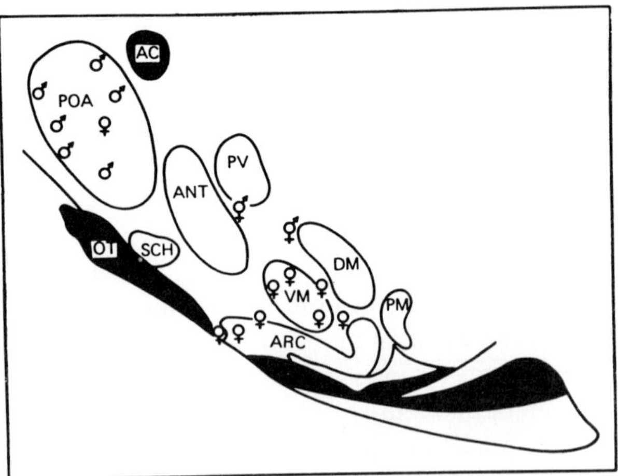

FIG. 12. Parasagittal diagram of rat hypothalamus showing sites at which estradiol implants were effective in elicitation of lordosis reflexes (F) or mounting attempts (M) in adult spayed females. Abbreviations: AC, anterior commissurae; ANT, anterior hypothalamic nucleus; ARC, arcuate nucleus; DM, dorsomedial nucleus; OT, optic chiasma; PM, premammillary nucleus; POA, preoptic area; PV, paraventricular nucleus; SCH, suprachiasmatic nucleus; VM, ventromedial nucleus. (Modified from Dörner *et al.*, 1968b).

vior in these animals. As shown in Fig. 12, the same holds true for estrogen implants (Dörner *et al.*, 1968a; Lisk, 1962).

The reverse experiment has also been done by Dörner (1967). He castrated newborn male rats or castrated and treated them with testosterone propionate. In adulthood, all groups, were equally treated with 1.0 mg testosterone propionate per day. In neonatally castrated male animals not treated with testosterone propionate, female sex behavior predominated over male behavior. Exactly the opposite situation was produced when neonatally castrated animals received testosterone propionate. In spite of equal replacement therapy, male sex behavior prevailed in adulthood in such animals (Fig. 13).

It should also be mentioned that estrogen administration during the first days of life has the same effect as testosterone treatment with regard to the future behavioral pattern (Harris and Levine, 1965; Levine and Mullins, 1967). Exhaustive investigations in this field were done by Whalen and co-workers (Whalen and Edwards,

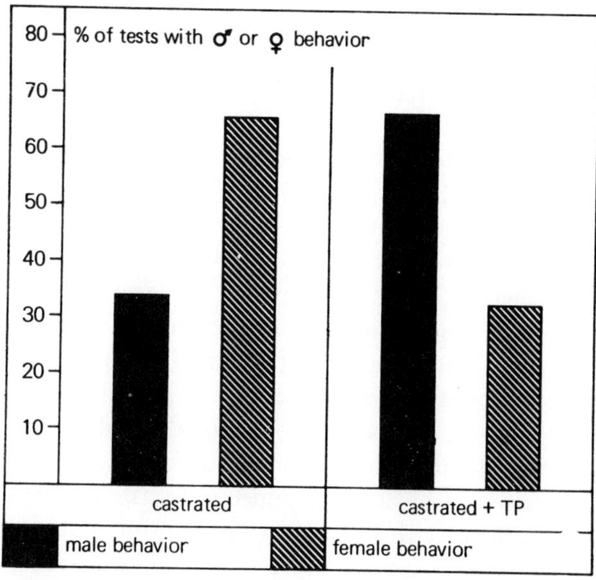

FIG. 13. Sexual behavior of castrated (day 1 of life) or castrated and androgen-treated (0.02 mg of testosterone propionate, TP, on the third day of life) male rats. Androgen treatment in adult life: 1 mg TP/day, for 10 days and then 0.25 mg TP/day. In each test the rat was exposed for 5 min to an estrous female or to a male, respectively (From Dörner, 1967).

1967; Whalen and Nadler, 1963, 1965). Normal female animals display predominantly lordosis reactions and little defensive behavior, whereas estrogenized female animals behave exactly the opposite way (Whalen and Nadler, 1963).

Even the effect of neonatal castration can be partly abolished in male animals by estradiol treatment (Levine and Mullins, 1967; Mullins and Levine, 1968a, b). With regard to sex behavior, the time at which androgens and estrogens become effective neonatally is also of crucial importance, similarly to the time element in the previously described differentiation of hypothalamic centers that regulate gonadotropin secretion (Goy *et al.*, 1962; Grady *et al.*, 1965). For example, the complete sequence of sex behavior cannot be induced by androgen substitution in male rats castrated on the fourth day of life.

2. *Guinea Pigs and Hamsters*

The role played by androgens in the differentiation of the hypothalamus in hamsters is as important as in rats (Crossley and Swanson, 1968).

In guinea pigs, the sensitive period of hypothalamic differentiation occurs during pregnancy. The treatment of pregnant guinea pigs with testosterone propionate leads to pseudohermaphroditism in the female offspring.

Extensive investigations on guinea pigs have been carried out by the team of Young, Phoenix and Goy (cited below). As it has been previously described for rat experiments, such guinea pig hermaphrodites were castrated in adulthood and then tested for female or male sex behavior after appropriate replacement therapy with estradiol benzoate and progesterone or testosterone propionate, respectively. The complete pattern of mating behavior in the castrated female can be restored by replacement therapy with female sex hormones at any time during adulthood (Wilson and Young, 1941).

Figure 14 demonstrates the intensification of male mating behavior and the suppression of lordosis response in female guinea pigs whose mothers have been treated with testosterone propionate in pregnancy (Goy *et al.*, 1961). The animals were treated with estradiol benzoate and progesterone or testosterone propionate in adulthood (Phoenix *et al.*, 1959; Young, 1960, 1963).

The greatest number of mounts was achieved by treatment of the mother from the 30th to the 65th day of pregnancy. In this group

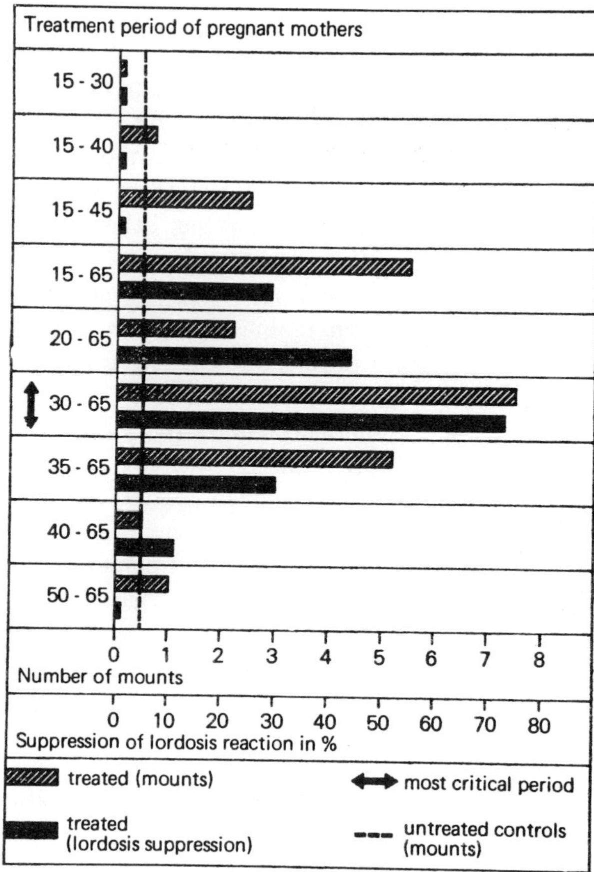

FIG. 14. The maximally effective period for behavioral modification of female guinea pigs treated prenatally with testosterone propionate. Numbers on the left are days of pregnancy. Experimental animals were spayed postpubertally and tested four times for mounting with a receptive female or male (From Goy *et al.*, 1961).

the lordosis response was also suppressed to the highest degree. Other experiments revealed that the androgen-sensitive period extends from the 30th to the 35th day of fetal life (Goy *et al.*, 1964). To masculinize the external genitals, higher androgen doses are necessary than for the suppression of the lordosis reaction (Phoenix *et al.*, 1959). The genital tissues appear to be sensitive to androgens somewhat earlier than the neural tissues that mediate mating behavior (Goy *et al.*, 1964).

Special attention should be given to an interesting point of view. Both male and female guinea pigs are known to display the lordosis response on the day of birth following appropriate tactile stimulation (Boling *et al.*, 1939). Boling and co-workers have also explored whether female guinea pig sensitivity to estrogens is more increased after birth than in male animals or female pseudohermaphrodites. This could be an explanation for the different sex behavior. Indeed, this was the case (Goy *et al.*, 1967). The conclusion was reached that adult pseudohermaphroditic females are less sensitive to estrogen than are normal females. When mounting behavior was studied in pseudohermaphroditic guinea pigs, the results suggested an increased sensitivity to androgens in these animals (Young, 1960, 1963).

Thus the adult female pseudohermaphroditic guinea pig has been characterized as showing both reduced sensitivity to estrogens and an increased sensitivity to androgens as compared with the sensitivity of the normal female (Goy *et al.*, 1964; Phoenix *et al.*, 1959). Androgens effective during the hypothalamic differentiation period in guinea pigs are not so much inhibitory with regard to the establishment of the mechanisms for lordosis reaction itself; rather they inflence those mechanisms that will be triggered later by estrogen and progesterone to mediate lordosis reaction.

3. *Monkeys*

Experiments done by Young and co-workers (1964) on rhesus monkeys indicate that patterns of behavior may be impressed on primates by androgens. Young and co-workers treated monkeys from the 46th to the 90th day of pregnancy with 10 to 25 mg testosterone per day. Thereafter the female young were strongly virilized. In these animals, some patterns of behavior were tested that are not immediately connected with sexual behavior. The pseudobehavior of these animals did not differ importantly from that described for normal males. These female pseudohermaphrodites threatened, initiated play, and engaged in a rough-and-tumble play pattern more frequently than did the controls. The authors conclude: " Analysis of the sexual behavior displayed by these pseudohermaphroditic females showed that it is not only in their patterns of withdrawing, playing and threatening that they display a bias toward masculinity. In special tests with pairs of females, one pseudohermaphroditic and one normal, the pseudohermaphrodites consistently displayed more frequent attempts to mount. Their attempts to mount, while infantile and poorly oriented, are beginning to be integrated with pelvic thrusting and even, on a few occasions, phallic erections ".

4. *Dogs*

Recently we started investigations in another species, namely, dogs. Pregnant dogs were treated from the 23rd day or pregnancy to term with 10 mg per kilogram of the antiandrogen cyproterone acetate injected intramuscularly. The offspring were reared and primarily the somatic sex differentiation was studied (Neumann *et al.*, 1969; Steinbeck *et al.*, 1970). We found that all androgen-dependent differentiation processes were abolished. Among other things, these animals had complete feminine external genitals, including a vagina.

Since the differentiation of the somatic sex was impaired, we assume that the effect of androgens on brain differentiation processes was also abolished. In preliminary investigations, such animals behaved bisexually; they accept mounting by other males and they themselves mount other dogs. The two pictures (Fig. 15a,b) show the same animal, at one time mounting and at another time being mounted. It should be emphasized that no replacement therapy was given and that the animals retained their testes. These investigations are continuing. However, the initial results support the assumption that androgens, rather than catalyzing the determination of male sex behavior, suppress the differentiation of centers that control female sex behavior.

B. Aggressive Behavior; Fighting Impulse

Isolated caged male mice readily fight when paired (Beeman, 1947). However, female mice do not fight even when administered testosterone (Tollman and King, 1956). Following castration of male animals, the fighting impulse is lost and after adequate androgen substitution it reappears. Similarly, fighting can be evoked in neonatally androgenized female mice after castration and testosterone propionate substitution in adulthood (Fig. 16). Figure 16 shows that testosterone propionate treatment is much more effective on the first day of life than on the tenth day in causing fighting in adulthood (Edwards, 1968).

Somewhat contrary to these findings are the results of Ulrich and

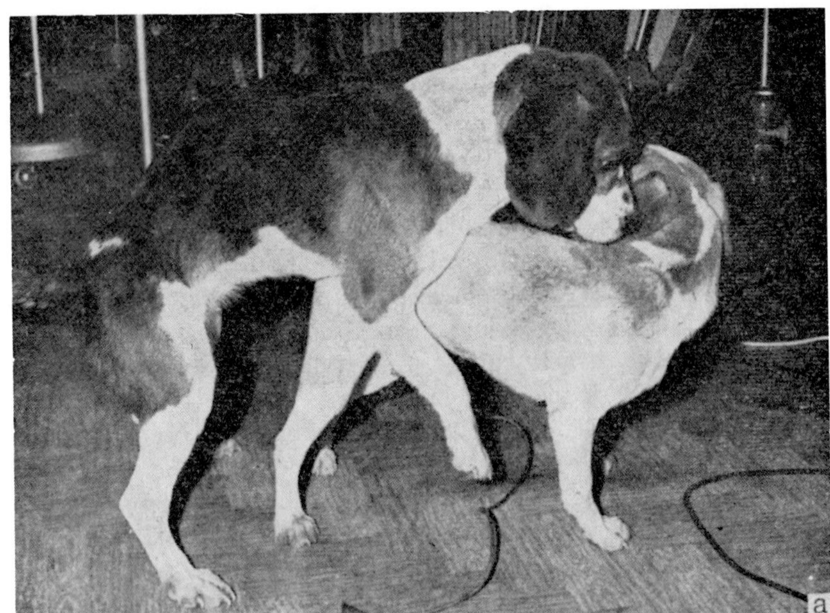

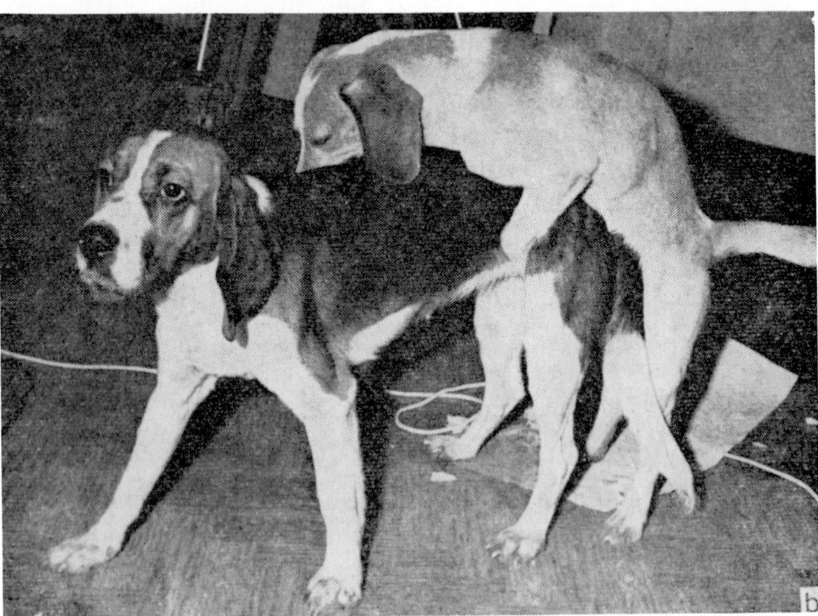

FIG. 15. Bisexual behavior of feminized male dogs. The light-colored dog's mother was treated with 10 mg/kg/day cyproterone acetate from the 22nd day of pregnancy to term. The dark-colored dog's mother was treated with 10 mg/kg/day cyproterone acetate from days 22 to 38 of pregnancy. Both dogs have a vagina.

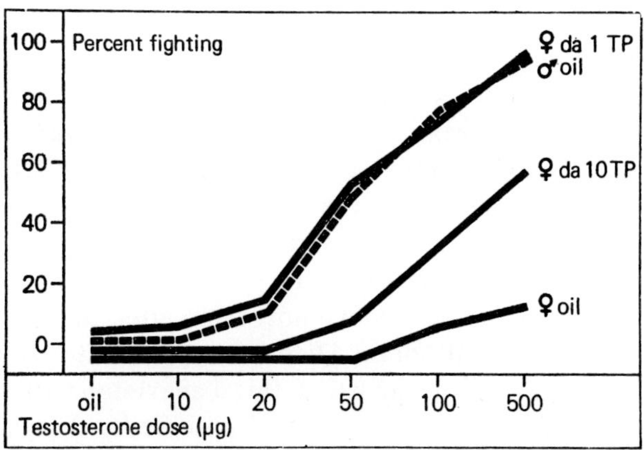

FIG. 16. Percent of adult castrated mice in each group fighting at different dose levels of testosterone propionate (TP). Abscissa: adult dose levels of TP. Neonatal dose: 0.5 mg (From Edwards, 1968).

Azrin (1962), who measured aggressive behavior after electrical stimulation. Neonatally castrated male rats showed levels of aggressive behavior that resembled those seen in normal intact females. Paradoxically, neonatal testosterone propionate injections to females suppressed aggressive behavior below the normal female level. With regard to aggressive behavior these animals were so to say "hyperfeminized." This is in agreement with the findings of Mullins and Levine (1968b). They reported that administration of testosterone propionate to female rats 120 hr after birth resulted in facilitation of the lordosis response in low-dose groups (5 to 50 μg testosterone propionate) and suppression of lordosis in high-dose groups (100 to 1.000 μg testosterone propionate). Injection of estradiol led to inhibition of lordosis reaction, regardless of dosage.

C. MATERNAL INSTINCT; RETRIEVE BEHAVIOR

There are almost no data as to whether these behavioral patterns are influenced by hormones at early stages of differentiation. Our own results seem to support this view. We attempted to induce lactation in feminized male rats by appropriate hormone treatment (Neumann *et al.*, 1966b). Pregnant rats received cyproterone acetate from the 13th to the 22nd day of pregnancy; the newborn were treated with the antiandrogen for the first 3 weeks of life.

More or less as a " fun " project, we observed whether these animals would be able to nourish newborn rats after lactation was successfully initiated. It was not possible to keep the newborn rats alive for more than a week; obviously our treatment was not optimal to maintain lactation. However, one observation was striking: the feminized male rats cared for the young and retrieved them when they had escaped from the nest (von Berswordt-Wallrabe and Neumann, unpublished observations). To our knowledge, normal male rats never display this behavior, not even after appropriate hormone treatment in adulthood. Therefore it is quite obvious that androgens also play a role in fixing this behavioral pattern. Before definitive statements are possible, more specifically designed experiments have to be done. Money (personal communication) relates that married women with the testicular feminization syndrome (i.e., chromosomal male individuals) have a much more distinct desire for children than women who are sterile from other causes. This is expressed statistically in the comparatively greater number of adoptions in this group. This observation indicates that in humans the maternal instinct may be influenced by hormones.

D. Activity

The activity of female rats varies markedly, depending on the state of the cycle. The highest activity is observed during estrus (see Harris 1963, 1964, where he reports measurements of running activity with the aid of a rotor and of a revolution counter). Neonatally castrated male rats implanted with an ovary in adulthood show the same cyclic activity patterns as normal females.

E. Emotional Behavior and Learning

Neonatal injection of either testosterone propionate or estradiol benzoate abolished the sex differences in activity and defecation frequency usually observed between male and female rats in the open-field situation. In this situation female rats tend to be more exploratory and to defecate less often than males. Females, androgenized after birth, behave like males (Gray *et al.*, 1965; Levine, 1966).

In addition, certain behavior patterns of adult rats are also influenced by administration of various pharmaceutical agents, such as norepinephrine or chlorpromazine, within the first 10 days of life. Animals injected with norepinephrine on day 2 or 4 of life were highly emotional in adulthood. They made more errors and took more time to solve a simple problem than did normal controls (Young, 1963). This indicates that the future ability to learn can be influenced by the action of certain pharmaceutical agents at an early stage of life.

III. Experiments in Humans

For obvious reasons, all attempts to apply observational data from animal experiments to human beings remain more or less speculative. However, it can be reasonably assumed that the most important regulatory mechanisms are at least similar in all mammals and that therefore disturbances during the period of hypothalamic center differentiation must have an influence on future behavior. Young and co-workers (1964), Levine (1966), and Dörner (1967, 1969), and others are concerned that in man certain sexual deviations are attributable to hormonal imbalances during the hypothalamic differentiation period rather than to early childhood experiences.

Some authors even claim homosexuality to be caused by disturbances during the differentiation period (Levine, 1966). According to Dörner (1967, 1969), male homosexuality results when an androgen deficiency had existed during the appropriate differentiation period. Female homosexuality ensues when androgens have been effective. Male hypo- or hyper-sexuality is also explained by Dörner to be due to relative androgen deficiency or androgen excess during the differentiation period. In some recent publications, Dörner (1969) has even proposed prophylactic prevention of male and female homosexuality by intrauterine testosterone treatment of fetuses to prevent male homosexuality or antiandrogen administration to prevent female homosexuality.

We believe that this interpretation of experimental data obtained from animals is still inconclusive, at least with regard to the genesis of homosexuality.

On the other hand, we are concerned that hormonal imbalances during the period of differentiation could result in transsexuality. Some observations in man are in accordance with this (Hinman,

1951), and mention is made of just one investigation done by Erhardt and Money (1967). These authors studied the behavior of ten girls between 3 and 14 years of age whose mothers had been treated during pregnancy with gestagens having androgenic side effects. Of the ten girls examined, nine were tomboyish. The common criteria of these girls were: preference in playing with boy's toys, athletic energy, outdoor pursuits, and minimal concern for feminine frills, doll play, baby care, and household chores.

Interesting statements were given by their mothers in reply to questions regarding their daughters' behavior and preferred toys. Here is a verbal quotation of some answers: " She would rather play baseball than dolls, she gets into fights, she likes guns and soldiers, she likes to play with boys, she likes to run and jump, her interests are guns and cowboys, she even wanted a baseball glove."

From these and other observations it may be assumed that, in principle, humans do not differ greatly from rats with regard to fixation of behavioral patterns.

References

Alloiteau, J. J. (1954). *C. R. Soc. Biol.* (Paris) **148,** 223.
Arai, Y. (1963). *J. Fac. Sci. Univ. Tokyo, Sect. IV,* **10,** 243.
Arai, Y., and Gorski, R. A. (1968). *Proc. Soc. Exp. Biol. Med.* **127,** 590.
Barraclough, C. A. (1955). *Am. J. Anat.* **97,** 493.
Barraclough, C. A. (1961). *Endocrinology* **68,** 62.
Barraclough, C. A. (1966a). *Recent Progr. Hormone Res.* **22,** 503.
Barraclough, C. A. (1966b). *Endocrinology* **78,** 1053.
Barraclough, C. A. (1967a). *In* " Neuroendocrinology " (L. Martini and W. F. Ganong, eds.), Vol. II, pp. 61–99. Academic Press, New York.
Barraclough, C. A. (1967b). *In* " Hormonal Steroids " (L. Martini, F. Fraschini and M. Motta, eds.), pp. 913–916. Excerpta Medica, Amsterdam.
Barraclough, C. A., and Gorski, R. A. (1961). *Endocrinology* **68,** 68.
Barraclough, C. A., and Gorski, R. A. (1962). *J. Endocrinol.* **25,** 175.
Barraclough, C. A., and Leathem, J. H. (1954). *Proc. Soc. Exp. Biol. Med.* **85,** 673.
Barraclough, C. A., Yrarrazaval, S., and Hatton, R. (1964). *Endocrinology* **75,** 838.
Beach, F. A. (1942). *Endocrinology* **31,** 673.
Beach, F. A. (1945). *Anat. Record* **92,** 289.
Beach, F. A., and Holz, A. M. (1946). *J. Exp. Zool.* **101,** 91.
Beeman, E. A. (1947). *Physiol. Zool.* **20,** 373.
Boling, J. L., Blandau, R. J., Wilson, J. G., and Young, W. C. (1939). *Proc. Soc. Exp. Biol. Med.* **42,** 128.
Boyd, R., and Johnson, D. C. (1968). *Acta Endocrinol.* (Kbh) **58,** 600.
Bradbury, J. T. (1941). *Endocrinology* **28,** 101.
Courrier, R., Guillemin, R., Justisz, M., Sakiz, E., and Aschheim, P. (1961). *C .R. Soc. Biol.* (Paris) **253,** 922.

Critchlow, V. (1958). *Am. J. Physiol.* **195**, 171.
Crossley, D. A., and Swanson, H. H. (1968). *J. Endocrinol.* **41**, xiii.
D'Angelo, S. A., and Kravatz, A. (1959). *Anat. Record* **113**, 370.
Dantchakoff, V. (1938). *Biol. Zbl.* **58**, 302.
Dörner, G. (1967). *Acta Biol. Med. Ger.* **19**, 569.
Dörner, G. (1968). *J. Endocrinol.* **42**, 163.
Dörner, G. (1969). *Deut. Med. Wochschr.* **94**, 390.
Dörner, G., and Hinz, G. (1967). *Germ. Med. Mth.* **12**, 281.
Dörner, G., and Staudt, J. (1968). *Neuroendocrinology* **3**, 136.
Dörner, G., Döcke, F., and Moustafa, S. (1968a). *J. Reprod. Fertility* **17**, 173.
Dörner, G., Döcke, F., and Moustafa, S. (1968b). *J. Reprod. Fertility* **17**, 583.
Edwards, D. A. (1968). *Science* **161**, 1027.
Erhardt, A. A., and Money, J. (1967). *J. Sex. Res.* **3**, 83.
Everett, J. W., and Radford, H. M. (1961). *Proc. Soc. Exp. Biol. Med.* **108**, 604.
Feder, H. H. (1967). *Anat. Record* **157**, 79.
Feder, H. H., Phoenix, C. H., and Young, W. C. (1966). *J. Endocrinol.* **34**, 131.
Flerkó, B., and Szentágothai, J. (1957). *Acta Endocrinol.* (Kbh) **26**, 121.
Flerkó, B., Petrusz, P., and Tima, L. (1967). *Acta Biol. Acad. Sci. Hung.* **18**, 27.
Goldzieher, J. M., and Axelrod, L. R. (1963). *Fertility Sterility* **14**, 631.
Gorski, R. A. (1966). *J. Reprod. Fertility* Suppl. **1**, 67.
Gorski, R. A. (1968). *Endocrinology* **82**, 1001.
Gorski, R. A., and Barraclough, C. A. (1962a). *Acta Endocrinol.* (Kbh) **39**, 13.
Gorski, R. A., and Barraclough, C. A. (1962b). *Proc. Soc. Exp. Biol. Med.* **110**, 298.
Gorski, R. A., and Barraclough, C. A. (1963). *Endocrinology* **73**, 210.
Gorski, R. A., and Wagner, J. W. (1965). *Endocrinology* **76**, 226.
Goy, R. W., Bridson, W. E., and Young, W. C. (1961). *Anat. Record* **139**, 232.
Goy, R. W., Phoenix, C. H., and Young, W. C. (1962). *Anat. Record* **142**, 307.
Goy, R. W., Bridson, W. E., and Young, W. C. (1964). *J. Comp. Physiol. Psychol.* **57**, 166.
Goy, R. W., Phoenix, C. H., and Meidinger, R. (1967). *Anat. Record* **157**, 87.
Grady, K. L., Phoenix, C. H., and Young, W. C. (1965). *J. Comp. Physiol. Psychol.* **59**, 176.
Gray, J. A., Levine, S., and Broadhurst, P. L. (1965). *Anim. Behav.* **13**, 33.
Greer, M. (1953). *Endocrinology* **53**, 380.
Halász, B., and Pupp, L. (1965). *Endocrinology* **77**, 553.
Halász, B., Pupp, L., Uhlarik, S., and Tima, L. (1965). *Endocrinology* **77**, 343.
Harris, G. W. (1963). *J. Physiol.* (London) **169**, 117 P.
Harris, G. W. (1964). *Endocrinology* **75**, 627.
Harris, G. W., and Jacobsohn, D. (1952). *Proc. Roy. Soc.* **139**, 263.
Harris, G. W., and Levine, S. (1962). *J. Physiol.* (London) **163**, 42 P.
Harris, G. W., and Levine, S. (1965). *J. Physiol.* (London) **181**, 379.
Hillarp, N. A. (1949). *Acta Endocrinol.* (Kbh) **2**, 11.
Hinman, F. (1951). *J. Clin. Endocrinol.* **11**, 477.
Hohlweg, W., and Junkmann, K. (1932). *Klin. Wochenschr.* **11**, 321.
Kawashima, S. (1960). *J. Fac. Sci. Univ. Tokyo, Sect. IV* **9**, 117.
Kurcz, M., and Gerhardt, V. J. (1968). *Endocrinol. Exp.* **2**, 29.
Kurcz, M., Kovács, K., Tiboldi, T., and Orosz, A. (1967). *Acta Endocrinol.* (Kbh) **54**, 663.
Kurcz, M., Nagy, I., Gerhardt, J., and Baranyai, P. (1968a). *Acta Biol. Acad. Sci. Hung.* **19**, 123.
Kurcz, M., Nagy, I., Gerhardt, J., and Baranyai, P. (1968b). *Kiserl. Orvostud.* **20**, 380.
Levine, S. (1966). *Sci. Am.* **214**, 84.

Levine, S., and Mullins, R., Jr. (1967). *In* " Hormonal Steroids " (L. Martini, F. Fraschini and M. Motta, eds.), pp. 925–931. Excerpta Medica, Amsterdam.

Lisk, R. D. (1962). *Am. J. Physiol.* **203**, 493.

Lloyd, C. W., and Weisz, J. (1967). *In* " Hormonal Steroids " (L. Martini, F. Fraschini and M. Motta, eds.), pp. 917–924. Excerpta Medica, Amsterdam.

Martinez, C., and Bittner, J. J. (1956). *Proc. Soc. Exp. Biol. Med.* **91**, 506.

Mazer, H., and Mazer, C. (1939). *Endocrinology* **24**, 175.

Meyerson, B. J. (1968). *Nature* **217**, 683.

Mullins, R. F., Jr., and Levine, S. (1968a). *Physiol. Behav.* **3**, 339.

Mullins, R. F., Jr., and Levine, S. (1968b). *Physiol. Behav.* **3**, 333.

Neumann, F., and Elger, W. (1965a). *Acta Endocrinol.* (Kbh) Suppl. **100**, 174.

Neumann, F., and Elger, W. (1965b). *Excerpta Med. Intern. Congr. Ser.* **101**, 168.

Neumann, F., and Elger, W. (1966). *Endokrinologie* **50**, 209.

Neumann, F., and Elger, W. (1969). *In* " Advances in the Biosciences " (G. Raspé, ed.), pp. 80–107. Pergamon Press-Vieweg, Braunschweig.

Neumann, F., and Kramer, M. (1967). *In* " Hormonal Steroids " (L. Martini, F. Fraschini and M. Motta, eds.), pp. 932–941. Excerpta Medica, Amsterdam.

Neumann, F., Elger, W., and von Berswordt-Wallrabe, R. (1966a). *J. Endocrinol.* **36**, 353.

Neumann, F., Elger, W., and Kramer, M. (1966b). *Endocrinology* **78**, 628.

Neumann, F., Hahn, J. D., and Kramer, M. (1967). *Acta Endocrinol.* (Kbh) **54**, 227.

Neumann, F., Elger, W., and Steinbeck, H. (1969). *J. Reprod. Fertil., Suppl.* 7, 9.

Palka, Y. S., and Sawyer, C. H. (1966). *Am. J. Physiol.* **221**, 225.

Petrusz, P., and Flerkó, B. (1965). *Acta Biol. Acad. Sci. Hung.* **16**, 169.

Petrusz, P., and Nagy, E. (1967). *Acta Biol. Acad. Sci. Hung.* **18**, 21.

Pfeiffer, C. A. (1936). *Am. J. Anat.* **58**, 195.

Phoenix, C. H., Goy, R. W., Gerall, A. A., and Young, W. C. (1959). *Endocrinology* **65**, 369.

Schuetz, A. W., and Meyer, R. K. (1963). *Proc. Soc. Exp. Biol. Med.* **112**, 875.

Segal, S. J., and Johnson, D. C. (1959). *Arch. Anat. Microscop. Morphol. Exp.* **48**, 261.

Selye, H. (1940). *Endocrinology* **27**, 657.

Shay, H., Gershon-Cohen, J., Paschkis, K. E., and Fels, S. S. (1939). *Endocrinology* **25**, 933.

Steinbeck, H., Elger, W., and Neumann, F. (1970). *J. Endocrinol.* (In press).

Swanson, H. E., and Van der Werff ten Bosch, J. J. (1963). *J. Endocrinol.* **26**, 197.

Swanson, H. E., and Van der Werff ten Bosch, J. J. (1964a). *Acta Endocrinol.* (Kbh) **45**, 1.

Swanson, H. E., and Van der Werff ten Bosch, J. J. (1964b). *Acta Endocrinol.* (Kbh) **45**, 37.

Takewaki, K. (1962). *Experientia* **18**, 1.

Taleisnik, S., and McCann, S. M. (1961). *Endocrinology* **68**, 263.

Tollman, J., and King, J. A. (1956). *Brit. J. Animal Behav.* **6**, 147.

Ulrich, R. E., and Azrin, N. H. (1962). *J. Exp. Anal. Behav.* **5**, 511.

Van Dyke, D. C., Simpson, M. E., Leprovsky, S., Koniff, A. A., and Brobeck, J. R. (1957). *Proc. Soc. Exp. Biol. Med.* **95**, 1.

Van Rees, G. P., Van der Werff ten Bosch, J. J., and Wolthuis, C. L. (1962). *Acta Endocrinol.* **40**, 95.

Weisz, J., and Lloyd, C. W. (1965). *Endocrinology* **77**, 735.

Whalen, R. E., and Edwards, D. A. (1966). *J. Comp. Physiol. Psychol.* **62**, 307.

Whalen, R. E., and Edwards, D. A. (1967). *Anat. Record* **157**, 173.

Whalen, R. E., and Nadler, R. D. (1963). *Science* **141**, 273.

Whalen, R. E., and Nadler, R. D. (1965). *J. Comp. Physiol. Psychol.* **60**, 150.

Wilson, J. G., and Young, W. C. (1941). *Endocrinology* **29**, 779.
Wilson, J. G., Hamilton, J. B., and Young, W. C. (1941). *Endocrinology* **29**, 784.
Wollman, A. L., and Hamilton, J. B. (1967). *Endocrinology* **81**, 350.
Wolthuis, O. L., Swanson, H. E., and Van der Werff ten Bosch, J. J. (1962). *Acta Physiol. Pharmacol. Neerl.* **11**, 313.
Yazaki, I. (1959). *Jap. J. Zool.* **12**, 267.
Yazaki, I. (1960). *Annot. Zool. Jap.* **33**, 217.
Young, R. D. (1964). *Science* **143**, 1055.
Young, W. C. (1960). *In* " Recent Advances in Biological Psychiatry" pp. 312–364. Grune and Stratton Inc., New York.
Young, W. C. (1961). *In* " Sex and Internal Secretions " (W. C. Young, ed.), Vol. II, pp. 1173–1239. Williams and Wilkins, Baltimore.
Young, W. C. (1963). *Canad. Psychiat. Ass. J.* **8**, 36.
Young, W. C., Goy, R. W., and Phoenix, C. H. (1964). *Science* **143**, 212.

Hypothalamo-Hypophyseal Relationships in the Fetus

A. JOST, J. P. DUPOUY, and A. GELOSO-MEYER

I. Introduction

In the discussion of any problem of fetal or of developmental endocrinology the importance of developmental chronology should be kept in mind. As a rule a phase of organogenesis precedes endocrine function, but it often remains difficult to correlate purely morphological data and onset of funtioning. This remark is especially valid with respect to the hypothalamo-hypophyseal system.

II. Development of the Hypothalamo-Hypophyseal System

The chronology of the development of definitive adult-like hypothalamo-hypophyseal relationships varies according to the animal species. Four examples will be summarized.

1. Rabbit: the general development of the rabbit pituitary gland has been described by Atwell (1918) and by Jost and Gonse (1953). Some major steps will be reviewed. On day 12 after insemination, the primordium of the anterior pituitary is made of a buccal evagination, which grows toward the diencephalon; the neurohypophysis appears as an outgrowth of the neural structure (Fig. 1). Three days later, on day 15, a flat Rathke's pouch is individualized; its lower

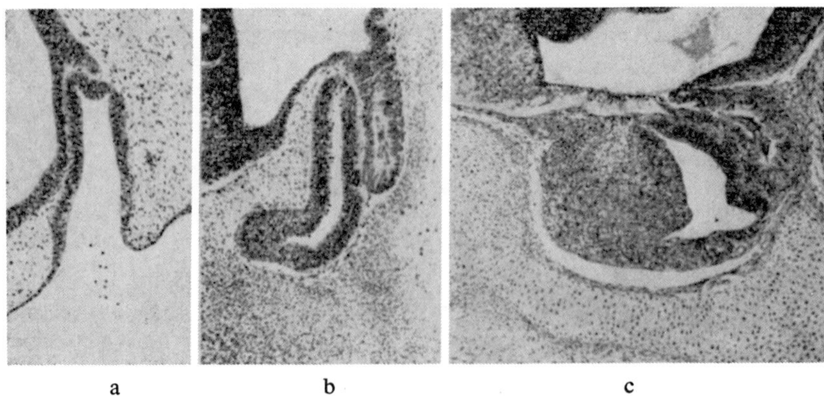

FIG. 1. Sagittal sections through the pituitary glands of rabbit fetuses, on days 12 (a), 15 (b), and 18 (c). The 12- and 15-day-old fetuses are litter mates taken from the uterus three days apart. Anterior part of the head, on the left side (x 68).

and anterior part is intensely proliferating and growing; it pushes an anterior process into the mesenchyme. On day 18 the anterior process has formed the major part of the pars distalis and it has grown upward beneath the tuber, giving the pars tuberalis. A large mesenchymal cleft still remains between the early pouch of Rathke and the tuberal process (fossa or Atwell's recess). Small blood vessels are entering the pars distalis, a few through its inferior part and others from vessels present in the fossa. The pars intermedia remains avascular until day 22. Campbell (1966) studied the development of the primary portal plexus in the median eminence in rabbit fetuses after the injection of Indian ink through the left ventricle of the heart. On day 17, the adenohypophysis still is avascular, but " a thin pial plexus lay along the base of the brain, becoming thicker in the region of the pars tuberalis to form the supratuberal region." On day 21, vessels from the pial plexus penetrate into the median eminence to form the dense network of capillary loops that in the adult is termed the primary plexus of the portal system.

It is worthwhile recalling that hypophyseal influence on the testis (Jost, 1951) and thyroid (Jost, 1953) becomes especially conspicuous at approximately the same stage.

2. Human: In the human fetus the morphological maturation of the hypothalamo-hypophyseal system is a prenatal event, as in the rabbit; the portal vessels are present on the fourth month, while the adult-like primary plexus develops somewhat later (Niemineva, 1950). Rinne (1963) studied hypothalamic neurosecretion.

3. Rat: According to Jost and Tavernier (1956), on day 14.5 the anterior process is growing on Rathke's pouch; on day 15.5 this process contains many cells stained with the Periodic Acid Schiff (PAS) technique and it is penetrated by the very first capillaries. Vascularity becomes increasingly important in the anterior lobe during days 16 and 17; the general shape of the pituitary complex is well outlined on day 17. Glydon (1957) studied the hypophyseal blood supply. On day 16 he observed an extensive infiltration of the tuberal process, except in its extreme anterior part, by connective tissue and blood vessels. On day 17 a thin layer of connective tissue is present between the lower part of the brain and the adenohypophysis. This layer is in continuity with the fossa and contains "many capillaries which derive blood from branches of the internal carotid arteries." Between days 17 and 19 an increasing number of vessels coming from this supratuberal plexus drain into the adenohypophysis. But these vessels do not penetrate the tuber cinereum.

The first hypophyseal portal vessels "represented in the caudal region of the plexus by one to three vessels of larger diameter than any others" were noted by Glydon (1957) on day 21 and on day 20 by Rinne and Kivalo (1965). The rich capillary network seen in the median eminence of the adult does not develop until the fifth day after birth (Glydon, 1957; Campbell, 1966) although some capillary loops penetrating the median eminence were seen as early as on day 3 by Rinne and Kivalo (1965).

Neurosecretory material stained with aldehyde-fuchsin could be detected in the infundibular process on day 19, and in a few fibers running toward the hypophyseal portal vessels only on and after day 20 of prenatal age (Rinne and Kivalo, 1965).

The postnatal completion of the primary plexus contrasts with the evidence indicating that during the prenatal period the pituitary influences the adrenals (Cohen, 1963) and the thyroid (Geloso, 1967) from day 18 onward.

4. Mouse: Enemar (1961) gives a careful description of the prenatal superficial network of capillaries on the surface of the prospective median eminence and of the late postnatal penetration of shallow loops into the median eminence. Björklund *et al.* (1968) found only very little fluorescent monoamines in the median eminence during the three last prenatal days, but they did observe a few fluorescent fibers in some hypothalamic centers. Prenatal hypophyseal adrenocorticostimulating activity has been studied by Eguchi (1961).

III. Experimental Techniques

The *in vivo* study of the hypothalamo-hypophyseal relationships necessitates reliable tests of hypophyseal activity; several such tests have been utilized in many experiments involving either fetal hypophysectomy by decapitation and hormonal replacement or hormonal feedback effects in the fetus (for review see Jost, 1966a).

In the experiments discussed here, comparisons were made between control fetuses, fetuses deprived of both their hypothalamus and pituitary by surgical decapitation, and fetuses in which the whole brain including the hypothalamus was removed while the pituitary was left *in situ* (technique described in Jost *et al.*, 1966). These fetuses will be referred to as encephalectomized fetuses (Fig. 2).

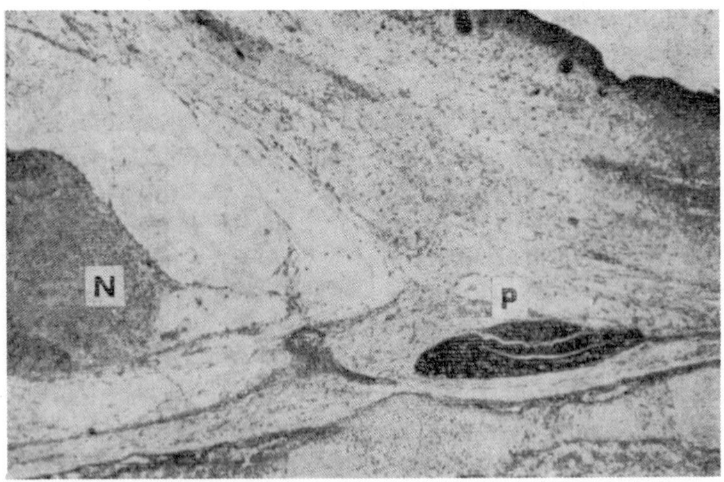

FIG. 2. Longitudinal section through the head of an encephalectomized rat fetus. Surgery done on day 18.5, sacrifice on day 21.5. The nervous structures normally covering the pituitary gland (P) were removed; the interrupted nervous stem (N) is seen on the left, in the posterior part of the head (From Jost and Geloso, 1967).

IV. Hypophyseal Corticostimulating Activity

In rat fetuses deprived of their pituitary by decapitation, growth of the adrenal glands is more or less completely stopped, and at birth these glands appear atrophic in comparison with those of control fetuses (Jost, 1966b; Jost and Cohen, 1966). Removal of the pituitary on or before day 18 also prevents the fetal adrenal cortex from influencing target organs; thus glycogen storage in the liver does not occur in decapitated fetuses of adrenalectomized mother animals (Jost and Jacquot, 1954; Jacquot, 1959).

The weight of the adrenals was studied in fetuses that were encephalectomized on day 16 or 19 and sacrificed on days 20 and 21 respectively (Jost, 1966b; Jost *et al.*, 1966). Despite the presence of the pituitary gland the weight of the adrenals was extremely low. In fetuses encephalectomized on day 19.5 the weight of the adrenal glands is similar to that found in decapitates (in other experiments done on day 18 the weight of the adrenals was slightly higher in encephalectomized than in decapitated fetuses).

Crude hypothalamic extracts prepared from rat hypothalami (acetonic powder was extracted by acetic acid and the lyophilized product was cleaned of its fat and extracted with methanol) or beef Corticotropin Releasing Factor (CRF) (purified and kindly sent to us by Dr. Schally) were given subcutaneously to fetuses encephalectomized on day 19.5. Subnormal or normal adrenal weights were obtained in encephalectomized fetuses, but the same treatment had no significant effect in decapitates (Fig. 3) (Jost, 1966b; Jost *et al.*, 1966).

These experiments definitely suggested that the corticostimulating activity of the fetal pituitary depends on the hypothalamus. The role of the hypothalamus was further explored in fetuses subjected to stress. It has been shown by Milkovic and Milkovic (1958) that a stress (injection of adrenaline into 20-day-old rat fetuses) produces a depletion of ascórbic acid in the adrenal gland within 1 hr unless the fetuses are decapitated (Milkovic and Milkovic, 1961). Similar experiments were conducted in rat fetuses that had been encephalectomized and injected with formalin on day 20 under ether anesthesia. Formalin is a more active stress-producer than adrenaline in intact rat fetuses (Cohen and Pernot, 1968); it produces a 25% reduction in ascorbic acid content of the adrenals. In encephalectomized fetuses a

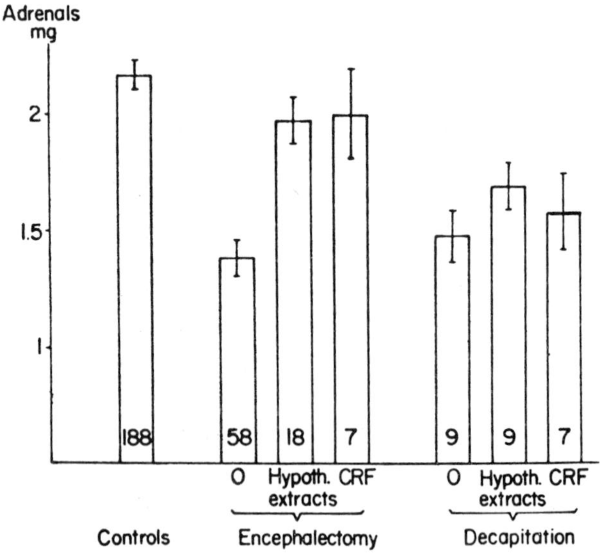

FIG. 3. Weight of one pair of adrenals in 21.5-day-old rat fetuses that were encephalectomized or decapitated on day 19.5; hypothalamic extracts or Corticotropin Releasing Factor (CRF) were given immediately. Confidence intervals and number of animals are indicated (From Jost, 1966b).

formalin injection has no effect on adrenal ascorbic acid (Cohen *et al.*, 1968). The hypothalamus is a probable link between the stressing agent and the hypophyseal response.

In the fetal pituitary-adrenocortical feedback system the maternal hormones play a role in one way or another, since maternal adrenalectomy in rats results in an increase in size of the fetal adrenals. Maternal hypercorticism has the opposite effect (for review see Jost, 1966a). Some experiments were made on pregnant rats that were adrenalectomized on day 14 of gestation; fetuses were encephalectomized (21 fetuses) or decapitated (22 fetuses) on day 19 and studied on day 21. The adrenals were slightly but significantly heavier in the encephalectomized than in the decapitated fetuses. This difference could result either from increased maternal CRF production and transfer to the fetus or from a diminished level of circulating corticosteroids in the fetuses of adrenalectomized pregnant females and the resultant diminished negative feedback directly at the level of the pituitary gland.

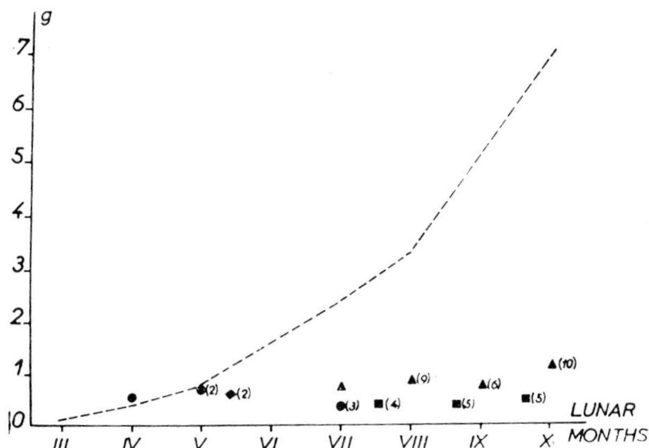

FIG. 4. Weight of the adrenals in normal controls (dotted line) and anencephalic human fetuses according to age in lunar months. The values were collected in the literature (see Jost, 1970a).

In another group of 5 fetuses encephalectomized on day 18 in adrenalectomized pregnant females, some glycogen storage had occurred in the liver on day 21 (one-third of controls), while glycogen remained low in the decapitates. It would appear that despite encephalectomy on day 18, some adrenocortical action upon liver glycogen storage had been present.

These data would suggest some hypophyseal functioning not under hypothalamic control and some direct feedback influence of corticosteroids upon the pituitary. But, as a whole, during the period of development studied so far in the rat fetus the hypophyseal corticostimulating activity has been mostly under hypothalamic control.

These results give an experimental model for studying the conditions prevailing in anencephalic human infants. It has long been known that their adrenal cortex is strikingly underdeveloped. An adenohypophysis is usually present, but it is either isolated or in contact with a neurohypophysis devoid of neurosecretory material (Tuchmann-Duplessis and Gabe, 1960). In anencephalic fetuses the growth of the adrenals (Fig. 4) is practically stopped after the fourth month as it is in rat fetuses decapitated before day 18 (Jost, 1966b; Jost and Cohen, 1966). The defective hypothalamic influence on the adenohypophysis is responsible for the defective corticostimulating activity.

Finally it should be noted that in decapitated chick embryos, pituitary grafts permit normal adrenocortical development in the absence of the hypothalamus (Betz, 1967).

V. Thyroid-Stimulating Activity

The effects of intrauterine hypophysectomy on the fetal thyroid were studied (Jost *et al.*, 1949; Jost, 1953) in fetuses that were partially decapitated (through the mouth so as to leave the lower jaw and the tongue *in situ* and to spare the thyroid any direct surgical trauma). Under such conditions the histological differentiation of the thyroid is delayed and its size at birth is reduced in rats (Jost, 1957a, b) and in rabbits (Jost, 1953); the uptake of radioiodine remains low in rabbits (Jost *et al.*, 1952) and in rats (Geloso, 1967); and the release of thyroxine in the fetal blood is practically abolished in the rat fetus (Geloso, 1967).

Another test for hypophyseal thyroid-stimulating activity is based on the disturbance of the pituitary thyroid feedback by antithyroid drugs: propylthiouracil (PTU) given to the mother reaches the fetus and causes fetal thyroid hypertrophy unless the fetus is decapitated (Jost, 1957a,b, 1959) or injected with thyroxine (Jost, 1959).

The role of the hypothalamus in the hypophyseal thyroid-stimulating activity was studied in rat fetuses that were encephalectomized on day 18.5; the pregnant female animal was untreated or given PTU by mouth (50 mg daily for 3 days). Three tests were studied on day 21.5: namely, thyroid histology, thyroid volume, and ^{131}I uptake in 1 hr; no significant difference was found between the intact or the encephalectomized fetuses; it is especially noteworthy that the latter showed a conspicuous response to the antithyroid drug (Jost and Geloso, 1967). The increase in the pituitary-thyroid-stimulating activity in the PTU-treated animals could be suspected of having resulted from maternal overproduction of Thyrotropin Releasing Factor (TSHRF) and transfer to fetus, but it seems to be more satisfactorily accounted for by changes in the pituitary-thyroid hormone feedback in the fetus.

Between days 18.5 and 21.5 the fetal hypophyseal thyroid-stimulating activity is largely independent of the hypothalamus. The same conclusion applies to very briefly reported experiments (section of pituitary stalk; decerebration) done by Noumura (1959). However, our most recent results suggest some hypothalamic influence on the thyroid-stimulating function: (1) whereas the ^{131}I uptake in 1 hr is the same in intact and in encephalectomized fetuses and is similarly reduced by PTU, after 24 hr, the uptake tends to be lower in encephalectomized fetuses; (2) the amount of circulating thyroxine was measured under conditions of isotopic equilibrium and expressed in nanograms of ^{127}I (technique of Geloso and Bernard, 1967); under these conditions it was only two-thirds of the control value on day 21 in fetuses that had been encephalectomized on day 17.

Preliminary experiments were done on rabbit fetuses encephalectomized on day 18 or 19 and studied on day 28. Some reduction of thyroid activity is suggested by the elevated total radioactivity of the fetal blood and the lowered ^{131}I uptake in 1 hr. Whether in rabbit fetuses the absence of the hypothalamus has a more pronounced effect on thyroid function than in rats remains to be assessed.

In human newborns the size of the thyroid gland is extremely reduced when the pituitary gland is completely missing (Brewer, 1957; Reid, 1960); on the contrary in anencephalics it is usually of normal size or somewhat hypertrophied (Tuchmann-Duplessis, 1959; Kind, 1962). Thyrotropic cells are well differentiated in the pituitary of the anencephalic newborn (Herlant, 1953) and the thyroid synthesizes thyroxine (Yamazaki *et al.*, 1959). However, it contains lower amounts of iodine and of thyronines than do the control glands (Biressi *et al.*, 1962). These observations are in good agreement with the experimental results on rat fetuses. One aspect deserves further comment. In both rat and human anencephalics some reduction of thyroid hormone production is likely in the absence of the hypothalamus. Nevertheless the thyroid gland is not strikingly hypertrophied; it could be speculated that the thyroid-pituitary feedback system is not deeply disturbed, perhaps because of changes in adrenocortical function.

In addition, it should be noted that permanent thyroid disturbances have been described in rats that were injected with thyroid hormones during the first postnatal days (Eayrs and Holmes, 1964; Bakke and Lawrence, 1966) or in dogs treated both prenatally and postnatally with PTU (Arosenius *et al.*, 1962). These observations suggest that the developing hypothalamus might be susceptible to an organizing action by thyroid hormones.

Some experiments suggest that the pituitary in chick embryos exerts thyroid-stimulating activity in the absence of the hypothalamus (Mess and Straznicky, 1964; Betz, 1967; Le Douarin and Ferrand, 1968). In tadpoles deprived of their hypothalamus, metamorphosis is prevented (Chang, 1957; Remy, 1962), but PTU still induces hypertrophy of the thyroid (Remy and Bounhiol, 1967).

VI. Other Pituitary Functions

Information about the role of the fetal hypothalamus in other pituitary functions is scarce and often indirect. In rabbit fetuses, glycogen deposition in the fetal liver requires a pituitary hormone that might be either growth hormone or prolactin (Jost, 1961). Glycogen deposition is prevented by decapitation (Jost and Hatey, 1949) but it is not affected (Bearn, 1968) or is only slightly reduced (Jost and Picon, 1970) in encephalectomized fetuses. This suggests that the pituitary produces the necessary metabolic hormone in the absence of the hypothalamus.

Testicular function in anencephalics has been discussed previously (Jost, 1966a, 1970b).

VII. Conclusions

Animal experiments as well as observations on human anencephalic fetuses concur and indicate that in the mammalian fetus the hypophyseal corticostimulating activity more strictly depends upon the hypothalamus than does the thyrostimulating function. In rats, hypothalamo-hypophyseal relationships are definitely established at a stage that precedes the complete maturation of the primary plexus. It remains to be determined whether the actual onset of hypophyseal functioning and hormonal release depends upon the hypothalamus.

References

Arosenius, K. E., Derblom, H., and Nylander, G. (1962). *Acta Endocrinol.* **40**, 430.
Atwell, W. J. (1918). *Am. J. Anat.* **24**, 271.
Bakke, J. L., and Lawrence, N. (1966). *J. Lab. Clin. Med.* **67**, 477.
Bearn, J. G. (1968). *Anat. Record* **160**, 311.
Betz, T. W. (1967). *Gen. Comp. Endocrinol.* **9**, 172.
Biressi, P. C., Scorta, A., and Patrito, G. (1962). *Folia Endocrinol.* **15**, 74.
Björklund, A., Enemar, A., and Falck, B. (1968). *Z. Zellforsch.* **89**, 590.
Brewer, D. B. (1957). *J. Path. Bact.* **73**, 59.

Campbell, H. J. (1966). *J. Anat.* **2**, 381.
Chang, C. Y. (1957). *Anat. Record* **128**, 531.
Cohen, A., (1963). *Arch. Anat. Microscop. Morphol. Exp.* **52**, 277.
Cohen, A., and Pernot, J. C. (1968). *C. R. Soc. Biol.* (Paris) **162**, 1675.
Cohen, A., Pernot, J. C., and Jost, A. (1968). *C. R. Soc. Biol.* (Paris) **162**, 2070.
Eayrs, J. T., and Holmes, R. L. (1964). *J. Endocrinol.* **29**, 71.
Eguchi, Y. (1961). *Endocrinology* **68**, 716.
Enemar, A. (1961). *Arkiv Zool.* **13**, 203.
Geloso, J. P. (1967). *Ann. Endocrinol.* (Paris) **28**, *Suppl.* **1**, 1.
Geloso, J. P., and Bernard, G. (1967). *Acta Endocrinol.* **56**, 561.
Glydon, R. St. J. (1957). *J. Anat.* **91**, 237.
Herlant, M. (1953). *Ann. Endocrinol.* **14**, 899.
Jacquot, R. (1959). *J. Physiol.* (Paris) **51**, 655.
Jost, A. (1951). *Arch. Anat. Microscop. Morphol. Exp.* **40**, 247.
Jost, A. (1953). *Arch. Anat. Microscop. Morphol. Exp.* **42**, 168.
Jost, A. (1957a). *C. R. Soc. Biol.* (Paris) **151**, 1295.
Jost, A. (1957b). *Rev. Suisse Zool.* **64**, 821.
Jost, A. (1959). *C. R. Soc. Biol.* (Paris) **153**, 1900.
Jost, A. (1961). *The Harvey Lectures* **55**, 201.
Jost, A. (1966a). *In* " The Pituitary Gland " (G. W. Harris and B. T. Donovan, eds.), Vol. II, pp. 299–323. Butterworths, London.
Jost, A. (1966b). *Recent Progr. Hormone Res.* **22**, 541.
Jost, A. (1970a). *In* " Hormones in Development " (M. Mamburgh and E. J. W. Barrington, eds.). Appleton Century Crofts, New York. (In press).
Jost, A. (1970b). *In* " Hermaphroditism, Genital Anomalies and Related Endocrine Disorders " (H. W. Jones and W. W. Scott, eds.). 2nd ed. Williams & Wilkins, Baltimore. (In press).
Jost, A., and Cohen, A. (1966). *Develop. Biol.* **14**, 154.
Jost, A., and Geloso, A. (1967). *C. R. Acad. Sci.* (Paris), **265**, 625.
Jost, A., and Gonse, P. (1953). *Arch. Anat. Microscop. Morphol. Exp.* **42**, 243.
Jost, A., and Hatey, J. (1949). *C .R. Soc. Biol.* (Paris) **143**, 146.
Jost, A., and Jacquot, R. (1954). *C. R. Acad. Sci.* (Paris) **239**, 98.
Jost, A., and Picon, L. (1970). *Advances in Metabolic Disorders* **5** (In press).
Jost, A., and Tavernier, R. (1956). *C. R. Acad. Sci.* (Paris) **243**, 1353.
Jost, A., Morel, F. F., and Marois, M. (1949). *C. R. Soc. Biol.* (Paris) **143**, 142.
Jost, A., Morel, F. F., and Marois, M. (1952). *C. R. Soc. Biol.* (Paris) **146**, 1066.
Jost, A., Dupouy, J. P., and Monchamp, A. (1966). *C. R. Acad. Sci.* (Paris) **262**, 147.
Kind, C. (1962). *Helv. Paediat. Acta* **3**, 244.
Le Douarin, G., and Ferrand, R. (1968). *C. R. Acad. Sci.* (Paris) **266**, 697.
Mess, B., and Straznicky, K. (1964). *Acta Biol. Acad. Sci. Hung.* **15**, 77.
Milkovic, K., and Milkovic, S. (1958). *Arch. Intern. Physiol. Biochem.* **66**, 534.
Milkovic, S., and Milkovic, K. (1961). *Proc. Soc. Exp. Biol. Med.* **107**, 47.
Niemineva, K. (1950). *Acta Paed.* **39**, 315.
Noumura, T. (1959). *Japan. J. Zool.* **12**, 301.
Reid, J. D. (1960). *J. Pediat.* **56**, 658.
Remy, C., (1962). *C. R. Acad. Sci.* (Paris) **254**, 567.
Remy, C., and Bounhiol, J. (1967). *Gen. Comp. Endocrinol.* **9**, 519.
Rinne, U. K. (1963). *Acta Neuroveget.* (Wien) **25**, 310.
Rinne, U. K., and Kivalo, E. (1965). *Acta Neuroveget.* (Wien) **27**, 166.
Tuchmann-Duplessis, H. (1959). *Ann. Endocrinol.* (Paris) **20**, 569.
Tuchmann-Duplessis, H., and Gabe, M. (1960). *Bull. Acad. Nat. Med.* (Paris) **144**, 102.
Yamazaki, E., Noguchi, A., and Slingerland, D. W. (1959). *J. Clin. Endocrinol.* **19**, 1437.

Some Effects of Gonadal Hormones on Sexual Behavior

F. A. BEACH

I. Introduction

My original assignment was to review the effects of hormones upon behavior. When I reviewed this same topic more than 20 years ago the result was a book with a bibliography of more than 80 pages. Research published in the enusing two decades could only be reviewed in three or four volumes the same size as the one published in 1948 (Beach, 1948). It is gratifying to observe that the study of hormones and behavior has burgeoned. Many new techniques have been invented. The number of species examined has increased greatly. The kinds of behavior studied have multiplied. Most important, the number of investigators has expanded from a very small band in the 1920s and 1930s to several hundred at the present time. In response to these changes a new journal was founded and the first issue appeared in April of 1969. Its title, appropriately enough, is "Hormones and Behavior".

Although one cannot possibly condense into one chapter a review of even the high points of current knowledge in the field of hormones and behavior it is feasible to present a general overview of one small portion of it. This article will deal exclusively with the single category of sexual behavior and the hormones produced by one type of gland, the gonads. In spite of these restrictions it should be possible to provide the reader with a general comprehension of the types of problems that are being studied, the investigational methods currently in use, the kinds of evidence presently available, and the theoretical interpretations currently entertained by experts in the area. Many of the points that will be made in discussing the hormonal control of sexual behavior are applicable to all studies of hormone-behavior relationships, regardless of the behavior or the particular hormones being considered.

II. Problems of Classifying Behavior

One of the very first problems in any study of behavior, as in any scientific endeavor, is the description and classification of the raw material or primary data, which in this case consists of behavioral responses.

When one mentions the adenohypophysis, adrenal cortex, corpus luteum, progesterone, or testosterone the referents of his terms are clear and definite. The words mean the same thing to all endocrinologists. In contrast, many of the terms used in classifying behavior are imprecise and ambiguous. Much has been written about hormones and stress, but the behavioral symptoms assigned to the so-called stress syndrome by different investigators are highly variable and often vaguely defined. This muddies the channels of communication and makes it difficult to compare the results of different studies.

Even so widely used a category as " sexual behavior " is subject to different implicit definitions, and the unconscious assumption of a mutual understanding has sometimes led to serious confusion and contradictory interpretations. Investigators of the effects of hormones on " sexual behavior " have dealt with such widely divergent phenomena as the content of women's dreams at different stages in the menstrual cycle (Deutsch, 1945) and genital reflexes in paraplegic male rats (Hart, 1967a). As long as primary behavioral data remain loosely, idiosyncratically, and intuitively defined, the category of " sexual behavior " can have no scientific validity and questions regarding its hormonal control will be meaningless.

In recent years there has developed an interesting and potentially fruitful trend which should lead to clarification of our concepts of sexual behavior. Biologically-oriented laboratory experimentalists have always treated sexual behavior as synonymous with heterosexual copulation. This definition seemed adequate to encompass the relatively simple and stereotyped mating patterns of guinea pigs, rats, and cats, which were about the only species studied for many years. At the same time more naturalistically-inclined students of behavior insisted that for many species the act leading to union of male and female gametes is only one brief segment of a much larger pattern of heterosexual interaction which is more properly identified as the sexual behavior of that species. They pointed, for example, to the fact that in some birds and fishes the act of fertilization is of necessity preceded by days or even weeks devoted to the formation of mated pairs, establishment of a breeding territory, building of one or more nests, etc. All these activities are properly classified as elements of sexual behavior. It is clear that definitions must be framed so as to allow for and encompass wide differences between species.

Another development in the conceptualization of sexual behavior and its hormonal correlates has been a tendency to set up categories of " sex-related " or " sexually dimorphic " behavior. In actual fact such categories have long been employed in other fields of behavior study. For decades child psychologists have been concerned with the process of sex typing or gender identification whereby the " psychological sex " of the individual becomes established within the first three or four years of life. Other psychologists have devised and employed with profit scales to measure " masculinity " and " femininity " in adult humans.

An interesting point of view relevant both to sex differences and to definitions of sexual behavior has been expressed by one clinician (Hampson, 1965). In studying patients presenting certain types of endocrine and/or hermaphroditic disorders, the following approach was adopted: " . . . we decided that from our point of view as psychiatrists the important dimension to be evaluated was the quality and pervasiveness of the *gender role* displayed by each individual. . . . As we have used the term, *gender role* is revealed by all those things a person says or does to disclose himself or herself as having the status of boy or girl, man or woman. It includes but is not restricted to sexual eroticism " (Hampson, 1965).

The activities and writings of psychologists and psychiatrists thus show clearly their recognition of the fact that there are important behavioral differences which can properly be classified as " sex-related. "

Similarly, naturalists involved in the study of freeliving animals always have recognized that male birds of some species sing but females of the same species do not; male sticklebacks build nests but females do not normally do so; adult male baboons join together and advance to threaten the approaching leopard while adult females flee or at least remain near the center of the troop.

There is nothing new about the recognition of sex differences in behavior, but what is noteworthy is the growing tendency for investigators dealing with the hormonal control of behavior to undertake the study of the endocrinological bases for various distinctively masculine and feminine behavioral characteristics other than those directly and immediately involved in reproduction. Results of this movement differ in important ways from the intuitive classifications of " sexual behavior, " which have led to confusion rather than clarification of issues concerning behavioral effects of hormones.

Growing recognition of the importance of sex-related behavior has been influenced by two lines of research. One line, proceeding principally from developmental studies, has involved demonstration of clear and stable sex differences in social behavior of sexually immature as well as adult animals. For example, Harlow (1965) has convincingly demonstrated that the behavior of juvenile male rhesus monkeys toward their peers differs in several respects from that of females in the same group. These differences appear in young animals reared separately from adults and plainly represent biologically rather than " culturally " determined tendencies.

III. Early Sexual Differentiation Affecting Behavior

A second line of inquiry leading to recognition of the importance of sexually dimorphic behavior has derived from studies of the contribution of certain hormones to sexual differentiation. The first experiments in this area dealt with factors influencing development or lack of development of the Wolffian and Müllerian systems and their derivatives. Lilly's classic description and analysis of the free-martin (1917), and early experiments on sexual differentiation in amphibians showed that testicular secretion contributes to growth of the Wolffian system and tends to inhibit development of the Müllerian derivatives.

Dantchakoff (1938) placed testosterone in the amniotic cavity of fetal guinea pigs at about 27 days of gestational age and noted that after birth the female offspring not only showed anatomical masculinization but also displayed male-like mounting activity (a pattern of behavior which, unknown to Dantchakoff, occurs spontaneously in normal females of this species). Even before this, some psychologists had suggested that "the gonads are absolutely necessary for completion of the developmental processes underlying overt expression of the sexual libido" (Stone, 1932), or that the sexual pattern may be "laid down" under the influence of gonadal hormones secreted before puberty (Ball, 1937).

Much later it was discovered that the cyclic gonadotropic activity of the female pituitary depends upon hypothalamic control and that the functional characteristics of the male and female hypothalamus are different. If a fetal or neonatal female is treated with androgen her hypothalamus is permanently "masculinized", in the sense that during adulthood it will stimulate steady secretion of pituitary gonadotropins but will not elicit cyclic function leading to ovulation.

The relevance to behavior of such experiments on sexual differentiation in the reproductive system and the brain lies in the fact that experimentally induced modifications of normal morphological and physiological development are accompanied by alterations in certain forms of sex-related behavior. Behavioral masculinization of female guinea pigs exposed to androgenic stimulation during the fetal period was, as already noted, reported by Dantchakoff as early as 1938, but interpretation of her findings is difficult because they do not allow for the fact that normal females of this species frequently mount conspecifics in a manner quite similar to that of the copulating male. In the late 1950s Young and his associates initiated a program of research that soon was taken up in other laboratories and which has led to important new insights concerning the ways in which gonadal hormones affect mating and other forms of sexually dimorphic behavior.

It is now generally believed that certain hormones act upon the developing brain to produce changes that have enduring effects upon behavior which will appear later in life. Young (1965) refers to these changes of the central nervous system as "organizational" effects induced by the hormones. At present the nature of the presumed modification in the brain is unknown, but the behavioral alterations produced by early hormone treatment seem incontrovertible.

These effects have been demonstrated in guinea pigs, rats, and to a certain degree in rhesus monkeys. If pregnant guinea pigs are injected with testosterone propionate during the second trimester, the female offspring are modified in several ways (Phoenix *et al.*, 1959). They possess recognizable ovaries and uteri, but the external genitalia are so masculinized that they cannot be distinguished from those of genetic males. When the pseudohermaphroditic females are ovariectomized and tested for their responsiveness to exogenous ovarian hormones, it becomes clear that they are refractory to the treatment that reliably induces sexual receptivity in spayed females. Injection of estrogen followed by progesterone regularly produces behavioral estrus in normal females after they have been ovariectomized, but when the same hormones are administered to masculinized females these animals show a lower percentage of tests positive for estrus, a shorter duration of heat and maximum lordosis plus an increase in the frequency of male-like mounting activity. The reduced behavioral responsiveness to ovarian homones appears to be permanent.

Female guinea pigs that have been exposed to androgenic stimulation during fetal life are more responsive than normal females to the behavioral effects of androgen administered in adulthood. Pseudohermaphrodites injected with testosterone propionate exhibit more frequent, intense, and complete male-like copulatory responses than can be elicited in normal females by the same treatment.

Female rats also are sensitive to the effects of early androgen treatment, although in this species the period of sensitivity extends beyond the prenatal limit (Goy *et al.*, 1962). Females injected with androgen during the first few days of life show hypertrophy of the clitoris. Females exposed to androgen both before and immediately after birth possess an essentially male phallus. Neonatally androgenized females do not show feminine copulatory responses in adulthood and are quite refractory to treatment that induces sexual receptivity in ovariectomized females. Masculinized female rats react to androgen injections in adulthood with much more frequent mounting behavior than can be elicited from normal females given similar treatment. If the pseudohermaphroditic females have been exposed to testosterone propionate in both the fetal and the neonatal phases of development, their pattern of masculine coital behavior in adulthood (when given exogenous androgen) is indistinguishable from that of a genetic male (Ward, 1969).

Male rats castrated in adulthood and injected with the combination of ovarian homones that induces receptivity in spayed females exhibit most of the elements of the estrous female rat's mating pattern, but these responses are difficult to elicit, occur infrequently, and are sluggishly performed (Beach, unpublished observations). In striking contrast is the coital behavior shown by males that were castrated at birth. When adult animals of this type are injected with the same dosages of estrogen and progesterone, they display the full complement of feminine mating reactions. To the observer their behavior seems identical to that of a genetic female in estrus, and to a normal male rat it is sufficiently similar to elicit repeated attempts to mate (Beach *et al.*, 1969).

Neumann and Elger (1965, 1966) administered the antiandrogen cyproterone acetate to pregnant female rats and found that the male offspring had feminized genitalia and showed mating behavior of the female type. Furthermore, the hypothalamus was "feminized" so that it stimulated the pituitary to cyclic secretion of gonadotropic hormones.

If male rats castrated at birth are given a single injection of testosterone propionate within the next two or three days, their behavioral responsiveness to ovarian hormones is as limited as that of adult castrates. However, if androgen treatment of neonatally castrated males is delayed until the tenth day of life, the capacity for feminine behavior in adulthood is not reduced (Beach *et al.*, 1969).

In summary, experiments on both sexes indicate that the presence of androgen in the rat during the first few days of life seriously interferes with functional development of mechanisms essential to the display of normal female copulatory reactions in adulthood.

Does androgen have any effect upon development of mechanisms for masculine behavior? When pregnant guinea pigs are treated with testosterone propionate, the exogenous androgen does not produce any striking morphological or behavioral effects on the male offspring. However, testicular hormone seems to play a very important role in differentiation or organization of the male system. Several experiments have yielded results suggesting that development of the neural basis for masculine copulatory behavior in male rats depends upon the effects of endogenous androgen secreted during the neonatal period. Males castrated at birth and injected with testosterone propionate in adulthood react to estrous females by mounting them vigorously and frequently but without displaying the ejaculatory response. The absence of this element in the normal copulatory pattern could indicate some deficiency in the central nervous system, but some investigators question this interpretation. They point to the fact that neonatal castrates given androgen as adults still possess a very small penis. This deficiency prevents the achievement of insertion during mounting and thus deprives a male of the stimuli that normally evoke ejeculation. According to this point of view, although androgen inhibits normal development or organization of the neural basis for feminine mating responses, it is not necessary for organization of brain mechanisms that mediate the masculine copulatory pattern. Results of several independent experiments appear to support this view, but for the present the solution of the dispute awaits more definitive evidence.

Long-term experiments on rhesus monkeys, which are currently in progress at the Oregon Primate Center, already have shown that females exposed to androgenic stimulation in fetal life exhibit anatomical and behavioral masculinization. At birth, such females possess a male-like phallus and scrotum and lack an external vagina. As juveniles the androgen-treated females exhibit social behavior that is intermediate between that of normal males and females or which tends to overlap with that of genetic males. These animals have not reached the age of sexual maturity, but even before puberty the pseudohermaphrodites mount their playmates much more frequently than do normal females of the same age. Male monkeys castrated shortly after birth display normal masculine behavior during the prepuberal periods.

Commenting on the work in progress at the Oregon Primate Center, Goy (1968) has written as follows: " in the rhesus monkey, as in certain rodents, the importance of early experience for the expression of normal patterns of sexual and sex-related behavior is well documented. Viewed in this context, the results from experiments with prenatal hormones suggest that the contributions made by experience act upon a substrate that is already biased either in a masculine or a feminine direction. As a result of some action of androgen on the developing nervous system, the individual whether genetically male or female is predisposed to the acquisition and expression of behaviors which normally characterize the genetic male ".

IV. Human Psychosexual Differentiation

Results of experiments on female rodents and monkeys have attracted the attention of clinicians who deal with girls and women showing different degrees of morphological masculinization. This sometimes results from administration of progestagens to the mother during pregnancy to prevent abortion. In other instances, masculinization is due to congenital hyperadrenalcorticalism which involves secretion of large amounts of androgen by the fetal adrenocortex. In contrast to patients of these types are individuals suffering from Turner's syndrome in which no gonads are present. If humans resembled other mammals, it would be expected that females exposed to androgenic stimulation during development would show more masculine behavioral traits than normal females, whereas the behavior of " agonadal " females would be completely feminine; according to Money and Ehrhardt (1968) this may actually be the case: " . . . girls affected by androgen *in utero* and diagnosed with either progesterone-induced hermaphroditism or hyperadrenocortical hermaphroditism showed more of a developmental tendency toward tomboyish behavior than girls with Turner's syndrome ".

In interpreting their findings, the same authors reach the following conclusions: " The present data suggest that there may well be a fetal hormonal effect on subsequent psychosexual differentiation. If so, it is limited in scope and does not induce anything approaching a complete psychosexual reversal in the genetic female. . . . Psychosexual differentiation in the human must, evidently, be understood as a composite or end result of various component factors, and not as a single or global entity ".

V. "Activational" Effects of Gonadal Hormones

A different approach to the study of endocrine control of organismic activities is represented by investigations into more immediate and transient effects of a given hormone or combination of hormones upon a particular pattern of behavior. In his discussions of the effects of gonadal hormones, Young (1965) dealt dichotomously with what he termed "organizational" versus "activational" effects. In the view of the present writer, organizational effects are characterized by the fact that they can occur only during certain fairly limited periods of development and that their consequences are irreversible. There is the additional characteristic that some behavioral consequences of organizational effects may not be manifested for a long time after the period of hormonal action.

Activational effects of gonadal hormones are reflected in the behavioral changes that follow orchidectomy or ovariectomy in adulthood and those produced by subsequent replacement therapy. Many animal experiments and some clinical studies dealing with both types of changes have been summarized and evaluated in various reviews, and some of these serve as reminders of the ever-present need for precise definitions of behavior. One author may conclude that neonatally castrated male rats exhibit "normal" copulatory behavior when treated with androgen in adulthood and another writer may conclude that the same experiment reveals a deficiency in coital performance. The contradictory interpretations are based upon different definitions. The first reviewer emphasizes the fact that the animals in question mount receptive females as often or even more often than do intact males, whereas the second stresses the fact that despite their frequent and vigorous mounting responses, neonatal castrates almost never ejaculate.

Any general discussion of hormonal effects upon behavior (or, more specifically, of effects of gonadal hormones on copulatory behavior) must include a phylogenetic dimension. That is to say it must take equal account of both differences and similarities between species. Generalizations concerning effects of castration on sexual behavior are sure to be invalid unless (1) sexual behavior is given a clear and objective definition and (2) the conclusions are restricted to one species or the nature and magnitude of species differences are explicitly stated.

A. Dependence of Male Behavior on Testicular Hormone

Castration of adult male rats is followed by a predictable series of changes in their coital reactions to estrous females. The first item in the behavior pattern to disappear is the ejaculatory response, which consists of considerably more than simple seminal emission (Beach *et al.*, 1966). This loss occurs within four weeks, according to most investigators, although postoperative survival for several months in some individuals has been described (Davidson, 1966a). Coincident with or relatively soon after disappearance of the ejaculatory response, castrated rats cease achieving intromission when they mount the receptive female. This change has been taken to signify the loss of erectile capacity. Mounting without insertion is the last response to deteriorate. It becomes progressively less frequent and usually reaches a baseline rate, which is very low but not necessarily zero, within a month or two after removal of the testes.

This general description of postcastrational changes in copulatory behavior is applicable, with minor modifications, to several species of rodents such as the guinea pig, hamster, and domestic mouse as well as to the male rabbit.

For every species of rodent in which the problem has been investigated, administration of exogenous androgen in sufficiently large amounts has induced reversal of the behavioral changes that follow castration in adulthood. It is important to add that the level of sexual performance does not depend simply and directly upon the amount of androgen given to castrated males. Quantitative measures of preoperative behavior in any species invariably reveal marked individual differences in the frequency and vigor of mating reactions. When guinea pigs or rats are castrated, coital behavior eventually declines to a very low level and individual differences are almost completely eliminated. At this point administration of equal amounts of testosterone propionate to all males evokes reappearance of the normal mating pattern, and the individual differences observed prior to operation reappear despite the fact that males are receiving identical hormone treatment (Grunt and Young, 1952; Beach and Fowler, 1959).

The effects of gonadectomy on male carnivores are in some respects similar to those seen in rodents, but there are important differences.

Some male cats with preoperative sexual experience are capable of intromission for at least three months after castration (Rosenblatt and Aronson, 1958). Mounting without insertion survives even longer, and in some individuals may do so indefinitely.

In the writer's laboratory sexually experienced male dogs castrated as adults have been shown to continue copulating with estrous bitches for as long as five years after operation (Beach, 1970). Tendencies to mount the female and execute copulatory thrusts are not influenced by loss of the testes. However, there almost always is a distinct postoperative deterioration in potency as reflected in the achievement and maintenance of full penile tumescence. This capacity is not lost but is definitely reduced within six months or less after the operation and remains at the new level for several years thenceforth. Adrenalectomy of castrated males does not eliminate the sexual performance that survives gonadectomy.

In both cats and dogs the change produced in coital behavior can be reversed by treatment with exogenous androgen. For dogs, such restorative therapy has been shown to be effective when applied more than four years after gonadectomy (Beach, 1970). No systematic studies of pre-and post-castrational behavior in adult male primates have been published, but the scattered reports that are available led Phoenix and associates (1967) to the following conclusion: " The effects of castration in the adult monkey and chimpanzee are unclear, but available evidence suggests that sexual behavior tends to persist for longer periods than those characteristic of nonprimate species ".

An important source of evidence concerning the importance of testosterone in the sexual activities of infrahuman male primates is provided by observations on male monkeys castrated before puberty. Goy (1964) describes the behavior of four rhesus males gonadectomized at three months of age and observed for the next nine months, during which time they showed normal sexual development. " The capacity for erection was maintained, and characteristic masculine patterns continued to develop normally in the complete absence of testicular androgen. " Four years later the same individuals were still displaying normal mounting and copulatory patterns typical of intact males of the same age (Phoenix *et al.*, 1967).

Comparisons among a few species of rodents, two carnivores, and one type of monkey scarcely provide a substantial basis for sweeping conclusions about phyletic differences and similarities. Nevertheless it is thought-provoking to observe that there is within this small series some indication of a progressive increase in the extent to which copulatory performance can survive in the absence of testosterone.

B. Dependence of Female Behavior on Ovarian Hormones

The next question deals with the influences of ovarian hormones on coital behavior of female mammals. Two kinds of evidence are of primary importance. The first deals with relationships between cycles of ovarian activity and rhythms of sexual receptivity. The second comes from studies of the effects of ovariectomy and subsequent replacement therapy.

Ever since the discovery of vaginal cycles in polyestrous rodents more than 40 years ago, there has been general recognition of a systematic relationship between changes in vaginal cytology and sexual responsiveness in female rats and guinea pigs. In general the condition of " vaginal estrus " (smears containing large amounts of cornified epithelium and few or no leukocytes) is associated with readiness to mate. There are individual exceptions to this general rule and they are of the utmost theoretical importance. In every species that has been carefully investigated a few females have been found who showed normal vaginal cycles but never became receptive to the male. Even for the majority of individuals that do exhibit both vaginal and behavioral cycles, individual differences exist. For example, some rats regularly become receptive in the proestrous stage whereas in others the onset of receptivity just as regularly occurs during vaginal estrus.

Female carnivores as exemplified by the domestic cat and dog are similar to female rodents in that sexual receptivity is totally dependent upon the support of ovarian hormones. In both species a female will cooperate with males in copulation only when she is physiologically in estrus. There is, however, an important difference between the two types of mammals. Among rodents the ovarian hormones usually are both necessary and sufficient for the display of sexual receptivity. In cats and dogs, on the other hand, although necessary, the hormones may not be sufficient to produce receptivity in every instance. This is shown quite clearly in the mating preferences exhibited by many female dogs (Le Boeuf, 1967).

Although she is in full physiological estrus, a particular bitch may respond to certain males by rejecting or even attacking them and then promptly and enthusiastically receive the copulatory attempts of other masculine partners. Individual preferences of this type are remarkably stable and have been observed to persist at least as long as five years (Beach and Le Boeuf, 1967). Qualitatively similar selectivity is said to be characteristic of some female cats in estrus, and the indicated conclusion would seem to be that physiological estrus and sexual receptivity in female carnivores are not identical. Extra-hormonal factors evidently play a role in determining the receptivity of a given female for a particular male.

In all rodents and carnivores that have been examined, removal of the ovaries results in prompt and permanent elimination of sexually receptive behavior. Mating performance can be revived by administration of exogenous ovarian hormones. Estrogen given in large amounts usually will induce copulatory responses, but for several species this priming with estrogen followed by one injection of progesterone produces higher degrees of receptivity than estrogen alone. The effectiveness of adding progesterone has been demonstrated in studies of guinea pigs, rats, hamsters, mice (Beach, 1948), and most recently, of dogs (Beach and Merari, 1968).

Among infrahuman primates, only catarrhine monkeys and apes have a true menstrual cycle. In the rhesus monkey the cycle is about 28 days and in the chimpanzee slightly longer. Laboratory studies have revealed systematic relationships between the female's cycle and rhythms of various types of heterosexual interaction (Michael, 1968). Average trends based on study of numerous pairs show that the frequency with which males mount females tends to increase at the approach of mid-cycle (ovulation) and to decrease during the luteal phase. When attention is directed to the behavior of individual pairs one is impressed with the differences that emerge.

In some pairs male mounting is highest near mid-cycle, declines sharply early in the luteal phase, and shows a secondary increase just before menstruation. A second common pattern involves high levels of mounting throughout the follicular phase, followed early in the luteal phase by a low level that persists until menstruation. Finally there are some pairs in which a low level of male mounting persists throughout the menstrual cycle.

The importance of individual preferences among female dogs and cats has been mentioned, and it appears that even greater differences characterize sexual relations in monkeys.

" Very striking individual variations and partner preferences exist which are not encountered to anything like the same extent in lower mammals. The absolute levels of activity of different pairs vary considerably: some animals show little interest in each other, the sexual interaction being minimal, yet when assigned a different partner the opposite is the case. It is a waste of time to persist with observation on pairs in which aggression has occurred or which are sexually ill-assorted, therefore our subjects are preselected for good levels of interaction; even so, clear-cut rhythms are observed in less than 50 per cent of cases and are not necessarily consistently present in a given pair in successive cycles. The dominant part played by individual differences and partner preferences is reminiscent of the human situation, and the three patterns of sexual interaction in the macaque also reminds one of the differences reported in the fluctuations in human sexual activity with the menstrual cycle " (Michael, 1968). As might be expected, gonadectomizing the female has different effects upon the behavior of different pairs of monkeys. In all pairs described by Michael (1968) there was a marked decrease in the male's mounting responses. Some males continued to mount at low frequencies whereas others ceased entirely. Injections of estradiol to ovariectomized females were followed by a prompt and pronounced rise in the number of mounts and by reappearance of the ejaculatory reaction, which was not shown in response to untreated female castrates. When estrogen-treated females were given progesterone, mounting and ejaculation by their male partners was suppressed.

Michael (1968) differentiates between the sexual " drive " of the female monkey, expressed by the frequency of her sexual presentations to the male, and her value as a stimulus, expressed by the proportion of her presentations that stimulate the male to mount. It appears that progesterone decreases both the female's drive and her attractiveness to the male. Probably the decline in male mounting of intact females early in the luteal phase is due to progestational activity. On the other hand, estrogen increases both feminine drive and excitatory value, although the two effects may well be mediated by different mechanisms.

VI. Mechanisms of Hormonal Action

Up to this point we have reviewed a very small part of the evidence pertaining to relationships between gonadal hormones and copulatory behavior in a few species of mammals. Limited as it necessarily has been, this survey suffices to illustrate in a general way the kinds of problems that have been attacked and the sorts of evidence currently available. Correlational studies like those described have raised many questions concerning the intervening mechanisms or processes by which hormones bring about changes in behavior. We cannot review the relevant evidence *in extenso*, but description of a few representative experiments will indicate how the problem is being approached and will provide at least a skeletal representation of the current state of knowledge in this area.

The earliest investigations of relationships between hormones and behavior led to the assumption that hormones produced their behavioral effects by inducing changes in the central nervous system; this undoubtedly is true, but other target organs may also be involved. A very clear example, which has already been mentioned, is androgenic control of development of the mammalian penis. If testosterone is lacking during a critical ontogenetic period, this organ cannot be stimulated to full growth no matter how much androgen is supplied in adulthood. As a consequence, mating behavior is never completely normal and the most apparent deficit is the absence of ejaculation *in copulo*.

A. Sensory Reception

Gonadal hormones exert two kinds of effects upon sensory cues involved in behavior. They change the stimulus qualities of animals and change responsiveness to certain forms of stimulation. As one very obvious example, ovarian hormones modify the chemical stimuli emanating from female mammals and testicular hormones alter the male's reactivity to such stimulation. Male dogs spend more time smelling urine taken (by catheter) from females in heat than urine from anestrous females (Beach and Gilmore, 1949). Beach and Merari (1968) compared the reactions of male dogs to small balls of cotton that had been inserted into the vaginae of spayed bitches before and after different kinds of hormone treatment. Cotton that had been in contact with the vulva of an estrogen-treated female was smelled for longer periods of time than cotton from an untreated, spayed bitch. Males reacted even more strongly to stimulus material from a female that had been injected first with estrogen and then with progesterone. The added stimulus value induced by progesterone became apparent within 4 hr after the hormone had been injected. Prepuberally castrated males did not discriminate between the three types of stimulus material.

The female monkey's hormonally induced sexual attractiveness to the male almost certainly involves chemical stimulation from the vagina (Michael, 1968). Some ovariectomized females were given estrogen by subcutaneous injection and others received injections of the same hormone directly into the vagina. Intravaginal administration resulted in a significantly greater increase in male mounting attempts than did subcutaneous injection, in spite of the fact that vaginally treated females were not receptive and did not present, whereas subcutaneously injected females showed an increase in presentation frequency and allowed males to copulate and ejaculate. The conclusion that estrogen-dependent changes in the vagina provide new olfactory cues to the male is supported by the demonstration that induction of temporary anosmia in males eliminated any change in behavior shown toward females when the latter were injected with enough estrogen to induce the vaginal condition that evokes frequent attempts to mount in normal males.

Normal male rats display a significant preference for the odor of estrous *vs.* that of diestrous females. Castrates ordinarily show no preference but will do so when injected with testosterone propionate (Le Magnen, 1952). Androgen clearly affects preference and not olfactory sensitivity because untreated, prepuberally castrated males will learn the discrimination if rewarded for doing so (Carr and Caul, 1962).

The problem of determining what kinds of changes hormones induce in the nervous system can be approached only when it is known where in the central nervous system those changes take place. Questions about functional localization are therefore of paramount importance.

B. Spinal Cord

Certain responses that normally constitute subunits of the total copulatory pattern of many mammalian species are known to consist of relatively simple spinal reflexes. For example, this is true of erection and ejaculation in male mammals. These reflexes survive after complete transection of the spinal cord in the thoracic region. There is evidence to show that the spinal mechanisms involved are sensitive to androgen.

For some time after spinal transection, paraplegic dogs react to genital massage with erection and ejaculatory reflexes (Hart, 1967b). However, the spinal lesion eventually results in testicular atrophy and this is associated with weakening of the genital reflexes. After this has occurred the reflexes can be restored to their original strength by systemic administration of testosterone propionate (Hart, 1968). Chronic spinal male rats also show penile reflexes in response to local stimulation, and these reactions are reduced by castration. They recover their normal strength if crystals of testosterone are inserted in the spinal canal caudal to the level of section (Hart and Haugen, 1968). The effect appears to be restricted to the cord because treated males give no evidence of systemic responses to androgen.

Female dogs in heat respond to unilateral tactile stimulation of the vulva by moving the hindquarters laterally toward the stimulated side. This clearly is a spinal reflex because it can be elicited in the bitch after bilateral cord transection. However, evocation of the response is possible only if the spinal female is treated with estrogen, which suggests that this hormone acts directly upon the spinal circuits involved (Hart, 1969).

C. Brain

Several methods have been used separately or in combination to identify the hormone-sensitive brain regions involved in sexual activities. The principal ones have been infliction of restricted lesions, recording from implanted electrodes, and direct application of hormones to selected brain areas.

Experiments using each of these techniques have been numerous in recent years and several excellent reviews are available (e.g., Sawyer *et al.*, 1966; Lisk, 1967a). In this presentation only a few representative studies in each category will be mentioned.

Much of the early work dealing with effects of brain lesions on mating behavior involved fairly massive ablations of cerebral cortical tissue, and although the results were of interest and significance they provided no insight into possible effects of hormones on behavior (for review see Beach, 1947). Exceptions were the report by Brookhart and Dey (1941) to the effect that lesions in the anterior hypothalamus were followed by loss or reduction of mating behavior in male guinea pigs, and the report by Bard (1940) that mating was eliminated in female cats by surgical invasion of the hypothalamus.

Subsequent investigations have confirmed the importance of the hypothalamus and adjacent regions to the mediation of hormonally dependent coital activities. Species differences are important and results from different laboratories have not always been in total agreement. Nevertheless, evidence has gradually accumulated to indicate that the preoptic-anterior hypothalamic area of the brain contains hormone-sensitive mechanisms that are essential for normal mating behavior. Medial preoptic lesions tend to eliminate copulation in male rats (Soulairac, 1963), although the secretion of testicular hormone appears to be undisturbed (Heimer and Larsson, 1964). Lesions of the ventromedial hypothalamus greatly reduce mating activity in male guinea pigs (Phoenix, 1961).

Female cats with injury to the anterior hypothalamus do not come into estrus and will not mate despite administration of exogenous estrogen (Sawyer and Robison, 1956). Other experiments on females have been summarized by Lisk (1967a) as follows: " . . . The lesion data for female mammals is consistent with the theory of dual hormone-sensitive systems within the hypothalamus. One system situated in the median eminence region appears to be the regulator of the pituitary-gonad axis. After its destruction, behavioral estrus is still possible, provided exogenous hormone is used. A second region, apparently hormone-sensitive, appears essential for the lordosis response in the female. This has been localized to the preoptic region and anterior hypothalamus for the cat, rat, guinea pig, the hamster, and ewe. Only the rabbit among the mammals so far investigated has a behavior-regulating region in the posterior hypothalamus ".

The frequency and patterning of electrically recordable activity in various parts of the brain are known to be influenced by certain gonadal steroids. For example, Sawyer and Kawakami (1959) noted in female rabbits a unique electroencephalographic (EEG) pattern which they termed an " afterreaction " because it followed coitus or artificial vaginal stimulation. This pattern, which involves changes in the hippocampus, limbic cortex, and frontal cortex, can be evoked only when the female has been injected with estrogen. Sawyer (1967) has also described an " EEG arousal threshold, " which involves " high frequency stimulation of the brain stem reticular formation including the posterior hypothalamus " and which is lowered by estrogen.

EEG recordings from the anterior lateral hypothalamus of the estrous cat show that vaginal stimulation is followed by several minutes of high-amplitude activity. This response can be obtained only when the female is in natural estrus or has been injected with estrogen (Porter *et al.*, 1957). Heightened activity in the anterior hypothalamus also occurs in the estrous rabbit during copulation (Green, 1954). Stimulation of the cervix produces EEG increases in the preoptic area of the female rat, provided she is in estrus at the time (Barraclough, 1960).

The technique of applying hormones directly to restricted areas of the brain is yet another method of localizing the sensitive regions. In general the results obtained are in agreement with the lesion data in implicating the anterior hypothalamus and adjacent tissues. Very small implants of estradiol benzoate placed in the lateral anterior hypothalamus-medial forebrain bundle area produce behavioral estrus in spayed female cats, whereas similar implants elsewhere in the hypothalamus are totally ineffective (Sawyer, 1963).

Also, in the female rat, estradiol implants to the anterior hypothalamus or the preoptic area will induce behavioral estrus (Lisk ,1962). A very important point to be made with regard to these studies of intracranial implantation is that the the vaginae of the treated animals remained in the anestrous state, which indicates that the hormone was not acting systemically but was directly stimulating neural tissue to produce behavioral changes.

Several experiments have shown that copulatory behavior can be induced in castrated male animals by androgenic stimulation of the brain. Using testosterone implants in male rats, Davidson (1966b) found that the most effective locus was in the medial preoptic area. Confirmatory results were reported by Lisk (1967b).

One more technique for identification of hormone-sensitive areas or cells within the brain involves systemic injection of radioactive hormone and subsequent autoradiography of tissue from different brain regions. Using this method with female cats, Michael (1965) has found that, following the administration of ^{3}H-hexestrol (an estrogen), highest concentrations of radioactivity are present in the anterior lateral hypothalamus.

If it is agreed that the past decade has been marked by significant advances in our knowledge of loci of hormone action within the brain, it must at the same time be admitted that very little is known concerning the ways in which gonadal steroids affect nerve cells and thus eventually produce changes in behavior. This represents perhaps the most pressing of the remaining problems in the hormone-behavior area.

It may yield to frontal attack, but at present the most promising approach would seem to be through the study of hormone action on nonneural tissue. For example, there are several advantages to investigation of the reactions of estrogen-sensitive cells in the uterus. Research already accomplished has shown that the physiological effects of estrogen in the uterus are mediated by increases in the synthesis of RNA and protein. It appears further that the early action of estrogen involves: first, binding of the hormone to nuclear chromatin; second, stimulation of RNA synthesis within the nucleus; and eventually transport of messenger RNA to the cytoplasm with the resultant modification of cytoplasmic properties for incorporation of amino acids and accompanying effects upon cytoplasmic protein. There are presently available several reasonable models and theories of estrogenic action (Hamilton, 1968), some of which may prove applicable to reactions of neurons to the hormone and eventually help to explain how hormones exert their effects upon behavior.

References

Ball, J. (1937). *J. Comp. Psychol.* **24**, 135.

Bard, P. (1940). *Res. Publ. Assoc. Res. Nervous Mental Disease* **20** 551.

Barraclough, C. A. (1960). *Anat Record* **136**, 159.

Beach, F. A. (1947). *Physiol. Rev.* **27**, 240.

Beach, F. A. (1948). "Hormones and Behavior." Hoeber, New York.

Beach, F. A. (1970). *J. Comp. Physiol. Psychol.* **70**, *Part.* **2**, 1.

Beach, F. A., and Fowler, H. (1959). *J Comp. Physiol. Psychol.* **52**, 50.

Beach, F. A., and Gilmore, R. W. (1949). *J. Mammal.* **30**, 391.

Beach, F. A., and Le Boeuf, B. J. (1967). *Anim. Behav.* **15**, 546.

Beach, F. A., and Merari, A. (1968). *Proc. Nat. Acad. Sci. U.S.* **61**, 442.

Beach, F. A., Westbrook, W. H., and Clemens, L. G. (1966). *Psychosomatic Med.* **28**, 749.

Beach, F. A., Noble, R. G., and Orndoff, R. K. (1969). *J. Comp. Physiol. Psychol.* **68**, 490.

Brookhart, J. M., and Dey, F. L. (1941). *Am. J. Physiol.* **133**, 551.

Carr, W. J., and Caul, W. F. (1962). *Animal Behav.* **10**, 20.

Dantchakoff, V. (1938). *C. R. Soc. Biol.* **127**, 1255.

Davidson, J. M. (1966a). *Animal Behav.* **14**, 266.

Davidson, J. M. (1966b). *Endocrinology* **79**, 783.

Deutsch, H. (1945). " Psychology of Women." 2 vols. Grune and Stratton, New York.

Goy, R. W. (1964). *In* " Human Reproduction and Sexual Behavior " (C. W. Lloyd, ed.), pp. 409–441. Lea and Febiger, Philadelphia.

Goy, R. W. (1968). *In* " Endocrinology and Human Behavior " (R. P. Michael, ed.), pp. 12–31. Oxford University Press, London.

Goy, R. W., Phoenix, C. H., and Young, W. C. (1962). *Anat. Record* **142,** 307.

Green, J. D. (1954). *Anat. Record* **118**, 304.

Grunt, J. A., and Young, W. C. (1952). *Endocrinology* **51**, 237.

Hamilton, T. H. (1968). *Science* **161**, 649.

Hampson, J. L. (1965). *In* " Sex and Behavior " (F. A. Beach, ed.), pp. 108–132. J. Wiley, New York.

Harlow, H. F. (1965). *In* " Sex and Behavior " (F. A. Beach, ed.), pp. 234–265. J. Wiley, New York.

Hart, B. L. (1967a). *Science* **155**, 1283.

Hart, B. L. (1967b). *J. Comp. Physiol. Psychol.* **64**, 388.

Hart, B. L. (1968). *J. Comp. Physiol. Psychol.* **66**, 726.

Hart, B. L. (1969). *Hormones Behav.* **1**, 65.

Hart, B. L., and Haugen, C. M. (1968). *Physiol. Behav.* **3**, 735.

Heimer, L., and Larsson, K. (1964). *Acta Neurol. Scand.* **40**, 353.

Le Boeuf, B. J. (1967). *Behavior* **29**, 268.

Le Magnen, J. (1952). *Arch. Sci. Physiol.* **6**, 295.

Lilly, F. R. (1917). *J. Exp. Zool.* **23**, 371.

Lisk, R. D. (1962). *Am. J. Physiol.* **203**, 493.

Lisk, R. D. (1967a). *In* " Neuroendocrinology " (L. Martini and W. F. Ganong, eds.), pp. 197–239. Academic Press, New York.

Lisk, R. D. (1967b). *Endocrinology* **80**, 754.

Michael, R. P. (1965). *In* "Hormonal Steroids" (L. Martini and A. Pecile, eds.). Vol. **II**, pp. 469–481. Academic Press, New York.

Michael, R. P. (1968). *In* " Endocrinology and Human Behaviour " (R. P. Michael, ed.), pp. 67–93. Oxford University Press, London.

Money, J., and Ehrhardt, A. P. (1968). *In* " Endocrinology and Human Behaviour " (R. P. Michael, ed.), pp. 32–48. Oxford University Press, London.

Neumann, F., and Elger, W. (1965). *Acta Endocrinol. Suppl.* **100**, 174.

Neumann, F., and Elger, W. (1966). *Endokrinologie* **50**, 209.

Phoenix, C. H. (1961). *J. Comp. Physiol. Psychol.* **54**, 72.

Phoenix, C. H., Goy, R. W., Gerall, A. A., and Young, W. C. (1959). *Endocrinology* **65**, 369.

Phoenix, C. H., Goy, R. W., and Young, W. C. (1967). *In* " Neuroendocrinology " (L. Martini and W. F. Ganong, eds.), pp. 163–196. Academic Press, New York.

Porter, R. W., Cavanaugh, E. B., Critchlow, B. V., and Sawyer, C. H. (1957). *Am. J. Physiol.* **189**, 145.

Rosenblatt, J. S., and Aronson, L. R. (1958). *Behaviour* **12**, 285.

Sawyer, C. H. (1963). *Anat. Record* **145**, 280.

Sawyer, C. H. (1967). *In* " Hormonal Steroids " (L. Martini, F. Fraschini and M. Motta, eds.), pp. 123–135. Excerpta Medica, Amsterdam.

Sawyer, C. H., and Kawakami, M. (1959). *Endocrinology* **65**, 622.
Sawyer, C. H., and Robison, B. (1956). *J. Clin. Endocrinol. Metab.* **16**, 914.
Sawyer, C. H., Kawakami M., and Kanematsu, S. (1966). *Res. Publ. Assoc. Res. Nervous Mental Disease* **43**, 59.
Soulairac, M. L. (1963). *Ann. Endocrinol. Suppl.* **24**, 1.
Stone, C. P. (1932). *J. Genet. Psychol.* **40**, 296.
Ward, I. L. (1969). *Hormones Behav.* **1**, 25.
Young, W. C. (1965). *In* " Sex and Behaviour " (F. A. Beach, ed.), pp. 89–107. Wiley, New York.

Pituitary Control of Avoidance Behavior

D. De Wied

I. Introduction

The pituitary-adrenal system plays an essential role in defence against or reaction to noxious stimuli. Neurogenic and psychic stresses induce the discharge of Adrenocorticotropic Hormone (ACTH) from the anterior pituitary and the subsequent secretion of glucocorticosteroids from the adrenal cortex. These secretions may in turn influence the Central Nervous System (CNS). Reports of mental aberrations and sometimes psychosis in patients on adrenocortical therapy have appeared, and electroencephalographic alterations, convulsions, and mental changes have been found in patients during hyper- as well as hypo-adrenocorticism. In fact both steroids (e.g., cortisol) and ACTH itself may alter brain excitability (Woodbury, 1954; Feldman *et al.*, 1961; Wasserman *et al.*, 1965).

The first demonstration of an effect of the pituitary-adrenal system on animal behavior was presented by Mirsky *et al.* (1953). These authors postulated that ACTH diminished the effectiveness of an anxiety-producing stimulus. Murphy and Miller (1955) found that ACTH delayed extinction of an avoidance response, and subsequently Miller and Ogawa (1962) showed that ACTH had a similar effect in adrenalectomized rats. The adrenals themselves apparently do not play a significant role in the maintenance of a conditioned avoidance response.

Other workers also found an effect of ACTH and adrenocortical hormones on spontaneous and conditioned behavior (Lissák and Endröczi, 1961; Bohus and Endröczi, 1965; Korányi *et al.*, 1966, 1967; Levine and Jones, 1965; Ferrari *et al.*, 1961, 1963; De Wied, 1966, 1967) (for detailed description of the literature, see De Wied, 1969).

II. Effect of Adrenocorticotropic Hormone and its Analogues on the Rate of Extinction of a Conditioned Avoidance Response

Our interest in the behavioral effect of ACTH originated from studies on the responsiveness of the adenohypophysis to emotional stress in posterior lobectomized rats (Smelik, 1960; De Wied, 1961). These animals had a behavioral deficiency, in that extinction of a conditioned avoidance response was markedly facilitated. Treatment of posterior lobectomized rats with vasopressin but also with ACTH normalized conditioned behavior of these animals (De Wied, 1965). Studies in intact rats revealed similar effects. Administration of ACTH as a long-acting zinc phosphate preparation during the extinction period delayed the rate of extinction of a shuttlebox avoidance response in a dose-dependent manner (De Wied, 1967). It was further found that pharmacological doses of ACTH (which induced hypercorticism) reduced the inhibitory effect of ACTH on extinction. It seemed as if the resultant high level of circulating corticosteroids had an effect on extinction of the avoidance response opposite to that of ACTH. Further proof of the fact that the inhibitory effect of ACTH was not mediated by the adrenal cortex was obtained when it was found that both α- and β-Melanocyte-Stimulating Hormone (MSH) inhibited extinction of the avoidance response in a way similar to that induced by ACTH (De Wied, 1966). In contrast, protamine zinc insulin had no effect.

In an attempt to determine the active part of the ACTH molecule responsible for the behavioral effect, a pole-jumping avoidance procedure was used in conjunction with a dosage of various ACTH analogues (De Wied, 1966). The amino acid sequence of a number of ACTH analogues is shown in Table I. It appeared that the peptide

TABLE I

Amino Acid Sequence of ACTH β 1-24, α-MSH, ACTH 1-10, and ACTH 4-10

		1	2	3	4	5	6	7	8	9	10	11	12	13	14			
ACTH β 1-24		Ser	Tyr	Ser	Met	Glu	His	Phe	Arg	Try	Gly	Lys	Pro	Val	Gly	Lys	Lys	Arg
												Arg	Pro	Val	Lys	Val	Tyr	Pro
α-MSH	CH_3CO –	Ser	Tyr	Ser	Met	Glu	His	Phe	Arg	Try	Gly	Lys	Pro	Val				
ACTH 1-10		Ser	Tyr	Ser	Met	Glu	His	Phe	Arg	Try	Gly							
ACTH 4-10					Met	Glu	His	Phe	Arg	Try	Gly							

ACTH 1-10 had an effect on extinction equipotent to that of MSH and ACTH β 1-24. The peptide ACTH 11-24 had no effect. Subsequent experiments showed that the heptapeptide ACTH 4-10 (which is common to ACTH, α- and β-MSH) was as active as its parent molecules in inhibiting the rate of extinction of the avoidance response (Greven and De Wied, 1967).

III. Opposite Effect of Two Decapeptides Derived from Adrenocorticotropic Hormone on the Rate of Extinction of a Conditioned Avoidance Response

During the studies of the shuttlebox behavior it was found that the sequence ACTH 1-10, in which the amino acid phenylalanine in the seventh position had been replaced by the D-isomer, had an opposite effect on extinction of the avoidance response (Bohus and De Wied, 1966). Administration of ACTH 1-10 in which phenylalanine had been replaced by the D-isomer (ACTH 1-10 (7D-phe) in the same dose as the L-form peptide facilitated the rate of extinction of a shuttlebox avoidance response, while the L-form peptide inhibited extinction. The effect of the D-form peptide could not be explained by a direct antagonistic action against naturally occurring hormones like ACTH and MSH, since it exhibited the same effect in hypophysectomized rats. The facilitatory effect of the D-form peptide on extinction of the avoidance response was not due to motor and/or sensory disfunction. Peptide administration for 14 days did not affect escape speed in a runway. Neither the D-form nor the L-form peptide affected gross behavior. The L-form or D-form decapeptide 18 hr after a single subcutaneous injection did not have a significant effect upon the rate of ambulation in an open field, rearing, grooming, or the number of fecal boluses produced during the 3-min period of observation. Later it was found that ACTH 4-10 (7-D-phe) was as potent as ACTH 1-10 (7-D-phe) in facilitating the rate of extinction of the shuttlebox avoidance response (De Wied and Greven, 1968). It is apparent, therefore that two closely related peptides are able to induce either inhibition or facilitation of extinction of a conditioned avoidance response. These two peptides differ only in the optical rotation of one amino acid and are devoid of gross behavioral effects or systemic influences. In this respect it may be worth mentioning that the D-form peptide affects conditioned avoidance behavior in the same way as do the major tranquilizers (e.g., chlorpromazine and other phenothiazine derivatives). The effect of the L-form peptide resembles that of amphetamine and related drugs that delay extinction of the avoidance response. The difference between the two peptides and these drugs is that the former accomplish their effect on conditioning without any other demonstrable effect on the organism whereas the tranquilizer drugs interfere with gross behavior, metabolism, temperature, and endocrine and autonomic nervous activity.

IV. Site of Behavioral Action of an Adrenocorticotropic Hormone Analogue in the Central Nervous System

CNS lesions were utilized in an attempt to localize the site of action of peptide analogues on extinction of the avoidance response (Bohus and De Wied, 1967a). Lesions were made in the posterior thalamic area; large lesions in this region produced severe deficits in avoidance acquisition and escape behavior. Smaller lesions also interfered with acquisition but not with escape behavior. Bilateral destruction of the nuclei parafascicularis did not affect avoidance acquisition, but extinction was markedly facilitated. Administration of long-acting α-MSH during the extinction period failed to delay the rate of extinction of the avoidance response in rats with lesions in the nuclei parafascicularis (Bohus and De Wied, 1967b). This is evidence that the CNS is involved in the behavioral effect of the ACTH-like peptides.

Cardo and Valade (1965) found that the impairment in avoidance behavior of rats with lesions in the nuclei parafascicularis could be partially restored by the administration of dexamphetamine. The failure of α-MSH to affect the rate of extinction in similar rats again suggests that the effect of ACTH-like peptides is of a more specific character than mere excitation of central nervous activity.

V. Effect of Adrenocortical Steroids on the Rate of Extinction of a Conditioned Avoidance Response

As mentioned above, adrenocortical hormones seem to have an effect opposite to that of ACTH and its analogues. In fact, these steroids appeared to facilitate extinction of the avoidance response (De Wied, 1967). Daily treatment of rats with corticosterone in various dose levels during the extinction period facilitates the rate of extinction of a shuttlebox avoidance response in a dose-dependent manner. Since daily treatment with glucocorticosteroids leads to inhibition of ACTH release, the results could be interpreted to indicate that facilitation of extinction is the result of blocking the ACTH release.

However, corticosterone also facilitates extinction of the avoidance response in hypophysectomized rats. This is in agreement with findings of Bohus (1968), who showed that cortisol implants in the mesencephalic reticular formation markedly facilitate extinction of the avoidance response but only slightly reduce pituitary ACTH release. In contrast, cortisol implants in the median eminence markedly reduce pituitary ACTH release but hardly affect avoidance behavior.

In order to determine the site of action of the corticosteroids in facilitating the rate of extinction of the avoidance response, a study was designed in which these steroids were implanted into several brain areas of the conscious rat during extinction sessions with the pole-jumping avoidance response (Van Wimersma Greidanus and De Wied, 1969).

It appeared that implantation of dexamethasone phosphate or corticosterone facilitates the rate of extinction of the avoidance response when implanted in the thalamic parafascicular area, the same area that had to be destroyed to prevent the inhibitory effect of α-MSH on the rate of extinction of a shuttlebox avoidance response. Preliminary experiments with ACTH 1-10, studied in the same way as the steroids, indicate that implantation of this peptide in the nucleus parafascicularis, as does systemic administration, delays extinction of the avoidance response. The site of action of ACTH analogues and of the glucocorticosteroids in the CNS therefore appears to be the same.

VI. Conclusions

The mode of action of the pituitary-adrenal system in its effect on conditioned avoidance behavior is not clear. Stimulatory and inhibitory effects on electrical activity of the CNS in various species (Krivoy and Guillemin, 1961; Korányi and Endröczi, 1967; Koltai and Minker, 1966; Feldman *et al.*, 1961; Wasserman *et al.*, 1965) indicate that the pituitary-adrenal system affects neural elements, either directly or by influence on the biochemical or vascular environment of these elements. In which way these neural processes are affected is unknown. The process may involve autonomic nervous transmission, since some behavioral effects of MSH may be counteracted by adrenergic blocking agents (Sakamoto and Prasad, 1968). It may also involve the synthesis of macromolecules involved in neuronal activity, since bockade of protein synthesis may impair learning and memory (Flexner *et al.*, 1967).

The biological meaning of the effect of the pituitary-adrenal system on conditioned avoidance behavior remains to be elucidated. Avoidance conditioning, according to Mowrer's theory, is motivated by fear and reinforced by fear reduction. It may be, therefore, that the pituitary-adrenal system modifies fear, as suggested by Mirsky *et al.* (1953). However, conditioned avoidance behavior contains other elements as well. The effect of the pituitary-adrenal system may be on learning, on memory, on motivation, on the drive state of the animal, etc.

The pituitary-adrenal system plays an essential role in adaptation when homeostasis of the organism is threatened. Learning, recognition, and experience are involved in the mechanism of adaptation. Modification of the rate of extinction of an avoidance response as induced by the pituitary-adrenal system may therefore reflect a modification of an adaptive process in the CNS.

References

Bohus, B. (1968). *Neuroendocrinology* **3**, 355.
Bohus, B., and De Wied, D. (1966). *Science* **153**, 318.
Bohus, B., and De Wied, D. (1967a). *Physiol. Behav.* **2**, 221.
Bohus, B., and De Wied, D. (1967b). *J. Comp. Physiol. Psychol.* **64**, 26.
Bohus, B., and Endröczi, E. (1965). *Acta Physiol. Acad. Sci. Hung.* **26**, 184.
Cardo, B., and Valade, F. (1965). *C. R. Acad. Sci.* (Paris) **261**, 1399.
De Wied, D. (1961). *Endocrinology* **68**, 956.
De Wied, D. (1965). *Intern. J. Neuropharmacol.* **4**, 157.
De Wied, D. (1966). *Proc. Soc. Exp. Biol. Med.* **122**, 28.
De Wied, D. (1967). *In* "Hormonal Steroids" (L. Martini, F. Fraschini, and M. Motta, eds.), pp. 945–951. Excerpta Medica, Amsterdam.
De Wied, D. (1969). *In* "Frontiers in Neuroendocrinology 1969". (W. F. Ganong and L. Martini, eds.), pp. 97–140. Oxford Univ. Press, New York.
De Wied, D., and Greven, H. M. (1968). Proc. 24 Congress Intern. Union Physiological Sciences, Vol. VII, p. 110.
Feldman, S., Todt, J. C., and Porter, R. W. (1961). *Neurology* **11**, 109.
Ferrari, W., Gessa, G. L., and Vargiu, L. (1961). *Experientia* **17**, 90.
Ferrari, W., Gessa, G. L., and Vargiu, L. (1963). *Ann. N. Y. Acad. Sci.* **104**, 330.
Flexner, L. B., Flexner, J. B., and Roberts, R. B. (1967). *Science* **155**, 1377.
Greven, H. M., and De Wied, D. (1967). *Europ. J. Pharmacol.* **2**, 14.

Koltai, M., and Minker, E. (1966). *Acta Physiol. Acad. Sci. Hung.* **29**, 410.
Korányi, L., and Endröczi, E. (1967). *Neuroendocrinology* **2**, 65.
Korányi, L., Endröczi, E., and Tárnok, F. (1966). *Neuroendocrinology* **1**, 144.
Korányi, L., Endröczi, E., Lissák, K., and Szepes, E. (1967). *Physiol. Behav.* **2**, 439.
Krivoy, W. A., and Guillemin, R. (1961). *Endocrinology* **69**, 170.
Levine, S., and Jones, L. E. (1965). *J. Comp. Physiol. Psychol.* **59**, 357.
Lissák, K., and Endröczi, E. (1961). *In* " Brain Mechanisms and Learning " (A. Fessard, R. W. Gerard and J. Konorski, eds.), pp. 293–308. Blackwell, Oxford.
Miller, R. E., and Ogawa, N. (1962). *J. Comp. Physiol. Psychol.* **55**, 211.
Mirsky, A., Miller, R., and Stein, M. (1953). *Psychosomat. Med.* **15**, 574.
Murphy, J. V., and Miller, R. E. (1955). *J. Comp. Physiol. Psychol.* **48**, 47.
Sakamoto, A., and Prasad, K. N. (1968). *In* " Protein and Polypeptide Hormones " (M. Margoulies, ed.), pp. 503–504. Excerpta Medica, Amsterdam.
Smelik, P. G. (1960). *Acta Endocrinol.* **33**, 437.
Van Wimersma Greidanus, Tj. B., and De Wied, D. (1969). *Physiol. Behav.* **4**, 365.
Wasserman, M. J., Belton, N. R., and Millichap, J. G. (1965). *Neurology* **15**, 1136.
Woodbury, D. M. (1954). *Recent Progr. Hormone Res.* **10**, 65.

Hypothalamic Control of Hypophyseal Function in Anurans

C. Barker Jørgensen

I. Introduction

It is the task of the comparative physiologist to elucidate the patterns of structural and functional relations that may exist between brain, hypophysis, and peripheral endocrine glands in representatives of the various systematic and ecological groups of vertebrates.

We already know that functional relations between brain, hypophysis, and peripheral endocrine glands (thyroids, adrenals or gonads) vary strongly throughout the vertebrate system. Table I summarizes available data on the dependency of adenohypophyseal functions upon hypothalamic control in some vertebrate groups (for references see Jørgensen, 1968 a, b; Rust and Meyer, 1968). Apparently, the dependence of adenohypophyseal functions on central nervous control varies from absolute dependence to more or less complete independence. Such extreme variation has been observed even within smaller systematic groups, which suggests the adaptive character of the functional relationships between brain and adenohypophysis.

Moreover the evidence suggests that hypophyseal functions are generally more dependent upon control by the brain in vertebrate groups placed higher in the phylogenetic hierarchy. Indeed, brain dependence of pars distalis functions has not been demonstrated in cyclostomes and selachians. This may be due in part to the fact

TABLE I

SECRETION OF HORMONES FROM THE ADENOHYPOPHYSIS DEPRIVED OF ITS NORMAL HYPOTHALAMIC CONTACT

			Gonadotropins				
	ACTH	TSH	♀	♂	GH[1]	LTH[2]	MSH
Cyclostomes			n	n			
Selachians		n	n				+
Teleosts		n to +	—		—	n	
Amphibians	— to n	— to n	— to n	— to n	n	+	+
Birds		n	—	—		—	
Mammals	— to n	— to n	—	—	—	+	+

Symbols: —, significantly reduced or absent; n, normal or only slightly reduced; +, enhanced (inhibitory control).

[1] GH, growth hormone.

[2] LTH, luteotropic hormone (prolactin).

that these groups have been less extensively studied than teleosts and tetrapods. It is noteworthy that the greatest variability has been found within the amphibians (i.e., the class in which more species have been studied than in any other class of vertebrates). It is desirable to extend studies to non domesticated and non laboratory-adapted mammalian species.

The brain may act either by stimulating or by inhibiting adenohypophyseal functions even though control by stimulation is more common. This type of control of a hypophyseal function may vary between groups. Thus the hypothalamus inhibits prolactin secretion in mammals and amphibians, but it seems to stimulate secretion in birds. Thyrotropin (TSH) secretion, too, is perhaps controlled by different mechanisms in different groups of vertebrates.

II. Brain Control of Toad Adenohypophysis

We have for some years been studying hypophyseal functions in the adult toad *Bufo bufo* with special reference to the central nervous control of such functions. Some results are summarized in Table II. It appears that all adenohypophyseal functions studied turned out to be more or less dependent upon normal structural connections between hypophysis and hypothalamus.

TABLE II

SECRETION OF HORMONES FROM THE ECTOPICALLY TRANSPLANTED HYPOPHYSIS IN THE TOAD (BUFO BUFO)

ACTH	TSH	Gonadotropins ♀	ICSH ♂	FSH	MSH
—	—	— to n	—	n	+

Symbols: As defined in Table I.

A. FUNCTION OF THE ECTOPICALLY TRANSPLANTED PARS DISTALIS

The ectopically transplanted pars distalis did not secrete measurable amounts of Adrenocorticotropic Hormone (ACTH), as judged by the absence of measurable amounts of corticosterone in blood (Spies, personal communication). Moreover, toads bearing an ectopic auto-transplanted pars distalis died with symptoms of adrenocortical deficiency as fast as did hypophysectomized controls (Jacobsohn and Jørgensen, 1956).

Ectopic transplantation of the pars distalis strongly reduces TSH secretion, as judged by the low rate of ^{131}I uptake by the thyroid gland (van Dongen *et al.*, 1966).

Gonadotropin secretion was evaluated on the basis of the functional state of the gonads and the secondary sex characters. In the male, the ectopic pars distalis secreted Follicle-Stimulating Hormone (FSH) in amounts that maintained a fairly normal spermatogenesis, whereas Interstitial Cell-Stimulating Hormone (ICSH) secretion did not maintain the thumb pads (a secondary sex character). In females only ovarian growth was evaluated; no secondary sex characters exist in the female toad. The dependence of gonadotropic function upon hypothalamic contact varied seasonally. During the months following the breeding season in contrast to results obtained in experiments performed in autumn (van Dongen *et al.*, 1966; Jørgensen, 1968a, b), there was little difference between ovarian growth in females with the pars distalis transplanted to an eye muscle and in females with pars distalis regrafted under the median eminence (See also Section II D).

B. FUNCTION OF THE DENERVATED PARS INTERMEDIA

Whereas the hypothalamic control of the amphibian pars distalis is, as in other tetrapod vertebrates, humorally mediated, the control of the pars intermedia is by direct innervation (Jørgensen and Larsen,

1963). The nerves are inhibitory (Etkin, 1962). Denervation of the gland causes darkening of the toads owing to uncontrolled release of Melanocyte-Stimulating Hormone (MSH), which causes dispersion of melanin granules within the melanophores. After a latency period of up to several months some denervated toads kept on a white illuminated background resumed the ability to concentrate the melanin granules, presumably concurrent with regeneration of the inhibitory innervation of the gland (Jørgensen and Larsen, 1963).

C. Localization of the Hypothalamic Structures Controlling the Hypophysis

We studied the question of which parts of the brain are involved in the control of the various hypophyseal functions by transecting the ventral brain stem at various levels in front of the hypophysis. The operations isolated the hypophysis along with a smaller or larger part of the hypothalamus from the rostral parts of the brain. Table III summarizes the effects of transections immediately caudal or immediately frontal to the optic chiasma on various hypophyseal functions.

It was found that transection caudal to the optic chiasma permitted normal gonadotropic and thyrotropic function, whereas the corticotropic function was abolished. MSH secretion was constantly high. After transection anterior to the optic chiasma, all hypophyseal functions studied were more or less normal except the thyrotropic function, which was significantly enhanced. Further details of the effects of the transections at the two levels on corticotropic and thyrotropic functions of the pars distalis are provided in Tables IV and V.

TABLE III

Effect of Transection of Ventral Brain Stem on Hypophyseal Functions in the Toad

Level of Transection	ACTH	TSH	Gonadotropins ♀	Gonadotropins ♂	MSH
Caudal to optic chiasma	—	n	n	n	+
Frontal to optic chiasma	n	+	n	n	n

Symbols: As defined in Table I.

TABLE IV

EFFECT OF TRANSECTION OF VENTRAL BRAIN STEM ON ADRENOCORTICOTROPIC FUNCTION OF PARS DISTALIS IN THE TOAD

Group	Corticosterone, μg/100 ml blood		
	Experiment 1 (Mean ± S.E.)	Experiment 2 (Mean ± S.E.)	Experiment 3 (Mean ± S.E.)
Unoperated controls	1.72 ± 0.18 (6) [1]	1.44 ± 0.22 (6)	1.76 ± 0.38 (5
Transection anterior to optic chiasma	1.74 ± 0.61 (7)		0.89 ± 0.22 (5)
Transection posterior to optic chiasma		0.07 ± 0.02 (6)	0.03 ± 0.02 (5)
Pars distalis extirpated			0.03 ± 0.02 (4)

[1] Figures in parentheses indicate number of animals in the group.

It thus appears that neurons which originate in the posterior hypothalamus between the hypophysis and optic chiasma can maintain normal thyrotropic and gonadotropic functions. This agrees with Dierickx's finding (1967) of a gonadotropic area located in the infundibular region of the hypothalamus in the female frog.

In order to secure normal corticotropic function the hypophyseal-hypothalamic complex should include the region of the optic chiasma (Spies and Vijayakumar, personal communication).

TABLE V

EFFECT OF TRANSECTION OF VENTRAL BRAIN STEM ON THYROTROPIC FUNCTION OF PARS DISTALIS IN THE TOAD

Group	Percentage Uptake of Tracer Dose ^{131}I by Thyroid	
	2 Weeks after Operation (Mean ± S.E.)	4 Weeks after Operation (Mean ± S.E.)
Unoperated controls	5.4 ± 0.6 (9) [1]	3.8 ± 0.5 (8)
Transection anterior to optic chiasma	13.6 ± 1.6 (10) [2]	10.0 ± 0.7 (9) [2]
Transection posterior to optic chiasma	7.5 ± 0.8 (10)	3.5 ± 0.5 (7)
Pars distalis extirpated	2.7 ± 0.2 (9) [3]	1.8 ± 0.2 (8) [3]

[1] Figures in parentheses indicate number of animals in the group.
[2] Values significantly higher than in unoperated controls; $p < 0.001$.
[3] Values significantly lower than in unoperated controls; $p < 0.001$ and $p < 0.01$, respectively.

The finding that the rates of ^{131}I uptake by the thyroids increased after transection of the hypothalamus in front of the optic chiasma, whereas it was low after ectopic transplantations of pars distalis, suggests that the control of TSH secretion is complex and perhaps includes inhibitory as well as stimulatory nervous components (Rosenkilde, 1969).

This pattern of interrelations among brain, hypophysis, and thyroids may be different in other amphibians. Thus, in the frog *Rana temporaria*, Rosenkilde (personal communication) found that transections anterior and posterior to the optic chiasma had no significant effect on the thyroidal ^{131}I uptake, which was high in both groups (Table VI). Moreover, in contrast to the toad, the transplanted pars distalis exhibits pronounced autonomous thyrotropic activity.

TABLE VI

EFFECT OF TRANSECTION OF VENTRAL BRAIN STEM ON THYROTROPIC FUNCTION OF PARS DISTALIS IN THE FROG

Group	Percentage Uptake of Tracer Dose ^{131}I by Thyroid 1–2 Weeks after Operation (Mean ± S.E.)
Unoperated controls	12.0 ± 2.2 (8)[1]
Transection anterior to optic chiasma	11.0 ± 1.8 (8)
Transection posterior to optic chiasma	9.0 ± 1.1 (12)
Posterior hypothalamus extirpated	7.0 ± 0.9 (10)[2]
Pars distalis extirpated	3.6 ± 0.5 (10)

[1] Figures in parentheses indicate number of animals in the group.
[2] Value significantly higher than in group with pars distalis extirpated, $p < 0.001$; and lower than in unoperated controls, $0.02 < P < 0.05$.

D. Neuroendocrine Control of Ovarian Cycle

In describing the pattern of relations between brain and hypophyseal functions we have to consider that these relations may vary with the periods of life. Thus, in amphibians, thyrotropic function is more dependent upon intact structural connections to the brain during the period of metamorphosis than before and after that period. In seasonally reproducing vertebrates the relations among brain, gonadotropic function of the hypophysis, and gonadal (especially ovarian) function may possibly vary with the phase of the reproductive cycle. However, nothing definite is known, and the study of the control mechanisms of ovarian cycles in the various groups of lower vertebrates is still at its beginning. The toad *Bufo bufo* is one of the better studied species with respect to the complex neuro-endocrine interactions among ovary, hypophysis, and brain in a group of annually breeding lower vertebrates.

Bufo bufo spawns in spring (in Denmark, ordinarily in April). Immediately after spawning, the ovaries contain only gonadotropin-independent oocytes of up to about 0.8 mm in diameter. A few weeks after spawning, a large number of these oocytes normally starts vitellogenic growth, which requires gonadotropin. Hypophysectomy prevents small oocytes from starting vitellogenesis. Recruitment to the vitellogenetic growth phase is usually complete within some weeks and the individual oocytes of the population grow at about equal rates until full size (1.6–1.8 mm in diameter) is reached in autumn before the toads commence hibernating. However, the oocytes do not mature until breeding time in the following spring when maturation, division and dissolution of the nuclear membrane occur just prior to ovulation (Schuetz, 1967).

We may distinguish three main controls operating during the ovarian cycle: (1) number of oocytes entering the vitellogenic growth phase, (2) rate at which the oocytes grow, and (3) ovulation.

1. *Number of Oocytes Entering the Vitellogenic Growth Phase*

The number of oocytes that commences vitellogenesis in spring is presumably regulated, since the final ovarian weight is always about one-sixth the total body weight. Normally, only few oocytes become atretic.

In order to get an idea about the mechanism of regulation of this recruitment we tested whether the number of oocytes recruited is predestined or whether some extra oocytes form a potential reserve. We studied the effect of extirpating part of the ovarian mass upon the size of the population that commences vitellogenesis. It was found that by reducing the ovarian mass to one-sixth of the original at the beginning of the growth phase in early May the remaining ovarian fragment would strongly increase the number of young oocytes that start vitellogenesis. By the end of June the ovarian fragments had on an average produced populations of oocytes undergoing vitellogenesis that were of about the same size as those produced by one whole ovary in the normal controls; that is, the ovarian fragments had increased their normal quota of oocytes to the vitellogenic growth phase by a factor of 3 (Vijayakumar, personal communication).

It can therefore be concluded that the presence of a normal population of growing oocytes in an ovary prevents further recruitment to the population. The likelihood that full-grown oocytes also inhibit recruitment appears from the observation that elimination of full-grown oocytes uniformly resulted in resumed recruitment to the vitellogenic growth phase. Experiments to demonstrate this may conveniently be made in late autumn or winter when the oocyte populations in the ovaries consist only of small gonadotropin-independent stages and full-grown oocytes. The population of full-grown oocytes may be eliminated by various means, for instance, by gonadotropin-induced ovulation or by extirpating and regrafting the pars distalis under the median eminence. The transient lack of gonadotropin, lasting from the operation until the gland has become revascularized, will result in degeneration and resorption of all large oocytes. Irrespective of the way in which the population of eggs is eliminated, young oocytes will regularly start growing and new oocytes will continue to enter the growth phase until a normal population of oocytes undergoing vitellogenesis has once more been attained (Van Dongen *et al.*, 1966).

The large oocytes may be thought to exert their effect locally in the ovary or centrally (e.g., on the hypothalamus or the hypophysis). If the effect is exerted directly in the ovary one should expect that elimination of the large oocytes in only one ovary would result in small oocytes commencing vitellogenesis in this ovary and not in the other one that contains an intact population of large oocytes. Such unilateral elimination of the population of large oocytes was achieved by ligating the blood vessels to the ovary for a few days. The transient lack of blood supply caused the large oocytes to die, whereas the gonadotropin-independent oocytes survived. The operation regularly resulted in small oocytes starting vitellogenesis, however, in equal numbers in both ovaries (Table VII; Vijayakumar, personal communication). It is thus suggested that the population of growing or full-grown oocytes does not prevent vitellogenesis in other oocytes by a local inhibitory mechanism. This conclusion is further supported by the finding that administration of exogeneous gonadotropin (purified mammalian FSH) stimulated recruitment of

TABLE VII

EFFECT OF UNILATERAL ELIMINATION OF THE POPULATION OF FULL-GROWN OOCYTES ON RECRUITMENT TO VITELLOGENIC PHASE

Full-grown Oocytes Eliminated by Means of Temporary Ligation of the Blood Vessels to One Ovary	Ligated Ovary		Non Ligated Ovary	
	No Recruitment	Recruitment	No Recruitment	Recruitment
Ovary ligated 4–5 days [1]	4	2	1	5
Ovary ligated 2–5 days	4	3	3	4

[1] Three-fourths of ovary extirpated.

small oocytes to the growth phase whether or not the ovaries had a population of large oocytes (Kirsten Kjœr, personal communication). The absence of recruitment to the growth phase in the presence of large oocytes therefore does not seem to be due to inability of the small oocytes to respond to gonadotropin.

It seems reasonable to suggest that recruitment to the vitellogenic growth phase may be regulated by a feedback mechanism in which secretion of hormone from the growing or full-sized follicles controls the gonadotropic activity of the hypophysis. Another mechanism can, however, be imagined. The removal of all (or a fraction of) large oocytes results in a reduction of the gonadotropin-requiring ovarian mass. If the large follicles represent significant consumers and metabolizers of gonadotropin, their elimination must leave more gonadotropin available to the small oocytes that might previously have been prevented from initiating vitellogenesis because of lack of gonadotropin. A self-regulatory mechanism for maintenance of a specific number of gonadotropin-requiring oocytes in the ovaries might thus operate even at constant autonomous levels of gonadotropin secretion (see McLaren, 1966). Further experiments are in progress to determine the type of mechanism that controls recruitment of oocytes to the vitellogenic growth phase in the toad ovary.

2. *Rate of Growth*

In the normal toad the oocytes that start growing in spring accumulate yolk relatively slowly and usually do not finish vitellogenesis until September. The normal growth rate seems to be adapted to the length of the annual period of feeding, which terminates when the toads enter hibernation. As mentioned, elimination of the population of large oocytes in autumn or winter is followed by immediate replenishment to the growth phase from the reserve of small oocytes. However, at this time of the year, vitellogenesis proceeds rapidly and a normal population of full-sized eggs can be restored in the ovaries in less than two months after the elimination of the original population. The rapid replenishment of the lost population of oocytes does not take place in toads with the pars distalis transplanted to an eye muscle (Van Dongen *et al.*, 1966; Jørgensen, 1968b).

The fast rate of growth of oocytes in autumn and winter thus appears to be dependent upon central nervous stimulation of gonadotropin secretion from the pars distalis, whereas in spring only slight stimulation of gonadotropin secretion could be demonstrated. As mentioned, there was only minimal difference in ovarian development in toads with pars distalis regrafted under the median eminence or autotransplanted to an eye muscle.

It is noteworthy that subtotal ovariectomy in spring resulted only in a greater number of small oocytes commencing vitellogenesis. The rate of growth did not increase. The regulation mechanism therefore seems to be adjusted for securing a complete population of oocytes that will be recruited to the growth phase and which will grow slowly. Later during the normal ovarian development the toad acquires an increased potential for rapid vitellogenesis, perhaps due to an increased potential for central nervous stimulation of gonadotropin secretion. However, the secretion is normally checked by the presence of the population of large oocytes and is maintained at a level necessary to maintain their survival. It is therefore suggested that one principal factor in the regulation of the annual ovarian cycle in the toad is an annual rhythm in the activity of gonadotropin-controlling structures in the brain. These structures may be located in the posterior hypothalamus.

3. *Ovulation*

The toads leave their hibernation quarters and migrate to the breeding ponds in spring, presumably in response mainly to increased temperature and rainfalls. Little is known about a possible hormonal control of the migration to water, e.g. whether prolactin plays a similar role in the water drive of anurans as it does in urodeles.

Ovulation in the toad seems primarily to be triggered by contact of the skin with water, whereas the clasping of the male is less important. Thus, Heusser (1963) observed that when unmated female toads were kept out of water they would not ovulate, but toads kept in water tanks ovulated at the same time as those breeding in nature. Occasionally the isolated females in water did not ovulate at the expected time. Such toads eventually ovulated when mated.

4. *Ovarian Dysfunction*

In some years freshly sampled toads have frequently been found to exhibit deficient ovarian development. It is also often difficult in the laboratory to produce conditions that will secure normal ovarian function. A number of factors that may cause ovarian dysfunction are known. Some of these laboratory observations have contributed to elucidating the mechanisms that control the normal annual cycle.

a. Nutritional state. Undernourished vertebrate females often exhibit deficient ovarian function. The female toad is no exception, but the effect of undernourishment has been found to vary with the experimental conditions. Proper gonadotropic function of the ectopically transplanted pars distalis has been observed only in well-nourished female toads. In undernourished toads bearing an autotransplanted pars distalis the ovarian development remained at the hypophysectomy level. Conversely, the pars distalis regrafted under the median eminence was capable of maintaining normal ovarian development in similarly undernourished toads. It thus appears that central nervous stimulation can to some degree counteract the depressing effect of starvation on the hypophyseal-ovarian function (Jørgensen, 1968b).

b. Prolonged exposure to room temperature. In nature, toads are exposed to falling temperature during the autumn, and hibernation is initiated. Toads that remain at room temperature in the laboratory do not hibernate, but continue their active life apparently normally. However, whereas the full-grown oocytes remain intact in the ovaries of hibernating toads, atresia sets in among the full-grown oocytes in the ovaries of toads kept at the higher temperature of the laboratory. Atresia begins to appear as early as November and December and it has usually included the whole population of full-grown oocytes by January and February. Concomitantly with the progressing atresia of the large oocytes, small oocytes start vitellogenesis eventually to substitute the loss. However, the regular pattern of the normal ovarian cycle is blurred (Vijayakumar, personal communication). Very likely the period of hibernation plays an important role in maintaining and synchronizing the normal annual ovarian cycle.

Moreover, normal ovulation seems to require exposure of the toads to the low temperatures of hibernation. The full-grown follicles thus do not respond to injections of gonadotropin by ovulation unless the toads have been kept for some time at low temperatures (Su and Yu-lan, 1963a, b; Jørgensen, 1968a).

c. Transection of the brain and regrafting of the pars distalis. Operations such as transection of the anterior hypothalamus or regrafting of the pars distalis under the median eminence have often led to failure of ovarian development, which does not seem to result from the interruption of the functional connections between the hypophyseal gonadotropic cells and gonadotropin-controlling structures in the brain. At any rate, apparently identical operations have regularly been found to be compatible with normal ovarian development (Vijayakumar, personal communication; Van Dongen *et al.*, 1966). Similar operations did not interfere with the thyrotropic and adrenocorticotropic function. Presumably, therefore, the integrative function of the ovary-hypophysis-brain complex is more easily suspended than those involving the thyroid or the adrenal glands. Perhaps the apparently high vulnerability of ovarian function serves to ensure that oocytes are recruited to the vitellogenic growth phase only when conditions are favorable for growth.

III. Conclusions

In the toad *Bufo bufo* all adenohypophyseal functions studied are more or less dependent upon hypothalamic control. Structures that control gonadotropic functions have been demonstrated in the posterior hypothalamus. Corticotropin-controlling structures have been located in the middle hypothalamus, whereas thyrotropin-controlling structures, which may be both inhibitory and stimulatory, seem to be more widely distributed.

The toad ovary normally exhibits one annual cycle. After spawning in spring, small oocytes enter the vitellogenic, gonodatropin-dependent growth phase in sufficent numbers to compensate for mature eggs lost at ovulation. The oocytes grow slowly and reach final size in autumn. It seems that hypothalamic stimulation of the gonadotropin secretion is low after spawning. The potential for stimulation becomes great in autumn, but is normally checked by the population of growing or full-grown follicles. The population of full-grown oocytes survive until the next breeding season only if the toads become exposed to the low temperatures of their hibernation quarters. Exposure to low temperature is moreover needed for the ovaries to ovulate in response to gonadotropin. The seasonal exposure of the toads to low temperature may play an important role in maintaining and synchronizing the normal annual ovarian cycle.

The function of the brain-hypophysis-ovary complex disintegrates easily.

Acknowledgments

I gratefully acknowledge permission to refer to the unpublished results of the following colleagues: Mrs. Kirsten Kjœr, Mr. Per Rosenkilde, Miss Ingrid Spies and Dr. S. Vijayakumar. The work has been supported by grants from the Danish Science Foundation and the Carlsberg Foundation.

References

Dierickx, K. (1967). *Z. Zellforsch. Mikrosk. Anat.* **77**, 188.
Etkin, W. (1962). *Gen. Comp. Endocrinol. Suppl.* **1**, 148.
Heusser, H. (1963). *Rev. Suisse Zool.* **70**, 741.
Jacobsohn, D., and Jørgensen, C. B. (1956). *Acta Physiol. Scand.* **36**, 1.
Jørgensen, C. B. (1968a). *In* "Perspectives in Endocrinology" (E. J. W. Barrington and C. B. Jørgensen, eds.). Academic Press, London, pp. 469–541.
Jørgensen, C. B. (1968b). *Arch. Anat. Histol. Embryol.* **51**, 356.
Jørgensen, C. B., and Larsen, L. O. (1963). *Gen. Comp. Endocrinol.* **3**, 468.
McLaren, A. (1966). *Proc. Roy. Soc.* **166**, 316.
Rosenkilde, P. (1969). *In* "La spécificité zoologique des hormones hypophysaires et de leur activités" (M. Fontaine, ed.). Éditions du C.N.R.S., Paris, Vol. 177, pp. 287–290.
Rust, C. C., and Meyer R. K. (1968). *Gen. Comp. Endocrinol.* **11**, 548.
Schuetz, A. W. (1967). *Proc. Soc. Exp. Biol. Med.* **124**, 1307.
Su, T., and Yu-lan, W. (1963a.) *Scientia Sinica* **12**, 1161.
Su, T., and Yu-lan, W. (1963b.) *Scientia Sinica* **12**, 1165.
Van Dongen, W. J., Jørgensen, C. B., Larsen, L. O., Rosenkilde, P., Lofts, B., and van Oordt, P. G. W. J. (1966). *Gen. Comp. Endocrinol.* **6**, 491.

The Hypothalamus as a Thermodependent Neuroendocrine Center in Urodeles

V. MAZZI

I. Introduction

In several species belonging to various vertebrate classes, external stimuli such as light, temperature, food availability, sounds, and social environment play an important role in the synchronization not only of endogenous circadian rhythms, but also of the circennian rhythms (Bünning, 1964; Solberger, 1965; Van Tienhoven, 1968).

In a restricted number of vertebrate species, including some urodele amphibians, an endogenous annual rhythm seems to be lacking. In this case, temperature is the synchronizing factor that takes on the role of the controlling mechanism of cyclic events. This article will examine experimental results affording evidence that the hypothalamus may be regarded as a thermosensitive center in which stimuli arising in the external environment are integrated with internal stimuli and transmitted to the adenohypophysis through the hypothalamo-hypophyseal system.

II. Hypothalamic Control of Adenohypophyseal Functions

A. GONADOTROPIC FUNCTION

In many species of urodele amphibians the development of the sexual cycle, and in particular of spermatogenesis (for review see Galgano, 1952), is more or less strictly determined by seasonal temperature variations. In this connection a typical example is provided by the crested newt (*Triturus cristatus carnifex* Laur.). In this species, according to the classification proposed by Galgano, spermatogenesis is rigostatic in that it undergoes a stasis during the cold period of the year; it is potentially continuous in that no autonomous factors other than those which are temperature-dependent are involved in the stasis determinism, since temperature decline alone is its effective cause. Finally, spermatogenesis is eureactive in that it responds in a predictable way to the gradual temperature decline by completely suppressing spermatogonial transformation into spermatocytes, thus concluding the spermatogenetic cycle.

In the crested newt the regular development of spermatogenesis depends upon the integrity of the hypothalamo-hypophyseal connections (Mazzi, 1950, 1952a,b; Mazzi and Peyrot, 1960, 1963). Mechanical lesions of the preoptic region and the hypothalamic floor, insertion in the median eminence of a barrier preventing the regeneration of the hypothalamo-hypophyseal tract and of the vessels coming from the hypothalamic floor, and heterotopic hypophyseal autotransplantation are all followed by a marked weight loss in the testes and by blockage of spermatogenesis. In addition, reactivation of spermatogenesis (which is known to occur in the winter animal when kept under optimal temperature conditions, 22–23 °C) is prevented by the interruption of the hypothalamo-hypophyseal connections, although in this case a moderate proliferative spermatogonial activity may be elicited.

On the other hand, the evaluation of the large number of experimental data we have collected has enabled us to ascertain that reactivation of spermatogenesis in animals bearing a hypothalamic lesion is possible only when some capillaries from the hypothalamic floor have regenerated. Spermatogenesis, moreover, is facilitated by the regeneration of a satisfactorily organized neurohemal organ as long as vascular connections with the adenohypophysis are realized. These observations together with the fact that the heterotopic pituitary autograft is in no case capable of supporting spermatogenesis in the crested newt, as is also the case for Pleurodeles (Pasteels, 1960), point to the importance of the vascular connections between the pituitary and the hypothalamus.

A particularly marked antigonadal effect is seen to occur in the case of permanent lesions involving the hypothalamic floor anteriorly to the median eminence. This is accounted for by both the special organization of the hypophyseal portal system to which blood is supplied by the cranial branches of the hypothalamic artery (Peyrot, 1960) and the organization of the hypothalamo-hypophyseal tract, whose fan-like distributed fibers converge upon the median eminence coursing in a loose arrangement upon the hypothalamic floor (Mazzi, 1953). This anatomical situation is a great obstacle to the restoration of the conditions that ensure the flow of neurohormonal information to the adenohypophysis.

The regressive phenomena occurring in the testes of animals bearing either irrevocable hypothalamic lesions or a pituitary autograft are due to the decrease in pituitary gonadotropin synthesis or to suppression of gonadotropin release. The cells responsible for either Follicle-Stimulating Hormone (FSH) or Luteinizing Hormone (LH) elaboration (type II and type III basophils, respectively) appear functionally inactive, although still identifiable at both the optical and ultrastructural levels. This has also been reported for other amphibian species belonging to the anurans and urodeles (Doerr-Schott, 1968; Masur, 1968). Replacement therapy with balanced doses of purified mammalian FSH and LH administered to the hypophysectomized animal and to that bearing a pituitary autograft it sufficient to secure the recovery of spermatogonial proliferation and to promote the onset of meiosis after the transformation of the late spermatogonia into spermatocytes I (Vellano, 1968a).

The blockage of spermatogenesis naturally occurring with the decline of temperature in autumn may also be due to decreased FSH production.

The declining phase of spermatogenesis is evidenced by the appearance of a degenerating band that involves the late spermatogonia, thereby impairing the production of new spermatocytes I, while meiosis is carried on by the cells that are already engaged in it. These cells achieve spermiogenesis independently of the environmental temperature. The degeneration band can be provoked in as short a time as 5 days, even in animals that have attained the culminating phase of spermatogenesis, by subjecting them to a drop in temperature of about 10 degrees or by treating them with prolactin (Mazzi *et al.*, 1966, 1967). Simultaneous administration of adequate FSH and prolactin doses prevents the degeneration band from forming, thus enabling normal spermatogenesis to proceed (Mazzi and Vellano, 1968).

On the one hand, a rise in the environmental temperature stimulates spermatogenesis, but on the other hand it inhibits the development of the secondary sex characters (Galgano, 1944). Most of these characters are testosterone-dependent, since they undergo regression in the gonadectomized animal and are restored by testosterone administration. In the crested newt the synthesis of testosterone by the steroidogenic tissue of the testis is supported by the pituitary LH, whose production is dependent in turn upon the integrity of the neurovascular connections between the pituitary and the hypothalamus. The testosterone-dependent secondary sexual characters consistently regress in the animal bearing adequately placed hypothalamic lesions and in those bearing a pituitary autograft. However, during the summer months when the secondary sex characters are at their lowest, LH production by the hypophysis is likely not to be completely suppressed. Some activity of the enzymes involved in steroidogenesis is histochemically detectable and the tissues responsible are identifiable (Vellano, 1968b). Moreover, some amount of testosterone seems to be essential for spermatogenesis to proceed regularly (Vellano, 1968a).

The majority of observations reported above support the assumption that a thermosensitive center is likely to be present in the hypothalamus and could be responsible for a "thermo-pituitary-sexual reflex," as previously suggested by Galgano and Mazzi (1951).

In the light of recent observations in other vertebrates this concept can be integrated by accepting the premise that the hypothalamic thermosensitive centers operate like endocrine transducers in which chemical messengers such as the Follicle-Stimulating Hormone Releasing Factor (FSHRF) and the Luteinizing Hormone Releasing Factor (LHRF) are elaborated and transferred to the hypophysis through the hypophyseal portal system. To date we do not know precisely where these hypothalamic gonadotropic centers are located. They may be in the preoptic region, since lesions there definitely impair the hypophyseal gonadotropic activity (Mazzi, 1952a) and since electrical stimulation in this region evokes both ovulation in the crested newt (Mazzi, 1952c) and spermiation in *Rana esculenta* (Stutinsky and Befort, 1962). The study of the biology of reproduction indicates that the centers controlling the production of FSHRF and LHRF are separate from one another and differ in their sensitivity to temperature variations. It may be reasonably deduced that the production of FSHRF is enhanced by high temperatures and inhibited by low ones, while the reverse is true for LHRF. However, LHRF would only be depressed, though not inhibited, by high temperatures.

Changes in the sexual behavior of the crested newt are also closely correlated to temperature variations. Thus far the study of the hormonal determinism of these changes has not been thoroughly carried out in the crested newt. However, on the basis of the observations of Galgano (1944), Von Gauss (1961) and Grant (1966) it may be suggested that both gonadotropic hormones (LH in particular) and prolactin are directly involved in these changes. LH not only controls the development and maintenance of the secondary sex characters and spermiation, but also possibly plays a role in the determinism of courtship. Prolactin is likely to promote at least one ambisexual seasonal character, namely tail heigth (Mazzi *et al.*, 1969), and is doubtlessly responsible for eliciting water drive (Tuchmann-Duplessis, 1949; Mazzi *et al.*, 1966). Further details are discussed in the following section.

B. Prolactin Function

In the crested newt, prolactin production by the pituitary also appears to be cyclic and temperature-dependent. This hormone would be stored in the competent pituitary cells during the summer months, then released massively into the circulation and coincidentally with the temperature decline in autumn, thus eliciting water drive; finally prolactin would be most actively produced and released into the circulation during the winter months, resulting in the animals being " saturated " with prolactin during this period.

The main evidence supporting this working hypothesis rests on the quantitative determination of pituitary prolactin content at different periods of the year (Mazzi *et al.*, 1966; Peyrot *et al.*, 1966) and on the capacity of the pituitary tissue to produce prolactin in short-term *in vitro* cultures, either in the presence or absence of acid hypothalamic extracts (Vellano *et al.*, 1968). On the basis of these observations, which are further supported by the determination of prolactin in the pituitary autotransplanted for varying periods of time (30 and 140 days; Peyrot *et al.*, 1969a), it was suggested that trolactin synthesis and secretion in the crested newt are controlled py two factors, an inhibitory one (Prolactin Inhibiting Factor, PIF and an activating one (Prolactin Releasing Factor, PRF) which are balanced and simultaneously produced by the hypothalamus. The release of these two factors into the circulation is believed to be controlled by seasonal temperature variations (Vellano *et al.*, 1968) (Table I). While the presence of PRF has not been verified experimentally so far, the presence of PIF, which is operating in the summer animal, has been confirmed by the observation that the heterotopic pituitary autograft supports water drive in the aestivating animal for over 4 months after operation and carries on prolactin production at a markedly higher rate as compared with that of the pituitary from control animals (Peyrot *et al.*, 1969a, b) (Table I).

Some fundamental aspects of the biology of the crested newt seem to be closely connected with the temperature-dependent cyclic variations of prolactin. In the first place the onset of the water drive (Tuchmann-Duplessis, 1949; Mazzi *et al.*, 1966, 1967), whereby the natural tendency of the crested newt to stay on land in summer converts to the tendency to migrate into water when temperature declines in the autumn. Water drive entails the shift from a dehydrating habitat to a hydrating one and is considered as a " second metamorphosis " by Wald (1960). Water drive is attended by marked alterations in water and electrolyte metabolism and in the texture and permeability of the skin, in which prolactin may possibly play a primary role.

TABLE I

PROLACTIN CONTENT IN THE CULTURE MEDIUM AND IN THE HOMOGENATE OF IN VITRO CULTURED PITUITARIES

Subject	Pituitary	Culture Medium	Total	Comparisons	
Controls [1]	39.12± 9.97	26.50± 4.92	65.62± 10.12 (*A*)	*A–C*	$P \leq 0.01$
Short-term operated animals [2]	64.44±24.47	242.46±154.81	306.90±159.35 (*B*)	*A–B*	Not significant
Long-term operated animals [2]	55.20±17.98	127.40± 47.79	182.60± 38.41 (*C*)		

(From Peyrot *et al.*, 1969a).

[1] Means ± S.E. of nine cultures of two pituitaries each, after 2-hr incubation at 20–22 °C, expressed in ImU.

[2] Means ± S.E. of five cultures of two pituitaries each, after 2-hr incubation at 20–22 °C, expressed in ImU.

In the hypophysectomized animal, prolactin was found to affect the sodium content of the skeletal muscle in a manner similar to that of aldosterone (Sampietro and Vercelli, 1968). In the castrated newt, prolactin enhances the mitotic activity of the epidermal cells and slows down keratinization, thereby inhibiting moulting and speeding up the penetration of the anesthetic MS222 (Mazzi *et al.*, 1969).

The variation in the weight of the fat bodies and in the glycogen content of the cardiac muscle are in part related to the cyclic temperature-dependent prolactin variations. Systematic investigations throughout the year have demonstrated that the cardiac glycogen drops to a minimum during summer then increases gradually until a maximum is reached in winter, and declines again with the rise in the environmental temperature. The opposite is true for the weight of the fat bodies (Scalenghe *et al.*, 1968; Andreoletti and Rotta, 1968) (See Fig. 1). In the summer animal, exogenous prolactin was found to reverse both phenomena, i.e. to promote an increase in cardiac glycogen and a weight loss in the fat bodies (Scalenghe *et*

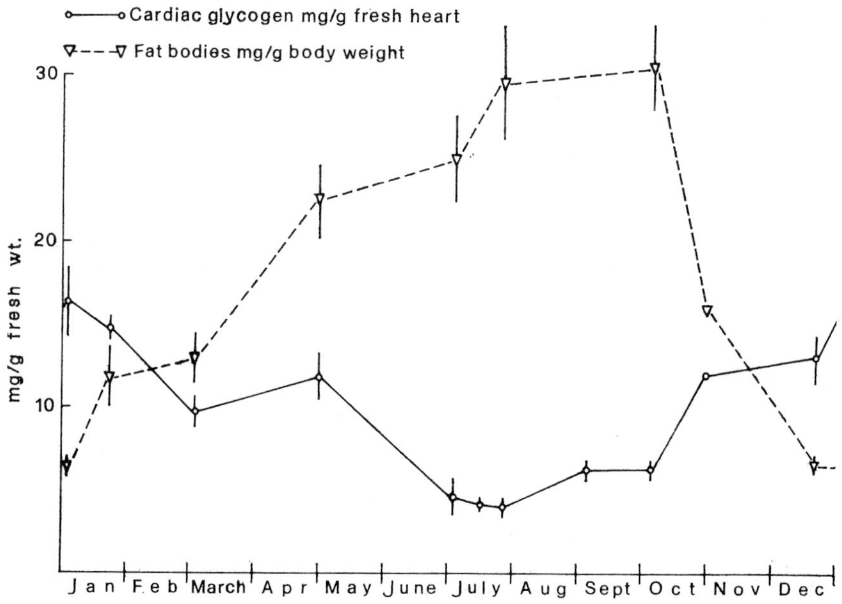

FIG. 1. Cyclic annual variations in cardiac glycogen and fat bodies weight.

TABLE II

EFFECTS OF PROLACTIN ON THE CARDIAC GLYCOGEN AND FAT BODIES WEIGHT

Subject	Group No.	Cardiac Glycogen, mg/g heart wet wt	Fat Bodies, mg/g body wt
Control aestivating animals (60 days at 22 °C)	I	5.5 ± 0.61 ($n=6$)[1]	15.99 ± 2.05 ($n=19$)
Prolactin-treated aestivating animals	II	13.5 ± 0.83 ($n=7$)	5.75 ± 1.62 ($n=9$)
Comparisons	I–II	$P \leq 0.001$	$P \leq 0.01$

(From Scalenghe *et al.*, 1968).

[1] Mean ± S.E. Valu *n* in parentheses gives number of animals used.

al., 1967a, b) (See Table II). By taking into account the prolactin year cycle it may be inferred that prolactin plays a primary role in the determination of both glycogen and fat bodies behavior, although other factors may operate synergically or permissively with prolactin. In the newt bearing a pituitary autograft and showing an enhanced prolactin production for over 4 months after operation (Peyrot *et al.*, 1969c), the heart glycogen content is lower than in controls, while the weight of the fat bodies is higher (Andreoletti, personal communication). In the hypophysectomized animal fasted for 22 days, in which both the weight of the fat bodies and the cardiac glycogen were seen to decline, prolactin treatment associated with Adrenocorticotropic Hormone (ACTH) (as well as Growth Hormone, GH, plus ACTH), though inducing an increase in cardiac glycogen, does not further affect the weight of the fat bodies (Sampietro *et al.*, 1969).

C. THYROTROPIC FUNCTION

In the crested newt the activity of the thyroid also follows a annual cycle that is dependent upon the seasonal temperature variations. Thyroid activity attains its maximum in winter and drops to a minimum during the warmer months in a manner similar to that reported by Miller and Robbins (1955) for *Taricha torosa*. These variations are less apparent in the female than in the male. In *Triturus viridescens*, maximum thyroid activity is coincidental with the breeding season during spring and minimum thyroid activity occurs in summer

(Morgan and Fales, 1942). A number of observations on different urodeles consistently indicate some degree of thyroid autonomy from the central nervous system. In the animal bearing a pituitary autograft, moulting is normal or almost normal, as reported for *Pleurodeles waltii* by Jørgensen (1968); for metamorphosed *Ambystoma mexicanum* by Jørgensen and Larsen (1963); for several *Triturus* species by Schotté and Tallon (1960), Dent (1966), Larsen and Rosenkilde (personal communication), and Mazzi (unpublished observations). Thyrotropic cells in the autotransplanted pituitary respond to thiouracil by hypertrophying (Mazzi and Peyrot, 1963). In the crested newt bearing cronic hypothalamic lesions, moulting does not seem to be severely affected (Mazzi, 1958). Histologically, the thyroid appears functionally more active than in the hypophysectomized controls and gives a positive response to thiourea (Mazzi and Peyrot, 1960). In several urodele species ^{131}I uptake by specimens bearing a heterotopic pituitary autograft is normal or almost normal, as reported for *Ambystoma mexicanum* by Jørgensen (1968) and for *Triturus viridescens* by Dent (1966), or even higher than in controls (reported for *Triturus cristatus* by Peyrot *et al.*, 1966). However, the activity pattern in the crested newt differs from that of controls, particularly in the monoiodotyrosine/diiodotyrosine ratio (which is nearer to that of hypophysectomized animals) and in the rate of thyroxine synthesis, which appears lower than in controls (Peyrot *et al.*, 1969c). As compared to normal controls these variations are more marked in short-term operated animals (4 ½ months) than in long-term (1 yr) operated ones (Peyrot *et al.*, 1969c). In animals bearing irreparable hypothalamic lesions the pattern of thyroid activity is wholly comparable to that of animals bearing a heterotopic pituitary autograft (Peyrot, 1969). Regeneration of the neurovascular connections between the hypothalamus and the pituitary however partial, is sufficient to restore the normal thyroid activity even in those specimens in which spermatogenesis has not been reactivated or those which show only partial reactivation under aestivating conditions (22 °C for 40 days; Peyrot, 1969) (Table III). The different degree of dependence of the hypophyseal gonadotropic and thyrotropic functions upon the central nervous system, previously noticed in the crested newt (Mazzi, 1958; Mazzi and Peyrot, 1960), seems therefore to be confirmed by the observations cited above.

TABLE III

PATTERN OF THYROID ACTIVITY UNDER VARYING EXPERIMENTAL CONDITIONS

Operations	% ^{131}I Thyroid Uptake	% ^{131}I Loss into Water	Thin-Layer Chromatography [1]				Electrophoresis	MIT/DIT Ratio
			MIT	DIT	T_3	T_4		
Controls	24.93±1.08[2]	56.38±2.22	17.19±0.89	29.85±1.43	2.50±0.35	4.50±0.30	8.15±1.13	0.59±0.04
Autografted	35.52±3.27	47.76±4.57	25.46±0.79	26.71±2.17	2.20±0.34	4.61±0.39	7.57±1.09	1.01±0.08
Hypophysectomized	15.01±2.51	75.01±3.30	38.96±1.59	52.41±2.76	1.98±0.26	3.76±0.37	9.89±2.68	0.76±0.06
Lesions								
Successful	37.28±1.02	47.18±2.30	18.48±3.70	23.74±4.44	2.29±0.44	5.50±0.98	7.34±1.66	0.75±0.04
Partially successful	34.12±3.04	50.82±4.14	17.27±0.90	34.75±3.33	2.33±0.57	5.02±0.57	11.87±3.52	0.51±0.05
Thiouracil	2.19±0.27	79.03±3.11	3.31±0.65	1.86±0.42	0.35±0.06	0.70±0.01	89.85±1.81	2.29±0.43

(From Peyrot, 1969 and Peyrot and Vellano, 1968).

[1] MIT, monoiodotyrosine; DIT, diiodotyrosine; T_3. triiodotyronine; T_4, thyroxine.

[2] Means ±S.E.

D. Corticotropic Function

As to the hypothalamic control upon the corticotropic function in urodeles, only controversial experimental results have been obtained so far.

If survival of the individual bearing a pituitary autograft is taken as a criterion for evaluating corticotropic activity, it should be concluded that the transplanted pituitary is able to secrete ACTH in all species so far studied (*Pleurodeles waltii*: Pasteels, 1960; *Triturus cristatus*: Mazzi and Peyrot, 1963; *Triturus viridescens*: Schotté and Tallon, 1960, and Dent, 1966; *Triturus hongkongensi*: Jørgensen, 1968; *Ambystoma mexicanum*: Jørgensen and Larsen, 1963). Obviously this does not rule out the occurrence of a central nervous system control of ACTH secretion (Jørgensen, 1968).

However, a more direct evaluation of the interrenal activity indicates that it declines at most in some of the animals bearing a pituitary autograft. In the interrenal gland of the crested newt, 4 months after operation (unlike Pleurodeles: Pasteels, 1960) the cells are much smaller than normal and show shrunken nuclei. Water increase in the skeletal muscle and an altered Na/K ratio are seen to occur in these animals (Ferreri *et al.*, 1966). Under the same experimental conditions the activity of Δ^5-3β-hydroxysteroid-dehydrogenase as well as the Schultz reaction are reduced (Vellano and Peyrot, 1966). It was reported by Schotté and Tallon (1960) and by Dent (1967) that in *Triturus viridescens* bearing a pituitary autograft the capacity for limb regeneration is maintained. This is likely to be due to the enhancement of prolactin production by the grafted pituitary rather than to normal ACTH production. Prolactin administration, either alone or in more efficient combination with thyroxine, secures limb survival and regeneration in the hypophysectomized newt (*Triturus viridescens*: Connelly *et al.*, 1968). No information is available to date on a possible annual interrenal activity in the crested newt. In *Taricha torosa* the increase in the interrenal activity was found to arise within 4-6 weeks prior to the breeding season (Miller and Robbins, 1955).

III. Conclusions

The analytical data reported in the foregoing sections indicate that in many urodele amphibians, and in particular in the crested newt (*Triturus cristatus carnifex* Laur.), the activity of the adenohypophysis follows a annual cycle that coincides with environmental temperature variations. The controlling influence of this external factor is exerted not directly but throught the central nervous system. As in the amniota the final common pathway mediating the flow of information from the central nervous system to the adenohypophysis is represented by the hypothalamo-hypophyseal system and the hypophyseal portal system. Besides temperature, internal factors inherent in the endocrine balance are also implicated in the qualitative and quantitative modulations of the many activities of the adenohypophysis. This endocrine interplay also varies according to the environmental temperature variations.

It was shown experimentally that disturbance of this balance after administration of an exogenous hormone impinges on the whole endocrine constellation. Negligible effects are exerted by the administration of 100–120 IU of prolactin to the crested newt during winter, when endogenous production of this hormone is very high. By contrast, dramatic effects are produced in the summer or aestivating animals. These effects consist in marked variations in the cytological pattern of the adenohypophysis, increase in thyroid activity, decrease of spermatogenesis, development of the secondary sexual characters, increase in cardiac glycogen, weight loss from the fat bodies, and behavioral changes (water drive, enhanced vivacity, and triggering of courtship after long-term treatment). During a short lapse of time the summer or aestivating animal assumes an endocrine balance similar to that which is normal to the winter animal (Vellano *et al.*, 1967). This readjustment of the endocrine balance is likely to be mediated through the hypothalamus, since prolactin does not elicit any effect on the thyroid of the animal bearing a heterotopic pituitary autograft or a successful hypothalamic lesion (Vellano *et al.*, 1969).

The sites of the thermodependent centers responsible for the elaboration of the various hypothalamic neurohormones in urodele amphibians are unknown. Localization of these centers is a fairly difficult task, owing to the fact that the neurons are distributed over a continuous periventricular gray column. Nor is any knowledge available about the role played by some extrahypothalamic areas (habenula, striated body and mesencephalon), as was recently brought to light in mammals (see critical survey by Mess and Martini, 1968), nor about the chemical nature of pituitotropins.

ACKNOWLEDGMENTS

I thank my co-workers and collegues Prof. A. ˎ 'ardabassi, Prof. M. Sacerdote, Prof. A. Peyrot, and Dr. C. Vellano for help and advice.

REFERENCES

Andreoletti, G. E., and Rotta, G. P. (1968). *Boll. Zool.* **35**, 411.
Bünning, E. (1964). "The Physiological Clock." Academic Press, New York.
Connelly, T. G., Tassava, R. A., and Thornton, C. S. (1968). *J. Morph.* **126**, 365.
Dent, J. N. (1966). *Gen. Comp. Endocrinol.* **6**, 401.
Dent, J. N. (1967). *Am. Zool.* **7**, 714.
Doerr-Schott, J. (1968). *Ann. Biol.* **7**, 189.
Ferreri, E., Mazzi, V., and Socino, M. (1966). *Gen. Comp. Endocrinol.* **6**, 156.
Galgano, M. (1944). *Arch. Ital. Anat. Embriol.* **50**, 1.
Galgano, M. (1952). *Boll. Zool.* **19**, 97.
Galgano, M., and Mazzi, V. (1951). *Riv. Biol.* **43**, 21.
Grant, W. C., Jr. (1966). *Am. Zool.* **6**, 354.
Jørgensen, C. B. (1968). *In* "Perspectives in Endocrinology" (E. J. W. Barrington and C. B. Jørgensen, eds.), pp. 469-541. Academic Press, New York.
Jørgensen, C. B., and Larsen, L. O. (1963). *Symp. Zool. Soc.* (London) **9**, 59.
Masur, S. (1968). *Gen. Comp. Endocrinol.* **12**, 12.
Mazzi, V. (1950). *Atti Accad. Naz. Lincei Rend. Classe Sci. Fis. Mat. Natur.* **9**, 280.
Mazzi, V. (1952a). *Arch. Ital. Anat. Embriol.* **57**, 1.
Mazzi, V. (1952b). *Atti Accad. Naz. Lincei Rend. Classe Sci. Fis. Mat. Natur.* **12**, 605.
Mazzi, V. (1952c). *Monit. Zool. Ital.* **59**, 68.
Mazzi, V. (1953). *Z. Zellforsch.* **39**, 298.
Mazzi, V. (1958). *Z. Zellforsch.* **48**, 332.
Mazzi, V., and Peyrot, A. (1960). *Arch. Ital. Anat. Embriol.* **45**, 295.
Mazzi, V., and Peyrot, A. (1963). *Monit. Zool. Ital.* **70–71**, 124.
Mazzi, V., and Vellano, C. (1968). *J. Endocrinol.* **40**, 529.
Mazzi, V., Vellano, C., and Toscano, C. (1966). *Ric. Sci.* **36**, 3.
Mazzi, V., Vellano, C., and Toscano, C. (1967). *Gen. Comp. Endocrinol.* **8**, 320.
Mazzi, V., Vellano, C., and Sacerdote, M. (1969). *Ric. Sci.* **39**, 676.
Mess, B., and Martini, L. (1968). *In* "Recent Advances in Endocrinology" (V. H. T. James, ed.), pp. 1–49. J. and A. Churchill, London.
Miller, M. R., and Robbins, M. E. (1955). *Anat. Record* **122**, 79.
Morgan, A. H., and Fales, C. H. (1942). *J. Morphol.* **71**, 356.
Pasteels, J. L. (1960). *Arch. Biol.* (Liège) **71**, 409.
Peyrot, A. (1960). *Arch. Anat. Microscop. Morphol. Exp.* **49**, 411.
Peyrot, A. (1969). *Gen. Comp. Endocrinol.* **13**, 525.
Peyrot, A., and Vellano, C. (1966). *Ric. Sci.* **36**, 1073.
Peyrot, A., and Vellano, C. (1968). Boll. Zool. **35**, 417.
Peyrot, A., Biciotti, M., and Pons, G. (1966). *Boll. Soc. Ital. Biol. Sper.* **42**, 1026.
Peyrot, A., Mazzi, V., Vellano, C., and Lodi, G. (1969a). *J. Endocrinol.* **45**, 525.
Peyrot, A., Vellano, C., and Mazzi, V. (1969b). *Gen. Comp. Endocrinol.* **12**, 179.
Peyrot, A., Pons, G., and Biciotti, M. (1969c). *C. R. Ass. Anat.* **142**, 1299.
Sampietro, P., and Vercelli, L. (1968). *Boll. Zool.* **35**, 419.
Sampietro, P., Andreoletti, G. E., Rotta, G. P., and Mazzi, V. (1969). *Ric. Sci.* **39**, 672.
Scalenghe, F., Sampietro, P., and Andreoletti, G. E. (1967a). *Boll. Soc. Ital. Biol. Sper.* **43**, 1334.

Scalenghe, F., Sampietro, P., and Andreoletti, G. E. (1967b). *Boll. Zool.* **34**, 172.
Scalenghe, F., Andreoletti, G. E., Sampietro, P., Rotta, G. P., and Mazzi, V. (1968). *Arch. Anat. Histol. Embryol.* **51**, 629.
Schotté, O. E., and Tallon, A. (1960). *Experientia* **16**, 71.
Solberger, A. (1965). "Biological Rhythm Research." Elsevier Publishing Co., Amsterdam.
Stutinsky, F., and Befort, J. J. (1962). *Gen. Comp. Endocrinol.* **2**, 621.
Tuchmann-Duplessis, M. (1949). *Arch. Anat. Microscop. Morphol. Exp.* **38**, 302.
Van Tienhoven, A. (1968). "Reproductive Physiology of Vertebrates." Saunders, Philadelphia.
Vellano, C. (1968a). *Boll. Zool.* **35**, 420.
Vellano, C. (1968b). *Boll. Soc. Ital. Biol. Sper.* **44**, 2062.
Vellano, C., and Peyrot, A. (1966). *Ric. Sci.* **36**, 1073.
Vellano, C., Peyrot, A., and Mazzi, V. (1967). *Monit. Zool. Ital. (N.S.)* **1**, 207.
Vellano, C., Peyrot, A., Lodi, G., Longo, S., and Mazzi, V. (1968). *Boll. Zool.* **36**, 177.
Vellano, C., Peyrot, A., and Mazzi, V. (1969). *Gen. Comp. Endocrinol.* **13**, 537.
Von Gauss, G. H. (1961). *Z. Tierpsychol.* **18**, 60.
Wald, G. (1960). *Circulation* **21**, 916.

Author Index

Numbers in italics indicate the pages on which the complete references are listed.

C

F

H

I

L

T

Y

Z

Subject Index

R

S

T

U

V

RANDALL LIBRARY-UNCW
3 0490 0023423 U